VERTEBRATE DISSECTION

EIGHTH EDITION

W9-BZM-532

Warren F. Walker

Oberlin College
Oberlin, Ohio

Dominique G. Homberger

Louisiana State University
Baton Rouge, Louisiana

HARCOURT COLLEGE PUBLISHERS

Fort Worth Philadelphia San Diego
New York Orlando Austin San Antonio
Toronto Montreal London Sydney Tokyo

ISBN 0-03-047434-5

Text Typeface: Century Oldstyle
Compositor: General Graphic Services
Acquisitions Editor: Julie Alexander
Developmental Editor: Dena Digilio-Betz
Managing Editor: Carol Field
Project Editor: Nancy Lubars
Copy Editor: Martha Colgan
Manager of Art and Design: Carol Bleistine
Art Director: Doris Bruey
Cover Designer: Louis Fuiano
Text Artwork: J&R Art Services, Inc.
Director of EDP: Tim Frelick
Production Manager: Charlene Squibb
Senior Marketing Manager: Marjorie Waldron

Cover Credit: © David Masse

Address for Domestic Orders
Harcourt College Publishers, 6277 Sea Harbor Drive, Orlando, FL 32887-6777
800-782-4479
(e-mail) collegesales@harcourt.com

Address for International Orders
International Customer Service
Harcourt, Inc., 6277 Sea Harbor Drive, Orlando, FL 32887-6777
407-345-3800
(fax) 407-345-4060
(e-mail) hbintl@harcourt.com

Address for Editorial Correspondence
Harcourt College Publishers,
Public Ledger Building, Suite 1250, 150 S. Independence Mall West,
Philadelphia, PA 19106-3412

Web Site Address
http://www.harcourtcollege.com

Printed in the United States of America

VERTEBRATE DISSECTION 8/E

Library of Congress Catalog Card Number: 91-050763

0 1 2 3 4 5 6 7 8 9   071   20 19 18 17 16 15 14 13 12 11 10

PREFACE TO THE EIGHTH EDITION

◆ ◆

Introduction to the Eighth Edition

Although our knowledge of the anatomy of vertebrates does not change between editions of this manual, our understanding of their anatomy and evolution continues to advance rapidly.

Dr. Dominique G. Homberger of Louisiana State University has joined the senior author in preparing the eighth edition of this book. The book is enriched by the perspectives of two of us, and the participation of an active teacher helps us meet the needs of current students. We have rewritten many of the directions for dissection to make them easier to follow and to clarify areas with which students have had trouble. We have also updated the background (nondissection) text so that students can appreciate the evolutionary and functional contexts into which the structures they are studying fit.

A few examples follow that give an indication of the scope and nature of the changes we have made throughout the book. We no longer include directions for the dissection of the mink because few instructors use this animal and its anatomy is very similar to that of the cat. We have rewritten the overview of vertebrate evolution, and included a synoptic classification to help students keep track of the evolutionary position of the species we discuss. Basic muscle structure and function are important and of interest to the students and we have expanded this section. The analysis of fish swimming is updated. New information is included on the nature and function of the sense organs, particularly those of fishes, and of the parts of the fish brain. We have expanded and clarified the sections on the development of coelomic divisions and the mesenteries. A different approach is used to expose the branchial arteries of the dogfish. We have rewritten and expanded the sections on the structure, function, and development of the kidneys, gonads, and reproductive ducts. The background text does not simply reiterate information found in textbooks of functional and compara-

tive vertebrate anatomy. It consists of carefully selected data from embryology, histology, physiology, ethology, and natural history, which help the students in their understanding of the structures they see during the dissections. If necessary to suit the time constraints of the course, some or all of the background text can be disregarded. We have added many new figures to the background sections of the text and have carefully checked all the other figures so that terms used in the text and figures are consistent.

Approach

The purpose and basic approach of this edition remain the same as those of previous ones. The central theme of this manual is a study of the major structural transformations that have occurred in the vertebrates during their evolution from the fish to the mammalian stage, with the view of making the anatomy of the mammal meaningful. We cannot, of course, examine the actual evolutionary sequence to mammals in living vertebrates, but we can simulate this sequential study to some extent by examining selected living vertebrates. For most organ systems, the more important stages can be seen by dissecting examples from three levels of organization—a lungless fish, an amphibian or reptile, and a mammal. The dogfish (*Squalus*) is used as an example of an early jawed fish; the mudpuppy (*Necturus*) or the turtle (*Chelydra*), as examples of an amphibian and reptile; and both the cat (*Felis*) and rabbit (*Oryctolagus*) as examples of the mammal. Directions are also included for certain sheep organs that are readily available.

These directions give the instructor not only a choice of mammal to be studied, but also the option of introducing students to comparisons at one major evolutionary level. Different students can be issued different mammal species, including some not specifically included in these directions. Natural inquisitiveness, or formal comparisons of dissections after an exercise, will acquaint students with the species his or her neighbors may be dissecting. This will serve to emphasize the basic similarity of animals at one level of organization. The student sees that a carnivore and a rabbit, for example, have much in common, and can better realize that human beings too share many of these features. At the same time, the student has the opportunity to make comparisons between animals at the same level and to see the diverse features that are superimposed upon a common pattern. Although most of the differences between a carnivore and the rabbit are the result of adaptation to different modes of life, some doubtless result from chance divergence. A few of the differences also illustrate primitive and advanced stages in the evolution of a structure within the mammalian level of organization. The duplex uterus of the rabbit and the bipartite uterus of carnivores are a case in point.

In addition to the jawed fish, early terrestrial vertebrate, and mammal, directions are included for the study of representative protochordates and an agnathous fish (the lamprey), for those who have time to supplement the major sequence by examining these more primitive types.

Flexibility has been an aim throughout, for the length and emphasis of comparative anatomy courses are subject to much variation. Enough material is included for an intensive course. On the other hand, the directions are written so that much can be omitted. For example, the protochordates and lamprey could be omitted, leaving the jawed fish, early tetrapod, and mammal. A still shorter course could also omit the amphibian or reptile. An alternate way of shortening the course would be by omitting parts of some organ systems—the muscles of the hind limb, certain of the sense organs, the lymphatic system, and so on. Certain dissections could also be replaced by demonstrations. Regardless of what, if anything, is omitted, the sections left are rather thorough; for in our opinion it is a better educational experience for the student to do a limited amount of material well than to cover a lot of ground superficially. A detailed table of contents is supplied to facilitate the instructor's selection of the most relevant parts for a particular course, or for studying all of the organ systems of one species, if the instructor prefers the systematic approach.

Some textual (background) material has been incorporated into the laboratory directions. These remarks are not intended to replace a textbook but rather are included to summarize the major structural changes that have occurred and to relate these to the functions of the organs. Structure and function are inseparable, and changes in them have enabled vertebrates to exploit the opportunities, and to meet the challenges, imposed by the chang-

ing environment during the course of their evolution. Other sections explore some of the principles by which structure and function are interrelated—that is, how the specific configuration of the structures meets the physical and other constraints imposed by the functions they must perform. "Functional anatomy" gives new insights to our understanding of the evolution of organisms. It is a subject receiving increasing research attention nowadays. Still other textual remarks point out the ways in which the species being studied resembles or departs from a generalized member of its group.

In describing mammals, we have continued to use the terminology of the *Nomina Anatomica Veterinaria,* and we have also applied these terms to nonmammalian vertebrates where appropriate. This terminology is very similar to that given in the *Nomina Anatomica* for human beings, differing only where the structure between humans and quadrupeds differs significantly (e.g., pectoralis superficialis muscle = human pectoralis major), or with respect to terms for direction (e.g., cranial vena cava = human superior vena cava). An additional useful guide has been the *Nomina Anatomica Avium* by Baumel et al. Although this text was written for ornithologists, most of the terms are applicable to reptiles and amphibians. These terminologies are becoming the standard, and many students will encounter them in later work.

The appendix of major Greek and Latin roots for anatomical terms (Appendix 1) will help students gain an understanding of the basis for the formation of the numerous terms they will encounter. As students begin to see the ways the same roots are used in different combinations, they should be able to infer the meanings of unfamiliar terms.

Acknowledgments
We are indebted to many for help in preparing this edition. Drs. Kurt Schwenk of the University of Connecticut, W. David Sissom of Elon College, and Mark L. Wygoda of McNeese State University reviewed the previous edition and Drs. Frank E. Fish of West Chester University and John H. Roese of Lake Superior State University reviewed the final draft. All have made suggestions that have helped guide us in preparing this edition. We thank them for taking time from busy schedules to help us in this way. Many students and colleagues have given us further suggestions. We thank them and earnestly hope that users of this edition will call our attention to any errors or deficiencies they find. The staff of Saunders College Publishing have been very supportive and helpful. Dena Digilio-Betz, our Development Editor, guided us as we prepared the manuscript, and Nancy Lubars, our Project Editor, helped us with all phases of produc-

tion. Many figures retained from earlier editions were originally prepared by former students of the senior author, some of whom have become professional illustrators. Once again we are indebted to them: Lisabeth Daly, Ginny Hull, H. Jon Janosik, J. Paul Nail, Daniel T. Magidson, and Molly E. Wing.

New art for this edition was prepared under the supervision of the art department of Saunders College Publishing. Tensy Walker, wife of the senior author, has continued to lend encouragement and help in the proofreading.

Warren F. Walker, Jr.
Dominique G. Homberger

V

Preface to
the Eighth
Edition

CONTENTS

◆ ◆

CHAPTER FIVE The Trunk Skeleton 95

CHAPTER SIX The Appendicular Skeleton 111

CHAPTER SEVEN The Muscular System 133

CHAPTER EIGHT   The Sense Organs   212

CHAPTER NINE   The Nervous System   237

CHAPTER TEN   The Coelom and the Digestive and Respiratory Systems   287

CHAPTER ELEVEN   The Circulatory System   336

CHAPTER TWELVE   The Excretory and Reproductive Systems   394

A NOTE TO THE STUDENT: ANATOMICAL TERMINOLOGY

◆ ◆

When studying anatomy for the first time, students are confronted with numerous unfamiliar terms that must be mastered, because effective communication requires their use. Some understanding of the derivation of these terms will help to fix their meaning and spelling in mind. Most are based on Latin or Greek roots, and as you become familiar with the more common roots you will recognize the terminology for the shorthand that it is. The root "chondro," for example, always means cartilage, and it is used in many combinations: Chondrichthyes (cartilaginous fish), chondrocranium (cartilaginous braincase), perichondrium (connective tissue around cartilage), chondrocyte (cell in cartilage). A sampling of the more important word roots used in anatomy and for the names of organisms is presented in Appendix 1. There are also certain conventions, described below, that make the terminology more rational.

Terms for Organs

Anatomists originally described most organs by their appearance in a human being, for human anatomy was one of the first to be studied. Over the centuries a plethora of names has been proposed, and many organs have a long list of synonyms. To bring some order to the chaotic situation, human anatomists have agreed upon a code of terminology known as the *Nomina Anatomica*. Terms are in Latin or Greek, but they are often translated into the vernacular of each language. Veterinary anatomists have agreed upon a *Nomina Anatomica Veterinaria* in which they have brought most of the terminology for other mammals into agreement with human terminology, the major exception being certain terms for direction which are naturally different between a biped and quadruped. Bird anatomists have agreed upon a *Nomina Anatomica Avium*, and many of these terms are applicable to earlier terrestrial vertebrates.

In writing this book we have used *Nomina Anatomica Veterinaria* terms for mammals. We have applied mammalian or avian terms to other vertebrates where this appears to be reasonable. Favored terms are placed in boldface when first used for each animal. In the few cases where this is not a *Nomina Anatomica* term, the official term is given in italics, e.g., **kidney** (*ren*). The official term in this case is seldom used in English as a noun, but it forms the basis for the common adjective, renal. We have avoided introducing synonyms unless they are in common use, in which case the synonym is given in regular type, e.g., **cleidobrachialis muscle** (clavodeltoid). Terms based on the names of individuals are avoided in official terminology, but some in common use have been included, e.g., **auditory tube** (eustachian tube).

Terms for Directions, Planes, and Sections

Terms for Direction

Many terms for direction are the same in comparative and human anatomy, but there are certain differences occasioned by our upright posture. A structure toward the head end of a quadruped is described as **cranial** (e.g., cranial vena cava); one toward the tail, as **caudal** (Fig. N-1). The term **rostral** replaces cranial for directions within the head. Comparable positions in human beings are described as **superior** (e.g., superior vena cava) and **inferior.**

A structure toward the back of a quadruped is **dorsal**; one toward the belly is **ventral.** The terms dorsal and ventral are not used in human anatomy; comparable directions are referred to as **posterior** and **anterior.** It is recommended that the terms anterior and posterior not be used in a quadruped for the cranial and caudal ends of the body because of the possible confusion with the different usage in human anatomy. (However, they are sometimes still used.) In general we have followed this recommendation except for well-

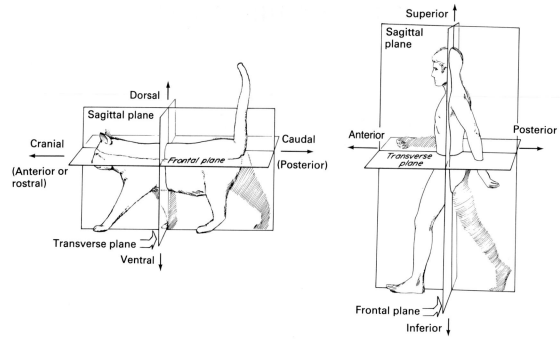

Figure N-1
Diagrams to show the planes of the body and the differences in the terms for direction in a quadruped and in a human being.

established terms for certain organs (e.g., posterior cardinal veins in a fish).

Other terms for direction are used in the same way in all animals. **Lateral** refers to the side of the body; **medial,** to a position toward the midline. **Median** is used for a structure in the midline. **Distal** refers to a part of some organ, such as an appendage or blood vessel, that is farthest removed from the point of reference, such as the center of the body or the origin of the vessel; **proximal,** to the opposite end of the organ, i.e., the part nearest the point of reference.

Left and **right** are self-evident, but it should be emphasized that in anatomical directions they always pertain to the *specimen's* left or right, regardless of the way the specimen is viewed by the observer.

Adverbs may be formed from the above adjectives by adding the suffix **-ly** or **-ad** to the root, in which case the term implies motion in a given direction. To say that a structure extends caudally, or caudad, means that it is moving toward the tail.

Planes and Sections of the Body
The body of a specimen is frequently cut in various planes to obtain views of internal organs. A longitudinal, vertical section from dorsal to ventral that passes through the median longitudinal axis of the body is a **sagittal** section (Fig. N-1). Such a section lies in the sagittal plane. Sections or planes parallel with, but lateral to, the sagittal plane are said to be **parasagittal.**

A section cut across the body from dorsal to ventral, and at right angles to the longitudinal axis, is a **transverse** section, and it lies in the transverse plane.

A **frontal** (coronal) section or plane is one lying in the longitudinal axis, and passing horizontally from side to side.

CHAPTER ONE

◆ ◆

HEMICHORDATES AND PROTOCHORDATES

Chordates, echinoderms (sea stars and their relatives), and several phyla of wormlike creatures (Pogonophora, Chaetognatha, Hemichordata) are grouped together as **deuterostomes** because a second opening invaginates from the larval surface into the gut cavity and becomes the mouth. The original larval opening into the gut cavity—the blastopore—becomes the anus, or this original opening closes, and a new anus invaginates nearby.

The phylum **Chordata** embraces about 43,000 contemporary species, ranging in complexity from sessile, soft-bodied sea squirts to the wide variety of backboned vertebrates. Primitive chordates, like many other animals, are filter feeders that trap food particles from the surrounding sea water, but their method of collecting plankton is unique. Food particles in a stream of water passing through the pharynx are trapped in a sheet of mucus secreted by an **endostyle,** a longitudinal groove in the floor of the pharynx that is lined with glandular and ciliated cells. Excess water escapes through a series of **pharyngeal slits.** Some endostyle cells also bind iodine to protein and are homologous to the vertebrate thyroid gland. Terrestrial vertebrates lack pharyngeal slits as adults but as embryos have pharyngeal pouches. Pharyngeal pouches and an endostyle or its derivative, the thyroid gland, are defining features of chordates. Chordates are also distinguished from other deuterostomes by the presence, at some stage of their life cycle, of two additional unique characters: a **notochord,** and a **nerve cord** that differs from that of other animals in being **unpaired, dorsal,** and **tubular.** The notochord and nerve cord are functionally related to the locomotor system. The nerve cord integrates the activities of longitudinal muscle fibers in the body wall, and the notochord, which is a noncompressible yet bendable rod of cells, prevents the body from shortening when the muscle fibers contract. Contraction of longitudinal muscle fibers thus results in a side-to-side bending of the body, rather than a telescoping of the trunk, as occurs in an earthworm. Variations in the degree of development of the pharyngeal slits, notochord, and nerve cord are among the criteria used to divide the phylum Chordata into its three subphyla: **Urochordata, Cephalochordata,** and **Vertebrata.** Urochordates and cephalochordates are often called **protochordates.**

Courses in comparative anatomy deal primarily with the vertebrates, but it is desirable for you to have some idea of the nature of protochordates, and of hemichordates, which are probably more closely related to chordates than any other deuterostome group. In particular, you should become acquainted with their general external features and understand how the fundamental chordate characters are represented. The cephalochordates, which are closer in structure to the vertebrates, are often studied in more detail.

PHYLUM HEMICHORDATA

The 85 species of present-day hemichordates include a few rare, colonial, deep-sea forms, the pterobranchs, and wormlike enteropneusts, or acorn worms, which range in length from a few centimeters to about 2 meters. Enteropneusts burrow in the sand and mud of tidal flats and shallow coastal waters, where they are sometimes abundant. *Balanoglossus* and *Saccoglossus* are common genera along North American coasts.

Examine the external features of one of the enteropneusts (Fig. 1-1A). You may have to use low magnification to see certain structures. The body is divided into three distinct regions: an anterior **proboscis,** a **collar,** and a long **trunk.** The proboscis attaches by a narrow stalk to the encircling collar located just posterior to it. The proboscis nesting in the collar often gives the appearance of an acorn in its cup—a fact that gives the common name to the group. The proboscis and collar assist the ciliated epidermis in burrowing. The collar coelom fills (Fig. 1-1B), thus inflating the collar and anchoring the worm in its burrow. Then the deflated proboscis is pushed forward into the substratum by the action of circular muscles within it; its coelom is filled, and it anchors the worm. Finally the collar coelom is emptied, and the worm is pulled forward toward the proboscis.

When the animal is feeding, the proboscis is extended out of the burrow, and cilia upon its surface carry minute food particles caudally to the **mouth,** which is situated inside the front of the collar ventral to the proboscis stalk (Figs. 1-1B and C). Food is carried into the ventral part of the pharynx, and excess water enters the dorsal part to escape through numerous pharyngeal slits. The two parts of the pharynx are partially separated by a longitudinal fold. The **external pharyngeal slits** can be seen dorsolaterally on the cranial portion of the trunk. As many as 150 pairs have been counted on a 40-centimeter specimen. Each external pharyngeal slit leads to a pharyngeal pouch, which connects with the pharynx by way of an unusual U-shaped internal pharyngeal slit. It is of interest that at one stage of development the pharyngeal slits of amphioxus (a cephalochordate) have an identical appearance. Food continues from the ventral part of the pharynx into a simple, straight intestine. Material not digested and absorbed leaves through the terminal **anus.**

In mature individuals, prominent **genital ridges** will be found just ventral to the caudal external pharyngeal slits and extending a short distance caudad. The sexes are separate but cannot be distinguished externally. In some but not all species, conspicuous **hepatic ridges** will be seen caudal to the genital ridges. When present, they are the outward manifestation of digestive glands, the hepatic caeca, that bud off the intestine.

A longitudinal middorsal ridge and a similar midventral ridge can also be seen on the trunk. Each contains a superficial and solid **nerve strand** that extends forward into the collar. Within the collar, the dorsal strand rolls up upon itself and thereby forms one or more cavities. Neurons from an echinoderm-like subepidermal nerve plexus connect with the nerve strands.

As to the diagnostic chordate characteristics, pharyngeal slits are well represented, but food is collected on the surface of the proboscis and not within the pharynx. Mucus is secreted by an endostyle. Nerve strands are present, but they are not in the form of a single, dorsal, tubular nerve cord. A diverticulum from the rostral end of the mouth cavity extends into and helps to stiffen the proboscis (Fig. 1-1B). Its wall consists partly of a chitinous plate and partly of vacuolated cells resembling those of a

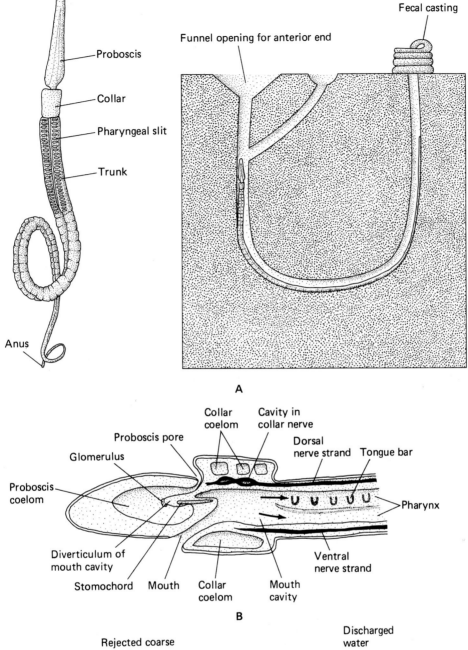

Proboscis

Collar

Pharyngeal slit

Trunk

Anus

Fecal casting

Funnel opening for anterior end

A

Collar
coelom

Cavity in
collar nerve

Proboscis pore

Dorsal
nerve strand

Tongue bar

Glomerulus

Proboscis
coelom

Pharynx

Diverticulum of
mouth cavity

Ventral
nerve strand

Stomochord   Mouth   Collar
coelom

Mouth
cavity

B

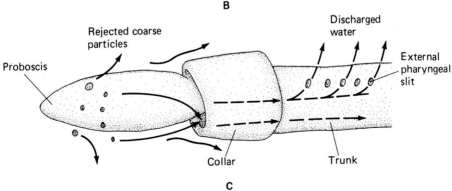

Rejected coarse
particles

Discharged
water

Proboscis

External
pharyngeal
slit

Collar

Trunk

C

Figure 1-1
Saccoglossus lives in burrows in the sea floor. **A,** An external view of a specimen, **B,** sagittal section
of the anterior end. **C,** Cilia on the proboscis sort fine from coarse particles and move the fine
particles to the mouth. *(From Dorit, Walker, and Barnes,* Zoology.*)*

notochord. Some investigators consider it to represent a rudimentary notochord (the hemichordates get their name from this "half notochord"), but others question this interpretation and prefer to avoid any implications of homology by calling it a **stomochord.** Since most of the chordate characteristics are not well represented at any stage of development, the Hemichordata are considered to be a distinct phylum distantly related to but not a part of the phylum Chordata.

PHYLUM CHORDATA, SUBPHYLUM TUNICATA

Members of the subphylum Urochordata or Tunicata are marine animals, most of which are encased in a leathery membrane called the tunic; hence, the animals are commonly referred to as tunicates. There are approximately 2000 species. Although a few species are pelagic, the most familiar tunicates are the sessile sea squirts belonging to the class **Ascidiacea.** They are found attached to submerged objects in coastal waters or occasionally partly buried in the sand or mud. Many sea squirts are colonial, but some are solitary. Their anatomy can be studied conveniently on one of the latter types, such as *Molgula,* which is one of the most abundant species along the Atlantic coast.

External Features

Examine a specimen of *Molgula* in a pan of water and notice its saclike appearance. Two spoutlike openings, or siphons, will be seen near the top of the animal (Fig. 1-2). Water enters the organism through the topmost aperture **(incurrent siphon)** and is discharged through the opening that is set off on one edge **(excurrent siphon).** The margin of the incurrent siphon bears six small tentacles. Also notice the external covering, or **tunic.** It contains a large amount of tunicin, a complex polysaccharide similar to the cellulose found in plant cell walls. The tunic is secreted by cells derived from the underlying body wall. Minute, hairlike processes extend out from the tunic and help to anchor the animal to its substrate. The lower part of the tunic may have sand grains adhering to the hairs, for *Molgula* often lies partly buried in the sand.

Although the shape of the animal appears rather asymmetrical, the sea squirt in fact has a modified bilateral symmetry. If you compare a larva with an adult (Fig. 1-2), you will notice that the region between the two siphons represents the dorsal surface, and the rest of the edge of the adult represents the ventral surface. The incurrent siphon lies anteriorly and the excurrent siphon posteriorly.

Dissection

To expose the inside of the animal, cut through the tunic beneath the excurrent siphon and extend the cut around the base of the sac to a point near the incurrent siphon. Reflect the tunic, observing that it is attached to the rest of the body only at the siphons. Detach the tunic at these points. A number of bundles of longitudinal and circular muscle fibers lie within the thin body wall, or **mantle,** and aid in expelling water. The mantle is nearly transparent, and many of the internal organs can be seen beneath it. To see them more clearly, carefully peel off the mantle without unduly injuring organs that may adhere to it. This part of the dissection should be done beneath water.

Study the dissection and compare it with Figure 1-2**A** and **B**. You may need low magnification to see certain structures. The incurrent siphon leads into a large, thin-walled, vascular **pharynx,** which occupies most of the inside of the body. The pharynx wall of *Molgula* is characterized by many **longitudinal bars** between which lie microscopic **pharyngeal slits.** To see the slits clearly, remove a piece of the pharynx wall, prepare a wet mount of it, and view it through a microscope. The slits appear as arclike slits arranged in spirals. Although they will not be seen in this type of preparation, the bars between the slits contain blood vessels in which gas exchange with the environment occurs; gills are not present. The bars are covered with cilia, which create the current of water that passes through the animal. The pharyngeal slits do not lead directly to the outside but into a delicate chamber, the **atrium,** located on each side of the pharynx. The lateral portions of the atrium may not be seen, but they converge posteriorly to form a more conspicuous median atrial chamber, which opens to the surface through the excurrent siphon.

The fold along the ventral surface of the pharynx is the **endostyle.** Certain of its cells are ciliated; some are glandular and secrete mucus that entraps minute food particles in the water; and some produce iodinated proteins as do the cells of the vertebrate thyroid gland. The food-containing mucous band is moved toward the incurrent siphon by the cilia. Near the anterior end of the pharynx it is carried to the dorsal side by lateral **peripharyngeal bands.** These may be difficult to see. Then the mucous string moves

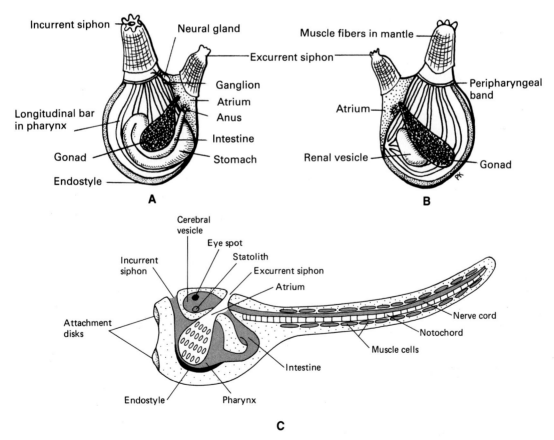

Figure 1-2
A and **B,** Left and right sides, respectively, of a dissection of *Molgula.* All of the tunic and most of the mantle have been removed. **C,** Diagramatic lateral view of a larval ascidian. (**C,** *From Walker, Functional Anatomy of Vertebrates.)*

posteriorly to the esophagus along the middorsal fold called the **dorsal lamina.** Thus, the pharynx of the sea squirt is primarily a food-gathering device, although some gas is exchanged between the blood and environment through its walls.

The rest of the alimentary canal of *Molgula* lies on the left side of the pharynx. A short **esophagus** leads from the posterior end of the pharynx to a slightly expanded portion of the gut, sometimes called the **"stomach,"** and this is followed by the **intestine** proper. The esophagus, stomach, and first part of the intestine form a C-shaped loop. Then the intestine doubles on itself, goes back beside the stomach and esophagus, and opens at the anus into the median portion of the atrial chamber. There is no stomach in the vertebrate sense. Little food is stored in this "stomach"; together with minute glandular folds evaginated from it, the stomach secretes enzymes that act upon carbohydrates and fats as well as upon proteins. The intestine appears to be primarily absorptive.

Sea squirts reproduce asexually by budding; they also reproduce sexually. In mature individuals, large **gonads** will be seen. In *Molgula* there is one on each side of the pharynx. Inconspicuous genital ducts lead from the gonads to the median portion of the atrium. Tunicates are hermaphroditic but generally not self-fertilizing. *Molgula*, however, can fertilize itself.

An oval **renal vesicle** lies ventral to the right gonad. Although some concretions of uric acid accumulate in the vesicle and stay there until the death of the animal, most nitrogen is excreted as ammonia and is lost by diffusion through the pharynx wall. The significance of the vesicle is not clear.

A small oval-shaped structure, seen on the dorsal edge of the pharynx between the two siphons, is the **neural gland complex.** It consists of a group of dense cells that are connected by a duct and coelomic funnel (probably not visible) to the pharynx cavity. Early investigators compared the group of dense cells to part of the vertebrate pituitary gland, but studies by Ruppert (1990) show that these cells are phagocytic and not secretory. Ruppert further demonstrated that sea water is drawn into the ciliated funnel, down the duct, and across the group of dense cells and enters the blood in adjacent pharyngeal blood vessels. The phagocytic cells clear the sea water of minute particles. Ruppert postulates that the neural gland complex restores blood volume, which may be reduced during periods when the body wall contracts vigorously and squirts water from the siphons. Carefully pull off the neural gland complex, and you will see beneath it an elongated nerve **ganglion.** Other internal organs are not usually seen in this type of dissection.

Apart from the endostyle and abundant pharyngeal slits, little about the appearance of an adult sea squirt suggests a chordate. The other diagnostic features of the phylum are present in the free-swimming, tadpole-shaped larva that disperses the population (Fig. 1-2**C**). The tail of the larva, which is its locomotive organ, contains longitudinal muscle fibers, a supporting notochord, and above the notochord the unpaired, dorsal, tubular nerve cord. The term "urochordate" derives from the position of the notochord. The nerve cord expands anteriorly into a sensory vesicle containing receptors for equilibrium and light that enable the larva to find a suitable substratum on which to settle. At metamorphosis, the larva attaches to the substratum by its anterior end, and its tail atrophies. The notochord and nerve cord are lost, and the sensory vesicle is reduced to the nerve ganglion.

PHYLUM CHORDATA, SUBPHYLUM CEPHALOCHORDATA

The subphylum Cephalochordata includes the lancelet or amphioxus (*Branchiostoma*[1]) and the related genus *Asymmetron*. The two genera include about 45 species of superficially fishlike animals that have an extremely long notochord extending beyond the nerve cord to the very front of the animal. This extreme extension of the notochord is probably correlated with the burrowing habits of the animal. It also gives the name Cephalochordata to the subphylum. These animals are found in coastal waters, usually lying partly buried in the sand with only their front end protruding because, like the sea squirt, amphioxus is a filter feeder (Fig. 1-3A). At times it actively swims to new feeding sites, but its locomotion is not very efficient because stabilizing fins are poorly developed. In the United States, they are found south from Chesapeake Bay and Monterey Bay. Since many of the features of amphioxus are believed to be very primitive for chordates and hence may throw light on the origin of vertebrate structure, it will be considered in more detail than the other protochordates.

External Features

Examine a preserved specimen of amphioxus in a pan of water. Mature specimens range from 4 to 6 centimeters in length. You will need low magnification to see certain structures. The shape of amphioxus is streamlined, or fusiform, being elongate, flattened from side to side (compressed), and pointed at each end (Fig. 1-3**B**). Segmental, V-shaped muscle blocks, the **myomeres,**[2] can be seen through the transparent epidermis. The

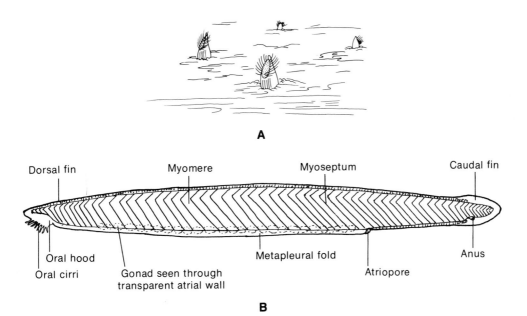

A

Dorsal fin   Myomere   Myoseptum   Caudal fin

Oral hood   Metapleural fold   Anus
Oral cirri   Gonad seen through   Atriopore
transparent atrial wall

B

Figure 1-3
A, Amphioxus in its habitat; **B**, lateral view of an adult amphioxus. (*A, From Walker,* Functional Anatomy of Vertebrates.*)*

---

[1]Although *Branchiostoma* (Costa, 1834) has priority over *Amphioxus* (Yarrel, 1836) as the technical name, the term amphioxus is so familiar that it is customary to retain it, at least as a common name.

[2]**Myotome** and **myomere** are terms for the primitive muscle segments. These terms are often used synonymously, but we follow the usage of *myotome* for an embryonic muscle segment and *myomere* for an adult segment.

apex of each V points cranially. Note that the **myomeres** extend nearly the length of the body. Their number is a specific character, ranging from about 55 to 75 in American species. The lines of separation between the myomeres are connective tissue partitions called **myosepta.**

A **dorsal fin** extends along the top of the body, and a **ventral fin** will be seen beneath the caudal quarter of the animal. Dorsal and ventral fins are continuous around the tail and expand slightly in this region to form a **caudal fin.** A pair of ventrolateral fins, or **metapleural folds,** continue forward from the rostral end of the ventral fins. (See also Fig. 1-5.)

The metapleural folds end a short distance from the front of the body. In front of them, and ventral to the cranial few myomeres, you will see a transparent chamber called the **oral hood.** The mouth is located deep within this chamber and will not be seen at this time, but the opening of the oral hood on the ventral surface can be seen. It is fringed with small tentacles called **cirri** (singular: **cirrus**), which are often folded across its opening. The cirri contain chemoreceptive cells and also aid in excluding large material, permitting only water and small food particles to enter.

Water that enters the pharynx passes through pharyngeal slits into an atrial chamber whose opening **(atriopore)** you will see between the caudal ends of the two metapleural folds. The intestine opens by an **anus** located on the left side of the caudal fin, so there is a **postanal tail.** A postanal tail characterizes cephalochordates and vertebrates; other groups of animals have a terminal anus.

If the specimen is mature, you will see on each side a row of whitish, square **gonads.** They lie just ventral to the myomeres in the cranial half of the body. Their number ranges from about 20 to 35, varying slightly with the species. The gametes are discharged directly into the atrium. The sexes are separate.

Whole Mount Slide

Study a stained microscope slide of a small specimen of amphioxus under the low power of the microscope (Fig. 1-4). Note its fusiform shape and find the structures described earlier: myomeres; dorsal, ventral, and caudal fins; metapleural folds; oral hood and cirri; atriopore; anus. The **myomeres** have been cleared to render them somewhat transparent, so they will not be seen as plainly as in the preserved specimen, but you should see indications of them, at least, just ventral to the dorsal fin.

Observe that the **dorsal** and **ventral fins** are supported by small, transparent blocks called **fin ray boxes.** In order to see the **metapleural folds** you will have to focus sharply on the surface. Each will appear as a horizontal line, parallel and slightly dorsal to the ventral edge of the body.

Since the small specimens used to make slides are not sexually mature, the **gonads** will not be fully developed or may even be absent. If present, they will appear as a row of lightly staining, oval structures close to the ventral edge of the body.

Notice the **notochord** located in the back, dorsal to the dark-staining alimentary canal. It extends nearly the entire length of the animal in the general position of the vertebral column of vertebrates.

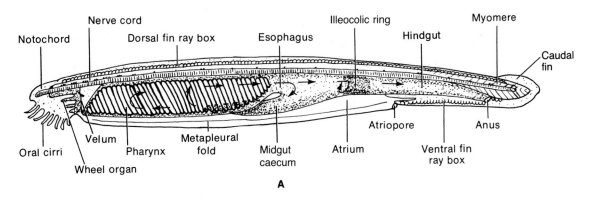

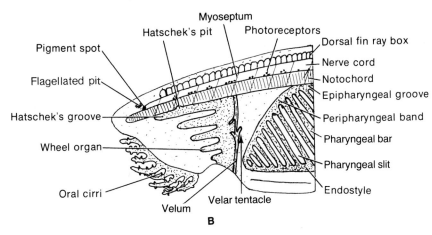

Figure 1-4
Anatomy of amphioxus. **A**, Lateral view of a whole mount slide of a oung specimen; **B**, Enlargement
of the anterior end. Arrows show the directions of mucus and food movement.

The notochord functions primarily to prevent the body from shortening when the
myomeres contract, so that their contractions are transformed into a series of lateral
undulations that sweep down the length of the body. The notochord also stiffens the
body enough for the animal to push forward or backward into the sand. Water provides
the primary support for aquatic animals as small as amphioxus.

As with other chordates, the notochord of amphioxus is a cellular organ encased in a
collagen sheath, but the notochord of amphioxus is unique in that most of the cells are
flattened discs of paramyosin muscle stacked like a column of checkers (Flood, 1975).
The cells themselves are vacuolated, and other fluid-filled extracellular spaces lie
between the paramyosin discs. Paramyosin muscle, which also occurs in the adductor
muscle of clam shells, is able to remain contracted for long periods with a minimal
energy expenditure. When muscles in the notochord contract, they decrease its
diameter and, because of the firm surrounding sheath, increase its turgidity. As a
result, the notochord of amphioxus is an adjustable hydroskeleton. The turgidity of the
notochord may be adjusted according to whether amphioxus is swimming or burrowing
forward or backward. Whichever direction the animal is going, the trailing end of the
body oscillates to a greater extent than the leading end.

The single, dorsal, tubular **nerve cord** will appear as a dark-staining band lying
dorsal to the notochord. Its position may be recognized by the dark pigment granules along
its ventral edge. Each granule represents parts of a simple **photoreceptor.** Notice that
the granules are particularly numerous near the front of the animal. The nerve cord ends in
a blunt point rostrally; there is no expanded brain. A prominent **pigment spot** of unknown

function will be seen in front of the nerve cord. Sharp focusing will also reveal a clear, saclike structure just dorsal to the front of the nerve cord. It is called the **flagellated pit** and is believed to be a chemoreceptor. It occurs only on the left side of the snout. In the embryo it connects with the nerve cord.

Examine the region of the **oral hood** in detail. The **cirri** have small processes along their edges, and each cirrus is supported by a skeletal rod of cartilage-like material. All the rods connect with a common basal rod. Ciliated grooves, or bands, are located on the inside of the lateral walls of the oral hood. In a lateral view they appear as large, dark-staining, finger-like bands extending forward from a common basal band. This complex is called the **wheel organ,** and it functions to draw a current of water into the organism. The dorsalmost band, which is called **Hatschek's groove,** is longer than the others. Slightly anterior to the middle of Hatschek's groove you will see a region where the groove is deeper and forms a pit that extends dorsally to overlap the right side of the notochord. This is **Hatschek's pit;** embryonically it connects with a part of the coelom. Besides aiding in the ciliary current, Hatschek's groove and pit secrete mucus that helps to entrap minute food particles in the water.

Caudal to the wheel organ you will notice a dark-staining line that is approximately in the transverse plane. This is the **velum**—a transverse partition that forms the posterior wall of the oral hood. The mouth, which cannot be seen in this view, is located in its center. The mouth is fringed with velar **tentacles,** which can be seen. They too act as strainers and probably contain chemoreceptive cells.

The mouth leads into a large **pharynx.** Most of the lateral walls of the pharynx are perforated by numerous elongate **pharyngeal slits** with ciliated **pharyngeal bars** between them. In favorable specimens, supporting rods may be seen within the pharyngeal bars. Mature specimens have over 200 bars. The pharyngeal bars provide a very large ciliated surface that plays the major role in moving water and food particles through the pharynx. Food is entrapped in mucus, and the water escapes through the pharyngeal slits. The pharynx is primarily a food-concentrating mechanism. Although blood vessels pass through the pharyngeal bars, no gills are present. Because of the great activity of the ciliated cells in this region, it is even possible that the blood leaving the bars contains less oxygen than blood entering them. The major site of blood aeration appears to be the general body surface. As in the sea squirt, in amphioxus water does not pass to the outside directly but through the pharyngeal slits into an **atrium.** The only part of the atrium to be seen in this view is the clear space ventral to the pharynx and continuing beneath the gut to the **atriopore.** The atrium is formed by the downgrowth of folds of the body wall around the pharyngeal slits. The encasement and protection of the delicate pharyngeal apparatus is essential in a burrowing species.

The longitudinal band that extends along the entire floor of the pharynx is the **endostyle.** Its function is the same as that of urochordates. It secretes mucus in which minute food particles become entrapped, and its cilia carry the mucus anteriorly. The string of mucus passes to the dorsal side of the pharynx along the pharyngeal bars and peripharyngeal bands. A **peripharyngeal band** is located on each side of the front of the pharynx and appears as a dark-staining line extending from the ventral edge of the velum diagonally dorsad and caudad just above the anterior pharyngeal slits. The mucus sheet is carried caudally along a middorsal **epipharyngeal groove.** Experiments have shown that certain endostylar cells also concentrate radioactive iodine, and Barrington has proposed that these cells are homologous to those of the vertebrate thyroid gland. In amphioxus the iodinated proteins are discharged into the gut and absorbed further caudad rather than into the blood as in vertebrates.

The caudal end of the pharynx floor extends diagonally dorsad and, just behind the last pharyngeal slit, the pharynx leads into a short, narrow **esophagus.** The outlines of the esophagus are often obscured by a large midgut caecum but can be seen by looking carefully. The top of the esophagus lies just ventral to the notochord, and its floor will appear as a longitudinal line extending caudally a short distance from the bottom of the last gill slit.

The diameter of the alimentary tract increases two- or threefold just caudal to the esophagus, for the large **midgut caecum** has evaginated at this point. The midgut caecum extends toward the front of the animal, lying along the ventral side of the esophagus and right side of the pharynx. It is located within the atrium.

Some smaller food particles are carried into the midgut caecum to be digested by enzymes secreted here and absorbed directly. Absorbed food is stored as glycogen and lipid in certain caecal cells; thus, the caecum has some functions (e. g., storage) in common with the vertebrate liver and others (e. g., enzyme secretion) not associated with the liver. Most food particles continue down the midgut, which becomes quite narrow caudal to the caecum. The deeply stained segment of the alimentary canal is called the **ileocolic ring.** At this point, cilia impart a rotary action to the cord of mucus and food in the entire midgut. This aids in the discharge of enzymes by the midgut caecum and in mixing them with the food mass. A still narrower **hindgut** follows the ileocolic ring and opens on the body surface at the **anus.** Remains of microorganisms, including the shells of diatoms, are often seen in the hindgut.

Other internal structures are not seen in this type of preparation.

Cross Sections

♦ **(A) COMMON FEATURES IN THE SECTIONS**

You will understand better the anatomy of amphioxus if you examine slides of representative cross sections. While studying such sections, compare them with the diagrams shown in Figures 1-5 and 1-6, and correlate the appearance of organs in this view with their appearance in the whole mount.

Many features will look much the same in any section. The body surface is covered by a simple **integument** consisting of an epidermis composed of a single layer of columnar epidermal cells underlain by a gelatinous subcutis containing a few connective tissue fibers. There are no gland or pigment cells. The **dorsal fin** will be recognized along with the hollow-appearing **fin ray box.** Actually, the fin ray box contains a gelatinous connective tissue. Cross sections of **myomeres** will appear as several oval chunks of tissue beneath the integument. They are separated from each other by the **myosepta.**

The **nerve cord** is the large tubular structure slightly ventral to the dorsal fin. Its cavity, the **neurocoel,** is very narrow. In favorable sections you will see lateral extensions of the cord that superficially resemble the dorsal and ventral roots of vertebrate spinal nerves. The dorsal extensions pass between the myomeres and are

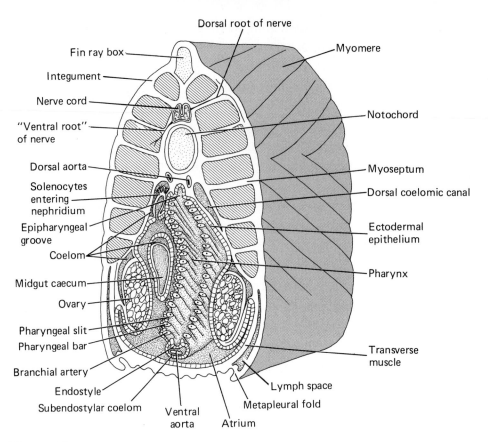

Figure 1-5
A diagrammatic cross section through amphioxus at the level of the posterior part of the pharynx.
The section is viewed from the front; thus, the right side of the drawing is on the left side of the
animal.

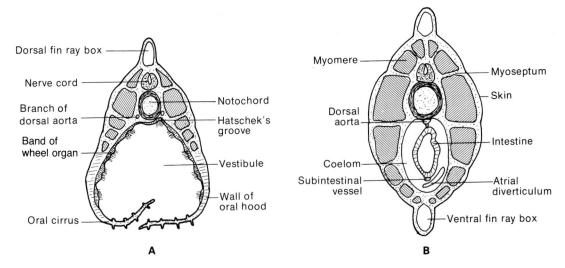

Figure 1-6
Diagrammatic cross sections through amphioxus at the level of the oral hood (**A**) and hindgut (**B**).
This specimen's right side is on the right side of the drawing.

indeed nerves carrying sensory fibers into the nerve cord and motor fibers to the ventral, nonmyomeral muscles. The apparent ventral roots are not nerves but are composed of delicate extensions of muscle cells, those of both the myomeres and the notochord, that make myoneural junctions within the cord. A similar pattern of innervation is found in round worms (nematodes) and some other invertebrates.

The **notochord** lies just beneath the nerve cord. It consists of vacuolated par-amyosin muscle cells that are distended with fluid and held tightly together by a firm connective tissue sheath. You will see the sheath, but the details of the cells will not be apparent.

◆ **(B) SECTION THROUGH THE ORAL HOOD**

In a section taken near the front of the animal (Fig. 1-6A), you will see a space, the vestibule, lying ventral to the notochord and myomeres and flanked laterally by the thin walls of the **oral hood.** The ciliated bands of the **wheel organ** will appear as thicker patches of epithelium on the inside of the wall of the hood. **Hatschek's groove,** the most dorsal of these, is located a bit to the right of the median plane. The oral hood opens ventrally, but some sections may be taken just caudal to the opening. Pieces of **cirri** will be seen in the section.

◆ **(C) SECTION THROUGH THE PHARYNX**

In a section through the pharynx (Fig. 1-5), the **metapleural folds** will be seen projecting from the ventrolateral portion of the body. There is a prominent **lymph space** in each. The wrinkled body wall between them contains a **transverse muscle** sheet that extends from the myomeres on one side to those on the other. This layer serves to compress the atrial cavity dorsal to it and thus aids in expelling water. The **pharynx** occupies most of the center of the section and is surrounded laterally and ventrally by the **atrium.** Note the numerous **pharyngeal bars** that form the wall of the pharynx and the **pharyngeal slits** between them. The deeply grooved **endostyle** will be seen in the floor of the pharynx, and a similar **epipharyngeal groove** in its roof.

In certain sections, pieces of the **gonads** protrude into the atrium from the body wall, carrying the lining of the atrium before them. An ovary consists of many large nucleated cells; testis tissue appears as small dark dots or fine tubules.

If the section is taken near the caudal end of the pharynx, the hollow, oval-shaped **midgut caecum** will be observed lying on the right side of the pharynx. It first appears to be completely within the atrium but is actually covered with a layer of atrial epithelium because it has pushed into the atrium from behind (see Fig. 1-4A).

A coelom is present in amphioxus but in a highly modified form. Close examination of the section will reveal certain of its subdivisions. A pair of **dorsal coelomic canals** are located slightly lateral to several of the most dorsal pharyngeal bars. The atrium in this region is a narrow space between the pharyngeal bars and the dorsal coelomic canals. Another coelomic canal will be found ventral to the epithelium of the endostyle. The **subendostylar coelom** connects with the dorsal coelomic canals by small coelomic passages within every other pharyngeal bar; portions of these passages may be found. The bars containing the coelomic passages are the primary pharyngeal bars; secondary bars (tongue bars) grow down between them from the roof of the pharnyx during embryonic development as they do in the hemichordates. Another portion of the coelom will be seen lateral to the gonads. It too connects with the dorsal coelomic canals, but this connection often disappears in the adult. Finally, a very narrow coelomic space may be seen between the cells of the midgut caecum and the surrounding cells of the atrial epithelium.

Excretion occurs via clusters of small cells that may be seen protruding into the dorsal coelomic canals (Fig. 1-5). These cells have been called **solenocytes** (although they are certainly not homologous to the solenocytes of certain primitive worms). The solenocytes connect to a short duct called a **nephridioduct** that opens into the atrium.

The excretory organs of amphioxus are unique (Ruppert and Smith, 1988). The solenocytes lie within the dorsal coelomic canals; each cell body has a series of footlike processes called **podocytes** that abut a glomerular artery, which branches from the dorsal aorta (Fig. 1-7). Long microvilli form a slitted, tubelike structure that extends through the dorsal coelomic canals to enter the nephridioduct. Excretory products are thought to pass from the glomerular artery between the podocytes and into the dorsal coelomic canal. From here the excretory products pass through the slits between the

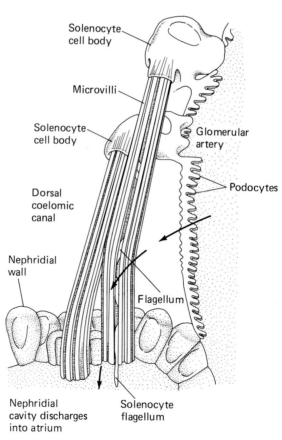

Figure 1-7
Diagram of two solenocytes of amphioxus based on electron microscope studies. The solenocytes lie within the dorsal coelomic canals and discharge into nephridia that open into the artrium. Arrows indicate the direction of flow of waste products. *(From Walker,* Functional Anatomy of Vertebrates, *after Brandenburg.)*

microvilli and are propelled by a long flagellum in the microvilli tube into the nephridioduct and on to the atrium.

Certain blood vessels may also be seen in the section through the pharynx, but we will not consider the circulatory system in detail. It is of interest, however, that the general course of blood flow is the same as in a vertebrate, i.e., from the tissues cranially to the ventral side of the pharynx, dorsally through the gill bars, and caudally in a dorsal aorta to the body. There is no well-developed heart, and many of the vessels are contractile. The tissues are supplied by open lacunae, for true capillaries are absent. The blood contains a few amoebocytes but no other cells or respiratory pigments.

♦ **(D) SECTION THROUGH THE INTESTINE**

In a section through the midgut or hindgut (Fig. 1-6**B**), the metapleural folds are absent, and the **ventral fin** is present instead. Such a section will also show the **intestine** lying within a large **coelomic space,** for the separate coelomic passages of the pharyngeal region have coalesced. A caudal diverticulum of the **atrium** lies on the right side of the intestine and coelom.

♦ **(E) SECTION THROUGH THE ANUS**

A section through the anus passes through the **caudal fin.** Note that this fin is narrower and higher than either the dorsal or ventral fins, which it replaces. The intestine opens at the **anus** on the left side of the fin.

All of the diagnostic characteristics of chordates are well represented in the cephalochordates. There can be no doubt about the affinities of the group. In addition, amphioxus has many features that are found in the vertebrates. Notable among these are the myomeres, a ventral, glandular diverticulum of the alimentary canal, and a postanal tail. On the other hand, amphioxus has certain peculiarities, including the atrial chamber, contractile notochord, and an extreme rostral extension of the notochord, which are adaptations for its burrowing mode of life and indicate that it is not a direct ancestor of the vertebrates. Although it is clear that there is a relationship among vertebrates, protochordates, hemichordates, and echinoderms, biologists are uncertain of the course that evolution has taken here.

CHAPTER TWO

◆ ◆

THE LAMPREY—A JAWLESS VERTEBRATE WITH SEVERAL PRIMITIVE FEATURES

There are nearly 43,000 described species of contemporary vertebrates; they are by far the largest of the chordate subphyla. Vertebrates are the most active of all chordates, and their distinctive, or derived, characters reflect this change in lifestyle. In the adults of most species, a segmented **vertebral column** replaces the notochord as a strut that resists compression. Sense organs and nervous tissue are concentrated in a well developed head; vertebrates have become **cephalized. Nasal cavities, eyes, ears,** and (in fishes) a **lateral line system** are present. A **brain** integrates the activities of these sense organs with the rest of the body. A braincase, or **cranium,** encases the brain and major sense organs. An increased level of metabolism is made possible by the **muscularization of the gut** and by the evolution of **visceral arches** between the pharyngeal pouches. Aquatic vertebrates (fishes and larval amphibians) can contract and expand their pharynx by means of muscles and draw in a larger volume of water and food than can protochordates with their ciliary currents. The larvae of ancestral vertebrates probably continued to be filter feeders, but the combination of a cranial array of sense organs and a muscular pharynx probably enabled the adults of ancestral vertebrates to find and ingest small prey organisms and tear off bits of soft tissue from plants and dead animals. **Pancreatic tissue** secretes an array of digestive enzymes, and a **liver,** among its many functions, stores excess food as lipids or glycogen. **Gills** in the pharyngeal pouches provide for increased gas exchange; a chambered **heart** pumps blood efficiently through a closed circulatory system. Many waste products are eliminated through **kidneys.**

The earliest vertebrates are jawless fishes of the class **Agnatha.** The earliest agnathans, in turn, were several orders of heavily armored fishes collectively called **ostracoderms.** Fragments of ostracoderm bony scales have been discovered in late Cambrian deposits over 500 million years old. These fishes thrived in the early Paleozoic era but died out nearly 300 million years ago. Although ostracoderms are extinct, agnathans are represented today by about 60 species of hagfishes (order **Myxiniformes**) and lampreys (order **Petromyzontiformes**). Lampreys and hagfishes share several features, including a round, jawless mouth. For this reason they are sometimes grouped as cyclostomes, but they differ in so many other ways from each other that they have probably had a long independent evolution from ostracoderm ancestors.

We will study the anatomy of vertebrates primarily by examining each organ system in turn in a representative series of species, but we will begin by studying the overall structure of a species that has retained a number of primitive characteristics. This exercise will acquaint you with the basic organization of vertebrates and will give you an idea of how various organs and structures might have appeared in an ancestral vertebrate. A favorable species for study is the large sea lamprey, *Petromyzon marinus.* This species enters fresh water to breed and has become established in the Great Lakes.

The lamprey shows a mixture of primitive and derived specialized characters. In studying the animal, you should try to separate one from the other, for the aim is to learn the anatomy of the lamprey not just for its own sake, but also for the information it may provide concerning the structure of ancestral vertebrates. Among the major specializations of the lamprey are its eel-like shape, the absence of an armor of dermal bone, and its mode of feeding. Lampreys attach to other fishes by means of a suctorial buccal funnel, rasp away the flesh of their prey, and suck its blood. A number of modifications of the digestive and respiratory systems are correlated with this lifestyle.

THE ADULT LAMPREY

External Features

Examine the external features of a lamprey, noting its eel-like shape and the scaleless, slimy skin. The body can be divided into **head, trunk,** and **caudal** regions. The head extends through the gill or branchial area; the caudal region, or tail, from the cloacal aperture to the tip of the tail. The body is rounded in cross section cranially, but, progressing caudally along the trunk and tail, the body becomes compressed, or flattened from side to side. This increases the surface area that can be pushed against the water as the body undulates during locomotion.

Observe that the only fins present are in the median plane—two **dorsal fins** and a symmetrical **caudal fin** (Fig. 2-1). Absent are the lateral (paired) fins, usually found in other groups of fishes. The median fins are supported by slender, cartilaginous **fin rays,** which you can see best if you cut across the fin in the frontal plane.

Notice the **buccal funnel** at the front of the head. It is fringed with **papillae** and lined with **horny teeth.** The papillae are sensory and also enable the lamprey to form a tight seal when it attaches to another fish. The teeth within the buccal funnel are composed of cornified cells and hence differ from the true teeth of jawed vertebrates. A protrusible

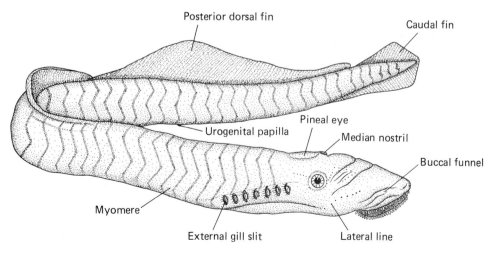

Figure 2-1
Lateral view of the sea lamprey, *Petromyzon marinus.*

tongue is situated near the center of the funnel. It is outlined by a ring of dark tissue and is provided with small, rasplike, horny teeth. The **mouth opening** is just dorsal to the tongue. No jaws are present. This is a primitive characteristic in vertebrates, but the horny teeth and the tongue are specializations for the animal's blood-sucking mode of feeding. A single, **median nostril** is located far back on the top of the head.

A median nostril is an unusual but ancient feature, for it is also found in certain of the ostracoderms. The manner in which the nostril is displaced during embryonic development from the more usual ventral position is shown in Figure 2-2. This condition derives from the tremendous enlargement of the upper lip of the embryo to form the buccal funnel.

18

Chapter 2
The
Lamprey—
A Jawless
Vertebrate
with Several
Primitive
Features

Just caudal to the nostril, you will see an oval area that is often slightly depressed and generally a lighter color than the rest of the skin. The **pineal eye,** an ancient feature also found in the ostracoderms, lies beneath this depigmented skin. Experiments have shown that the pineal eye detects changes in light and initiates diurnal color changes in larval lampreys, probably by its effect upon the hypothalamus and hypophysis, or pituitary gland. It is possible that other physiological activities are adjusted to the diurnal cycle in a similar way. A pair of conventional but lidless **lateral eyes** are on the sides of the head. Seven pairs of oval **external gill slits** are located behind the eyes.

If you let the head dry a bit and examine it with low magnification, you may be able to detect groups of pores, or little bumps, arranged in short lines. One group is found caudal to the top of the lateral eye; another extends from the underside of the eye rostrally and dorsally. A third group is located on the ventral side of the head, caudal to the buccal funnel. These, together with other less conspicuous pores, are parts of the **lateral line system,** a group of sense organs associated with detecting vibrations and movements in the water (p. 216).

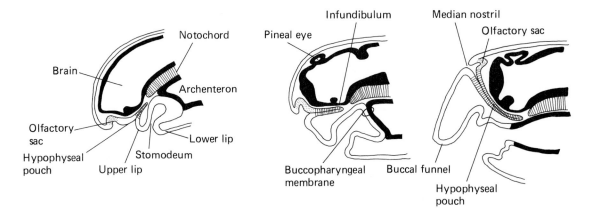

Figure 2-2
Diagrams in the sagittal plane of three stages in the development of the head region of *Petromyzon*. Note in particular how the originally independent hypophyseal pouch and olfactory sac become crowded together and are pushed on the top of the head by the enlargement of the upper lip. This region also differs from gnathostomes in that the hypophysis invaginates anterior to the stomodeum rather than from within the stomodeum. *(From Parker and Haswell, after Dohrn.)*

The body musculature consists chiefly of segmented **myomeres,**[3] whose outlines can be seen through the skin of the trunk and tail, especially in smaller specimens. Each myomere is roughly W-shaped, the top of the W being cranial. Furthermore, each is continuous from its dorsal to ventral end, with no interruption near its middle as in jawed vertebrates.

On the underside of the caudal end of the trunk you will see a shallow pit called the **cloaca.** It receives the excretory and genital products, which leave through the tip of a small **urogenital papilla** and, just cranial to the papilla, the **anus,** or opening of the intestine. The opening of the cloaca to the surface is called the **cloacal aperture.**

Sagittal and Cross Sections

Study the internal structure of the lamprey by examining a midsagittal section (Fig. 2-3) and a series of cross sections (Fig. 2-4). It is desirable to work in groups for this portion of the work. Certain students should prepare cross sections, and the rest, sagittal sections. A good series of cross sections is as follows: (1) through the pineal and lateral eyes, (2) through a pair of external gill slits near the middle of the branchial region, (3) just caudal to the last external gill slit, (4) about 3 centimeters behind the branchial regions, (5) near the middle of the trunk, (6) about 3 centimeters cranial to the cloaca, and (7) through the tail. In preparing the sagittal sections, use a large knife and be particularly careful to cut the head and branchial region as close as possible to the sagittal plane. If the halves are unequal in size, take the larger one and dissect it down to the sagittal plane.

♦ **(A) THE SKELETAL AND MUSCULAR SYSTEMS**

Study the better sagittal section, correlating the appearance of structures in this view with their appearance in the cross sections. Notice that the main skeletal axis is a long **notochord** extending from the caudal end of the body to a point beneath the middle of the **brain.** It has a gelatinous texture but is enclosed in a strong fibrous sheath. It is firm yet flexible and probably serves primarily to prevent the body from telescoping when the myomeres contract. In favorable cross sections, cartilaginous blocks **(arcualia)** will be seen above the notochord on either side of the **spinal cord.** They constitute the rudiments of vertebral arches. Other cartilages surround parts of the brain and extend into the roof of the buccal funnel. These are parts of the primary braincase, or **chondrocranium.** A long median **lingual cartilage** extends into the tongue. Still other cartilages will be found beneath and lateral to the gill region and posterior to the heart. They form a **branchial basket** that supports the gill region. The branchial basket of the lamprey and other agnathans lies lateral to the gill pouches rather than medial to the gills as do the visceral arches of all other vertebrates. Many zoologists believe that the branchial basket of the lamprey and the visceral arches of other vertebrates evolved independently and are not homologous.

---

[3]See footnote 2, p. 7.

20

Chapter 2
The
Lamprey—
A Jawless
Vertebrate
with Several
Primitive
Features

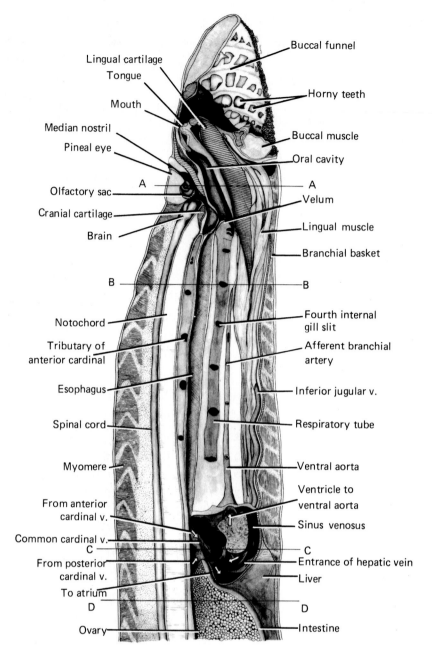

Figure 2-3
A sagittal section through the anterior part of a lamprey. The direction of blood flow through the heart is shown by arrows. The planes of four of the transverse sections shown in Figure 2-4 are indicated by the arrows A to D.

Examine a special preparation of the skeletal system, if available, to appreciate this system better (Fig. 2-5).

The muscular system consists primarily of the segmented myomeres already observed. Note their appearance in the sections. Each myomere consists of bundles of longitudinal muscle fibers that attach onto connective tissue **myosepta** between the myomeres. Waves of contraction passing anteroposteriorly along the two sides of the body cause the lateral undulations of the trunk and tail. Jets of water expelled from the gill slits may also aid locomotion. The movements of the buccal funnel and tongue during feeding are caused by the intricate musculature you see associated with these structures.

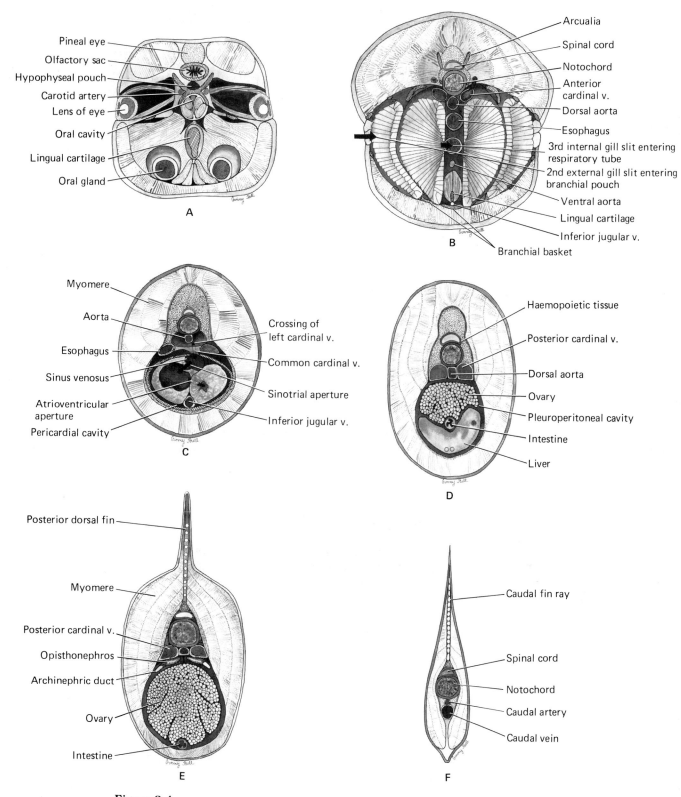

Figure 2-4

Representative transverse sections of the lamprey: **A**, through the pineal and lateral eyes (plane A in Fig. 2-3); **B**, through an external gill slit (plane B); **C**, through the heart (plane C); **D**, through the cranial end of the liver (plane D); **E**, through the caudal part of the trunk; **F**, through the tail.

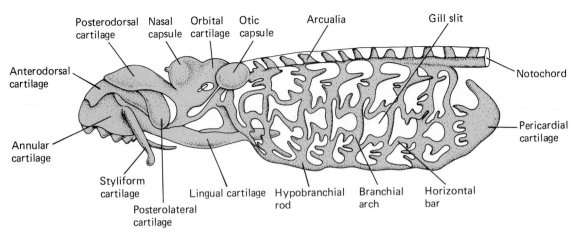

Figure 2-5
Lateral view of the cranium and branchial basket of the lamprey. *(From J. Z. Young, after Parker.)*

22

Chapter 2
The
Lamprey—
A Jawless
Vertebrate
with Several
Primitive
Features

♦ **(B) THE NERVOUS SYSTEM AND SENSE ORGANS**

The brain and the spinal cord have been seen lying above the notochord. The lamprey's brain has poorly developed acousticolateral centers, and the cerebellum is small, but in other major respects it is similar to the dogfish's brain (p. 241).

Notice the connections of the nostril in the sagittal section. It first leads into a dark **olfactory sac** located rostral to the brain. The internal surface of the sac is greatly increased by numerous folds. Then a **hypophyseal pouch** continues from the entrance of the olfactory sac and passes ventral to the brain and the rostral end of the notochord. Respiratory movements of the pharynx squeeze the end of the hypophyseal pouch much as we squeeze the bulb of a medicine dropper, thereby moving water in and out of the olfactory sac. The adenohypophysis, a part of the pituitary gland, is derived from an embryonic hypophyseal pouch, but in the jawed vertebrates the hypophysis invaginates from the roof of the mouth (stomodeum, Fig. 2-2). The pituitary gland is difficult to see in this type of dissection.

The **pineal eye** is represented by the clear area caudal to the olfactory sac. Actually, there are two median eyes here, for the pineal eye lies dorsal to a smaller parietal eye, but details cannot be seen. The well-developed **lateral eyes** show in one of the cross sections (Fig. 2-4A). The eye of fishes will be studied in more detail later (p. 220), but you will recognize the large spherical **lens** characteristic of fishes, a dark **pigmented layer,** and possibly the whitish **retina** between the pigmented layer and the lens. If you make another cross section just caudal to the lateral eye, you will see part of the **inner ear.** It will appear as a bit of tissue imbedded in a cavity of the chondrocranium lateral to the brain.

The ear of fishes will be studied later, but only the inner part is present. As you will learn (Fig. 8-13), the inner ear normally has three semicircular ducts. The lamprey, however, has only the two vertical ducts. This condition was also found in certain of the ostracoderms.

In the sagittal section, you will see that the mouth opening leads into an **oral cavity,** which extends caudad to the level of the rostral end of the notochord. A pair of **oral glands,** which can be seen in the first cross section (Fig. 2-4**A**), lie ventral and lateral to the oral cavity. Their secretion, which is discharged beneath the tongue, acts as an anticoagulant, preventing the blood of the prey from clotting. The secretions of the oral glands also contain hemolytic and cytolytic enzymes that initiate the digestion of blood cells and proteins. The caudal end of the oral cavity leads into two tubes—an **esophagus** dorsally and a **respiratory tube** ventrally. The esophagus can be recognized by the numerous oblique folds in its lining. At the level of the heart it passes to the left of the large common cardinal vein that enters the heart (Fig. 2-4**C**) and leads to the **intestine,** the first part of which lies between the liver and gonad (Figs. 2-3 and 2-4**D**). The intestine continues as a long, straight tube to the cloaca. Its internal surface area is increased by longitudinal folds, one of which is particularly prominent; it has a somewhat spiral course and is called the **spiral valve.** Otherwise there is little differentiation to the intestine.

A large and, in the sagittal section, triangular **liver** lies just caudal to the heart. It develops embryonically as a ventral outgrowth of the intestine. During larval life a gall-bladder (see Fig. 2-7) and bile duct are present, but these are lost at metamorphosis.

> The **pancreas** of jawed vertebrates is not present as a gross organ in the lamprey, but Barrington (1945) has found patches of cells comparable to the enzyme-secreting cells of the pancreas in the wall of that part of the intestine adjacent to the liver, and other groups of cells comparable to the endocrine portion of the pancreas (islets of Langerhans) imbedded in the intestinal wall and liver. Experimental destruction of the latter cells causes a rise in blood sugar.

The intestine and liver lie within a division of the **coelom** known as the **pleuroperitoneal cavity.** The intestine of vertebrates is usually supported by a long dorsal mesentery, but this is reduced in the lamprey to a few strands surrounding blood vessels that pass to the intestine.

Returning to the respiratory tube, you will see that its entrance is guarded by a series of tentacles that constitute the **velum;** its wall is perforated by seven **internal gill slits.** By looking at a cross section, and using a probe, you will note that each internal gill slit leads into an enlarged gill pouch, or **branchial pouch** (Fig. 2-4**B**), lined with gill lamellae. The pouches open to the surface through the **external gill slits.**

> Both the respiratory tube and the esophagus of the adult develop from a longitudinal division of the larval pharynx. The resulting separation of digestive and respiratory tracts is correlated with the lamprey's mode of feeding. When attached to its prey, the lamprey respires by pumping water in and out of the external gill slits. Some water may seep into the respiratory tube, but this would not interfere with feeding. The most active phase of respiration is expiration, for then the branchial muscles constrict the pouches, and water is forcibly expelled. Inspiration results primarily from the

elastic recoil of the branchial basket. When the lamprey is not feeding, some water may enter the gill pouches through the mouth and internal gill slits—the usual situation in fishes.

24

Chapter 2
The
Lamprey—
A Jawless
Vertebrate
with Several
Primitive
Features

♦ **(D) THE CIRCULATORY SYSTEM**

Study the **heart** in sagittal and transverse sections (Figs. 2-3 and 2-4**C**). Its structure is also shown diagrammatically in Figure 2-6. The heart lies in another division of the coelom, the **pericardial cavity.** The heart consists of a thin-walled, tubular **sinus venosus** located between the atrium and ventricle. The sinus venosus receives blood low in oxygen content from the body and leads into the large atrium through a **sinuatrial aperture.** The **atrium** is generally filled with hardened blood and hence is dark in color. It is located lateral to the sinus venosus and fills most of the left side of the pericardial cavity. It leads by an **atrioventricular aperture** into a muscular **ventricle** located in the right ventral portion of the pericardial cavity. Valves present in the apertures keep blood moving in the correct direction. A conus arteriosus, found in most fishes, is not present, so the **ventral aorta** leaves directly from the ventricle.

The pericardial cavity is separated from the pleuroperitoneal cavity by a **transverse septum.** In the lamprey this septum is stiffened by a large cartilage of the branchial basket. Since the pericardial wall is stiff and will not collapse, contraction of the ventricle probably results in the increase of the "open" space within the pericardial cavity and, hence, a reduction of pressure around the atrium and sinus venosus. This reduced pressure would help "suck" blood in from the veins. Such a mechanism has been demonstrated in the dogfish (p. 356), where it plays an important role in the return of venous blood. Blood pressure in the veins is very low in fishes.

The ventral aorta extends forward beneath the respiratory tube, giving off eight paired **afferent branchial arteries** that lead to capillaries in the gills. The branchial arteries may not be seen; the first pair leads to a pair of vestigial branchial pouches located rostral to the first pair of well-developed pouches. The **dorsal aorta** can be found in the cross sections just ventral to the notochord. It receives oxygenated blood from the gills by way of **efferent branchial arteries** and continues caudally, supplying the body musculature and viscera. In the tail it is called the **caudal artery** (Fig. 2-4**F**). The head is

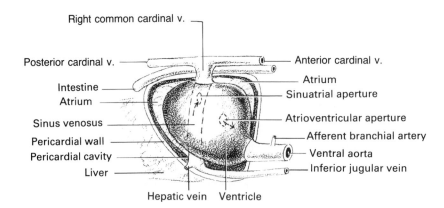

Figure 2-6
Lateral view of the right side of the heart of a lamprey. The position of the sinus venosus, which lies between the atrium and ventricle, is shown by dotted lines.

supplied chiefly by a pair of **carotid arteries** (Fig. 2-4**A**) that leave from the front of the dorsal aorta.

A **caudal vein** is located ventral to the artery. The characteristic renal portal system of fishes is absent, for the caudal vein goes not to the kidneys but directly to the paired **posterior cardinals.** These veins can be seen in cross sections of the trunk on either side of the dorsal aorta (Fig. 2-4**D** and **E**). At the level of the heart the right posterior cardinal unites with the right anterior cardinal coming from the head to form a **right common cardinal** (duct of Cuvier) that passes ventrally on the right side of the esophagus to enter the sinus venosus (Fig. 2-3). Most fish have a comparable left common cardinal, but in the lamprey the left posterior and anterior cardinals curve medially to join the right common cardinal (Fig. 2-4**C**). An inconspicuous hepatic portal system runs from the alimentary canal to the liver. From here blood passes to the sinus venosus through a **hepatic vein** (Fig. 2-3).

The head and the branchial region and their numerous venous sinuses are drained by a pair of **anterior cardinals** (Fig. 2-4**B**), which lead caudally to the right common cardinal, and by a median **inferior jugular vein** that enters the ventral part of the sinus venosus. The anterior cardinals are located lateral to the notochord; the inferior jugular vein is found ventral to the prominent tongue musculature in the floor of the branchial region.

Lampreys have no spleen. Blood-forming, **haemopoietic tissue** is represented by the dark, specialized connective tissue surrounding the spinal cord and notochord (Fig. 2-4**D**). The excretory organs consist of paired **opisthonephric kidneys** that appear as flaps suspended from the dorsal wall of the caudal half of the pleuroperitoneal cavity (Fig. 2-4**E**). You may have to push the gonad aside to see them. Each kidney is drained by an **archinephric duct,** which runs along its free border. Cut away the lateral body wall in the cross-section segment that contains the cloaca and observe that the two archinephric ducts unite at the caudal end of the pleuroperitoneal cavity to form a **urogenital sinus,** which opens at the tip of the urogenital papilla. You may be able to pass a bristle into an archinephric duct and out the urogenital papilla.

Although the **ovary** or **testis** (gonads) develop from paired primordia, that of the adult is a large median organ that fills most of the pleuroperitoneal cavity. It is suspended by a **mesentery.** Genital ducts are absent in both sexes, the gametes being discharged directly into the coelom. They leave the coelom through paired **genital pores** located on either side of the urogenital sinus at the extreme posterior end of the pleuroperitoneal cavity. Search for the pores by carefully probing this area.

THE AMMOCOETES LARVA

In the spring of the year, some adult lampreys leave the ocean and large lakes where they have been living and ascend streams to spawn and die. In about three weeks the eggs hatch into larvae that, when first discovered, were thought to be a distinct species, named *Ammocoetes.* The ammocoetes larvae burrow into the sand, much like amphioxus, and live as filter feeders for five to seven years. After attaining a length of about 12 centimeters, the larvae metamorphose. Their sense organs become better developed, their oral region changes, and the pharynx becomes divided longitudinally into the "esophagus" and respiratory tube. These changes adapt the animal to its new lifestyle as a predaceous blood sucker. Adults then descend to the lakes or ocean where they live for one or two years before ascending the rivers to spawn.

26

Chapter 2
The
Lamprey—
A Jawless
Vertebrate
with Several
Primitive
Features

♦   (A) WHOLE MOUNT SLIDE

Examine a whole mount slide of a small ammocoetes larva with the low power of a microscope (Fig. 2-7). Note its fusiform shape and the **dorsal** and **caudal fins,** which are continuous with each other. The fine dark lines, or specks, on the body surface are pigment cells (**chromatophores**). If the pigment is dispersed, you can see that each chromatophore consists of a central area from which branching processes radiate. Although the **myotomes** have been rendered transparent in preparing the slide, their outlines may show as faint lines on the surface.

The **spinal cord** appears as a dorsal, dark-staining band that is enlarged rostrally to form the **brain.** In favorable specimens, you can see that the brain is composed of several lobes separated by constrictions, the largest and most caudal of which is the hindbrain (**rhombencephalon**); the next, the midbrain (**mesencephalon**); and the most rostral (sometimes subdivided), the forebrain (**prosencephalon**). The **notochord** appears as a light-staining, longitudinal band ventral to the spinal cord and the caudal two divisions of the brain.

The small surface protuberance rostral to the brain is the **median nostril,** which leads into the **hypophyseal pouch** lying ventral to the brain. If apparent, each **lateral eye** is represented by a round, dark spot lying between the mesencephalon and prosencephalon. An evagination from the caudal portion of the roof of the prosencephalon is the primordium of the **pineal eye;** the large, clear vesicle overlapping the front of the rhombencephalon is the primordium of the **inner ear.**

Many of the specializations of the adult digestive and respiratory systems are absent in the larva because it feeds by sifting minute food particles from the water. Observe that the **upper lip** has already enlarged to form the primordium of the buccal funnel. The **lower lip** appears as a transverse shelf. The **mouth opening** is surrounded by a series of **oral tentacles** that function as strainers and as sensory organs. Behind the mouth is a

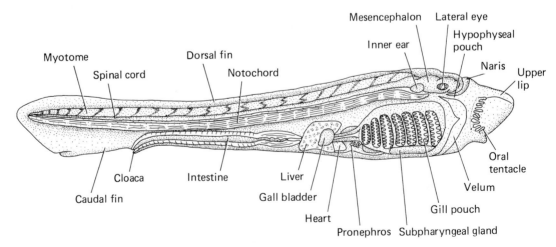

Figure 2-7
Lateral view of a whole mount slide of the ammocoetes larva.

clear chamber, the **oral cavity,** bounded caudally by a pair of large, muscular flaps, the **velum.** Movements of the velum bring a current of water and food into the **pharynx** behind it. It is important to note that the feeding current is caused by muscular action of the velum and pharynx rather than by ciliary action as in protochordates. This is more efficient and is undoubtedly a factor that permits the ammocoetes larva to attain a considerably larger size than any of the protochordates. Observe the seven large **gill pouches** in the pharyngeal region. They are lined with **gill lamellae** and open through small, round, **external gill slits,** which may be seen by sharp focusing on the surface. Ventral to the pharynx you will find a large, dark-staining, elongate body called the **subpharyngeal gland.** In young larvae the subpharyngeal gland may secrete mucus, which is discharged through a duct to the pharynx. Additional mucus is produced by the lining of the pharynx and gill pouches. The mucus forms a longitudinal cord in the pharynx that traps food particles, and the cord is carried back into the esophagus. The subpharyngeal gland has many of the characteristics of the protochordate endostyle and may be its homologue. Certain of its cells also produce iodoproteins, and these cells are transformed into the thyroid gland of the adult lamprey. Posterior to the esophagus the alimentary canal widens to form the **intestine,** which continues to the **cloaca.**

The **liver** is located adjacent to the caudal portion of the larval esophagus and contains a large, clear vesicle, the **gallbladder.** Notice the **heart** lying ventral to the esophagus in front of the liver. Between the heart and esophagus you will see a few bell-shaped or finger-like processes, or tubules, which are parts of the larval **pronephric kidney.**

♦ **(B) CROSS SECTION THROUGH THE PHARYNX**

Examine a cross section through the pharynx with the low power of a microscope and correlate the appearance of structures in this view with their appearance in the whole mount slide. A somewhat flattened **spinal cord** will be seen dorsally lying above the large **notochord** (Fig. 2-8). Notice that the spinal cord contains the characteristic central canal of chordates and that the notochord is composed of vacuolated cells encased in a dense sheath. The **dorsal aorta** lies ventral to the notochord, and the paired **anterior cardinal veins** flank the notochord ventrolaterally. Bits of cartilaginous **arcualia** may be seen beside the spinal cord and notochord.

A large central **pharynx** and lateral **branchial pouches** occupy most of the slide ventral to the dorsal aorta. The branchial pouches and the septa separating them slant posteriorly as they extend from the pharynx toward the body surface; hence, a transverse section may cut through parts of two pouches. The relationship of the feathery **gill lamellae** to the septum to which they attach is complex. Lamellae on the more anterior surface of a septum point toward the body surface (left side of Fig. 2-8), whereas those on the posterior surface are directed toward the pharynx (right side of Fig. 2-8). Many blood vessels will be seen in the gills. **Internal gill slits** lead from the pharynx to a branchial pouch, and each pouch discharges on the body surface through a single **external gill slit.**

The large **subpharyngeal gland** forms a ridge in the floor of the pharynx. It may appear as two or three glandular masses, depending on the level of the section. A **ventral aorta** and **hypopharyngeal ridge** overlie the subpharyngeal gland. An **epipharyngeal**

28

Chapter 2
The
Lamprey—
A Jawless
Vertebrate
with Several
Primitive
Features

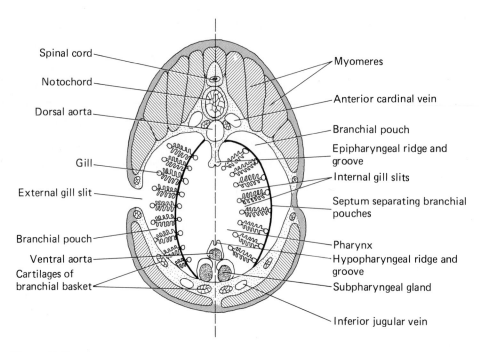

Figure 2-8
A diagrammatic section through the pharynx of an ammocoetes larva. A section taken at the level of an external gill slit is shown on the left, and one between external gill slits is shown on the right.

ridge extends into the pharynx from its roof. Cartilages of the **branchial basket** and associated muscles will be seen in the body wall lateral to the branchial pouches.

The ammocoetes larva is of great phylogenetic interest and may give some idea of the nature of ancestral vertebrates. It resembles amphioxus in that it has all of the chordate characteristics but lacks such specializations as the atrium surrounding the pharynx and the extreme rostral extension of the notochord. It also lacks many of the specializations of the adult lamprey for blood sucking yet has all of the essential vertebrate features. Its well-developed sense organs, brain, myomeres, and caudal fin enable it to be more active than any protochordate. It uses efficient muscular movements of the velum and pharynx in filter feeding, and muscular contractions of the gut carry food posteriorly. Being a larger and more active animal than amphioxus, it no longer depends on cutaneous gas exchange but has evolved gills at a site where there is a good flow of water, i.e., in the gill pouches. These gill pouches, therefore, are now part of both the feeding and the respiratory mechanisms.

◆ ◆

THE EVOLUTION AND EXTERNAL ANATOMY OF VERTEBRATES

Building on our knowledge of the anatomy of the lamprey, a representative of the jawless vertebrates **(Agnatha),** we can now begin to investigate the anatomy of the jawed vertebrates **(Gnathostomata).** A general history of vertebrate evolution will provide a framework for later discussions and interpretations of the structures in the various animals we will study. If possible, you should supplement the following history of vertebrate evolution by touring a museum, aquarium, or synoptic collection.

VERTEBRATE EVOLUTION

The past and present diversity of vertebrates is the result of evolutionary processes extending over millions of years. This evolutionary history is best described in a narrative text and illustrated by an evolutionary (or phylogenetic) tree (Figs. 3-1 and 3-2). This type of description, however, is also lengthy and complex and is therefore usually summarized and condensed in the form of a classification (Tables 3-1 and 3-2). In translating the complex, multidimensional process of evolution into a listlike classification, however, we lose much information and the arrangement of the various vertebrate groups is therefore the result of compromises. The classification presented here is no exception; it is a compromise between the latest classifications presented in various contemporary textbooks and our goal to provide you with a didactically useful list of vertebrates that are relevant for the study of comparative anatomy. For example, we have tried to avoid listing monotypic groups of uncertain relationships and have included the Acanthodii with the Osteichthyes, and the Chelonia with the Anapsida. We also left out some intermediate groupings that we felt were not relevant in the context of this manual, such as the Neopterygii, which comprise the Holostei and the Teleostei. Finally, we retained several groups that some scientists regard as artificial but are didactically useful, such as the Ostracodermata, the Cyclostomata, the Holostei, and the Reptilia.

   The earliest fossils of vertebrates are found in late Cambrian deposits over 500 million years old. They are representatives of the **Ostracodermata,** a group of extinct jawless fishes that includes several orders. They were usually heavily armored with bony plates. The paired spines or lobes found in the pectoral region of some species were not true fins, having no internal skeleton or musculature. Ostracoderms are thought to have been mostly bottom-dwelling filter feeders. All ostracoderms were extinct by the end of the Devonian. The only jawless vertebrates that have survived into the present are the **Cyclostomata,** comprising the lampreys **(Petromyzontiformes)** and the poorly known marine hagfishes **(Myxiniformes).**

(text continues on p. 33)

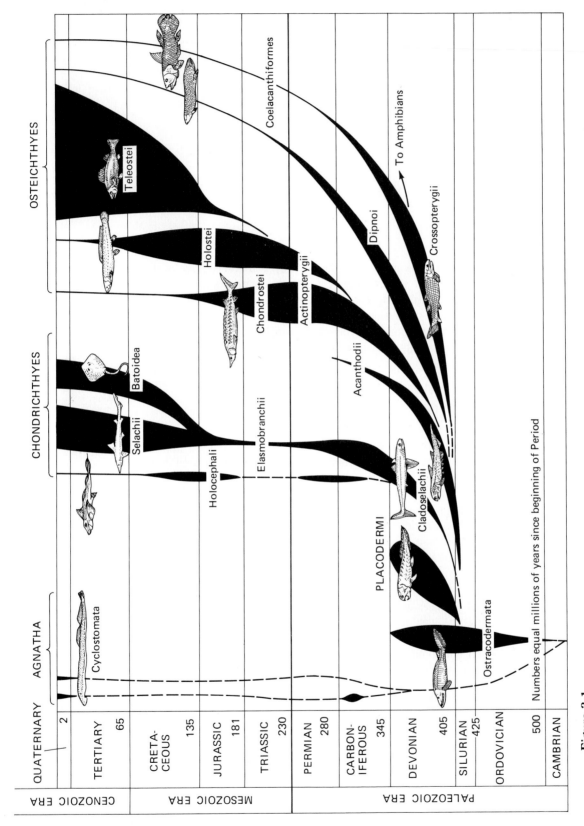

Figure 3-1
An evolutionary tree of fishes, showing the relationships of the various groups, their relative abundance, and their distribution in time.

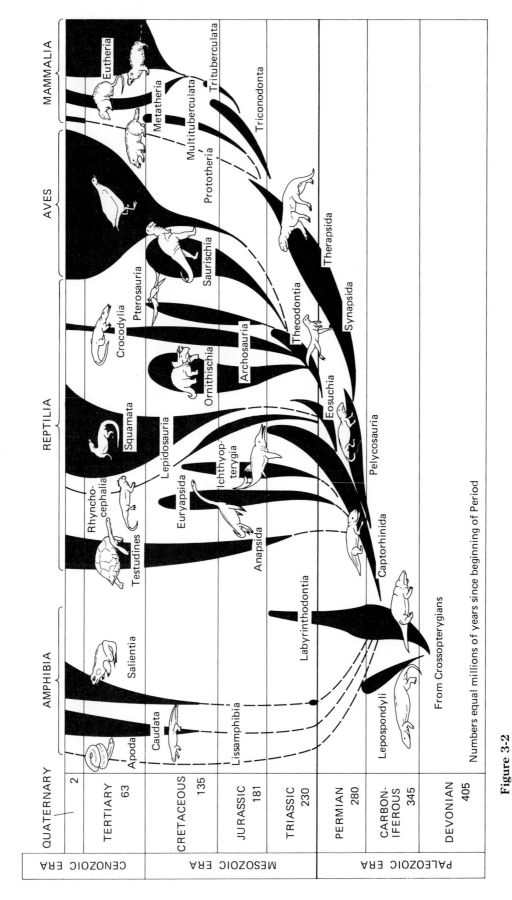

Figure 3-2

An evolutionary tree of terrestrial vertebrates, showing the relationship of the various groups, their relative abundance, and their distribution in time.

Table 3-1   Synoptic Classification of the Vertebrates (Subphylum Vertebrata)

1. **Class Agnatha (jawless vertebrates)**
    Ostracodermata (informal assemblage of extinct agnathans)
    Cyclostomata (informal assemblage of living agnathans)
        Order Petromyzontiformes (lampreys)
        Order Myxiniformes (hagfishes)
2. **Class Placodermi**
        Order Arthrodira
        Order Antiarchi
3. **Class Chondrichthyes (cartilaginous fishes)**
    Subclass Elasmobranchii
        Order Selachii (Squaliformes) (sharks)
        Order Batoidea (Rajiformes) (skates, rays)
    Subclass Holocephali (chimaeras)
4. **Class Osteichthyes (bony fishes)**
    Subclass Acanthodii
    Subclass Actinopterygii (ray-finned fishes)
        Superorder Chondrostei
        Order Palaeoniscoidea
        Order Polypteroidea
            Family Polypteridae (bichirs)
        Order Acipenseroidea
            Family Acipenseridae (sturgeons)
            Family Polyodontidae (paddlefishes)
        Superorder Holostei (*Lepisosteus* species, garpikes, and *Amia calva,* mudfish or choupique)
        Superorder Teleostei
    Subclass Sarcopterygii (lobe-finned fishes)
        Order Crossopterygii
        Suborder Rhipidistia
        Suborder Coelacanthiformes *(Latimeria chalumnae)*
        Order Dipnoi (lungfishes)
5. **Class Amphibia**
    Subclass Labyrinthodontia
        Order Ichthyostegalia
        Order Temnospondyli
        Order Anthracosauria
    Subclass Lepospondyli
    Subclass Lissamphibia
        Order Urodela (Caudata) (salamanders, newts)
        Order Gymnophiona (Apoda) (caecilians)
        Order Anura (Salientia) (frogs, toads)
6. **Class Reptilia**
    Subclass Anapsida
        Order Captorhinida (stem reptiles)
        Order Chelonia (Testudines) (turtles, tortoises)
    Subclass Diapsida
        Superorder Lepidosauria
        Order Eosuchia
        Order Rhynchocephalia (*Sphenodon* species, tuatara)
        Order Squamata
            Suborder Lacertilia (lizards)
            Suborder Ophidia (Serpentes) (snakes)
            Suborder Amphisbaenia
        Superorder Archosauria
        Order Thecodontia
        Order Saurischia (dinosaurs)
        Order Ornithischia (dinosaurs)
        Order Pterosauria
        Order Crocodylia (alligators, crocodiles)
        Superorder Euryapsida
        Superorder Ichthyopterygia

Subclass Synapsida (mammal-like reptiles)
   Order Pelycosauria
   Order Therapsida
7. Class Aves (birds)
8. Class Mammalia
 Subclass Prototheria
   Order Triconodonta
   Order Monotremata
     Family Tachyglossidae (spiny anteaters)
     Family Ornithorhynchidae (duck-billed platypus)
 Subclass Allotheria
   Order Multituberculata
 Subclass Theria
  Infraclass Trituberculata
  Infraclass Metatheria (Marsupialia)
  Infraclass Eutheria (Placentalia)

Table 3-2   Informal Vertebrate Groups

Group	Classes
Agnatha	Agnatha
Gnathostomata	Placodermi, Chondrichthyes, Osteichthyes, Amphibia, Reptilia, Aves, Mammalia
Pisces (Fishes)	Placodermi, Chondrichthyes, Osteichthyes
Tetrapoda	Amphibia, Reptilia, Aves, Mammalia
Anamniota	Agnatha, Placodermi, Chondrichthyes, Osteichthyes, Amphibia
Amniota	Reptilia, Aves, Mammalia
Sauropsida	Reptilia, Aves

Except in their lack of a jaw, the cyclostomes differ from the ostracoderms in almost every aspect. They lack a bony armor and paired appendages, and their skeleton is purely cartilaginous. Cyclostomes have a very poor fossil record; the only pre-Quaternary fossil was found in Carboniferous deposits and looks already very much like a contemporary lamprey. For this reason, cyclostomes are thought to have had a long independent evolution. They may have evolved not from any known ostracoderm but rather from an earlier jawless vertebrate ancestral to both the ostracoderms and cyclostomes.

In the late Silurian and especially in the early Devonian, a variety of fishes appeared, all of which possessed jaws and paired pectoral and pelvic appendages. These fishes are generally grouped into the three classes Placodermi, Chondrichthyes, and Osteichthyes and are hypothesized to have evolved independently from one another from an earlier ancestral jawed fish.

The **Placodermi,** a diverse assemblage of heavily armored fishes with a variety of lifestyles, appeared in the early Devonian, flourished during the Devonian, and were extinct by the early Carboniferous. The best-known placoderms were large fusiform fishes up to 10 meters long. Their head and thorax were covered by bony shields, which were interconnected by a movable joint; hence, these fishes are named **Arthrodira.** They had formidable jaws and large eyes, and many are thought to have been fast-swimming open-water predators. The **Antiarchi** were generally smaller and flat-bodied placoderms. They had a small mouth on the ventral side of the head and peculiar pectoral appendages that were ensheathed by an exoskeleton. They are thought to have been bottom-dwelling fishes that fed on small invertebrates or plant matter.

The **Acanthodii,** which appeared already in the Silurian and survived into the early

Permian, form another group of extinct jawed fishes. These were smaller, fusiform, large-eyed, unarmored fishes covered with small scales, and they had several pairs of long, hollow ventral spines. The "spiny sharks," as the acanthodians are sometimes called, are thought to have been agile open-water predators.

Because the placoderm and acanthodian fishes have so many structures and features that are unique to them and not found in other fishes, they are believed to have been a side branch of fish evolution and not ancestral to either the Chondrichthyes or Osteichthyes, both of which appeared later in the fossil record than the placoderms and acanthodians. The acanthodian fishes, however, do share certain similarities with the Osteichthyes and are, therefore, usually classified in this class (Table 3-1).

The first chondrichthyan, or cartilaginous, fish is known from late Devonian deposits. In the Carboniferous, the **Chondrichthyes** split into two groups, the **Holocephali** and the **Elasmobranchii.** The holocephalans, or chimaeras, have survived into the present and are deep-sea dwellers; they are rarely caught, and very little is known about their biology. The elasmobranchs became especially successful during the Jurassic at which time they split into two groups. The **Batoidea** include the extremely flattened skates and rays; these are bottom-dwelling fishes and usually disguise themselves with sand or mud or by mimicry coloration. They feed mostly on hard-shelled invertebrates. The **Selachii** include the usually fusiform sharks and dogfishes, some of which are pelagic, fast-swimming predators and some of which live near the bottom and shores of the sea.

Except for the placoid scales in their skin, chondrichthyan fishes do not have the bony armor found in other early fishes. Their skeleton is entirely cartilaginous. With very few exceptions, the chondrichthyans living today are marine. Although the chondrichthyans surviving today have changed considerably from their ancestors in the Paleozoic, the sharks in particular have retained several primitive anatomical characters, such as externally visible gill slits, a relatively unmodified visceral skeleton, a simple circulatory system, an absence of air-breathing organs, and a spiral valve in the intestine. This and the relative ease of their procurement make sharks the animals of choice to serve as models with which to learn about ancestral vertebrates.

All remaining fishes belong to the **Osteichthyes,** the class with the largest number of species and individuals among the vertebrates. Although osteichthyan fishes are very diverse, they are characterized by a bony operculum which covers the external gill slits, a skeleton that is usually at least partly ossified, and lungs or a swim bladder. The first fossil osteichthyan, or bony, fishes are found in Devonian deposits, and already at this early point of their evolution two groups can be distinguished, namely the Actinopterygii and the Sarcopterygii.

The **Actinopterygii,** or ray-finned fishes, form the larger and more diverse group of bony fishes. From the Carboniferous onward, they were the dominant fishes on earth. They are recognized by their fan-shaped paired fins, which are supported by numerous delicate bony rays, and by their nostrils, which end in blind nasal sacs. Three major groups are distinguished among the actinopterygian fishes: the **Chondrostei, Holostei,** and **Teleostei.**

The earliest actinopterygian fishes in the Devonian, the **Palaeoniscoidea,** were small fresh water chondrostean fishes with ganoid scales and a heterocercal tail. They became the dominant fishes in the lakes and rivers of the late Paleozoic but were extinct by the end of the Mesozoic. Today only a few species in three families of the Chondrostei survive as descendants of their palaeoniscoid ancestors. The **Polypteridae** from the Nile region of Africa have retained from the earliest fishes many primitive structures, such as a spiracle, lungs, a bony skeleton, ganoid scales, and an intestine with a spiral valve. The caviar-producing sturgeons (**Acipenseridae**) from the entire Northern Hemisphere and the paddlefishes (**Polyodontidae**) from the Mississippi River and eastern China still have a spiral valve in their intestine and a heterocercal tail, but their scales are reduced or lost, and their skeleton is cartilaginous.

In the late Paleozoic, fishes belonging to the **Holostei** appeared and became the dominant fish group by the middle of the Mesozoic, but by the end of the Cretaceous most holostean fishes were extinct. Today only two freshwater holostean genera survive in North America, namely several species of garpikes *(Lepisosteus)* and the bowfin, also called choupique *(Amia calva)*. These holostean fishes have a bony skeleton and have retained a skull bone configuration that is primitive for osteichthyan fishes. In addition, *Lepisosteus* has retained well-formed primitive ganoid scales.

Fishes belonging to the **Teleostei** appeared first in the early Mesozoic; since the Cretaceous they have been the dominant fish group to this day and have been able to enter virtually all aquatic niches. Practically all fishes that are commercially caught for food or the pet industry are teleosts. Teleost fishes that live today have an extensively modified anatomy compared with that of their ancestors; therefore, we cannot use them to learn about the anatomy of ancestral vertebrates.

The **Sarcopterygii,** or lobe-finned fishes, make up the other osteichthyan subclass. They are characterized by their lobe-shaped fins, which are supported and moved by internal bones and muscles, and by their nostrils, which extend into nasal cavities that open in the roof of the mouth cavity. Two major groups are distinguished within the sarcopterygian fishes: the **Crossopterygii** and the **Dipnoi.** The first crossopterygian fishes were the **Rhipidistia** in the Devonian. They were streamlined freshwater fishes and dominated the lakes and rivers before the heyday of the palaeoniscoid fishes, but by the end of the Permian they were extinct. Before that, however, the **Coelacanthiformes** had branched off from the rhipidistians and had invaded the oceans. Coelacanth fishes were once believed to have been extinct since the end of the Mesozoic. In 1939, however, a living coelacanth, *Latimeria chalumnae,* was discovered off the coast of Madagascar. *Latimeria* is a deep-sea fish that has evolved many adaptations for its particular environment and thus has diverged greatly from the ancestral rhipidistians. The Dipnoi, or lungfishes, evolved probably from a rhipidistian ancestor and appeared first in middle Devonian deposits, before the first amphibians appeared (see the following paragraph). They became rare in the Triassic and today comprise only three genera. These lungfishes have retained several primitive structures from their ancestors, such as lungs for air-breathing, an intestine with a spiral valve, and nasal cavities that communicate with the mouth cavity. During their long independent evolution, however, they have also acquired many unique features, especially in their skulls. Therefore, their anatomy is not really representative of that of the rhipidistians, which also gave rise to the tetrapods.

The first terrestrial tetrapods on earth were the **Amphibia,** in particular the **Labyrinthodontia,** so called because of their unique tooth structure, which is found otherwise only in the rhipidistians. (The **Lepospondyli** were a group of small early amphibians that lived only during the Carboniferous). Labyrinthodont amphibia were often quite large, up to 1 meter long. Several anatomical features distinguished them from their fishlike ancestors and allowed them to live at least partially on land. Their pectoral girdle is not attached to the skull, and their pelvic girdle articulates directly with the vertebral column. Their limbs have the basic configuration known for all land tetrapods, and their thorax is supported by massive ribs. An ear ossicle (i.e., columella or stapes) connects the auditory region of the skull with the cheek region of the body surface.

The first labyrinthodont amphibians, the **Ichthyostegalia,** appeared in the late Devonian, but by the late Carboniferous these early amphibians had radiated into a great variety of forms that can be grouped into several orders. The **Anthracosauria,** for example, include the mostly aquatic **Embolomera** as well as the mostly terrestrial **Seymouriamorpha.** The latter are thought to have given rise to the the reptiles (see the following section), but the anthracosaurian amphibians themselves were extinct by the late Permian. The **Temnospondyli,** another group of the labyrinthodont amphibians, survived until the late Triassic and comprised various life forms from long-snouted marine fish-eaters to small terrestrial animals. It is from these that the amphibians living today, the **Lissamphibia,** are thought to have evolved.

The first lissamphibian—an early form of frog—appeared in the Triassic. Unfortunately, there are no fossils with structures that are intermediate between those of the labyrinthodonts and lissamphibians. The lissamphibians differ very much from the labyrinthodonts in having very reduced and often highly modified skeletons. Thus, representatives of lissamphibians, such as *Necturus,* can provide us with only a limited insight into the anatomy of the ancestral tetrapods. Lissamphibians, however, are thought to have retained many physiological features from their ancestors, such as certain aspects of their circulatory, respiratory, and reproductive systems. Lissamphibians are therefore good models from which we can learn much about the evolutionary changes that took place during the transition from aquatic to terrestrial life. The lissamphibians, usually simply called amphibians (as long as it is understood that this term refers only to the amphibians living today) are classified into three orders: the **Urodela** (or **Caudata**), the salamanders and newts, which are found first in late Jurassic deposits; the **Gymnophiona,** or caecilians, which are little-known limbless animals and appear in the fossil record as late as the early Tertiary; and the **Anura** (or **Salientia**), the frogs and toads, which appear as fossils already in Triassic deposits.

The first reptiles on earth were representatives of the **Captorhinida** (stem reptiles). They evolved probably from a seymouriamorph amphibian ancestor and appeared in the late Carboniferous, flourished during the Permian, and were extinct by the end of the Triassic. They still closely resembled their amphibian ancestors in general body shape and differed from them only in a few anatomical details. They are believed, however, to have been fully adapted to terrestrial environmental conditions. The captorhinids gave rise to a multitude of reptiles, which are divided into subclasses or superorders according to the structural configuration of their skulls: Anapsida, Lepidosauria, Archosauria, Ichthyopterygia, Euryapsida, and Synapsida.

The **Anapsida** have a massive "anapsid" skull without fenestration, like the ancestral captorhinids. The only anapsid reptiles living today are the **Chelonia** (or **Testudines**), the tortoises and turtles, which appear as fossils for the first time in the Triassic. Turtles differ very much from their captorhinid ancestors in that they have lost their teeth and have evolved a shell of bony and keratinous plates, which encases and protects their body. In connection with the evolution of a shell, the anatomy of their internal organs and skeletomuscular system has been extensively modified. Turtles can therefore teach us much about the skull structure but little about the rest of the anatomy of ancestral reptiles.

The **Lepidosauria** are characterized by a "diapsid" skull with two fenestrae in the temporal region. The earliest lepidosaurian reptiles, the **Eosuchia,** appeared during the Permian and survived into the Triassic. They gave rise to two groups of diapsid reptiles: the **Rhynchocephalia** and the **Squamata.** The rhynchocephalians flourished from the Triassic to the lower Cretaceous but survive today only with a single genus in New Zealand. The tuataras *(Sphenodon),* rare and endangered lizard-like reptiles, have retained many anatomical features from their Mesozoic ancestors, such as a third, pineal eye. The Squamata appeared for the first time during the Jurassic and have flourished since then to become the dominant reptiles living today. They have been able to adapt to almost all terrestrial habitats and, in some cases, even to some aquatic environments. Partly as a consequence of this radiation, the anatomy of their skull and various organ systems has become highly modified from the primitive condition found in ancestral reptiles. They are classified into three major groups: the **Lacertilia,** or lizards, the **Ophidia,** or snakes, and the **Amphisbaenia,** peculiar burrowing and mostly limbless reptiles.

The **Archosauria** are also characterized by a diapsid skull, but this skull configuration may have evolved independently from the one found in lepidosaurian reptiles. The first archosaurian reptiles, the **Thecodontia,** lived from the late Permian and throughout the Triassic, and gave rise to several well-known groups of reptiles, one of which survives to this day. The **Saurischia** and **Ornithischia,** collectively called dinosaurs, include some of the largest terrestrial tetrapods that ever lived. They appeared first in the Triassic and became the dominant reptiles on earth before they disappeared at the end of the Cretaceous. The **Pterosauria** were flying reptiles, of which some species were very large (i.e., with a wingspan of up to 12 meters). They appeared in the Jurassic and were extinct by the end of the Cretaceous like the dinosaurs. The **Crocodylia,** the alligators and crocodiles, appeared in the Triassic and have remained a successful group up to the present. They are mostly large reptiles with an amphibious ecology and a highly developed brain that allows a complex social and parental behavior. The Aves, or birds, are also believed to have evolved from an archosaurian ancestor and have become one of the most successful vertebrate classes today (see a following section).

The **Euryapsida** and the **Ichthyopterygia** are peculiar descendants from captorhinid ancestors. Each group is characterized by a special skull configuration and first appeared in the late Permian and early Triassic, respectively, but became extinct during the Cretaceous.

The **Synapsida,** or mammal-like reptiles, are characterized by a "synapsid" skull with only one, lower fenestra in the temporal region. The earliest synapsid reptiles, the **Pelycosauria,** branched off from the ancestral captorhinid reptiles at the end of the Carboniferous, earlier than did the other major groups of reptiles discussed previously. They were the dominant reptiles during the Permian before they became extinct by the end of this period. The **Therapsida** evolved from a pelycosaurian ancestor in the Permian and flourished into the Jurassic when they gave rise to the mammals (see a following section).

As we discuss the reptiles in our attempts to understand the evolution of vertebrates from their piscine beginnings, we have to keep in mind that, anatomically, the ancestral captorhinid reptiles did not differ greatly from their labyrinthodont ancestors. In fact, it is often difficult to assign newly found fossils to either the amphibian or reptilian class. It is easiest to see the division between these two vertebrate classes when comparing living representatives, because the most clear-cut differences are found in their physiology and in the anatomy of their soft tissues and organs, none of which are preserved in fossils. Reptiles are much more adapted to terrestrial conditions than amphibians because reptiles have evolved, for example, a much thicker keratinized epidermis, which is largely impermeable and thus prevents the evaporative loss of body fluids. Reptiles, in contrast to fishes and amphibians, lay cleidoic eggs. In these eggs, the developing embryo is surrounded by several layers of extraembryonic membranes, which enclose fluid between them, and by a shell that allows gas exchange between the embryo and the environment. This modification in the design of the vertebrate egg allowed the reptiles to reproduce out of water and was retained and further developed by the birds and mammals. To emphasize the significance of the egg structure for the evolution of vertebrates, reptiles, birds, and mammals are often called **amniote** vertebrates, referring to the presence of a particular extraembryonic membrane, the **amnion.** In contrast, the **anamniote** vertebrates, which comprise the fishes and amphibians, lack an amnion.

The first **bird,** *Archaeopteryx lithographica,* is known from Jurassic deposits. Although it already possessed feathers typical of contemporary birds, it had also retained several anatomical structures typical of its reptilian ancestor. It is thought to have lived in trees and to have been capable of gliding but probably not of powered flight. During the Cretaceous a great variety of birds evolved. Most Cretaceous avian species are quite different from contemporary birds (many had teeth in their jaws), but they show that birds adapted very quickly to almost all ecological conditions present on earth. By the beginning of the Tertiary, the ancestors of all the contemporary avian orders had already evolved, and today birds, together with mammals, represent the vertebrates that acquired the most highly developed nervous system and, concomitant with this, the most refined social and parental behavior. Birds have evolved other features that are also found in mammals, such as highly complex respiratory, circulatory, digestive and locomotory systems, but any resemblances between these two classes are only superficial and have been acquired independently. It is important to remember that the avian and mammalian evolutionary lineages have diverged since the earliest time of reptilian evolution (see the foregoing discussion).

The first fossil mammals are known from the late Triassic. They were representatives of the **Triconodonta;** these small to cat-sized mammals evolved from a therapsid ancestor and became extinct during the lower Cretaceous. In the middle Jurassic a new group of mammals, the **Trituberculata,** appeared. They may have evolved from a triconodont ancestor and were small, insectivorous animals with a fully developed mammalian jaw articulation. The trituberculate mammals gave rise to the two main groups of mammals living today, namely the **Metatheria** (or **Marsupialia**) and the **Eutheria** (or **Placentalia**). Both the marsupial and placental mammals first appeared at the same time in the early Cretaceous and together have become the dominant land tetrapods since the Palaeocene of the Tertiary, that is, after the demise of the dinosaurs.

Both the marsupial and placental mammals bear live young, suckle their young with milk after giving birth, and closely resemble each other in most morphological, physiological, and behavioral features. The most fundamental difference between them is found in their reproductive biology. In placental mammals, the zygote develops a trophoblast, a layer of cells that surround the developing embryonic cells and that later, as they become part of the placenta, form an immunological barrier between the placental tissue of the mother and that of the embryo. Because of the trophoblast, the embryo of a placental mammal can remain within the body of the mother throughout its entire embryonic development, and the body of the mother does not reject the embryo as if it were foreign tissue. In marsupial mammals, however, no trophoblast is formed, and the embryo is born at a stage that would be considered grossly immature for placental mammals. The marsupial embryo then crawls into the pouch (marsupium) of the mother, where it attaches itself to a nipple and completes its embryonic development. Marsupial mammals are not "more primitive" than placental mammals, as one might sometimes hear; marsupials and placentals have simply evolved alternative approaches to being a viviparous mammal.

In addition to the marsupials and placentals, two additional groups of mammals have evolved from therapsid reptiles. One group, the rodent-like or rabbit-like **Multituberculata,** appeared first in the Jurassic and were extinct by the early Tertiary. The other group, the **Monotremata**, is known only from living representatives, namely the **Tachyglossidae** (or spiny anteaters) from New Guinea and Australia, and the purely Australian **Ornithorhynchidae** (or duck-billed platypus). These mammals have so many unique features that they may have evolved from some therapsid or triconodont ancestor independently from the metatherians and eutherians. The prototherians possess a curious mix of primitive reptilian characters, such as ovipary, and derived mammalian characters, such as the suckling of young with milk.

Several characters, such as the suckling of young with milk produced by modified skin glands, distinguish the mammals from the reptiles. The evolution of a secondary palate, muscular tongue, and heterodont dentition with limited tooth replacement permitted the mastication of food and thereby the exploitation of fibrous vegetarian or very large food or prey items. The evolution of a secondary jaw articulation between the squamosal and dentary bones paralleled the transformation of the articular and quadrate bones into ear ossicles. The circulatory, respiratory, excretory, nervous, and digestive systems were greatly modified to support a higher metabolic rate. Keep in mind, however, that all of these highly derived mammalian characters evolved gradually and at varying rates, as the fossil species and the prototherians with their mix of reptilian and mammalian characters amply demonstrate.

Species To Be Studied

The preceding review of the evolutionary history of the vertebrates allows us to appreciate the opportunities and limitations of a basic course in comparative anatomy for which this manual is intended. We understand now, for example, that contemporary reptiles did not evolve from the amphibians we know today as living animals. To grasp fully how today's variety of reptiles evolved from ancestral vertebrates, we would need to study mostly fossil material. Fossils, however, represent only very incomplete remains of organisms. Therefore, to learn about the evolutionary transformation of all the anatomical structures, including the soft tissues and organs, we resort to the comparative study of preserved material from living vertebrates.

We recognize also that all vertebrates living today can look back on an equally long evolutionary past, though some vertebrates have modified their anatomy in more fundamental ways than others. Thus, no living vertebrate is "more primitive," "lower," "more derived," "more advanced," or "higher" than any other contemporary vertebrate. When we compare the anatomy of an amphibian, such as *Necturus,* with the anatomy of a mammal, we do not do so because *Necturus* is "more primitive" than a mammal, but because *Necturus* has retained a greater number of primitive features from the ancestral tetrapod than have mammals and therefore is a good model for an ancestral tetrapod.

We also realize that no particular evolutionary lineage is inherently more interesting than any other, a fact to which all ichthyologists, herpetologists, ornithologists, mammalogists, and paleontologists can attest. Traditionally, however, the evolutionary line leading from the earliest fishlike vertebrate to the mammals has held special fascination for us. Therefore, this manual considers mainly the anatomy of animals that represent distinctive stages in this evolutionary line, namely, the anatomy of a lamprey as a model of an ancestral jawless vertebrate, that of a shark as a model of an ancestral jawed vertebrate, that of a salamander as a model of an ancestral tetrapod, and that of a mammal.

The choice of the particular species for each evolutionary stage has been dictated by purely practical considerations, namely abundance and ease of procurement.

A good example of a shark is the spiny dogfish of the North Atlantic and northern Pacific, *Squalus acanthias.* Adult males range in length from 0.7 to 1 meter; females are slightly larger. They prefer water temperatures ranging from 6°C to 15°C and, hence, migrate north in the spring and south in the fall. A migratory school includes thousands of fishes. They are voracious and prey upon most species of fish smaller than themselves. They are considered edible in Europe, but North American fishermen consider them only a nuisance. They drive more favorable fish away and are rather destructive to fishing gear and to hooked or netted fish.

The mudpuppy *(Necturus)* is a satisfactory example of a caudate amphibian. Unfortunately, it is unusual in one respect, for it is paedomorphic and retains certain larval features into adult life. **Paedomorphosis** evolves in animals either by an acceleration of sexual development relative to somatic development, called **progenesis,** or by a slowing down of somatic development relative to sexual development, called **neoteny.** *Necturus* is neotenic. Larval features retained by *Necturus* should not be confused with primitive adult characteristics. *Necturus* is distributed throughout most of the eastern half of the United States. The most widespread species is *N. maculosus.* It is most abundant in clear waters of lakes and larger streams, but it is also found in weed-choked, turbid, and smaller bodies of water. It is most active at night, when it forages for small fish, crayfish, aquatic insect larvae, and mollusks.

Since all placental mammals resemble each other in more ways than they differ, the basic features of a mammal can be seen by studying any convenient one. Directions in this manual are written in such a way that they can apply to the domestic cat *(Felis catus),* which belongs to the order **Carnivora,** or to the rabbit *(Oryctolagus cuniculus),* which belongs to the order **Lagomorpha.** If different students are provided with a cat and a rabbit, comparisons can be made between divergent species at the mammalian evolutionary level. This will illustrate the essential uniformity of mammalian anatomy and, at the same time, reveal important differences that have been superimposed upon a common structural pattern through a long independent evolution of these species and their adaptation to carnivorous and herbivorous modes of life.

EXTERNAL ANATOMY

The general body shape of vertebrates reflects mainly their particular modes of locomotion. At one end of the spectrum of vertebrate shape, we find the purely aquatic fishes with their streamlined body, adapted to reduce the resistance of the surrounding water during swimming. Undulations of their trunk and tail propel them forward, while they usually use their appendages, or fins, only for steering. At the other end of the spectrum are the terrestrial mammals, in which the paired appendages have become powerful limbs that not only support the body above the ground but also push or propel the body forward. Most mammals do not need a streamlined body, and a pronounced neck allows the head to be moved independently from the trunk. Between these two extremes in body shape, we find the terrestrial caudate amphibians and lizard-like reptiles, in which the neck region is poorly defined and the limbs contribute to the locomotion mainly by bracing the undulating body against the ground and preventing it from backsliding. In this manual we will consider the three vertebrate body shapes just described, but keep in mind that many other body shapes and locomotory modes have evolved among vertebrates, such as in limbless amphibians and reptiles, in birds, in mammals, and in specialized fishes.

The surface of vertebrates is covered by the **integument,** or skin *(cutis)*. It protects the body in many ways: against abrasion, undue exchanges of water and salts with the external environment, and ultraviolet radiation. But the skin does not isolate the body from the outside world. Because it contains many sensory receptors, it also establishes a connection between the organism and its environment. The skin consists of two layers of tissue—an outer **epidermis** of stratified epithelial cells derived from the embryonic ectoderm, and an inner **dermis** *(corium)* of dense connective tissue derived from the mesoderm (Figs. 3-4, 3-7, and 3-10). In fishes and amphibians, the epidermis is thin and permeable to water and gases. As vertebrates adapted to a terrestrial environment, however, the epidermis became thicker, and its outer cells became impregnated with keratin, a protein that contains sulfur. The keratinized, or horny, cells are hardened and dead and render the epidermis less permeable to gases and water and more resistant to mechanical injuries. They form a distinct layer, the superficial **stratum corneum.** These dead cells of the stratum corneum are eventually sloughed off by abrasion or shed by molting and are continually replaced by mitosis of the basal cells in the **stratum basale** or **stratum germinativum** of the epidermis.

Accessory structures of the integument that are found in various vertebrates are pigment cells, glands, and a variety of bony and horny structures such as scales, feathers, and hair. The pigment of fishes, amphibians, and reptiles is contained within specialized cells, the **chromatophores,** which, although derived from the embryonic neural crest, are located in the dermis. The most common chromatophores are stellate (star-shaped) **melanophores** containing the dark pigment **melanin.** The pigment moves within the melanophores of lower vertebrates under the influence of nerves and hormones. When the animal is dark, the pigment is dispersed throughout the cell; when the animal is light, the pigment is withdrawn to the center. The melanophores of birds and mammals transfer their pigment to epithelial cells. Pigments play an important role in many animals, serving for species recognition, for concealment, and sometimes for advertisement.

All skin glands are derivatives of epidermal cells, but the large glands invaginate into the dermis. The glands of fishes and amphibians produce primarily a protective coat of mucus, but the skin glands of mammals have a variety of functions, such as scent production, secretion of sweat for thermoregulation, and secretion of oily sebum to maintain the physical properties of healthy hair.

Feathers and hair are also derivatives of epidermal cells, but scales (of which there are many types) may develop from either or both layers of the skin. Bony formations in the skin have sometimes been called an "exoskeleton," but the term **integumentary skeleton** is more appropriate. These formations are cellular structures that develop within the skin and are not an acellular secretion on the skin surface, as is the exoskeleton of invertebrates.

Fishes

♦ **(A) GENERAL EXTERNAL FEATURES**

Examine a specimen, noting the streamlined, or **fusiform,** shape that enables the animal to move easily through the water (Fig. 3-3). The body regions are not as well demarcated as they are in tetrapods; they all blend into one another. Nevertheless, the body can be

40

Chapter 3
The Evolu-
tion and Ex-
ternal Anat-
omy of
Vertebrates

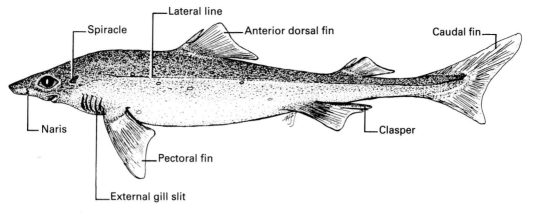

Figure 3-3
A lateral view of the dogfish, *Squalus acanthias.*

divided into a **head,** which includes the gill region, a **trunk,** which continues to the cloacal aperture, and a **tail** caudal to this. The tail is a powerful organ of locomotion. The cloaca is a chamber on the ventral side, which can be probed through the cloacal aperture and which receives the intestine and urinary and genital ducts. The urinary ducts, and in the male the genital ducts as well, open at the tip of a **urinary papilla,** which can be seen inside the cloaca. The **anus**—the opening of the intestine into the cloaca—lies cranial to the urinary papilla.

Note that the body is a dark color above and light beneath. Such a distribution of pigment, referred to as **counter shading,** is common in vertebrates, especially aquatic forms. Optically, it tends to neutralize the effect of natural lighting, which highlights the back and casts a shadow on the belly; thus, counter shading renders the organism less conspicuous when seen from the side.

There are two **dorsal fins** (cranial and caudal), with a large **spine** in front of each. The spines are defensive and are associated with modified skin glands that secrete a slightly irritating substance. Ringlike markings on the spines give an indication of the fish's age; the life span ranges from 25 to 30 years. Large, paired **pectoral fins** will be seen just behind the gill region and paired **pelvic fins** at the caudal end of the trunk. Males have stout, longitudinally grooved copulatory organs, called **claspers,** on the medial side of their pelvic fins. The tail ends in a large **caudal fin** of the **heterocercal** type; that is, the fin is asymmetrical, for the body axis turns up into its dorsal lobe, and most of the fin rays are ventral to the axis. All the fins are supported by fibrous fin rays (**ceratotrichia**). Cartilages, which will be seen later, lie within the base of each fin and provide further support. In well-preserved specimens, a lateral keel will be seen on the trunk on each side of the base of the caudal fin.

The **mouth,** which is supported by jaws, is located on the underside of the head and is bounded laterally by deep **labial pockets.** There is a **labial fold,** containing a cartilage, between the mouth and pocket. A pair of large **eyes** will be seen set in deep sockets on each side of the head. The rim of each socket forms immovable **eyelids.** Paired external nostrils, called **nares,** are on the underside of the pointed snout. The opening of each is partially subdivided by a little flap of skin, which separates the stream of water that flows through the nostril into and out of each **olfactory sac.** Pass a probe into a naris and notice that the olfactory sac does not communicate with the mouth cavity.

A row of five **external gill slits** is located in front of the pectoral fin. The term *elasmobranch* (the subclass to which sharks belong) means "platelike gills" and refers to the tissue between the gill slits. (In bony fishes the gills are covered by a common opercular flap.) Caudal to the eye you will see a large opening called the **spiracle.** Little parallel ridges, representing a reduced gill called the **pseudobranch,** can be seen on a fold of tissue that is separated from the rostral wall of the spiracular passage by a deep recess. Probe to determine the extent of this recess. This fold of tissue is a **spiracular valve,** which can be closed.

The spiracle is actually the reduced and modified first gill slit of ancestral gnathostomes. Most fishes respire by taking water into the pharynx through the mouth and discharging it through the gill slits. When a spiracle is present, water may also enter through it. In the bottom-dwelling skates and rays, the spiracle is enlarged and situated on the dorsal side of the flattened body. Water for breathing enters the pharynx only through the spiracle and leaves it through the gills. The breathing water thus is less likely to contain hard particles, which could injure the delicate gill surface, than if it entered through the mouth, which is usually buried in the mud or sand of the bottom.

If you look with low magnification at the top of the head between the spiracles, you will see a pair of tiny **endolymphatic pores,** one pore on each side of the midline. They communicate with the inner ear, which in most fishes is an organ of both equilibrium and hearing (see Fig. 8-13, p. 230). Vibrations of low frequency and movements in the water are detected by the lateral line system. The position of one canal of this system (the **lateral line** in a restricted sense) is indicated by a fine, light-colored, horizontal stripe extending along the side of the body. It is nearer the dorsal than the ventral surface. You will also see patches of pores on the head through which a jelly-like substance extrudes if the area is squeezed. They are the openings of the **ampullae of Lorenzini.** These ampullae are a modified part of the lateral line system and are electroreceptors.

♦ **(B) ACCESSORY STRUCTURES OF THE INTEGUMENT**

The skin of sharks contains chromatophores; glands, usually in the form of simple, scattered mucous cells; and hardened, dermal structures. In the dogfish these are minute **dermal denticles,** or **placoid scales,** which cover the animal and which you can feel by moving your hand cranially over the surface. To see them you will have to use low magnification, or, better still, observe a special microscopic preparation. (**Melanophores** may be seen at the same time.) Each scale consists of a basal plate imbedded in the dermis from which a spine perforates the epidermis and projects caudad (Fig. 3-4). Recent evidence indicates that the spiny scales reduce turbulence in the flow of water next to the skin and, hence, reduce drag during swimming. (At one time elasmobranch skin, sold as shagreen, was used as an abrasive for polishing wood.)

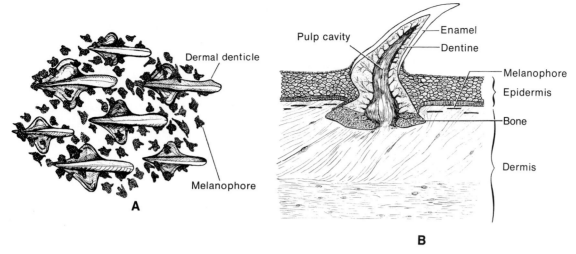

Figure 3-4
Skin and placoid scales of a shark (the tail end of the shark points to the right). **A,** Magnified surface view, with melanophores; **B,** magnified vertical section through a placoid scale and the skin. (**B,** *Redrawn from Dean.*)

Ancestral fishes had an extensive armor of bony scales and plates that consisted of three histologically distinct layers. Contemporary fishes have lost these primitive bony armors, but they usually still have scales imbedded in their skin.

Elasmobranchs have so-called placoid scales, the structure of which resembles that of teeth in many ways. Each scale consists of a base of **acellular bone** to which a cone of **dentine** is attached. The superficial part of the dentine, which breaks through the surface of the epidermis, is covered by a layer of **enamel.** The bony base of each scale is anchored within the dermis with connective tissue fibers. The center of a scale is hollow, forming the **pulp cavity,** which contains connective tissue, blood vessels, and nerves. Dentine is produced by mesenchymal cells in the dermis, which in turn are derived from ectodermal neural crest cells in the embryo. Dentine consists of inorganic crystals of hydroxyapatite with about 30 percent of organic material. Enamel, which is glasslike and harder than dentine, is produced by epidermal cells and consists almost entirely of hydroxyapatite. It is not clear whether the enamel of placoid scales is the same as the enamel of mammalian teeth (see Herold, Graver, and Christner, 1980). Therefore, the enamel of placoid scales is sometimes called **enameloid.** Although the placoid scales in the skin of sharks are very similar to the teeth in sharks, except in size and shape, it is important to realize that scales are not "derived" from teeth, or vice versa. They are similar in structure because both are formed by the same integument; the differences in their final shapes reflect differences in their location and function.

The hard dermal structures of the osteichthyans are not tooth-shaped but rather flat, platelike or scalelike structures. Nevertheless, the scales of osteichthyans consist also of three distinct layers, namely of the bonelike **isopedine,** the dentine-like **cosmine,** and the enamel-like **ganoine.** The scales of the actinopterygians tend to reduce the thickness of their cosmine layer and are called **ganoid scales.** For example, the scales of *Lepisosteus* comprise only the basal plate of isopedine and a thick superficial layer of ganoine. Furthermore, the scales of *Amia* and of teleosts have even lost the ganoine layer, and the remaining bony layer has become so thin that the scales are translucent and flexible. The scales of the sarcopterygians have a complex structure, and their cosmine layer is thicker than their ganoine layer; they are called **cosmoid scales.**

Other hard accessory structures of the integument are the ceratotrichia and spines already observed (p. 41). The spines are essentially modified, enlarged, and elongated placoid scales. The **ceratotrichia** are fibrous, flexible, and unsegmented rays. They are composed of elastoidin fibers, a special kind of connective tissue fibers, which are arranged in concentric layers. In the osteichthyans, the fins are supported by bony rays called **lepidotrichia,** which are segmented and distally branched. Both the ceratotrichia and lepidotrichia are formed in the dermis and have evolved from scales.

Amphibians and Reptiles

◆ (A) GENERAL EXTERNAL FEATURES

Examine a specimen of *Necturus* and compare it with *Squalus*. The body is elongate, with a modest-sized, flattened **head;** an incipient **neck** region; a long **trunk;** and a powerful, laterally compressed **tail** (Fig. 3-5). There are no median fins, except for traces on the tail, and these lack fin rays. The paired fins of fish have become transformed into pectoral and pelvic limbs. Each limb consists of three segments. In the pectoral appendage these are upper arm (**brachium**), forearm (**antebrachium**), and hand (**manus**). The elbow joint (**cubitus**) is between the brachium and antebrachium; the wrist (**carpus**) is in the proximal part of the manus. Corresponding parts of the pelvic appendage are the thigh (**femur**)[4], shank (**crus**), and foot (**pes**); corresponding joints are the knee (**genu**) and ankle (**tarsus**). Only four toes are present, the most medial toe of the ancestral pentadactyle foot having been lost. If the entire leg is pulled out to the side at right angles to the body with the palm of the hand (or sole of the foot) facing ventrally, the cranial border is said to be **preaxial;** the caudal, **postaxial.** Since *Necturus* is aquatic, its locomotion is still fishlike, with lateral undulations of the trunk and tail playing an important role.

In terrestrial salamanders, the rather weak legs are used mainly to prevent backsliding during the lateral undulations of the trunk, which propel the animal forward. When used, the proximal segment of the leg projects horizontally, the distal segment extends vertically downward, and the manus and pes point forward (Fig. 3-6). In such a position the preaxial border of the entire hind leg is still mostly cranial, and there is a simple

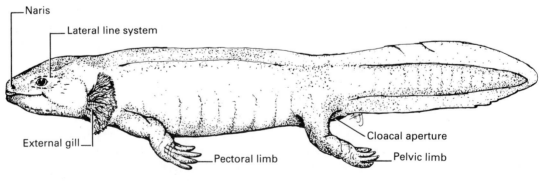

Figure 3-5
A lateral view of the mudpuppy, *Necturus maculosus.*

---

[4]The term *femur* can be used for the thigh and for the bone within it. The *Nomina Anatomica* term for the thigh is *femur* and for the bone, *os femoris.*

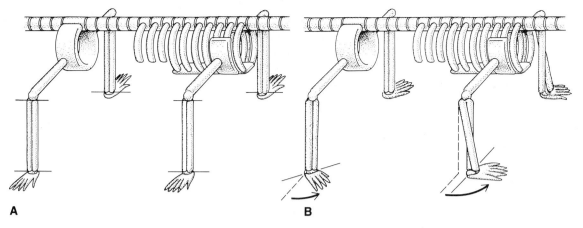

A   **B**

Figure 3-6
Models of the axial and appendicular skeleton, illustrating the evolutionary change in limb position
and orientation. **A,** Ancestral tetrapod *(redrawn from Starck, 1979)*; **B,** metamorphosed, terrestrial
caudate amphibian *(modified from Starck, 1979).*

hinge joint at the knee. In the front leg the preaxial border of the brachium is cranial, but
there has been a torsion, or rotation, at the elbow that brings the foot forward, so that the
preaxial border of most of the antebrachium is medial. The position of limbs in caudate
amphibia and lizard-like reptiles is such that the body is only slightly raised off the ground
and usually only during locomotion, and the humerus and femur move back and forth close
to the horizontal plane.

The **mouth** is terminal and is bounded by **lips.** Just above the upper lip is a pair of
widely spaced external nostrils **(nares).** As will be seen later, they communicate with the
front of the oral cavity by way of internal nostrils **(choanae),** thus permitting air to be
taken into the mouth through them. Small **eyes** are present but are devoid of lids. The
absence of eyelids is a larval feature in amphibians; movable eyelids are not necessary in
aquatic animals because the external surface of the eye (the cornea) is not in danger of
desiccation. Metamorphosed, terrestrial amphibians, like most tetrapods, have fully
developed movable eyelids, which help to protect, cleanse, and moisten the eye. Unlike
most amphibians and reptiles, salamanders also lack an external eardrum. There is an
internal ear, however, and vibrations reach it primarily by way of skull bones. The **lateral
line system** of fishes is retained and appears, when the specimen has dried a bit, as rows
of depressed dashes above and below the eyes, on the cheek, and on the ventral surface of
the head. A less obvious row of dashes also extends caudad along the side of the trunk.
The lateral line system, too, is a larval feature that is found only in aquatic amphibians;
metamorphosed terrestrial amphibians have lost it.

There are three pairs of prominent **external gills** at the caudal end of the head.
Although some gas exchange takes place through the highly vascularized skin, and the
animal occasionally comes to the surface to gulp air, these gills are the major respiratory
organ. External gills are larval structures and are lost by adults of those salamander
species that metamorphose into terrestrial adults. The gills of larval amphibians are
supported by elements of the visceral skeleton as are the gills of fishes (p. 71), but the
amphibian gills project outward from the body surface, whereas the fish gills lie within gill
pouches or are covered by an operculum. *Necturus* also has two **gill slits,** which can be
seen between the bases of the external gills. The fold of skin extending across the ventral
surface of the head between the gills is called the **gular fold.**

Finally, observe the **cloacal aperture** at the caudal end of the trunk. It is bounded by lips bearing tiny papillae in the male and small folds in the female.

♦ (B) ACCESSORY STRUCTURES OF THE INTEGUMENT

Very early amphibians (labyrinthodonts) still retained small cosmoid scales from their piscine ancestor, but most modern amphibians have no scales in their skin. Only the peculiar caecilians have small bony scales imbedded in their dermis. The skin of amphibians is only slightly keratinized, but it contains numerous **mucous glands** and **poison glands** (Fig. 3-7). The mucus helps to keep the skin surface moist and thus permeable to gases and water. Poison glands secrete a substance that is usually only slightly irritating but helps to keep some predators at bay. (In certain South American tree frogs, the secretion of the poison glands is extremely toxic; the indigenous people use it to prepare poison darts.)

The reptilian integument, as exemplified by that of a lizard or turtle, has acquired various features that have evolved in connection with the completely terrestrial biology of most reptiles and has lost several features typical of the skin of amphibians, such as mucous and poison glands. The reduction of skin glands in reptiles—they retain only a few scent glands—is correlated with an extensive cornification, or keratinization, of the epidermis that reduces its permeability to water and gases. The skin is covered with **horny scales** that are formed by a thickened stratum corneum. Sometimes these horny scales are underlaid by bony plates in the dermis. Between the individual horny scales, the epidermis is thinner, less keratinized, and flexible. On the head and on the external surface of the lips, the horny scales are enlarged to form horny plates. In turtles, the shell is formed by horny scales that have been modified to form large horny plates that are supported by plates of bone. Most of these plates represent new formations of bony tissue within the dermis, but some of the ventral plates (plastral plates) include remnants of the

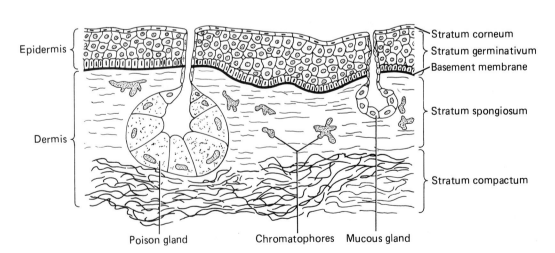

Figure 3-7
Diagram of a vertical section through the skin of an amphibian. The stratum corneum is not keratinized in *Necturus*. *(Adapted from Rabl in Bolk et al., Vol. 2, and from Starck, 1982.)*

original bony armor of ancestral fishes. The tips of the toes of reptiles bear **claws,** which are also keratin structures.

Mammals

◆ **(A) GENERAL EXTERNAL FEATURES**

Examine either a cat or a rabbit and compare it with *Necturus.* The diagnostic **hair** of mammals is at once evident (Fig. 3-8), and it will be seen that the evolutionary trends that began in ancestral tetrapods have continued. The **head** *(caput)* is large and separated from the **trunk** *(corpus)* by a distinct and movable **neck** *(collum).* The trunk itself can be divided into the **back** *(dorsum),* **thorax, abdomen,** and **pelvis.** A **tail** *(cauda)* is typically present in mammals but, except in the whales and their allies, is greatly reduced in size in comparison with that of an ancestral tetrapod. In some terrestrial mammals it is used as a balancing organ or to scare away insects. In some other mammals, such as the rabbit, it is vestigial and used mainly as a social signal; in a few, such as humans and apes, it has been lost as an external structure.

The paired appendages consist of the usual parts—**brachium, antebrachium,** and **manus** in the pectoral; thigh or **femur**[5], **crus,** and **pes** in the pelvic appendage. In both the cat and the rabbit the most medial, or first, toe of the manus is vestigial, and the corresponding toe of the pes has been completely lost as an externally visible structure.

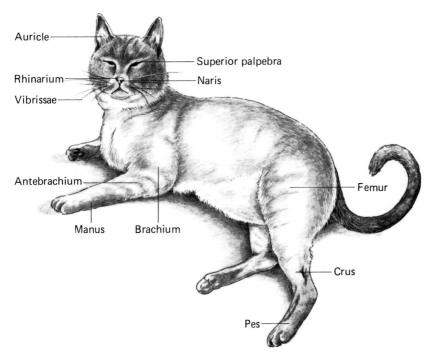

Figure 3-8
External features of a cat, *Felis catus. (From Walker,* A Study of the Cat.*)*

---

[5]See footnote, 4, p. 44.

Observe either on a mounted specimen or on a skeleton that a carnivore walks on its toes with the wrist and heel raised off the ground. This method of locomotion is referred to as **digitigrade,** in contrast to **plantigrade** (in humans), in which the entire sole of the foot is flat on the ground, or **unguligrade** (in ungulates such as the horse), in which the animal walks on the tips of its toes. The rabbit is also digitigrade with respect to the front feet, but the hind legs are modified for hopping. Just before the leap, the pes is in the plantigrade position. The terminal segment of each toe has a **claw;** in the cat this segment is hinged in such a way that the claw can be retracted or extended (p. 128).

In the evolution toward mammals, the elbow and knee are brought closer to the trunk so that the legs move back and forth closer to the vertical plane than is the case in ancestral tetrapods. This provides better support and makes a longer step and stride possible.[6] The degree to which the limbs rotate beneath the body varies in different groups of mammals (Jenkins, 1971). In-fast moving cursorial types, such as carnivores and lagomorphs, all parts of the limbs move in the same vertical plane. The front limb has rotated caudally so that the elbow points backward, but the hind limb has rotated cranially so that the knee points forward (Fig. 3-9). Both manus and pes are directed forward. The original preaxial border of the hind limb is now medial. The original preaxial border of the brachium is now lateral, but, because of a continued torsion at the elbow seen beginning in ancestral tetrapods (Fig. 3-6), the preaxial border of the antebrachium shifts from lateral to medial as it progresses distally. (In order to understand these changes, place your own appendages in the ancestral tetrapod position and then rotate them into the cursorial mammal position.)

The length of a step, and hence speed of travel, is also increased by a shift from the primitive plantigrade to a digitigrade foot posture. Lateral undulations of the trunk are no longer significant in locomotion, but in rapidly moving mammals vertical trunk undulations help increase stride length.

Examine the head. Notice that the **mouth** *(os)* is bounded by fleshy **lips** *(labia),* the upper one being deeply cleft in the rabbit (harelip). The paired external nostrils **(nares)** are close together on the nose surrounded by moist, bare skin known as the **rhinarium.** The **eyes** *(oculi)* are large and are protected by movable upper and lower eyelids **(palpebrae).** Spread the palpebrae apart and observe a third lid, called the **nictitating membrane,** in the medial corner of the eye. The nictitating membrane can be drawn across most of the eye, thus helping to moisten and cleanse this organ. Mammals have a prominent external ear consisting of a conspicuous external flap, called the **auricle,** or pinna, and an external ear canal that extends into the head from the base of the auricle. The eardrum **(tympanum)** is located at the bottom of the external ear canal. It will not be seen at this time. The part of the head that includes the jaws, mouth, nose, and eyes is referred to as the **facial region;** the rest, containing the brain and ears, as the **cranial region.**

The cloaca of nonmammalian vertebrates has become divided in therian mammals, so that the intestine and urogenital ducts open independently at the surface. The opening of the intestine, called the **anus,** will be found just ventral to the base of the tail. If the animal

---

[6]A step is the distance an animal is carried forward when one foot is on the ground; a stride, the distance an animal travels between the placement of one foot on the ground (left hind, for example) and the next placement of the same foot. A stride includes the step distance of all feet as well as the distance an animal is moved forward when one or more feet are off the ground.

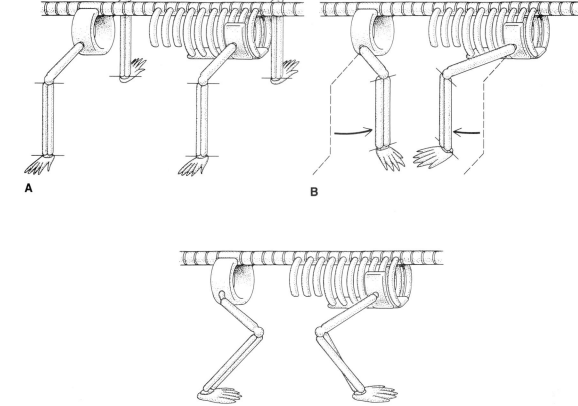

Figure 3-9
Models of the axial and appendicular skeleton, illustrating the evolutionary change in limb position
and orientation. **A,** Ancestral tetrapod; **B,** hypothetical intermediate stage between **A** and **C;**
C, mammal. (**B** and **C** shown with limbs only on one side). *(Redrawn from Starck, 1979.)*

is a female, the combined opening of the urinary and reproductive ducts will appear as a
second passage, the **vaginal vestibule,** bounded by small folds, ventral to the anus. If
the animal is a male, the urogenital duct opens at the tip of a **penis.** Associated with this
you will see the sac-shaped **scrotum** containing the testes. In the rabbit, the testes may
be retracted into the abdominal cavity. The entire area of the anus and external genitals is
called the **perineum** in both sexes. Further discussion of the details of the external
genital organs will be deferred until Chapter 12.

Carefully feel along the ventral surface of the thorax and abdomen on each side of the
midline and you will find two rows of teats **(papillae mammae)** hidden in the fur. These
bear the minute openings of the mammary glands. The teats are more prominent in
females, but rudiments can sometimes be found in males. There are usually four or five
pairs in the cat and six in the rabbit, but the number is subject to variation. The mammary
glands can be felt only in lactating females.

♦   **(B) ACCESSORY STRUCTURES OF THE
INTEGUMENT**

The integument of mammals is rich in accessory structures, many of which can be seen or
demonstrated in the laboratory (Fig. 3-10). A few pigment-producing melanophores (also

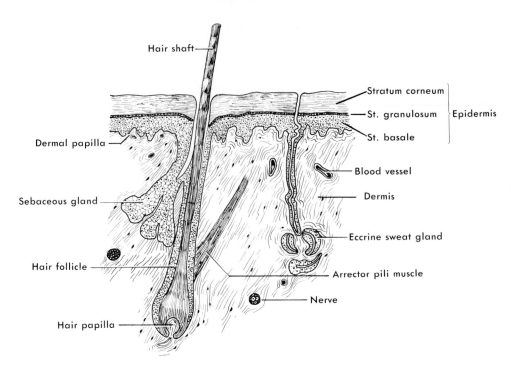

Figure 3-10

Diagram of a vertical section through the skin of a mammal. *(From Walker,* A Study of the Cat.*)*

called **melanocytes**) are present beneath the epidermis, but most of the melanin they produce is transferred to epithelial cells. There are many glands, which can be grouped into three categories: alveolar-shaped **sebaceous glands,** tubular **sweat glands** *(sudoriferous glands),* and **mammary glands.** Sebaceous glands usually discharge their secretions, sebum, into the hair follicles. Sebum appears to help lubricate, waterproof, and condition the hair. There are two types of sweat glands: apocrine and eccrine. **Apocrine glands,** which are abundant in the arm pits and in the genital area, discharge into the hair follicles and produce secretions responsible for body odors. Scent glands seem to be modified apocrine glands. **Eccrine glands** produce a more watery solution that is important in cooling the body in some mammals, such as human beings and horses. In heavily furred mammals, they tend to be limited to the snout, tail base, or soles of the feet. Openings of some can be seen with low magnification on your finger tips. Although alveolar in shape, the mammary glands resemble sweat glands in having contractile myoepithelial cells peripheral to the secretory cells.

The most conspicuous integumentary derivative is the protective and insulating covering of **hair** *(capillus).* Hair replaces the horny scales of reptiles in most mammals, but scales may still be found on the tails of certain rodents, and they have redeveloped over the bony plates of the armadillo shell. Although hair is composed of keratinized cells, it is a new development of the epidermis. It is not considered to be homologous with either horny scales or feathers, since details of its embryonic development are different. Moreover, the distribution of hair, as seen, for example, on the back of one's hand, leads to the conclusion that hairs evolved in small clusters between the horny scales in the ancestor of furry mammals, and that subsequently the scales were lost. The simultaneous presence of scales and hairs can be seen on the tails of some rodents. In most mammals the hair forms a dense fur over the body, being modified in certain places such as the eyelashes *(cilia)* and tactile whiskers **(vibrissae)** on the heads of carnivores and rabbits. But there are many departures from this pattern. Hair is reduced in humans and lost in the adults of such highly aquatic mammals as the whale. In some other mammals, hair has become adapted for very specialized purposes. The quills of a porcupine are a case in point, and the "horn" of a rhinoceros consists of a compact mass of hairlike elements.

Reptilian **claws** are retained in most mammals but have been transformed into **nails** *(ungulae)* in certain primates and into **hoofs** in the ungulates. Other common integumentary structures are the **foot pads** *(tori)* on the feet of most mammals. These are simply thickenings of the stratum corneum.

Aside from the widely distributed structures just mentioned, some mammals have still other hard accessory structures of the skin. Bony plates form in the dermis of the armadillo. The toothless baleen whales have large, fringed, keratinized plates of **baleen** that hang down from the roofs of their mouths and entrap plankton. Sheep, antelopes, and cattle have **horns** that consist of a core of skull bone covered with very heavily keratinized epidermis, that is, horn. Horns of this type should not be confused with the **antlers** found in the deer group. Antlers are bony outgrowths of the skull that are covered with skin (the velvet) only during their growth. In contrast to horns, they generally are restricted to the male, and they branch and are shed annually.

CHAPTER FOUR

◆ ◆

THE HEAD SKELETON

The next organ systems we will study are those concerned with the general functions of support and locomotion, namely, the skeleton, muscles, sense organs, and nervous system. It is appropriate to consider the skeleton first, as it is a fundamental system about which the body is built.

The vertebrate skeleton is internal, for it develops within body tissues rather than being a secretion on the surface, as is the exoskeleton of crayfish and many other invertebrates. Depending on the group of vertebrates, the notochord, cartilage, or bone may contribute to the adult skeleton. Skeletons resist shortening when the muscles contract and form lever arms that transfer muscle forces to some point of application such as the jaws or feet. Skeletons also protect the brain and many other internal organs and often store important mineral ions such as calcium and phosphorus. In terrestrial vertebrates, skeletons support the body against the pull of gravity.

Divisions of the Skeleton

The vertebrate skeleton consists of two basic parts—the **integumentary** or **dermal skeleton** and the **endoskeleton.** Although these two become united to various degrees, they are distinct in their ontogenetic and phylogenetic origins. Bone of the integumentary skeleton develops embryonically directly from the mesenchyme in, or just beneath, the dermis of the skin. This type of bone is called either **dermal** or **membrane bone.** It follows that the dermal skeleton is superficial. Bony scales and plates, and their derivatives, are dermal in nature. Among their derivatives are the dermal plates in the head region, teeth, and the dermal portions of the pectoral girdle. In addition, dermal bone has evolved independently of bony scales in the dermis of certain animals. Portions of the shell of the turtle and the armadillo are familiar examples.

The endoskeleton, on the other hand, arises in deeper body layers and consists of cartilage or bone that develops within and around cartilaginous centers of growth. Although such bone has the same histological structure as dermal bone, it is convenient to differentiate it as **cartilage replacement bone.** The endoskeleton may be subdivided into somatic and visceral portions. The **somatic** portion of the endoskeleton is associated with the "outer tube" of the body (body wall and appendages). It may be further broken down into axial and appendicular subdivisions. The **axial skeleton** includes those parts of the somatic skeleton located in the longitudinal axis of the body—vertebrae, skeleton of the median fins, ribs, sternum, and those portions of the braincase composed of cartilage or cartilage replacement bone. The **appendicular skeleton** consists of the more laterally placed portions of the somatic skeleton—the skeleton of the paired appendages and those portions of their girdles composed of cartilage or cartilage replacement bone. The **visceral**

Table 4-1   Divisions of the Skeleton

Integumentary Skeleton	Endoskeleton
(Dermal Bone)	(Cartilage Replacement Bone)
Bony scales	Somatic skeleton (in body wall)
Dermal plates (become associated with parts of endoskeleton)	Axial skeleton (chondrocranium, vertebrae, ribs, sternum)
Teeth	Appendicular skeleton (girdles, bones of paired appendages)
	Visceral skeleton (in gut wall)
	Visceral arches

skeleton, as the name implies, is associated with the "inner tube" (gut) of the body. It consists of skeletal arches (visceral arches) that form in the wall of the pharynx, contribute to the formation of the jaws and tongue skeleton, and support the gills in aquatic vertebrates. Aside from its location, the visceral skeleton differs from the somatic skeleton in developing from ectodermal mesenchyme that is derived from the neural crest rather than from mesodermal mesenchyme.

The divisions of the skeleton are summarized in Table 4-1.

Evolutionary Tendencies in the Integumentary Skeleton and Endoskeleton

Ancestral fishes such as the ostracoderms had an extensive integumentary skeleton consisting of thick, bony scales of the cosmoid type over the trunk and tail, and larger bony plates over the head. An endoskeleton, although present, was neither completely ossified nor so conspicuous. From these early ancestors the evolutionary tendency has been one of reduction of the integumentary skeleton and increased development of the endoskeleton. As seen in the preceding chapter, the primitive, heavy, bony scales have become much thinner in living fishes and are lost as such in recent tetrapods. However, the deeper parts of the original cephalic plates persist in osteichthyans and terrestrial vertebrates and become associated with the endoskeleton as integral parts of the head skeleton and pectoral girdle.

We must be continually aware of the difference between the endoskeleton and the integumentary skeleton, but in studying the evolution of the entire skeleton, we cannot always follow this dichotomy. Bony scales were studied with the integument, and the rest of the integumentary skeleton will be considered along with those portions of the endoskeleton with which it becomes associated. When we study the skeleton it is convenient to study groups of bones that have common functions as well as recognizing the divisions outlined previously. Those elements that encase the brain and major sense organs and form the jaws are collectively called the **skull.** Parts of the first two visceral arches become incorporated in the skull, but most of the visceral skeleton remains independent and participates in breathing and feeding movements.

FISHES

The head skeleton of *Squalus* shows more clearly than that of most fishes some of the major components of the head skeleton, but the skull of *Squalus* does not represent that of an ancestral fish. It is atypical in having an endoskeleton that is entirely cartilaginous. Cartilage forms the embryonic endoskeleton in all vertebrates, but in most species bone replaces most of the cartilage during development. *Squalus* also lacks the dermal plates that covered the head in ancestral vertebrates and in other groups of fishes. We will use the bowfin, *Amia* (class Osteichthyes, subclass

Actinopterygii), to illustrate the dermal cephalic bones that were present in fishes ancestral to tetrapods.

The head skeleton is a mixture of three groups of elements that are distinct in certain vertebrates but become confusingly united and mixed in other vertebrates. These are (1) the **chondrocranium,** (2) the **visceral skeleton** (splanchnocranium), and (3) associated **dermal bones.** The chondrocranium[7], which is the anterior end of the axial skeleton, surrounds a variable amount of the brain and forms protective capsules about the olfactory sacs and inner ears. The visceral skeleton is composed of visceral arches, which originally supported the gills but which in other vertebrates become involved in the jaw and ear apparatus and even help to encase the brain, to mention but some of their transformations. The chondrocranium and visceral arches are, of course, part of the endoskeleton, but the associated dermal bones are part of the integumentary skeleton. Primitively they covered the chondrocranium and visceral arches. Some are lost in terrestrial vertebrates, but many persist to help form the braincase, the jaws, and the facial portion of the skull.

◆ **(A) COMPOSITION AND STRUCTURE OF THE CHONDROCRANIUM**

The chondrocranium is a complex box of cartilage, or cartilage replacement bone, that we can best understand by describing briefly the major features of its embryonic development (Fig. 4-1). The basic elements in its formation in all vertebrates are two pairs of longitudinal cartilages that lie beneath the brain. The caudal pair, called **parachordals,** are located on each side of the cranial end of the notochord. (The notochord itself extends only as far forward as the pituitary gland, or hypophysis.) The cranial pair, called **trabeculae,** are situated in front of the hypophysis. The parachordals develop from mesodermal mesenchyme[8] and hence are closely related to the somatic skeleton. But the trabeculae, like the visceral arches, develop from mesenchyme derived from the neural crest. The parachordals enlarge and unite to form the **basal plate.** The trabeculae form the **ethmoid plate,** which extends rostrally between the nasal sacs. Together the basal and ethmoid plates form the floor of the chondrocranium. A variable number of **occipital arches,** which are serially homologous to developing vertebrae, unite with the posterior end of the basal plate, and surround the caudal end of the brain. An **otic capsule** develops around each inner ear, and a nasal capsule develops around each nasal sac. An **optic capsule** begins to form in the wall of the developing eye and may ossify as sclerotic plates, but it does not unite with the rest of the chondrocranium. The lateral walls of the chondrocranium between the eyes develop from a complex set of **orbital cartilages** that coalesce with each other and with other parts of the chondrocranium, leaving foramina for cranial nerves that pass through the orbit. In most vertebrates, dermal bones cover the brain dorsally, and only two rods of cartilage form the roof of the chondrocranium: the occipital arches and the **synotic tectum** between the otic capsules. Since dermal bones are absent in cartilaginous fishes, the chondrocranium is complete dorsally.

Study a preparation of the chondrocranium of *Squalus* (Fig. 4-2**A**,**B**, and **C**). The pointed, trough-shaped **rostrum** is at the front end; the caudal end is squarish. A pair of large sockets for the eyes, **orbits,** lies on the sides. The ventral surface of the chondrocranium is very narrow between the orbits; its dorsal surface is much wider in this

---

[7]The term *chondrocranium* is sometimes used to include all parts of the head skeleton derived from cartilage (chondrocranium proper and visceral arches), but it is here used in its narrower sense, which excludes the visceral arches. Some authors use the term *neurocranium* for the chondrocranium in this restricted sense.

[8]Mesenchyme is an embryonic connective tissue composed of stellate-shaped, migratory cells.

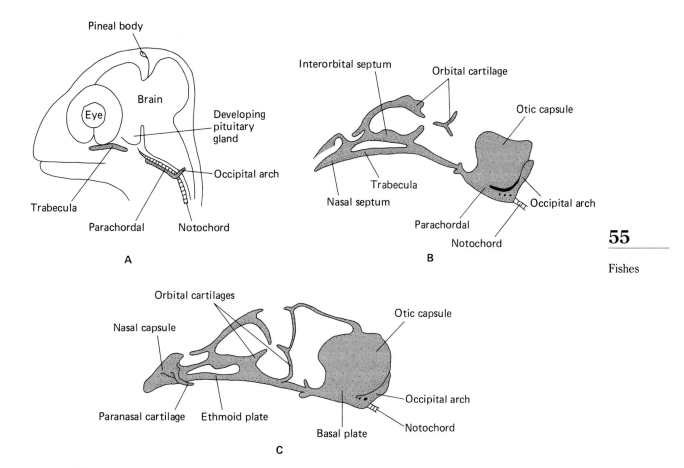

Figure 4-1
Lateral views of three stages in the embryonic development of the chondrocranium of a lizard. **A,** At
2.25-mm body length; **B,** at 5-mm body length; **C,** at 5.2-mm body length. *(After DeBeer.)*

region. The chondrocranium can be divided into several regions, which are not clearly
demarcated in the adult but do have distinct embryonic origins: the occipital region (from
occipital arches and synotic tectum), otic capsules (from otic capsules), basal plate (from
parachordals), orbital region (from orbital cartilages and caudal portions of the
trabeculae), and nasal region (from rostral portions of trabeculae and nasal capsules).

The **occipital region** is the very caudal portion of the chondrocranium lying in the
midline. It surrounds a large hole, the **foramen magnum,** through which the spinal cord
enters the **cranial cavity** within the chondrocranium. A pair of bumps, the **occipital
condyles,** will be seen ventral to the foramen magnum on each side of a centrum-like
area. They develop from parts of a vertebra that has been incorporated in the occipital
region, and they help to articulate the chondrocranium with the vertebral column. There is
little movement at the occipitovertebral joint in any fish, but a double occipital condyle of
this type is unusual; most have a single, rounded condyle located directly ventral to the
foramen magnum.

The paired **otic capsules** are the large, squarish caudolateral corners of the
chondrocranium that extend from the occipital region to the orbits. Between them, on the
dorsal side, is a large depression called the **parietal fossa.** Within the fossa you will see
two pairs of openings that communicate with the inner ear. The smaller, cranial pair is the
endolymphatic foramina for the endolymphatic ducts; the larger, caudal pair is the
perilymphatic foramina. On each capsule you may find ridges, two dorsally and one
laterally, beneath which lie the **semicircular ducts** of the ear. Finally, there are two

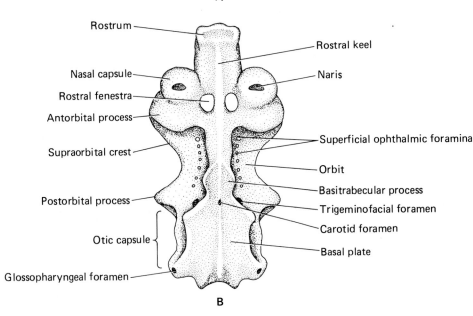

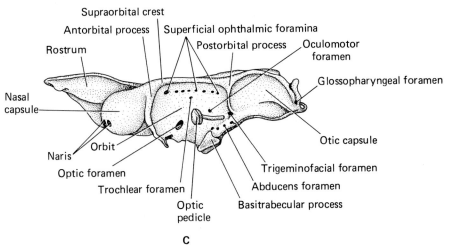

Figure 4-2
The chondrocranium of *Squalus*. **A,** Dorsal view; **B,** ventral view; **C,** lateral view.

large foramina on the caudal edge of the chondrocranium lateral to the occipital condyles. The more medial is the **vagus foramen** for the vagus nerve; the more lateral, the **glossopharyngeal foramen** for the glossopharyngeal nerve.

Ventrally, the otic capsules are connected by the flat, broad **basal plate.** Some of the notochord persists in the chondrocranium of the adult dogfish, and its position is indicated by a white strand of calcified cartilage that surrounds the notochord. This strand can be seen through the hyaline cartilage along the midventral line of the basal plate. The small hole in the midline, anterior to the strand of calcified cartilage, is the **carotid foramen,** which allows for the passage of internal carotid arteries.

The **optic region** is that area that includes and lies between the orbits. The cranial, dorsal, and caudal portions of each orbit are formed by walls of cartilage called the **antorbital process, supraorbital crest,** and **postorbital process,** respectively. Ventrally, the orbit is open. Most of the floor of the chondrocranium is narrow between the two orbits, but near the caudal part of the orbit, the floor is wider and bears a pair of prominent lateral bumps called the **basitrabecular processes.** As will be seen later, orbital processes of the upper jaws have a movable articulation with the floor of the chondrocranium just rostral to these processes. Note that the roof of the chondrocranium between the orbits is complete in *Squalus.* Primitively it was incomplete. The small, median hole near the rostral end of the roof is the **epiphyseal foramen.** In life, it contained a stalk of the same name that represents a vestige of the pineal eye. The series of foramina that perforate the supraorbital crest are the **superficial ophthalmic foramina** for the passage of the superficial ophthalmic nerve and its branches. Many other foramina can be seen in the medial wall of the orbit. The large rostral foramen is the **optic foramen** for the optic nerve; the large caudal one, the **trigeminofacial foramen** for the trigeminal and facial nerves. The remaining foramina are for smaller cranial nerves and blood vessels (Fig. 4-2C). A small cartilaginous stalk resembling a golf tee is left in the orbit in some preparations. This is the **optic pedicle;** it helps to support the eyeball.

The entire chondrocranium rostral to the antorbital processes may be called the **nasal region.** It consists of a pair of round **nasal capsules** attached to the front of each antorbital process, and a long, median **rostrum** that helps to support the snout. The wall of each nasal capsule is very thin and is generally broken. If it is complete, you will be able to see its external opening **(naris).** The opening within the capsule is for the passage of the olfactory tract. The rostrum is trough-shaped dorsally and keeled ventrally. Its dorsal concavity, called the **precerebral cavity,** communicates with the **cranial cavity** by way of a large opening, the **precerebral fenestra.** In life, the precerebral cavity is filled with a gelatinous material. Two other large openings **(rostral fenestrae)** into the cranial cavity will be seen ventrally lying between the nasal capsules on each side of the rostral keel.

57

Fishes

◆ **(B) SAGITTAL SECTION OF THE CHONDROCRANIUM**

Examine the inside of the cranial cavity in a sagittal section of the chondrocranium, if such a preparation is available. Certain structures seen earlier can be noted again in this view, but there are additional features of interest. Notice the depression in the floor dorsal to the basitrabecular processes. This is the **sella turcica,** a recess for the hypophysis. The large opening in the medial wall of the otic capsule, just caudal to the trigeminofacial foramen, is the **internal acoustic meatus.** Part of the glossopharyngeal nerve enters here to pass beneath the ear and emerge through the glossopharyngeal foramen, but most of the passage is occupied by the vestibulocochlear (statoacoustic) nerve from the inner ear.

◆   **(C) THE VISCERAL SKELETON**

The gill pouches and gills of jawless (agnathous) vertebrates are supported by a series of cartilaginous arches that lie *lateral* to the pouches, only a short distance beneath the skin. These cartilages form the branchial basket of the lamprey (p. 19). Comparable arches in extinct ostracoderms were attached to the overlying dermal bones. Some investigators believe that the branchial basket of agnathous vertebrates is a visceral skeleton. It may be, but it does not appear to be homologous to the visceral skeleton of jawed vertebrates (gnathostomes). The latter consists of arches of cartilage or cartilage replacement bone that are located *medial* to the gill pouches and next to the cavity of the pharynx. It is possible that the evolution of cartilaginous supports for the gills occurred independently and followed a different course in agnathans and early gnathostomes.

The jaws of cartilaginous fishes are formed by an enlarged arch of cartilage in the wall of the mouth cavity. This arch is in series with the other visceral arches and is regarded as the first visceral arch, or **mandibular arch.** In other fishes, dermal bones attach to the underlying mandibular arch and contribute to the jaws. The evolution of jaws allowed the size of the mouth opening to be controlled. The mouth can be opened and shut forcefully to seize prey, or the pharynx can be expanded and the mouth quickly opened widely to suck in prey. It seems probable that an anterior visceral arch, which would have been in a strategic position to open and close the mouth, was transformed into jaws in the course of evolution. We are uncertain whether the mandibular arch was the very first in a series of visceral arches, or whether one or more premandibular arches lay anterior to the present mandibular arch and were completely reduced.

The upper half of the mandibular arch, known as the **palatoquadrate cartilage,** must be braced some way so that its caudal end can act as a stable fulcrum for movements of the lower jaw. In ancestral jawed vertebrates it probably was braced by connections between the palatoquadrate and the chondrocranium. This type of jaw support is termed **autostylic** (Fig. 4-3). A complete gill slit may have been present between the mandibular arch and the second arch, which is called the **hyoid arch.** An ancestral shark has been discovered with this condition. In some fishes, including the crossopterygians, the dorsal part of the hyoid arch, called the **hyomandibular,** also extended as a prop between the otic capsule and the caudal end of the palatoquadrate. A connection also remained between the anterior part of the palatoquadrate and the chondrocranium. This is the **amphistylic** type of jaw suspension. The postmandibular gill slit is reduced to a spiracle or lost entirely. In the terrestrial vertebrate that evolved from crossopterygians, the hyomandibular is excluded from jaw suspension, and the palatoquadrate is again supported only by connections to the chondrocranium. This autostylic condition evolved secondarily from an amphistylic condition and is not a retention of the ancestral autostyly. In most fishes, including *Squalus,* supporting connections between the chondrocranium and palatoquadrate are lost, and the jaws are braced solely by the hyomandibular. This is the **hyostylic** type of jaw suspension. It evolved from an amphistylic condition and is the most advanced type among fishes. It permits the jaws to swing forward or downward during feeding.

Examine a preparation of the visceral skeleton of *Squalus* (Figs. 4-4 and 4-5). The seven **visceral arches** that compose it show clearly. The first, which is modified to form

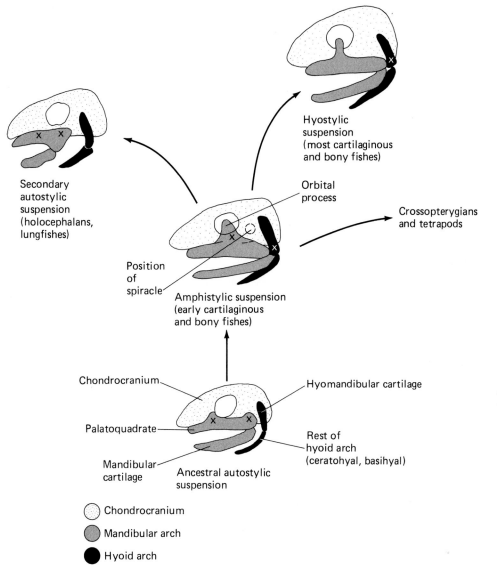

Figure 4-3
The probable evolution of the suspension of the palatoquadrate cartilage in fishes. The embryonic cartilaginous elements are shown, but many of these become ossified in bony fishes. X = points of palatoquadrate suspension.

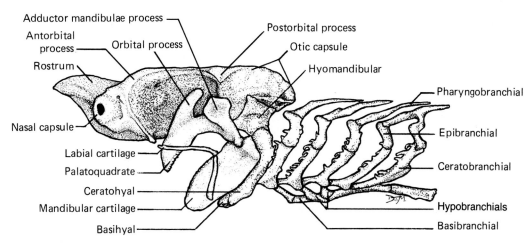

Figure 4-4
Lateral view of the chondrocranium and visceral skeleton of *Squalus*.

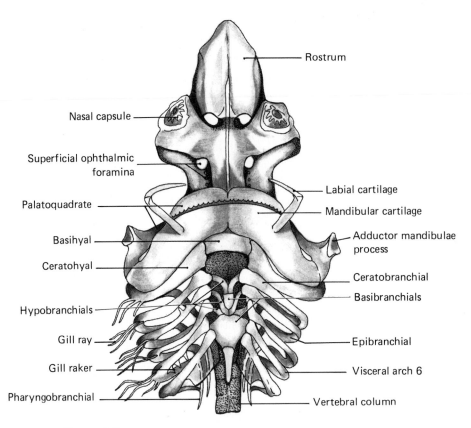

Figure 4-5
Ventral view of the chondrocranium and visceral skeleton of *Squalus*.

the upper and lower jaws, is the **mandibular arch.** The second arch, which extends from the otic capsule to the angle of the jaws and ventrally into the floor of the mouth, is the hyoid arch. The spiracle would be located between the dorsal portion of the hyoid arch and the upper jaw. The last five visceral arches, called **branchial arches,** support the interbranchial septa[9] and hence pass between the remaining gill slits. Note that the third visceral arch and first branchial arch are synonymous. In addition to the arches, small **labial cartilages** may be seen on the lateral surface of the jaws. They are located in the labial folds previously observed (p. 41) and are too superficial to be a part of the visceral skeleton.

Study the mandibular and hyoid arches in more detail. Each side of the upper jaw is formed by a **palatoquadrate cartilage,** each side of the lower jaw by the **mandibular cartilage** (Meckel's cartilage). Both cartilages bear several rows of sharp teeth, which are loosely attached to the surface of the jaws and are similar to one another (Fig. 10-9; p. 303). A dentition in which the teeth are essentially the same is referred to as **homodont.** The teeth of selachians differ from those of other fishes primarily in their triangular shape

---

[9]The tissue lying between successive gill slits supports the gill filaments and contains a number of structures associated with the gills—the skeletal arches, muscles, nerves, and blood vessels. There is some confusion as to what the entire complex should be called. It has often been called *gill, branchial,* or *visceral arch,* using the term *arch* in a broad sense. To avoid confusion we prefer to limit the term *arch* to the skeletal elements and to use the term *interbranchial septum* for the entire complex.

and sharp edges. *Squalus* and other sharks use these teeth to cut up prey. In many other fishes, the teeth are simple cones used to seize and hold prey, which is swallowed whole. Two prominent processes extend dorsally from the palatoquadrate. The one above the angle of the jaw is the **adductor mandibulae process** for the attachment of mandibular muscles. The one that extends up into the orbit **(orbital process)** passes lateral to the braincase and just rostral to a basitrabecular process. Orbital processes permit the jaws to move up and down slightly as the mouth is protruded and retracted during feeding, but they prevent lateral motion. The single midventral piece of the hyoid arch is called the **basihyal.** A **ceratohyal** extends from this element to the angle of the jaw, and a **hyomandibular** continues to the otic capsule. Ligaments unite the caudal end of the palatoquadrate cartilage and hyoid arch, so this arch connects the upper jaw to the chondrocranium.

Each of the branchial arches ideally consists of a midventral **basibranchial** and a chain of four additional elements extending dorsally on each side—a short, ventral **hypobranchial;** a longer **ceratobranchial** extending to the height of the angle of the jaws; an **epibranchial** continuing beyond this; and, finally, a **pharyngobranchial,** which overlies the pharynx and points caudally but does not unite with the vertebrae. All these elements can be seen in the first branchial arch, but there has been some fusion and loss of certain of the elements in the caudal branchial arches. Also notice that each branchial arch, and the hyoid arch as well, bears a number of laterally projecting cartilaginous **gill rays** that stiffen the interbranchial septa. Median projections from the branchial arches, called **gill rakers,** prevent food in the pharynx from entering the gill pouches.

Certain of the relationships of the visceral skeleton to other organs of the head can be seen clearly in transverse and sagittal sections of the head (Fig. 10-8, p. 302, and Fig. 11-13, p. 356).

♦   **(D) DERMAL BONES**

> The third component of the head skeleton is the dermal bones associated with the cranial portions of the endoskeleton as far caudally as the pectoral girdle. Although cartilaginous fishes have small dermal denticles in their skin (p. 42), dermal bones are not associated with the head skeleton or pectoral girdle. Recent evidence suggests that large dermal plates may never have been present in the ancestors of the chondrichthyans, although they were present in other fish groups, including the ancestors of terrestrial vertebrates. Interrelationships of the three components of the head skeleton (chondrocranium, visceral skeleton, dermal bones) are diagrammed in Figure 4-6. An appreciation of the extent of the dermal bones present in most fishes can be gained by studying them in the bowfin, *Amia calva.*

Examine the skeleton of the head of *Amia,* noting the groups, or series, of dermal bones that sheathe the chondrocranium and visceral arches (Fig. 4-7**A**). In some types of preparations the dermal bones have been removed on one side, exposing the endoskeletal structures. A **dermal roof** covers the top of the head and most of the cheek region. It is pierced by tiny nares (two on each side) and by the large orbits. There is also a large gap,

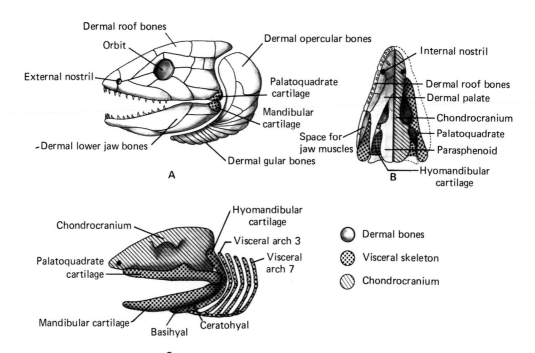

Figure 4-6
Diagrams of the components of the head skeleton of a generalized bony fish: **A,** Lateral view of the skull showing the superficial dermal bones that cover most of the other components. **B,** Ventral view of the skull with the dermal bones removed from the right side of the drawing. **C,** Lateral view of the skull after the removal of all of the dermal bones, leaving the chondrocranium and visceral arches.

which is not found in more ancestral fishes, between the cheek and upper jaw. The dermal bones on the margin of the jaw bear a single row of small, conical **teeth,** which are essentially similar to one another **(homodont).**

The caudal part of the palatoquadrate cartilage ossifies as the **quadrate,** but the part of it rostral to this is lost in the adult. A **palatal series** of dermal bones, on some of which there are several rows of teeth, forms much of the roof of the mouth in the general region of the missing portion of the palatoquadrate (Fig. 4-6). A third "series" (actually composed of but a single element, the **parasphenoid**) lies in the midline of the roof of the mouth directly beneath the chondrocranium. It bears a number of very small teeth.

A **lower jaw series** covers the lateroventral and medial surfaces of the mandibular cartilage, the caudal end of which usually ossifies as an **articular** bone. The jaw joint of all jawed vertebrates, except mammals, lies between the quadrate and articular, both derivatives of the mandibular arch. As with the upper jaw and palate, there is a single row of marginal teeth and several rows of more medial teeth. The medial teeth in both the palate and lower jaw occupy the same general position as the teeth on the mandibular arch of *Squalus* and are considered their homologues.

The remaining visceral arches are covered laterally by an **opercular series** and ventrally by a **gular series** of dermal bones. Finally, the endoskeletal portions of the pectoral girdle are covered laterally by a **pectoral series** of dermal bones.

The individual bones in these series are identified in Figure 4-7**B.** Most of these bones have the same names as those in terrestrial vertebrates. They occupy the same relative positions in fishes and terrestrial vertebrates, but the homologies are not entirely certain.

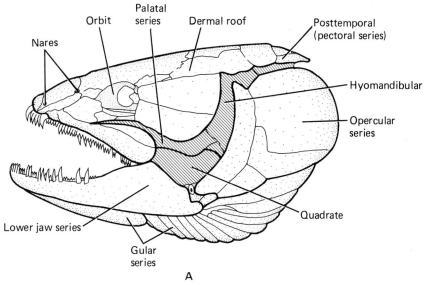

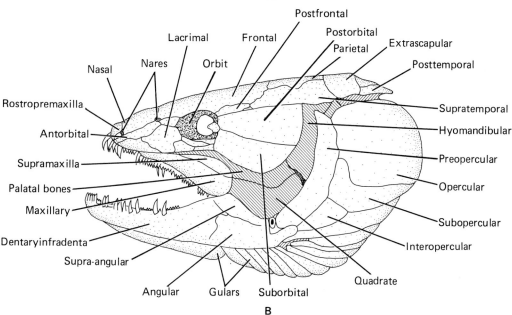

Figure 4-7
Lateral views of the skull of *Amia calva,* a bony fish with a primitive skull configuration. **A,** Groups of dermal bones are shown in heavy outline. **B,** Individual bones are identified. The quadrate and hyomandibular bones are parts of the visceral skeleton. *(After Goodrich.)*

AMPHIBIANS AND REPTILES

The skeletons of living amphibians and reptiles that are available in biological supply houses are specialized in many ways and therefore are not good representatives of the skeletons of ancestral terrestrial vertebrates. We can best understand the components of the tetrapod skull by first describing briefly the skull of *Palaeogyrinus,* an extinct labyrinthodont, one of the early amphibians. Then, when you examine the skull of *Necturus* or of the snapping turtle, *Chelydra serpentina,* you will appreciate how their skulls compare with that of a truly ancestral terrestrial vertebrate. *Necturus* is widely used as a representative of the class Amphibia, but you must remember that *Necturus* is neotenic and possesses some larval features in its skull. The turtle skull is primitive in some respects, but teeth have been lost, and the face is exceptionally short for a reptile. A lizard or alligator skull, if available, will show some primitive features not present in the turtle, but these skulls are specialized in the temporal region.

The Labyrinthodont Skull

The three groups of elements present in the head region of fishes (chondrocranium, visceral skeleton, and dermal bones) are also represented in amphibians and reptiles, but the opercular and gular series of dermal bones of fishes have been lost, and the visceral arches have been greatly reduced. These changes are correlated with the reduction of the gills and the beginning of neck formation as adaptations to terrestrial life. The elements that remain may be grouped, for purposes of description, into four units: (1) the skull in its restricted sense (that is, bones encasing the brain and sense organs and forming the upper jaw), (2) the lower jaw, (3) the teeth, and (4) the hyoid apparatus.

The skull of a labyrinthodont (Figs. 4-8 and 4-9) is composed of the chondrocranium, the palatoquadrate of the visceral skeleton, and the dermal bones that more or less encase these two. The chondrocranium covers the back, the underside, and most of the lateral surfaces of the brain. It does not cover the top of the brain, and there is a gap in each of its lateral walls. It is usually ossified to a large extent, as shown in Figures 4-8 and 4-9 and Table 4-2, but the rostral ethmoid region and nasal capsules

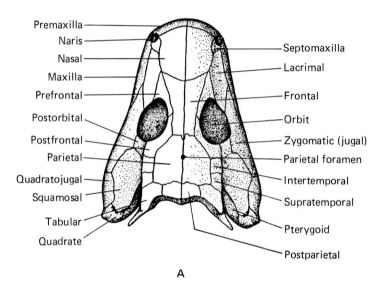

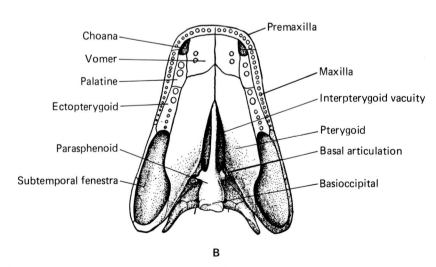

Figure 4-8
The skull of an ancestral terrestrial vertebrate based primarily on the Carboniferous labyrinthodont, *Palaeogyrinus*. **A**, Dorsal view; **B**, palatal view. *(Modified from Romer and Parsons.)*

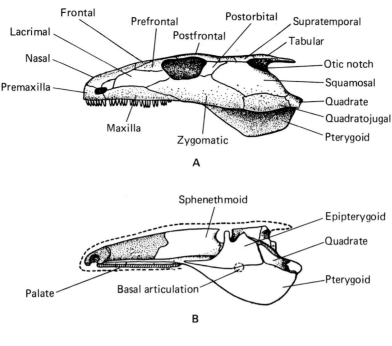

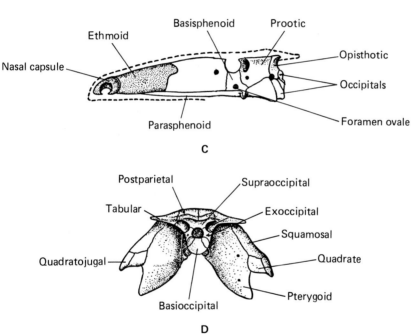

Figure 4-9
The skull of *Palaeogyrinus*. **A,** Lateral view; **B,** lateral view after removal of the dermal skull roof; **C,** lateral view of the chondrocranium; **D,** posterior view. *(Modified from Romer and Parsons.)*

are unossified. As in most fishes, but not *Squalus,* the occipital condyle is single. The chondrocranium is perforated by foramina for nerves and vessels and by an **oval window,** located on the lateral surfaces of the otic capsule, for the stapes (see a later section).

The posterior portion of the palatoquadrate ossifies as the **quadrate** and articulates with the lower jaw. The portion of it adjacent to the basitrabecular process of the chondrocranium generally ossifies as the **epipterygoid.** This bone articulates with the braincase and helps to fill in the gap in its side (see Fig. 4-19A). The rest of the palatoquadrate is lost in the adult.

(text continues on p. 68)

Table 4-2 Components of the Tetrapod Skull and Lower Jaw

The components of the skull and lower jaw of a labyrinthodont, together with the part of the skeleton to which they belong, are shown in the left hand column. The homologies between these elements and those of certain other tetrapods are shown in the right hand column. An X indicates that an element is present; O, that it is absent; Unos., that the region is unossified. All the elements are paired unless indicated to the contrary by a number in parentheses: (1) indicates a median element; (2) or (3) that two or three of these are present on each side.

Labyrinthodont	Necturus	Turtle	Lizard	Alligator	Mammal
SKULL					
Chondrocranium					
(Cartilage replacement bone)					
Basioccipital (1)	Unos.	X (1)	X (1)	X (1)	X (1)
Exoccipital	X	X	X	X Fused with opisthotic	X ⎱ Occipital
Supraoccipital (1)	Unos.	X (1)	X (1)	X (1)	X (1) ⎰
Opisthotic	X (Operculum)	X	X	X	X ⎱ Petrosal part of
Prootic	X	X	X	X	X (1) ⎰ temporal
Basisphenoid (1)	Unos.	X (1)	X (1)	X (1)	X (1) Body of basisphenoid
Sphenethmoid (1)	Unos.	Unos.	X Orbitosphenoid	X Laterosphenoid	X (1) Presphenoid
Unossified ethmoid region	Unos.	Unos.	Unos.	Unos.	X Ethmoid
Unossified nasal capsule	Unos.	Unos.	Unos.	Unos.	X Turbinates
Visceral arches					
(Cartilage replacement bone)					
Palatoquadrate					
Quadrate	X	X	X	X	X Incus
Epipterygoid	O	X	X	X	X Wing of basisphenoid
Hyomandibular					
Stapes	X	X	X	X	X
Dermal bones					
Roof					
Tooth-bearing marginal bones					
Premaxilla (incisive)	X	X	X	X	X
Maxilla	O	X	X	X	X
Median series					
Nasal	O	O	X	X	X
Frontal	X	X	X	X	X
Parietal	X	X	X	X	X
Postparietal	O	O	O	O	X Interparietal, often a part of occipital

Circumorbital series				
Lacrimal	O	O	X	X
Prefrontal	O	X	X	O
Postfrontal	O	O	X	O
Postorbital	O	X	X	O
Zygomatic (jugal)	O	X	X	X
Temporal series				
Intertemporal	O	O	O	O
Supratemporal	O	O	X	O
Tabular	O	O	O	X ? Part of occipital
Cheek bones				
Squamosal	X	X	X	X Squamous part of temporal
Quadratojugal	O	O	O	O
Palate and underside of chondrocranium				
Parasphenoid	X	X Fused with basisphenoid	X Fused with basisphenoid	X ? Part of basisphenoid
Vomer	X	X (1)	X	X (1)
Palatine	O	X	X	X
Ectopterygoid	O	O	X	X ? Part of basisphenoid
Pterygoid	X	X	X	X Pterygoid process of basisphenoid
LOWER JAW				
Visceral arches				
(Cartilage replacement bone)				
Mandibular cartilage				
Articular	Unos.	X	X	X Malleus
Dermal bones				
Lateral series				
Dentary	X	X	X	X
Splenials (2)	X	O	X	O
Surangular	O	X	X	O
Angular	X	X	X	X Tympanic part of temporal (endotympanic; a new cartilage replacement bone)
Medial series				
Coronoids (3)	O	X	X	O
Prearticular	O	X Fused with articular	X Fused with articular	X Anterior process of malleus

The dermal bones constitute the largest component of the skull. A great many were present in labyrinthodonts, but the evolutionary tendency since then has been one of reduction. Also notice certain general features of importance in the dermal roof and palate (Fig. 4-8). The external nostrils (**nares**) are widely separated, and the internal nostrils (**choanae**) enter the mouth at the very front of the palate. A **parietal foramen** is present between the parietal bones for the pineal or parietal eye. An **otic notch,** of unknown function, is located on the caudodorsal part of the skull roof. Aside from this notch, the temporal portion of the roof is solid (**anapsid** condition). Caudally, a pair of **posttemporal fenestrae** pass between the dermal roof and otic capsule. Mandibular muscles arise from the underside of the roof on each side and pass through a pair of lateral palatal openings (**subtemporal fenestrae**) to insert on the lower jaw. A **basal articulation** between the pterygoid and basitrabecular processes permits some movement between the palate and braincase. **Interpterygoid vacuities** separate the palate and braincase cranial to the basal articulation, and a **cranioquadrate passage,** which transmits important vessels and nerves, is situated just caudal to the basal articulation.

The **stapes** or **columella,** which evolved from the hyomandibular cartilage of fishes, may be added to these major components of the skull. In a frog or turtle it is a small, delicate bone that transmits high frequency sound waves from the tympanic membrane, across the middle ear cavity, to the otic capsule. It is doubtful that labyrinthodonts had a tympanic membrane, but they had a large stapes that abutted instead on the quadrate bone. Their stapes was too massive to have responded to high frequency air borne vibrations, but may have conducted lower frequency vibrations of sufficient intensity by bone conduction.

The lower jaw of early terrestrial vertebrates consists of a **mandibular cartilage** surrounded by a sheath of dermal bones. The caudal end of the mandibular cartilage (the region articulating with the quadrate) generally ossifies as the **articular bone.** The rest of the mandibular cartilage either remains cartilaginous or disappears.

The **teeth** of labyrinthodonts are similar in shape to those of fishes, for they are small, conical, numerous, and similar to one another (homodont). They are usually arranged in two series—a lateral series on the margins of the jaws and a medial series on the palate and medial side of the lower jaw.

As stated, the mandibular arch becomes incorporated into the skull and lower jaw, and the dorsal part of the hyoid arch (hyomandibular) becomes the stapes. The rest of the hyoid arch unites with variable portions of the more cranial branchial arches to form the **hyoid apparatus.** In contemporary tetrapods, this apparatus lies in the base of the newly evolved muscular tongue. Ligaments extend from the hyoid to the skull base. The hyoid apparatus forms a sling for the support of the tongue and pharynx floor and for the attachment of muscles associated with feeding and swallowing. Portions of the more caudal branchial arches form the cartilages of the larynx.

Necturus

♦   (A) ENTIRE SKULL

Study a skull of *Necturus*. Unfortunately, it has retained fewer of the primitive features and elements than many other amphibians or reptiles. First examine the top of the skull (Fig. 4-10). Most of the bone that you see belongs to the dermal roof, although other groups have been exposed through the loss of some of the original elements of the roof. The V-shaped, tooth-bearing bone on each side of the front of the upper jaw is the **premaxilla.** The **nasal cavity** is located in the notch posterior to its lateral wing.

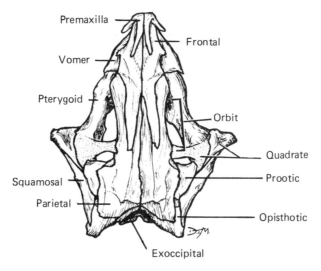

Premaxilla

Frontal

Vomer

Pterygoid

Orbit

Quadrate

Prootic

Squamosal

Parietal

Opisthotic

Exoccipital

Figure 4-10
Skull of *Necturus,* dorsal view.

Continuing caudally along the middorsal line, the next elements are a large pair of **frontal** bones. Paired **parietals** extend from the frontals nearly to the **foramen magnum.** Each parietal also has a narrow process that extends cranially lateral to the frontals. The **orbits** are located ventral to these processes.

Lateral to the posterior half of the parietal, you will see two small bones. One forms the very caudal angle of the skull and continues forward to a tiny window of cartilage. The other is cranial to this window. These bones, the **opisthotic** and **prootic,** respectively, form the otic capsule of the chondrocranium. The thin sliver of bone on the margin of the skull, lateral to the otic bones, is the **squamosal**—the last element of the original dermal roof. Cranial and ventral to it, at the point where the lower jaw articulates, you will see the partly ossified **quadrate**—the only part of the palatoquadrate present in *Necturus.*

The caudal end of the skull, lateral and ventral to the foramen magnum, is formed by the paired **exoccipitals.** Each bears a condyle. There are therefore two **occipital condyles** in *Necturus,* unlike the single condyle of more ancestral tetrapods.

Look at the underside of the skull (Fig. 4-11A). The palate consists of two pairs of dermal bones. The cranial pair, **vomers,** are located caudal to the premaxillae and, like the premaxillae, bear a row of teeth. The caudal pair, **pterygoids,** continue back from the vomers to the otic region. The fronts of the pterygoids also have a row of teeth. Portions of the vomers and pterygoids can be seen from the dorsal side. The large median bone on the ventral surface, lying between the vomers and the pterygoids and continuing to the exoccipitals, is the dermal **parasphenoid.** A bit of the unossified **ethmoid plate** of the chondrocranium is exposed rostral to the parasphenoid. The two otic bones can also be seen in the ventral view lying caudal to the pterygoids and lateral to the back of the parasphenoid. A cartilaginous area, containing the **oval window,** separates them. The oval window may be covered by a tiny disc-shaped bone bearing a little stem (stylus). This represents the **stapes** combined with the **operculum.** The opercular bone develops from the wall of the otic capsule and is a unique part of the amphibian auditory apparatus (p. 233). It is connected by its stylus and by ligaments to the squamosal.

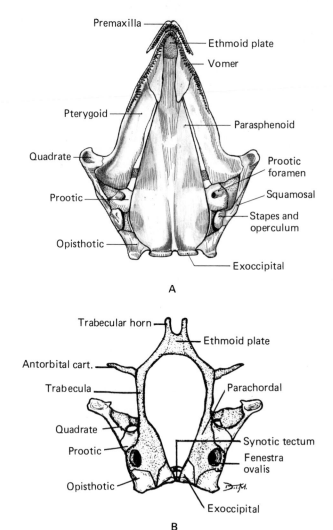

Figure 4-11
Skull of *Necturus*. **A**, Ventral view of entire skull; **B**, ventral view of chondrocranium and quadrate after removal of the dermal bones shown in **A**.

♦ **(B) THE CHONDROCRANIUM**

Although you can see parts of the chondrocranium in the complete skull, you will see them more clearly by examining a preparation in which the dermal bones have been removed. The chondrocranium of *Necturus* is not a good representative of the primitive adult stage since it is largely in the embryonic condition. (Compare what follows with the description of the development of the chondrocranium on page 54.) The pair of large, round caudolateral swellings are the **otic capsules.** You will see two ossifications in each—a cranial **prootic** and a caudal **opisthotic** (Fig. 4-11B). The hole on the side of the capsule is the oval window, which may be covered by the stapes. The only other ossifications are a pair of ventral **exoccipitals,** which form in the **occipital arch.** They are connected by a delicate cartilaginous bridge, the **basioccipital arch.** Dorsally, the otic capsules are connected by another bridge, the **synotic tectum.** Often the quadrate (a part of the mandibular arch) is left on preparations of the chondrocranium and will be seen cranial and lateral to the prootic.

The shelves of cartilage united with the medioventral edge of each otic capsule are the **parachordals.** The pair of cartilaginous rods that continue forward from the parachordals are **trabeculae.** They are united cranially to form an **ethmoid plate,** from which a pair of **trabecular horns** continue between very delicate nasal capsules. The nasal capsules are usually destroyed. A small **antorbital cartilage,** representing one of the orbital cartilages of other vertebrates, extends laterally from each trabecular cartilage.

◆ (C) THE LOWER JAW

Compare the lower jaw of *Necturus* with Figure 4-12. Only three dermal bones cover the mandibular cartilage in this animal. The largest of these is the **dentary,** which forms most of the lateral surface of the jaw and a small portion of the medial surface near the front of the jaw. It bears the long rostral row of teeth. The short caudal row of teeth is on the **splenial,** most of which is on the dorsomedial surface of the jaw, but a bit of the bone shows laterally. Most of the medial surface of the jaw is formed by the **angular.** This bone is widest caudally and then tapers to a point, which passes ventral to the splenial and continues rostrally to the dentary. A small portion of the angular shows laterally at the caudoventral corner of the jaw. The surface of the lower jaw that articulates with the quadrate is formed by the **mandibular cartilage,** which is unossified in *Necturus.*

◆ (D) TEETH

Notice that the teeth of *Necturus* are essentially fishlike. That is, they are small, conical, and all of the same type **(homodont).** They are also relatively numerous, and you can see parts of a **lateral** and **medial series.** The lateral series are on the premaxillae and dentaries; the medial, on the palatal bones and splenials. The teeth attach loosely to the jaws.

◆ (E) HYOID APPARATUS

Preserved skeletal material is better than dried material for studying the hyoid apparatus. Notice that it is made up of parts of four visceral arches—the hyoid arch and the first three branchial arches (Fig. 4-12). The hyoid arch is the most cranial and the largest component. It consists on each side of a short **hypohyal,** which lies just lateral to the midventral line, and a longer **ceratohyal,** which extends toward the angle of the jaw. A median **basibranchial 1** extends caudally from the hypohyals to the first branchial arch. This arch, too, is composed of two cartilages on each side, the more medial being **ceratobranchial 1;** the more lateral, **epibranchial 1.** The next two arches (branchial arches numbers 2 and 3) are greatly reduced and at first sight appear to consist only of **epibranchials 2 and 3.** On closer examination, however, you will see a small cartilage **(ceratobranchial 2)** that connects them with the first branchial arch. The branchial arches support the three external gills. The two gill slits pass on either side of the second branchial arch. A small, median **basibranchial 2,** which is generally partly ossified, extends caudad from the base of the first branchial arch.

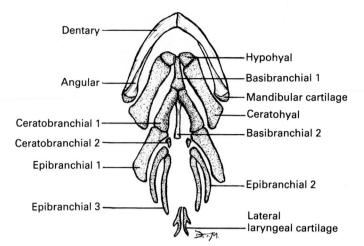

Figure 4-12
Ventral view of the lower jaw, hyoid apparatus, and laryngeal cartilages of *Necturus*. Cartilage is stippled.

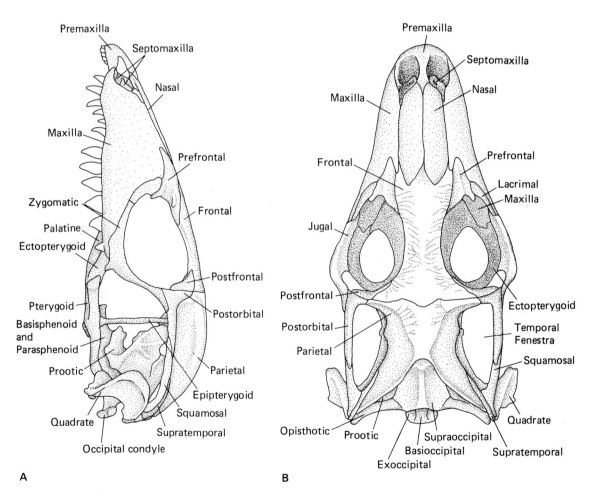

A

B

Figure 4-13
Skull and mandible of the common tegu lizard, *Tupinambis nigropunctatus,* which ranges from the southern United States to South America. **A,** Lateral view; **B,** dorsal view.

The last two branchial arches do not contribute to the hyoid apparatus and will not be seen. There is some doubt as to their fate, but traces of them may persist. A raphe in one of the branchial muscles, which contains a few cartilage cells, may represent a part of the fourth branchial (sixth visceral) arch. It is also probable that the rest of this arch plus the last arch have contributed to the lateral cartilages of the laryngotracheal chamber.

Reptiles

In many ways the skulls of certain living reptiles are better examples of an ancestral tetrapod skull than is that of *Necturus*. The skull of the snapping turtle, *Chelydra*, is described and illustrated in Figures 4-15 to 4-17, and illustrations of a lizard's and an alligator's skull (Figs. 4-13 and 4-14) are included for comparison.

♦   **(A) GENERAL FEATURES OF THE SKULL**

Examine a skull of *Chelydra* (Figs. 4-15 to 4-17). Teeth are absent and are replaced by a horny beak sheathing the jaw margins. The horny sheathing is often missing on dried

(*text continues on p. 76*)

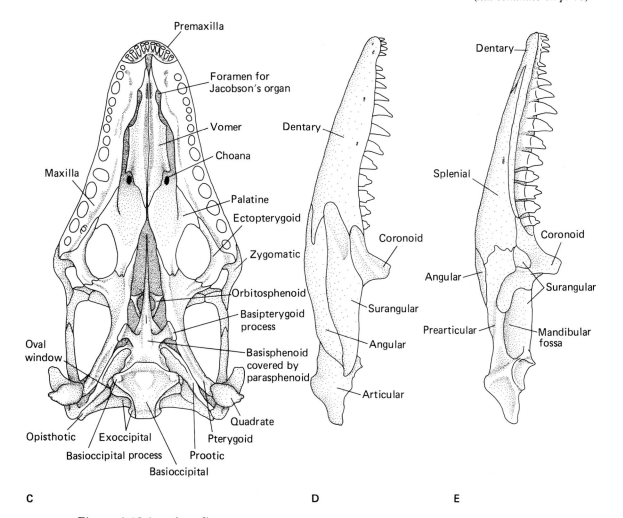

C   D   E

Figure 4-13 (continued)
C, palatal view; **D,** lateral view of the mandible; **E,** medial view of the mandible. Notice that one temporal fenestra lies dorsal to the postorbital-squamosal bar. Another lies ventral to this bar, but the bar of bone that bounded it ventrally in the ancestors of lizards is now lost. Most of the original temporal skull roof has also been lost. (*Redrawn from M. Jollie.*)

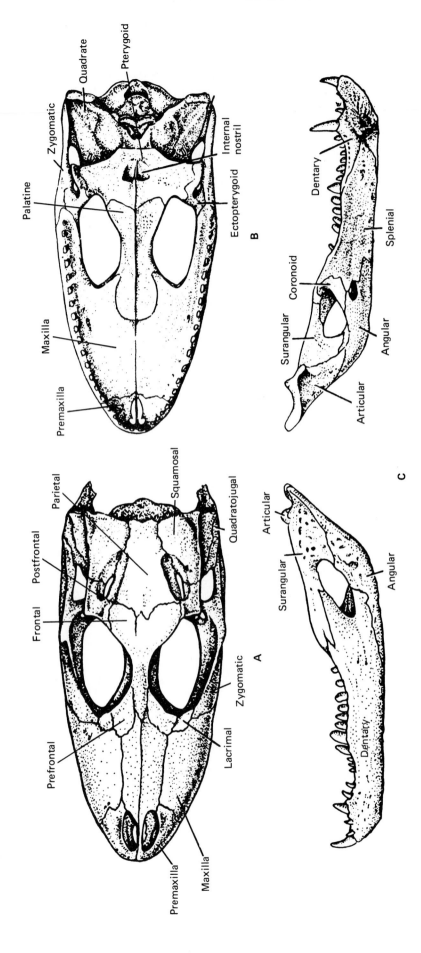

Figure 4-14
The alligator skull. **A**, Dorsal view; **B**, ventral view; **C**, lateral and medial views of the lower jaw. *(From Romer and Parsons.)*

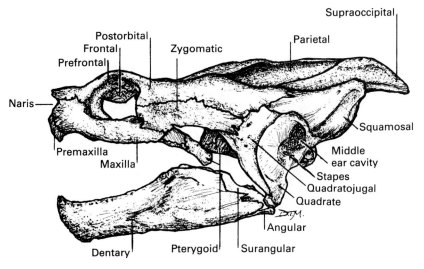

Figure 4-15
Lateral view of the skull and lower jaw of the snapping turtle, *Chelydra.*

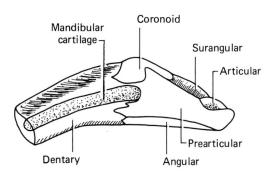

Figure 4-16
Snapping turtle skull and lower jaw. *Upper,* dorsal view of the skull; *lower,* median view of the right lower jaw.

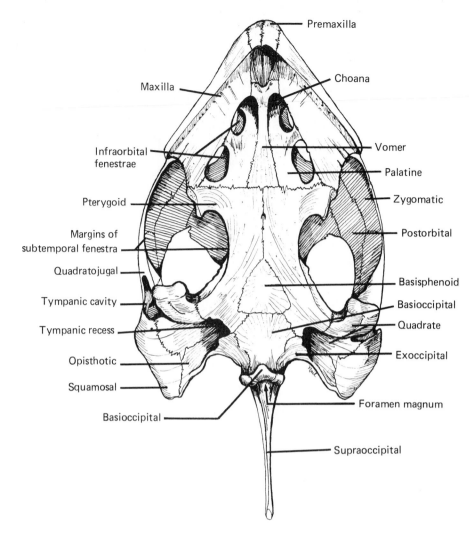

Figure 4-17
Ventral view of the skull of the snapping turtle.

skulls. The **dermal roof** and **palate** can be seen surrounding the small, partly ossified **chondrocranium.** The originally paired **nares** have united to form one opening at the surface of the dermal roof, but the nasal cavities remain distinct; they open into the mouth by paired internal nostrils **(choanae)** located at the front of the palate. The **orbits** are situated far forward, for the snout is short. The dermal roof in the temporal area has been "eaten away," or emarginated, from behind in most turtles but is complete in the sea turtles. Although this emargination is related to the attachment and bulging of the powerful jaw muscles, it is not comparable in its relationship to the temporal fenestrae of other reptiles. Turtles are usually considered to have the complete **(anapsid)** temporal region of ancestral amphibians and reptiles. In any case, you can easily visualize the nature of such a roof by looking at the turtle. A pair of large **posttemporal fenestrae** can be seen in a caudal view of the sea turtle skull between the dermal roof and otic capsule. They are present in other turtles, too, but the loss of the overlying dermal roof in this region makes them less apparent.

The palatal bones have united solidly with the underside of the braincase, so interpterygoid vacuities are absent. The pair of large **subtemporal fenestrae** can be seen between the palatal bones and the lateral margins of the dermal roof. Two pairs of **infraorbital fenestrae** can be found in the palate beneath the orbits.

Examine the back of the skull and note the position of the middle ear, or **tympanic cavity.** The reptile external and middle ears are located caudal to the quadrate. Often, as in turtles, the quadrate partly encases and constricts the tympanic cavity so that it appears as an hourglass-shaped cavity divided into medial and lateral portions.

♦ **(B) COMPOSITION OF THE SKULL**

The dermal elements along either lateroventral margin of the roof (Figs. 4-15 to 4-17) are a very small **premaxilla** ventral to the naris, a large **maxilla** continuing beneath the orbit, a **zygomatic** (jugal) caudal to this, and a **quadratojugal** just cranial to the middle ear cavity. The zygomatic enters only the caudoventral corner of the orbit. The middorsal elements are a pair of large **prefrontals** caudal to the naris and dorsal to much of the orbit; a pair of small **frontals** that do not enter the orbit in *Chelydra;* and a pair of large **parietals** that extend to the prominent, middorsal, occipital crest. Each parietal also sends a wide flange ventrally that covers a part of the brain not covered by chondrocranial elements. Two other dermal elements complete that portion of the roof situated between the dorsal and marginal bones. A large **postorbital** lies between the orbit and temporal emargination, and a **squamosal** forms a cap on the extreme caudolateral corner of the skull.

Dermal bones also make up the palate. The very front of this region is formed by palatal processes of the premaxillae and maxillae. A median **vomer** (a fusion of originally paired elements) is located caudal to the premaxillae, and **palatines** lateral to the vomer. The choanae enter on each side of the front of the vomer in most turtles. But in the sea turtles, and a few others, the vomer and palatines have sent out medial, shelflike extensions that unite with each other along the midline to form a small **secondary palate** ventral to the **primary palate.** This pushes the openings of the choanae caudad. The rest of the primary palate is formed by a pair of large **pterygoids.**

The chondrocranium surrounds the base, some of the sides, and the caudal portion of the brain, but it fails to cover the rest of the brain, which is covered instead only by dermal bones of the roof and by the epipterygoid (see later). Four bones surround the **foramen magnum**—dorsally the **supraoccipital,** which forms the occipital crest; laterally the paired **exoccipitals;** and ventrally the **basioccipital.** The **occipital condyle** has begun to shift from its primitive position on the basioccipital to the exoccipitals. Since distinct portions of it are born on all three of these bones, the condyle is tripartite. In a dorsal view an **opisthotic** can be seen extending laterally from the supraoccipital and exoccipital to the squamosal. A **prootic** is situated rostral to this. The **basisphenoid** can be seen ventrally lying between the pterygoids rostral to the basioccipital. The dermal **parasphenoid** has united with it. The interorbital, ethmoid, and nasal capsules of the chondrocranium remain unossified and are usually missing in dried skulls.

The caudal part of the palatoquadrate has ossified as the **quadrate.** This bone occupies the caudoventral corner of the skull, extending from the jaw joint dorsally to the squamosal. Most of it passes rostral to the middle ear, but a part of it extends caudally to the ear. A very slender, rodlike **stapes** may be seen extending from the lateral part of the tympanic cavity through a small hole in the quadrate to the otic capsule. Its medial end is enlarged to form a foot plate that fits into the **oval window.** The central portion of the palatoquadrate has ossified as the **epipterygoid.** This bone helps to fill in a gap in the side of the chondrocranium. The epipterygoid is a distinct element in a sea turtle skull, but it has fused to adjacent elements in *Chelydra*. It occupies a small triangular area between the pterygoid and ventral flange of the parietal, rostral to a large **prootic foramen** for certain cranial nerves (the trigeminal and the abducens).

◆ (C) THE LOWER JAW

Compare the lower jaw of *Chelydra* with Figures 4-14 and 4-15. The caudal end of the mandibular cartilage has ossified as the **articular** and can be recognized by its smooth articular surface. The rest of the cartilage remains unossified, but in the adult it can often be found lying in a groove on the medial surface of the jaw. Most of the lateral surface of the jaw is formed by the **dentary,** but a small **surangular** lies on the lateral surface between the dentary and the articular. The front half of the medial surface is also formed by the dentary, but three additional elements form the medial surface caudally—a **coronoid** dorsally; a **prearticular** ventral to it and bridging the groove for the mandibular cartilage; and an **angular** beneath the prearticular and forming the ventral border of the jaw. A bit of the coronoid and the angular can also be seen from the lateral surface.

◆ (D) HYOID APPARATUS

Study the hyoid apparatus of the turtle (Fig. 4-18). Its median body, the **corpus,** is lodged in the base of the tongue and supports the larynx. Its rostral cartilaginous portion is derived from the basihyal; its caudal ossified part, from the first two basibranchials. A

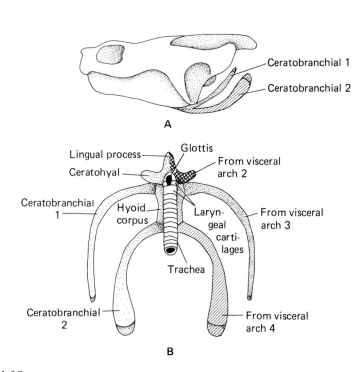

Figure 4-18
A, Lateral view of the hyoid apparatus of a snapping turtle in place beneath and behind the skull.
B, Dorsal view of the hyoid apparatus with the associated larynx and trachea. The derivation of the various parts is shown by different hatching on the right side.

lingual process extends forward from the corpus into the tongue. Cartilaginous and bony horns, the **cornua,** project laterally and dorsally from the corpus around the side of the neck, as visceral arches would in a fish. The first pair of horns are short, cartilaginous **ceratohyals,** and they are connected by ligaments to the base of the skull. The next two pairs of ossified horns represent the **first** and **second ceratobranchials** (visceral arches three and four). More caudal visceral arches have contributed to the **cartilages of the larynx.**

MAMMALS

The following directions for study are based on the cat, but they also apply to many other mammals. If material is available, compare the skeleton of the cat with that of human beings and other species.

The touchstone of the evolution of mammals from ancestral reptiles has been the development of an increased level of activity, a high level of metabolism, endothermy, and a larger and more varied repertoire of behavioral responses than in nonmammalian vertebrates. All this would not be possible without concomitant structural changes in all organ systems. The skull has been affected primarily by changes in certain sense organs, by the great enlargement and increased complexity of the brain, and by changes in breathing and feeding mechanisms needed by an endothermic animal. Major skull transformations related to these changes can be summarized at this time; details are set forth in Table 4-2 and will be seen during the examination of the cat skull.

Mammals have evolved a large and strong skull that houses the brain and resists the forces generated by the large and powerful jaw muscles. This has contributed to the reduction in the number of individual bones throughout the skull. Some bones present in ancestral tetrapods have been lost, and many others have fused with neighboring bones to form complex elements.

The median eye was lost, and the senses of hearing and smell became well developed in ancestral mammals. The cochlear region of the otic capsule expanded, and other changes that accompanied jaw changes increased the sound-transmitting capability of the middle ear. The nasal cavities enlarged, and the nasal capsules ossified as scroll-shaped **turbinate bones,** or **conchae,** which increase the surface area within the nasal cavities.

As the brain enlarged, the braincase expanded. The original chondrocranium in a sense has flattened so that it forms only the front, floor, and back of the mammalian braincase (Fig. 4-19). Its rostral part, which remains cartilaginous in ancestral tetrapods, ossifies as the **ethmoid** in mammals. Dorsally and laterally the brain is encased primarily by sheetlike extensions of dermal bone that grow inward from the original dermal skull roof, in the same way that the parietal of the turtle extends down from the roof. It can be seen from Figure 4-19 that the original dermal skull roof lay lateral to the jaw muscles, and that these extensions lie medial to the muscles. Although derived from the dermal skull roof, these extensions do not constitute the original skull roof. The braincase is completed laterally by the epipterygoid, which becomes the wing of the basisphenoid (alisphenoid). The epipterygoid is derived from that part of the palatoquadrate cartilage that in ancestral tetrapods forms a movable articulation between the palatoquadrate cartilage and the chondrocranium.

A high level of metabolism requires mammals to eat a great deal of food and also exchange a large volume of air in their lungs. Mammals must continue to breathe when they are feeding and chewing food. This is made possible by the evolution of a bony **hard palate,** which in life is extended caudally by a fleshy **soft palate.** Shelflike

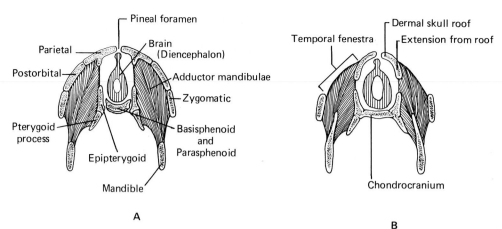

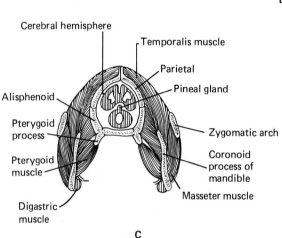

Figure 4-19
Evolution of the tetrapod skull as seen in cross sections caudal to the orbit of an ancestral reptile **A,** a mammal-like reptile **B,** and a mammal **C.** Development of the temporal fenestrae, the jaw musculature, and the braincase is shown. *(From Walker,* A Study of the Cat.*)*

processes of the premaxillary, maxillary, and palatine bones extend toward the midline and unite with each other ventral to the original palate to form the secondary hard palate (Fig. 4-20). This displaces the openings of the internal nostrils caudally, thus permitting the animal simultaneously to breathe and manipulate food within its mouth. Compare this with the secondary palate of the turtle (p. 77). The rostral portion of the primitive palate, now situated dorsal to the hard palate, regresses to some extent, and this area is occupied by enlarged nasal cavities.

Important changes in dentition and feeding mechanisms are correlated with increased energy needs. Mammals do not simply seize and swallow their food; they also cut it up and chew it. This increases the efficiency of chemical digestion by exposing more surface area of the food to digestive enzymes. On the line of evolution toward mammals, the teeth become firmly set in sockets (**thecodont**) and differentiated (**heterodont**) into different types of teeth that crop, cut, and crush the food (Fig. 4-21). A precise occlusion is needed, and this has affected both tooth replacement and jaw muscles. Approximately 80 percent of jaw growth occurs before the first set of teeth has fully erupted. Subsequent tooth replacement is not continuous, as it is in amphibians and reptiles, but limited. "Milk" teeth are lost and replaced by permanent teeth as the jaws enlarge.

Changes in the feeding mechanism also were accompanied by significant changes in the jaw musculature and the parts of the skull to which these muscles attach. The relatively simpler jaw musculature of reptiles became subdivided, larger, and stronger. The muscle mass that originally extended from the underside of the temporal region of

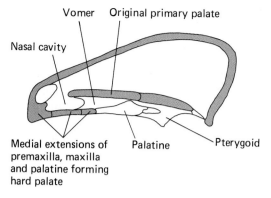

Vomer   Original primary palate

Nasal cavity

Medial extensions of   Palatine   Pterygoid
premaxilla, maxilla
and palatine forming
hard palate

Figure 4-20
A diagrammatic sagittal section through the skull of a late mammal-like reptile showing an early stage
in the evolution of the hard palate. *(After Romer and Parsons.)*

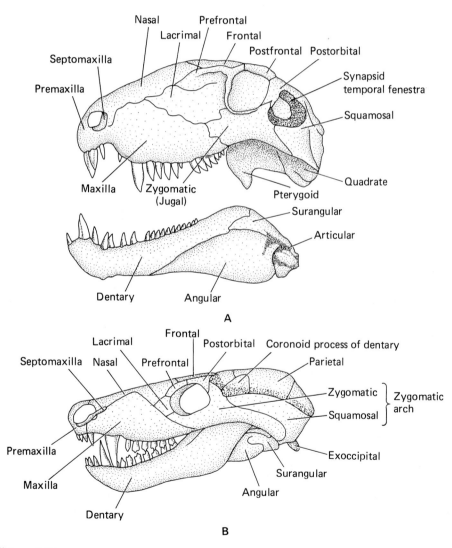

Nasal   Prefrontal
Lacrimal   Frontal
Postfrontal   Postorbital
Septomaxilla
Synapsid
temporal fenestra
Premaxilla
Squamosal

Quadrate
Maxilla   Zygomatic
(Jugal)   Pterygoid
Surangular
Articular

Dentary   Angular

A

Frontal
Lacrimal   Postorbital   Coronoid process of dentary
Septomaxilla   Nasal   Prefrontal   Parietal
Zygomatic   Zygomatic
arch
Squamosal
Premaxilla
Exoccipital
Maxilla   Surangular
Angular
Dentary

B

Figure 4-21
Evolution of dentition and the synapsid fenestra. **A,** An early mammal-like reptile, *Dimetrodon,* in
which the teeth are homodont, and the temporal fenestra is small. *(A, After Kemp.)* **B,** A late mammal-
like reptile, *Thrinaxodon,* in which the teeth are becoming heterodont, and the temporal fenestra
has enlarged but not yet merged with the orbit. **(B,** *After Romer and Parsons.)*

the skull roof to the lateral surface of the lower jaw became subdivided in mammals into a **temporalis** and a **masseter muscle. Pterygoid muscles** extend from the palate to the medial side of the lower jaw. Contraction of these muscles generates a bite force as the lower jaw is pulled upward toward the upper jaw. Smaller forces move the lower jaw fore and aft and from side to side. These movements bring the teeth together very precisely, cutting and crushing the food between them.

Stresses in the temporal skull roof of ancient reptiles from the pull of the temporal jaw muscles would have been less in the regions where roofing bones articulate (Fig. 4-18A), and an opening began to appear in this region (Fig. 4-18**B** and **C**). As the opening enlarged, the boundaries of the opening would have provided firmer attachment for jaw muscles than a flat surface of bone. Also, the opening enabled jaw muscles to bulge through it when they contracted. On the line of evolution toward mammals, a single temporal fenestra evolved that originally lay ventral to the postorbital and part of the squamosal bones. A fenestra in this position is known as a **synapsid fenestra** (Fig. 4-21). At first the fenestra was small, but it gradually enlarged. In early mammals, most of the original lateral surface of the dermal skull roof has disappeared in the temporal region so that the temporal fenestra merges with the orbit. What appears to be the lateral dermal skull roof in a mammal is actually the extension of the dermal bones contributing to the braincase. All that is left of the original dermal skull roof in this region is a strip of bone in the middorsal line, another bordering the occipital region, and the handle-like **zygomatic arch** lying ventral to the temporal fenestra and orbit.

The muscular forces that pull the lower jaw upward towards the upper jaw are opposed, of course, by equal and opposite reaction forces that push down upon the lower jaw. These forces are distributed in reptiles in such a way that the teeth come

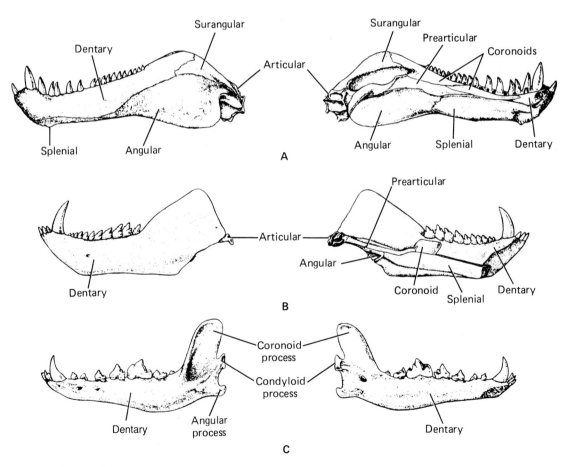

Figure 4-22
The left lower jaw of an early and a late mammal-like reptile (**A** and **B**) and a dog (**C**). The lateral surfaces are shown on the left side of the drawing and the medial on the right.

together in a moderately strong bite force. Forces acting on the jaw joint are also strong, and they are resisted by a large quadrate and articular, as you have seen in a turtle skull. In the course of evolution toward mammals, forces acting on the jaws were redistributed. The bite force increased considerably, while reaction forces acting at the jaw joint were greatly reduced. Because of the changing distribution of forces, the postdentary bones and quadrate became progressively smaller (Figs. 4-21 and 4-22). The dentary, on which the jaw muscles attach, enlarged. At the reptile-mammal transition, the dentary reached the squamosal bone of the skull and established a new jaw joint just lateral and cranial to the old one. The reduced quadrate and articular separated from the jaw apparatus; it entered the middle ear and became additional auditory ossicles, the **incus** and **malleus,** respectively. (Ear evolution, which is closely related to jaw changes, is discussed on page 234.)

The delicate auditory ossicles of mammals (malleus, incus, stapes) are protected in eutherian mammals by the formation of a plate of bone beneath the middle ear or tympanic cavity. Two elements contribute to this encasement in most instances—a cartilage replacement **endotympanic,** having no homologue in other vertebrates, and a dermal **tympanic,** homologous to the angular of the lower jaw of amphibians and reptiles.

♦   **(A) GENERAL FEATURES OF THE SKULL**

Examine the skull of a cat (Figs. 4-23, 4-24, and 4-25). As with other vertebrates, it can be divided into a **facial region,** containing the jaw, nose, and eyes, and a **cranial region**

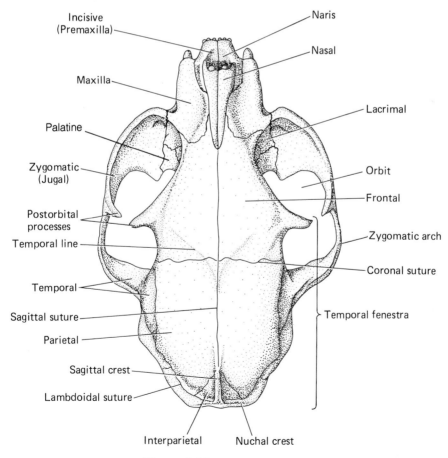

Figure 4-23
Dorsal view of the skull of a cat.

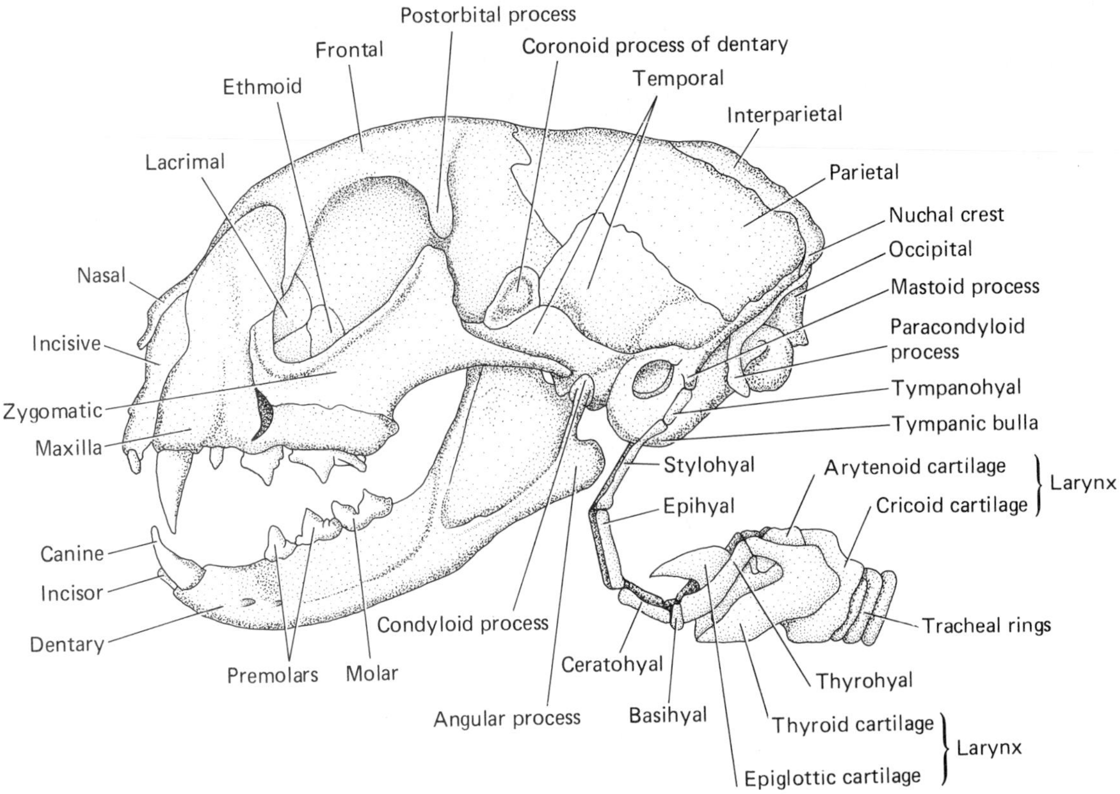

Figure 4-24
Lateral view of the skull, lower jaw, hyoid apparatus, and larynx of the cat.

housing the brain and ear. Notice the **nares** at the rostral end of the skull and the large, circular **orbits.** Although the two nares may appear contiguous, they are separated in life by a fleshy and cartilaginous septum.

The **foramen magnum** can be seen in the occipital region at the caudal end of the skull. There is an **occipital condyle** on each side of it. The division of the originally single occipital condyle and its shift to the sides of the foramen magnum are correlated with the greater head mobility of mammals. Vertical movements ("yes nods") of the head occur as rotations around the horizontal axis through the condyles. Side to side movements ("no shakings") of the head occur as rotations around a longitudinal axis between the first two cervical vertebrae (see also p. 108). In this way movements can occur with little stress upon the spinal cord, which enters the skull at this point. The large, round swelling on the ventral side of the skull, cranial to each occipital condyle, is the **tympanic bulla,** which encases the underside of the tympanic cavity. In life the tympanum is lodged in the opening **(external acoustic meatus)** on the lateral surface of the bulla. The tongue-shaped bump of bone on the lateral surface of the bulla just caudal to the external auditory meatus is the **mastoid process.** A similarly shaped **paracondyloid process** is located on the caudal surface of the bulla.

The handle-like bridge of bone on the side of the skull, extending from the front of the orbit to the external auditory meatus, is the **zygomatic arch.** All of it is included within the facial portion of the skull. A **mandibular fossa** for the articulation of the lower jaw appears as a smooth groove on the ventral surface of the caudal portion of the arch.

The orbit is bounded caudally by **postorbital processes,** one of which projects

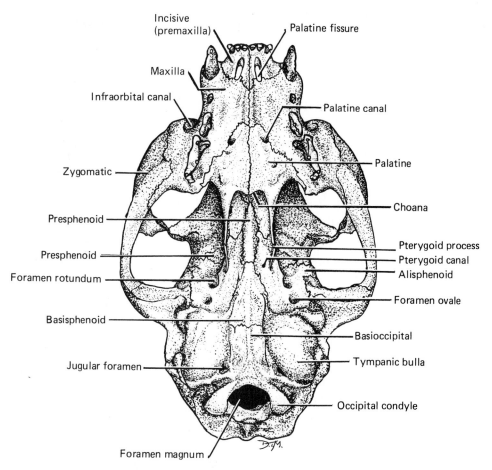

Figure 4-25
Ventral view of the skull of a cat.

Labels on figure:
Incisive (premaxilla)
Palatine fissure
Maxilla
Infraorbital canal
Palatine canal
Zygomatic
Palatine
Presphenoid
Choana
Presphenoid
Pterygoid process
Pterygoid canal
Foramen rotundum
Alisphenoid
Foramen ovale
Basisphenoid
Basioccipital
Jugular foramen
Tympanic bulla
Occipital condyle
Foramen magnum

down from the top of the skull and the other up from the zygomatic arch. A faint ridge, the **temporal line,** curves caudally from the base of the dorsal postorbital process to a middorsal ridge of bone, the **sagittal crest.** A **nuchal crest** extends from the caudal end of the sagittal crest ventrally to the mastoid process. The postorbital process, temporal line, sagittal crest, nuchal crest, and caudal half of the zygomatic arch form the boundary of a greatly enlarged **temporal fenestra,** an opening that leads into the **temporal fossa.** The floor of the temporal fossa is formed by the lateral wall of the braincase. The temporal fossa lying between the braincase and the boundary of the temporal fenestra is filled in life by the temporalis muscle. From its origin on the skull the temporalis muscle passes medial to the zygomatic arch to insert onto the lower jaw. The sagittal crest increases the area for the origin of this muscle. Neck muscles supporting the head attach onto the caudal surface of the skull and nuchal crest.

The shelf of bone that extends across the palatal region from the teeth on one side to those on the other is the **hard palate.** In life it is extended caudally by a fleshy **soft palate.** Together hard and soft palate constitute the **secondary palate.** The remains of the primitive primary palate lie dorsal to the hard palate. Internal nostrils, or **choanae,** will be seen dorsal to the caudal margin of the hard palate. A nearly vertical plate of bone, the **pterygoid process,** continues caudally from each side of the hard palate. The point of bone at its caudal end is called the **hamulus.** A small **pterygoid fossa,** for the attachment of pterygoid jaw muscles, appears as an elongated groove just lateral and posterior to the hamulus.

♦ **(B) COMPOSITION OF THE SKULL**

To visualize better the elements of the skull, you should examine a set of disarticulated bones along with the entire skull. Most of the top and lateral sides of the skull are formed by dermal bones. A small, tooth-bearing **incisive** (premaxilla) is located ventral and lateral to each external nostril. It also contributes a small, shelflike process to the hard palate. A **maxilla** completes the upper jaw. It, too, contributes a process to the rostral portion of the hard palate, sends one up rostral to the orbit, and forms the most rostral portion of the zygomatic arch. A small, delicate **lacrimal** is located in the medial wall of the front of the orbit just caudal to the dorsal extension of the maxilla. It is often broken. The portion of the zygomatic arch ventral to the orbit is formed by the **zygomatic** (jugal). A large **temporal** forms the caudal portion of the arch and the adjacent cranial wall, and encases the internal and middle ear.

The mammalian temporal bone has evolved through the fusion of a number of bones that are independent in amphibians and reptiles (Table 4-2). Its zygomatic process and the portion of it that helps to encase the brain **(squamous portion)** represent the squamosal. Of course, the part encasing the brain is an inward extension of the original squamosal of the dermal roof. The mastoid process and its inward extension into the cranial cavity, which will be seen in the sagittal section of the skull, represent the opisthotic and prootic of the chondrocranium. This part of the temporal is the **petrosal portion,** and the inner ear is contained within it. The thick part of the tympanic bulla adjacent to the external auditory meatus is the **tympanic** portion and is homologous with the dermal angular of the lower jaw of nonmammalian tetrapods. Finally, a new cartilage replacement **endotympanic** forms the rest of the bulla and completes the encasement of the tympanic cavity. It has no homologue in other vertebrates.

You may be able to glimpse the auditory ossicles by looking into the external acoustic meatus, but you should examine a special preparation to see them clearly. The **malleus** is roughly mallet-shaped. It has a long, narrow handle that attaches to the tympanic membrane and a rounded head that articulates with the incus. The **incus** is shaped like an anvil. It has a concave surface for the reception of the malleus and two processes that extend from the main surface of the bone. One of these articulates with the head of the stirrup-shaped **stapes.** A pair of narrow columns of bone extend from the head of the stapes to a flat, oval foot plate that fits into the oval window of the otic capsule. A stapedial artery passes between the columns of the stapes in the embryo and also in the adult of some species.

Dermal bones along the top of the skull are a pair of small **nasals** dorsal to the nares, a pair of large **frontals** dorsal and medial to the orbits, and a pair of large **parietals** posterior to the frontals. The parietals have a long suture with the squamous portion of the temporals. As was the case with the squamosal, much of each parietal and frontal represents shelflike extensions of the original skull roofing bones that grew down medial to the temporal muscle and helped to encase the brain.

A large, median **occipital** surrounds the foramen magnum and forms the caudal surface of the skull. The paracondyloid processes and most of the nuchal crest are on this element. Ventrally the occipital forms the floor of the braincase between the tympanic bullae. It is a compound bone containing the four separate occipital elements of the chondrocranium of nonmammalian tetrapods and, in most mammals, certain dermal elements (Table 4-2). In the skulls of young cats a triangular **interparietal** will be seen in the middorsal line in front of the occipital and between the caudal part of the two parietals. It represents a fusion of the paired postparietals present in the dermal skull roof of early tetrapods. In older individuals it often fuses with the occipital and parietals.

The occipital bone forms the caudal portion of the chondrocranium, and the otic elements are incorporated in the temporal. The rest of the primitive chondrocranium is represented by a part of the sphenoid, the ethmoid, and the turbinate bones. Although a tiny portion of the **ethmoid** can sometimes be seen in the medial wall of the orbit caudal to the lacrimal, most of this bone, and the turbinals too, can be seen best in a sagittal section described later.

The sphenoid can be observed in the entire skull. Roughly it forms the floor and part of the sides of the braincase rostral to the tympanic bullae and caudal to the hard palate. It is a single bone in human beings but is divided in the cat and many other mammals into a rostral presphenoid and a caudal basisphenoid. You should view these in a disarticulated skull to appreciate their extent. The **basisphenoid** includes the plate of bone on the underside of the skull just rostral to the basioccipital portion of the occipital; the caudal portion of the pterygoid process; the hamulus; the three caudal foramina of a row of four at the back of the orbit; and a winglike process extending dorsally between the squamous portions of the temporal and the frontal. Most of the bone is homologous to the primitive basisphenoid. That portion of it that includes the three foramina and dorsal wing is homologous to the epipterygoid of more primitive vertebrates and is sometimes called the wing of the basisphenoid, or the **alisphenoid.** There is some doubt as to the homologies of the pterygoid process and its hamulus, but the consensus is that this region includes the pterygoid of the primitive palate plus either the ectopterygoid or parasphenoid.

The **presphenoid** includes a narrow midventral strip of bone lying between the bases of the pterygoid processes, and lateral extensions that pass dorsal to the pterygoid processes to enter the medial wall of the orbits. The lateral portion contains the rostral foramen in the row of four referred to previously and has a common suture with a ventral extension of the frontal bone. The presphenoid is homologous to the sphenethmoid of nonmammalian tetrapods.

Certain of the original dermal palatal bones are incorporated in the sphenoid; two others can be seen in more rostral parts of the skull. The originally paired vomers have united to form a single element (the **vomer**), which appears as a midventral strip of bone rostral to the presphenoid and dorsal to the hard palate. You will have to look into the internal nostrils to see it, for it helps to separate the nasal cavities. Paired **palatine** bones form the caudal portion of the hard palate, the rostral portion of the pterygoid processes, and a bit of the medial wall of the orbits caudal and ventral to the lacrimals.

Most of the bones of the skull are joined to one another by a type of fibrous joint called a suture. **Sutures** are immovable joints (apart from the elasticity of the interconnecting connective tissue) in which the bones come close together and sometimes interlock by complex foldings of their margins, but the bones are separated by connective tissue called a **sutural ligament.** During growth, new bone forms in this connective tissue at the periphery of the bones. The surfaces of the bones are covered by a dense connective tissue, the **periosteum,** that crosses the suture and is anchored to the sutural ligament. In mature individuals, the sutural ligament may ossify and the bones fuse together. Sutures usually carry the names of the adjacent bones (e.g., frontomaxillary suture). A few have special names. The suture between the occipital and parietal bones is the **lambdoidal suture;** that between the frontals and parietals, the **coronal suture;** and that between the parietals, the **sagittal suture.** The frontal (coronal) and sagittal planes of the body pass through the comparable sutures in human beings.

Bones of the chondrocranium, which are cartilage replacement bones, are interconnected by cartilage. Such joints, which are also immovable apart from the elasticity of the cartilage, are called **synchondroses.** An example is the sphenooccipital synchondrosis, which you can see on the ventral surface of the skull between the basisphenoid and the occipital.

♦   **(C) INTERIOR OF THE SKULL**

Examine a pair of sagittal sections of the cat skull cut in such a way that the nasal septum shows on one half and the turbinates on the other. Note the large **cranial cavity** for the brain (Fig. 4-26). It can be divided into three parts—a **caudal cranial fossa** in the occipital–otic region for the cerebellum; a large **middle cranial fossa** for the cerebrum; and a small, narrow **rostral cranial fossa,** just caudal to the nasal region for the olfactory bulbs of the brain. In the cat, a partial transverse septum of bone (the **tentorium**) separates the middle and caudal cranial fossae. The internal part of the petrosal portion of the temporal, containing the inner ear, can be seen in the lateral wall of the caudal cranial fossa dorsal to the tympanic bulla. It is perforated by a large foramen, the **internal acoustic meatus.** The small fossa dorsal to the internal acoustic meatus lodges a lobule of the cerebellum. The saddle-shaped notch **(sella turcica)** in the floor of the middle cranial fossa lodges the hypophysis (Fig. 10-17, p. 317). It is in the basisphenoid. A large

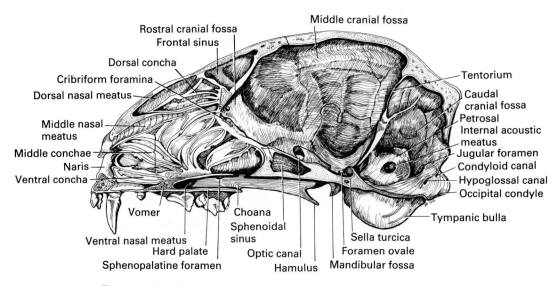

Figure 4-26
Sagittal section of the skull of a cat. The nasal septum has been removed.

sphenoidal air sinus can be seen in the presphenoid and a **frontal air sinus** in the frontal bone dorsal to the rostral cranial fossa.

The rostral wall of the rostral cranial fossa is formed by a sievelike plate of bone whose foramina communicate with the nasal cavities. You can also see this plate of bone, the **cribriform plate** of the **ethmoid,** by looking through the foramen magnum of a complete skull. A **perpendicular plate** of the ethmoid, which will be seen on the larger section, extends from the cribriform plate vertically and rostrally between the two nasal cavities. It connects with the nasal bones dorsally and with the vomer ventrally and forms much of the nasal septum. The very front of this septum is cartilaginous. Most of the ethmoid is shaped like the letter T, the top of the T being the cribriform plate and its stem the perpendicular plate. In addition, the ethmoid has small lateral processes that pierce the medial wall of the orbits caudal to the lacrimal and often can be seen in a lateral view of the skull (Fig. 4-24).

Each nasal cavity is largely filled with thin, complex scrolls of bone called the **conchae** (turbinates). They show best on the section that lacks the perpendicular plate of the ethmoid (Fig. 4-26). Although the turbinates ossify from the nasal capsules, they become attached to bones surrounding the nasal cavities. A small **dorsal concha** (nasoturbinate) lies rostral to the frontal sinus and lateral to a perpendicular septum of the nasal bone. A **ventral concha** (maxilloturbinate) lies in the rostroventral part of the nasal cavity. You can see its attachment to the maxillary bone by looking into the cavity through the naris. The large and very complexly folded **middle concha** (ethmoturbinate) lies between the others and fills most of the nasal cavity. Some of the air entering a nasal cavity passes directly caudally through an uninterrupted passage, the **ventral nasal meatus,** that lies between the turbinates and the hard palate, but much of the air travels back between and among the turbinates, which are covered by a moist nasal epithelium, and which serve to increase the surface area available for olfaction and for heating, cleansing, and moistening the inspired air. Turbinate bones are associated with endothermy. Their presence in ancestral mammals is evidence that these animals were warm-blooded. Further details of the nose are described on page 215.

♦   **(D) FORAMINA OF THE SKULL**

Numerous foramina for nerves and blood vessels perforate the skull. These may be considered either at this time or after the nerves have been studied. In any case, the foramina should be studied on a sagittal section of the skull so that both their external and internal aspects can be seen. Certain of the passages should be probed to determine their course.

First, study the foramina for the 12 cranial nerves. The first cranial nerve, the olfactory, consists of many fine subdivisions that enter the cranial cavity through the **cribriform foramina** in the cribriform plate of the ethmoid. The **optic canal,** for the second cranial nerve, is the most rostral of the row of four foramina in the caudomedial wall of the orbit (Fig. 4-27). Next caudad, and largest in this row, is the **orbital fissure** for the third, fourth, and sixth nerves going to the muscles of the eyeball, and for the ophthalmic division of the fifth nerve. The third and smallest foramen in this row **(foramen rotundum)** transmits the maxillary division of the fifth nerve, and the last in this row (foramen ovale) the mandibular division of this nerve. The **internal acoustic meatus** (Fig. 4-26), seen in the petrosal portion of the temporal, transmits both the seventh and eighth nerves. The eighth (vestibulocochlear) nerve comes from the inner ear, but the seventh (facial) nerve continues through a facial canal within the petrosal and emerges on the undersurface of the skull through the **stylomastoid foramen** located beneath the tip of the mastoid process (Fig. 4-27). The ninth, tenth, and eleventh nerves, in company with the internal jugular vein, pass through the **jugular foramen,** which can be seen internally in the floor of the caudal cranial fossa caudal to the internal acoustic meatus. The jugular foramen opens on the ventral surface of the skull beside the posteromedial edge of the tympanic bulla (Fig. 4-25). The **hypoglossal canal** for the twelfth cranial nerve can be seen in the floor of the caudal cranial fossa caudal and medial to the jugular foramen (Fig. 4-26). Probe to see that the hypoglossal canal extends cranially and emerges on the ventral surface in common with the jugular foramen.

As the ophthalmic and maxillary divisions of the fifth (trigeminal) nerve are distributed to various parts of the head, certain of their branches pass through other foramina, often in company with blood vessels. A branch of the ophthalmic division traverses a small **ethmoid foramen,** or series of foramina, in the medial wall of the orbit to enter the nasal cavity (Fig. 4-27). The ethmoid foramen lies in the frontal bone very near its suture with the presphenoid. One of the branches of the maxillary division emerges through an **infraorbital canal** located in the rostral part of the zygomatic arch. But before reaching the infraorbital canal, this branch has subsidiary branches that pass through several small foramina near the front of the orbit to supply the teeth of the upper jaw. A second branch of the maxillary division leaves the rostral portion of the orbit through a **sphenopalatine foramen** to enter the ventral part of the nasal cavity. The sphenopalatine foramen is the

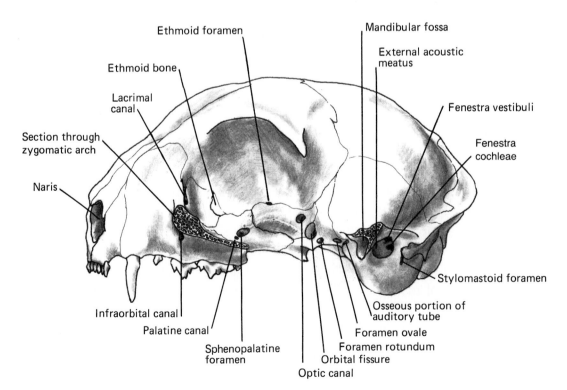

Figure 4-27
Lateral view of the skull of the cat after removal of the zygomatic arch. Major foramina are shown.

larger and more medial of two foramina that lie close together in the orbital process of the palatine bone. After entering the nasal cavity, part of this nerve continues forward through the nasal passages and finally drops down to the roof of the mouth through a **palatine fissure,** which is located just lateral to the midline at the rostral end of the hard palate (Fig. 4-25). In certain mammals the palatine fissure also carries an incisive duct that leads from the mouth to the vomeronasal organ (Jacobson's organ, p. 216) located in the nasal cavity. The small foramen lateral and rostral to the sphenopalatine is the caudal end of the **palatine canal.** A third branch of the maxillary division runs through this canal to the roof of the mouth. The rostral end of the canal can be seen on the ventral surface of the hard palate in or near the suture between the palatine and maxillary bone. A fourth branch of the maxillary backtracks to enter the orbital fissure. It then passes into a small **pterygoid canal,** whose rostral end may be seen in the floor of the orbital fissure. The caudal end of the canal appears as a tiny hole on the ventral side of the skull near the base of the pterygoid process of the basisphenoid. After emerging from the pterygoid canal, this branch enters the tympanic cavity through the osseous portion of the **auditory tube**—a large opening at the rostral edge of the tympanic bulla. This opening serves also as the point of entrance into the skull of the ascending pharyngeal artery (p. 379). A summary of the cranial nerves and the foramina through which they pass is given in Table 4-3.

The major foramina that remain do not carry nerves. A **lacrimal canal** for the nasolacrimal, or tear, duct will be seen in the lacrimal bone extending from the orbit into the nasal cavity. You may also be able to find parts of the small **carotid canal** for the vestigial (in the cat) internal carotid artery and ascending pharyngeal artery (p. 379). The

Table 4-3 Mammalian Cranial Nerves and Their Foramina

The 12 cranial nerves of mammals are listed together with the foramina through which they pass. In those cases in which a nerve goes through two or more foramina before reaching the organ it supplies, the foramina are listed in sequence from the brain.

Nerve	Foramen
I. Olfactory	Cribriform foramina
II. Optic	Optic canal
III. Oculomotor	Orbital fissure
IV. Trochlear	Orbital fissure
V. Trigeminal	
Ophthalmic division	Orbital fissure
one branch	Ethmoid foramina
Maxillary division	Foramen rotundum
one branch	Infraorbital canal
one branch	Sphenopalatine foramen, palatine fissure
one branch	Palatine canal
one branch	Pterygoid canal
Mandibular division	Foramen ovale
one branch	Mandibular foramen, mental foramina
VI. Abducens	Orbital fissure
VII. Facial	Internal acoustic meatus, facial canal, stylomastoid foramen
VIII. Vestibulocochlear	Internal acoustic meatus
IX. Glossopharyngeal	Jugular foramen
X. Vagus	Jugular foramen
XI. Accessory	Jugular foramen
XII. Hypoglossal	Hypoglossal canal, jugular foramen

caudal end of the canal appears as a tiny hole in the rostromedial wall of the jugular foramen. From here the canal extends forward, dorsal to the tympanic bulla, and enters the cranial cavity. Its point of entrance can be seen in the caudal cranial fossa rostral to the petrosal and ventral to the tentorium. A **condyloid canal** for a small vein can be found in the caudal cranial fossa dorsal to the hypoglossal canal.

If you have access to a specimen in which the tympanic bulla has been removed, you will be able to see two openings on the underside of the petrosal. The more dorsal is the oval window **(fenestra vestibuli)** for the stapes; the more ventral, the round window **(fenestra cochleae)** for the release of vibrations from the inner ear (Fig. 4-27).

◆   **(E) THE LOWER JAW**

> With the transfer of certain of the original lower jaw bones to the ear region, and the loss of others, the pair of enlarged dentaries are left as the sole elements in the mammalian lower jaw.

Examine the lower jaw, or **mandible,** of the cat (Fig. 4-28) and notice that the **dentary bones** are firmly united anteriorly by an **intermandibular suture.** The horizontal part of the dentary that bears the teeth is its **body;** the part caudal to this, its **ramus.** A large, triangular depression, the **masseteric fossa,** occupies most of the lateral surface of the ramus. Part of the masseter, one of the mandibular muscles, inserts here. Caudally the ramus has three processes—a dorsal **coronoid process,** to which the

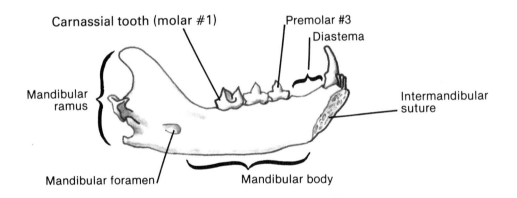

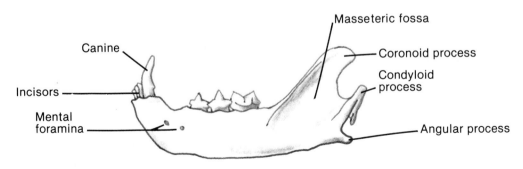

Figure 4-28
Medial (*upper*) and lateral (*lower*) views of the mandible of the cat.

temporalis muscle attaches; a middle, rounded **condyloid process** for the articulation with the skull proper; and a ventral **angular process** to which the rest of the masseter and pterygoid muscles attach.

A large **mandibular foramen** will be seen on the medial side of the ramus, and two small **mental foramina** on the lateral surface of the body near its rostral end. A branch of the mandibular division of the trigeminal nerve, supplying the teeth and the skin covering the lower jaw, enters the mandibular foramen and emerges through the mental foramina. Blood vessels accompany the nerve.

◆ **(F) TEETH**

The teeth of mammals are quite different from those of other vertebrates for they are limited to the jaw margins, are set in deep sockets **(thecodont),** are differentiated into various types **(heterodont),** and their replacement is limited **(diphyodont).** Most adult mammals have in each side of each jaw a series of nipping incisors at the front, a large piercing canine behind these, a series of cutting premolars behind the canine, and finally a series of chewing, or grinding, molars. The number of each kind of teeth present in a particular group of mammals may be expressed as a dental formula. For ancestral placental mammals this was

$$\frac{3.1.4.3}{3.1.4.3} \times 2 = 44$$

Such an animal had three incisors, one canine, four premolars, and three molars in each side of each jaw. Milk incisors, canines, and premolars are replaced by larger permanent ones as the jaws increase in size during the juvenile period. The molars appear sequentially later in life and are not replaced. The number of teeth, and their structure, are adapted to the animal's mode of life. The molars especially are subject to much divergence among the various groups of mammals.

Examine the teeth of the cat (Figs. 4-24 and 4-28). In each side of the upper jaw there are normally three **incisors** attached to the incisive bone. These are followed by one **canine,** three **premolars,** and one very small **molar,** all of which are in the maxilla. In the lower jaw there are three incisors, one canine, two premolars, and one large molar. The dental formula is therefore

$$\frac{3.1.3.1}{3.1.2.1}$$

During its evolution, the cat has lost the first premolar in the upper jaw, the first two in the lower, and all molars posterior to the first. The gap left between each canine and the premolars is called a **diastema.** The first of the remaining premolars and the molar of the upper jaw are more or less vestigial. But the last premolar of the upper jaw (phylogenetically premolar number four) and the lower molar have become large and complex in structure. Articulate the jaws and note how these two teeth, which are known as the **carnassials,** intersect to form a specialized shearing mechanism. Carnassials are restricted to carnivores, and in contemporary species have the formula Pm 4/M 1.

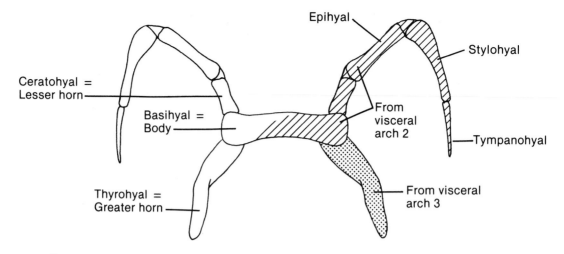

Figure 4-29
Dorsal view of the hyoid apparatus of the cat. The derivation of the various parts is shown by hatching on the right side.

♦   **(G) THE HYOID APPARATUS**

A hyoid apparatus is present in mammals but is not so large as in nonmammalian tetrapods, since it is formed from only the ventral parts of the hyoid and first branchial arches. More caudal arches are incorporated in the laryngeal cartilages (p. 320).

Study the hyoid apparatus of the cat (Figs. 4-24 and 4-29). It may be in place on the skeleton or removed and mounted separately. It consists of a transverse bar of bone, the **body of the hyoid,** composed of a single bone **(basihyal)** from which two pairs of processes (horns) extend cranially and caudally. The caudal or **greater horns of the hyoid** are the larger horns, and each consists of but one bone, which is called the **thyrohyal,** for it extends caudally to attach to the thyroid cartilage of the larynx (see also Fig. 10-20, p. 321). Each of the cranial or **lesser horns of the hyoid** consists of a small **ceratohyal.** A chain of ossicles connects the ceratohyal with the skull, attaching to the tympanic bulla medial to the stylomastoid foramen. From ventral to dorsal these are the **epihyal, stylohyal,** and **tympanohyal.** These ossicles, the ceratohyal, and the body of the hyoid are derivatives of the hyoid arch. The thyrohyal develops from the first branchial arch.

In some mammals (human beings, for example) the hyoid apparatus consists of a single bone, the hyoid, having a body, a lesser horn, and a greater horn. A stylohyoid ligament, which extends from a styloid process at the base of the skull to the lesser horn, replaces the chain of ossicles seen in the cat as the support for the apparatus. The styloid process itself presents a part of the hyoid arch fused onto the temporal bone.

CHAPTER FIVE

◆ ◆

THE TRUNK SKELETON

The trunk skeleton consists of the vertebral column, ribs, and sternum if one is present. All are parts of the axial skeleton. Usually these structures are represented by cartilaginous rudiments in the embryo and are replaced by bone as development continues in most species, but some parts of the vertebrae may ossify directly within the mesenchyme.

FISHES

The vertebral column surrounds and protects the spinal cord in all vertebrates. In addition to this, the vertebral column and ribs of fishes are part of the locomotor apparatus. The vertebral column acts as a compression strut, enabling the fish to push through a relatively dense medium, the water, and the vertebrae and ribs together participate in the undulatory movements of swimming (p. 148). Since all vertebrae perform nearly the same function and are subjected to similar stresses, there is little regional differentiation of the vertebral column. The first trunk vertebra articulates with the skull, and the caudal vertebrae are modified to protect the caudal artery and vein from being compressed and closed by the powerful locomotor muscles of the tail. The vertebral column, of course, also provides some support against gravity, but support is not a major problem in aquatic animals because of the buoyancy of water. Individual vertebrae are not as strong nor as securely linked together as they are in terrestrial vertebrates.

◆ **(A) RELATIONSHIPS OF THE VERTEBRAL COLUMN**

Make a fresh cross section of the tail of your specimen and observe that the vertebral column lies at the intersection of several connective tissue septa that separate the surrounding muscles (Fig. 5-1). A **dorsal skeletogenous septum** extends from the top of the vertebrae to the middorsal line of the body; a **ventral skeletogenous septum,** from the bottom of the vertebrae to the midventral line; and a **horizontal skeletogenous septum,** from each side of the vertebrae to the lateral surface of the body. The last is the hardest to see and may be confused with portions of the **myosepta** that separate the muscle segments. The horizontal septum forms an arc that reaches the skin at the position of the lateral line, which appears as a small hole in the skin. It divides

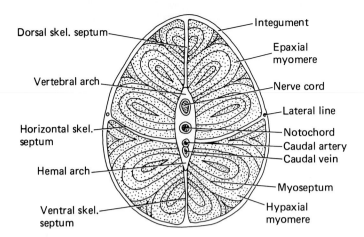

Figure 5-1
Cross section through the tail of the dogfish to show the relationship of the vertebral column to surrounding tissues. The section passes through the center of a caudal vetebra.

the myomeres horizontally into dorsal **epaxial** and ventral **hypaxial** portions. Although it will not be seen at this time, the relationships in the trunk are much the same. The only difference is that the ventral septum has split, so to speak, and passes on each side of the body cavity (coelom) (Fig. 11-11, p. 353).

♦ **(B) VERTEBRAL REGIONS AND CAUDAL VERTEBRAE**

The vertebral column of fishes consists of only **trunk vertebrae** and **caudal vertebrae,** but these become somewhat modified at the attachment of the chondrocranium and in the vicinity of the median fins. Study the structure of the caudal vertebrae of *Squalus* on a special preparation, or by dissecting the tail of your own specimen. If you dissect the tail, cut out a piece about 5 centimeters long, and carefully expose the vertebrae by cleaning away all the surrounding muscle and connective tissue. Do not select a piece adjacent to either the posterior dorsal or caudal fin.

Make a cross section through the joint between two vertebrae near one end of your piece and examine it along with the lateral aspect of other vertebrae. Each vertebra consists of a cylindrical central portion, called the centrum, or **vertebral body,** which bears a dorsal and ventral arch of cartilage. The dorsal arch, which protects the spinal cord, is the **vertebral arch** (neural arch); its cavity is the **vertebral canal** (Fig. 5-2**A**). The ventral arch, which surrounds and protects the caudal artery and vein, is the **hemal arch;** the passage through it is the **hemal canal.** You may see the vessels. The artery lies dorsal to the vein. The top of each arch may extend a short distance into the dorsal and ventral septa as a **spinous process** and a **hemal spine,** respectively.

Make a sagittal section through several caudal vertebrae and study it along with the lateral aspect of others (Fig. 5-2**B**). Notice that each side of the vertebral arch is composed of two roughly V-shaped blocks of cartilage. That block located directly on top of the vertebral body and having its apex pointing dorsally is called the **neural plate.** The other block located above the joint between vertebral bodies, its apex directed ventrally, is the **dorsal intercalary plate.** Each plate of the vertebral arch is perforated by a small hole, or **foramen,** for either the dorsal or the ventral root of a spinal nerve. A continuous

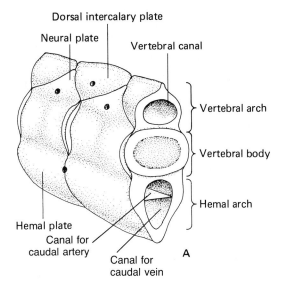

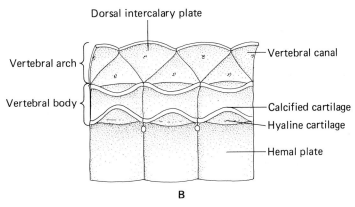

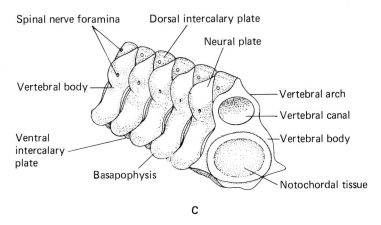

Figure 5-2
Vertebrae of *Squalus*. **A,** Three-dimensional view of two caudal vertebrae; **B,** sagittal section through three caudal vertebrae; **C,** three-dimensional view of several trunk vertebrae.

vertebral arch is peculiar to cartilaginous fishes; such a condition is possible because cartilage is much more flexible than bone. In other vertebrates there is some space between successive vertebral arches, and this facilitates lateral bending of the vertebral column. The vertebral arch of other vertebrates probably is homologous to the neural plate, for it develops primarily from a comparable group of cells.

The hemal arch is composed on each side of just one block, called a **hemal plate.** A small foramen for intersegmental branches of the caudal artery and vein can be found between the bases of the hemal arches.

Examine the vertebral body in the sagittal section. It is shaped like a biconcave spool, a shape termed **amphicoelous.** The concavity at each end contains a gelatinous substance, which is **notochordal tissue** that has persisted from the embryonic stage. A small strand of notochord also perforates each vertebral body. The cartilage immediately adjacent to the notochord has become calcified and thus appears white. Calcification of cartilage involves the deposition of calcium salts in the matrix. This process strengthens and stiffens the cartilage, but it is not the same as ossification. The area of calcified cartilage has an hourglass shape (Fig. 5-2**B**). The rest of the cartilage is glasslike, or hyaline.

♦   **(C) TRUNK VERTEBRAE**

Compare the caudal vertebrae with a special preparation of the trunk vertebrae (Fig. 5-2**C**). The structure is basically the same, but the trunk vertebrae lack hemal arches. Instead, they have short, ventrolateral processes, termed **basapophyses,** that project from the sides of the vertebral body. Basapophyses are serially homologous to the proximal part of the hemal arches. Small **ventral intercalary plates** lie between successive basapophyses, but they are often lost as the vertebrae are being prepared. They are the ventral counterparts of the dorsal intercalary plates.

Each vertebra develops embryonically from sclerotomal mesenchymal cells that migrate from the segmented, embryonic somites and accumulate around the nerve cord and notochord in an intersegmental position. A vertebra, thus, comes to lie between two segmental myomeres, which in turn have developed directly from the somites. Much of the vertebral mesenchyme forms cartilaginous vertebral arches, **arcualia.** Certain of these form the vertebral arch around the nerve cord; others form the hemal arch around the caudal vessels. The vertebral body develops between these arches and largely replaces the notochord. Its mode of development varies widely in vertebrates, but in fishes it appears to derive in part from arcualia bases and in part from the direct deposition of cartilage or, in some cases, of bone in concentric rings around the notochord and even within the notochord sheath.

♦   **(D) RIBS**

Notice on a mounted skeleton that short ribs articulate with the basapophyses and extend into the horizontal skeletogenous septum at the point where it intersects with the myosepta. Ribs in this position are called dorsal or **intermuscular ribs,** for they lie between the epaxial and hypaxial parts of the myomeres. Many fish have ventral or **subperitoneal ribs** in the myosepta adjacent to the body cavity. Some have both types of ribs. Ribs strengthen the myosepta onto which the muscle fibers of the myomeres attach.

♦   **(E) MEDIAN FIN SUPPORTS**

Look at the dorsal fins on a mounted skeleton and notice that each is supported proximally by several cartilages and distally by fibrous rods, the **ceratotrichia,** described earlier (p.

44). The cartilages, which collectively may be called **pterygiophores,** are in two series. Those that rest on the vertebral column are **basals;** the more distal ones are **radials.** There is some question as to whether the pterygiophores develop phylogenetically from spinous processes of the vertebrae or independently within the dorsal septum. The caudal fin, which is of the heterocercal type (p. 41), is supported proximally by enlarged spinous processes and hemal spines and distally by ceratotrichia.

AMPHIBIANS

Since the body of terrestrial vertebrates is not supported by water, their vertebral column is very important in both locomotion and maintaining posture. Individual vertebrae are more completely ossified than in most fishes, and successive vertebral arches articulate by pairs of overlapping zygapophyses (Fig. 5-3). The zygapophyses unite the vertebrae firmly; they resist vertical bending, yet permit lateral bending. Considerable regional differentiation evolves in the vertebral column and ribs, correlating with the changing forces that act on these parts of the skeleton. With the loss of gills and the bony connection between the skull and pectoral girdle, a neck region appears. The number of **cervical vertebrae** varies, but there is always at least one, the **atlas. Trunk vertebrae** follow the cervical vertebrae. At least one **sacral vertebra** and **rib** articulates with the pelvic girdle and transfers forces between the trunk and hind legs. The number of **caudal vertebrae** depends on tail length.

The composition of the vertebrae varies considerably, but the most primitive type seems to have been the **rhachitomous vertebrae** of primitive labyrinthodonts and captorhinids (Fig. 5-3, bottom). Such vertebrae consisted of a vertebral or neural arch beneath which was a notochord partly constricted by three blocks of bone that made up the vertebral body or centrum. A single U-shaped **intercentrum** lay beneath the notochord at the front of each vertebra. A pair of small **pleurocentra** was located dorsal and caudal to the intercentrum. Hemal arches (chevron bones) were also present in the caudal region, where they attached to the intercentra. In the evolution from ancestral amphibians through reptiles (Fig. 5-3) to birds and mammals, the intercentrum decreased in size and the pleurocentrum enlarged to become the definitive vertebral body. The notochord disappears in the adult.

Most of the vertebrae bear distinct ribs to which trunk muscles attach. Although legs are present, fishlike lateral undulations of the trunk contribute to locomotion in early terrestrial vertebrates. In most reptiles, rib movements also are important in ventilating the lungs. The cranial trunk ribs usually unite ventrally with a **sternum** of cartilage or cartilage replacement bone. Tetrapod ribs develop somewhat differently than do fish ribs. In tetrapods, the major portion of a rib occupies the same position as the subperitoneal rib of a fish, but its attachment more dorsally on the vertebra resembles an intermuscular rib. Most tetrapod ribs also have two processes that articulate with the vertebrae: a dorsal **tuberculum,** which articulates with the transverse process, or **diapophysis,** on the vertebral arch, and a ventral **caput,** which primitively articulates with the intercentrum (Fig. 5-3). Sometimes there is a small lateral process in the latter region called a **parapophysis.** With the reduction and loss of the intercentrum, the caput has an intervertebral articulation.

◆ (A) VERTEBRAL REGIONS

Examine the vertebral column and ribs on a skeleton of *Necturus*. There is more differentiation into regions than in the skeleton of *Squalus*. The most cranial vertebra, called the **atlas,** lacks distinct ribs and is otherwise specialized for articulation with the

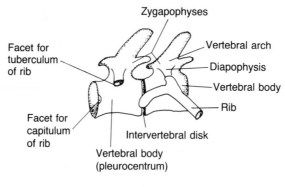

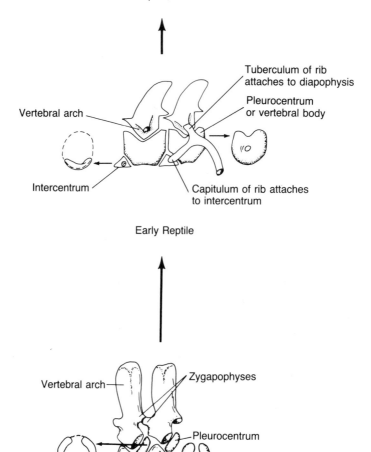

Figure 5-3
Diagrams showing the evolution of typical trunk vertebrae from a labyrinthodont to a mammal. The attachment of a rib is also shown. *(From Walker,* Functional Anatomy of the Vertebrates, *Saunders College Publishing, 1987.)*

skull. It is, of course, located in the incipient neck region and is the only one that can be called a **cervical vertebra. Trunk vertebrae** extend from here to the level of the pelvic girdle. All have free ribs. A single **sacral vertebra** (usually the nineteenth) and its pair of ribs are modified for the support of the pelvic girdle. The remaining are **caudal vertebrae;** most lack ribs and bear **hemal arches.**

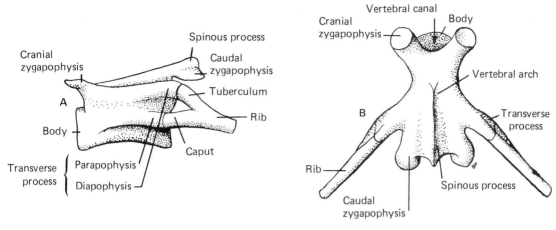

Figure 5-4
A, Lateral view, and **B,** dorsal view of a trunk vertebra and rib of *Necturus.*

♦ **(B) TRUNK VERTEBRAE**

Study a trunk vertebra on the mounted skeleton and also examine an isolated specimen (Fig. 5-4). Dorsally it consists of a **vertebral arch** that overlies the spinal cord located in the **vertebral canal.** A small **spinous process** projects caudad from the top of the arch. Two pairs of small processes for the articulation of successive vertebrae project laterally from the front and back of the vertebral arch. The cranial pair of processes, whose smooth articular facets face dorsally, are called the **cranial zygapophyses;** the caudal pair, whose facets face ventrally, are called the **caudal zygapophyses.** Note how the zygapophyses of adjacent vertebrae overlap.

Ventrally the vertebra consists of a biconcave (amphicoelous) **vertebral body.** Notochordal tissue persists in the concavities. A prominent **transverse process** projects laterally and caudally from each side of the vertebral arch and body. Observe that it is rather high from its dorsal edge to its ventral edge. The dorsal part represents a **diapophysis;** the ventral part is a **parapophysis.**

♦ **(C) RIBS**

The ribs of *Necturus* are shorter than the ribs of labyrinthodonts or captorhinids but have the two heads (the bicipital condition) characteristic of terrestrial vertebrates. The ventral **caput** attaches to the parapophysis; the dorsal **tuberculum** attaches to the diapophysis. The distal portion of the rib is its shaft, or **body.** In contemporary amphibians, ribs do not participate in lung ventilation.

♦ **(D) STERNUM**

Necturus does not have the small, cartilaginous sternum found in most urodeles just caudal to the pectoral girdle. However, traces of cartilage, which may represent an incompletely developed sternum, are located in the ventral portions of the myosepta dorsal and caudal to the pectoral girdle, and in the midventral connective tissue septum. You may notice them during the dissection of the muscles (p. 161).

MAMMALS

The trunk skeleton of mammals functions as a suspension girder for the body and transfers forces between the trunk and appendages. In the evolution toward mammals, as the appendages assume a more important role in locomotion, lateral undulations of the trunk and tail play less and less of a role. Except for whales, other aquatic mammals, and kangaroos, the tail seldom has a role in locomotion. It is important in balancing in arboreal species, in clearing insects from the body in larger herbivorous mammals, and it sometimes has a social role (e.g., a dog wagging its tail). The tail is lost as an external organ in humans and some other species. A change in the plane of articulation on some zygapophyses in cats and some other species that run rapidly allows for extension and flexion of the back. This increases stride length. Head movements are much freer in mammals than in amphibians and reptiles. Movable ribs are limited to the thoracic region and, together with the newly evolved diaphragm, encase the thoracic organs and help ventilate the lungs. Small embryonic ribs are present on other vertebrae, but during development they fuse onto the sides of the vertebrae, forming complex transverse processes. The rib component of a transverse process is known as a **pleurapophysis.** In an adult it is not always possible to distinguish the part of the transverse process that developed from a rib from the part representing the diapophysis and/or parapophysis, so the entire transverse process is often called a pleurapophysis. Because of the many functions of the trunk skeleton and the changing stresses along its length, the vertebral column displays more regional differentiation than is seen in early tetrapods. Indeed, no two vertebrae are exactly alike.

The individual vertebrae are similar to those of contemporary reptiles. The vertebral body evolved from the pleurocentrum, and the intercentrum was lost. **Intervertebral discs** of fibrocartilage lie between vertebral bodies. Traces of notochordal tissue may be found within them.

♦   **(A) VERTEBRAL GROUPS**

Examine a mounted skeleton of the cat and other mammals that are available, and also a string of disarticulated vertebrae, noting the five vertebral regions (Figs. 5-5 and 5-6). The most cranial group of vertebrae is the **cervical vertebrae.** There are more of them than in early tetrapods since, with few exceptions, mammals have seven cervical vertebrae, all of which lack ribs. **Thoracic vertebrae,** bearing ribs, follow the cervical vertebrae, and **lumbar vertebrae** follow the thoracic vertebrae. The lumbar vertebrae lack ribs but have very large transverse processes. The number of vertebrae in these two regions varies among mammals. The cat normally has 13 thoracic and 7 lumbar vertebrae; rabbits, 12 thoracic and 7 lumbar vertebrae; humans 12 thoracic and 5 lumbar vertebrae. **Sacral vertebrae** follow the lumbar vertebrae and are firmly fused together to form a solid point of attachment, the **sacrum,** for the pelvic girdle. The number of vertebrae contributing to the sacrum varies among species. Many mammals, including the cat, have three; rabbits, four; humans, five. The remaining are **caudal vertebrae.** Their number varies with the length of the tail, but all mammals have some. Even in human beings, there are three to five small caudal vertebrae that are fused together to form a **coccyx** to which certain anal muscles attach.

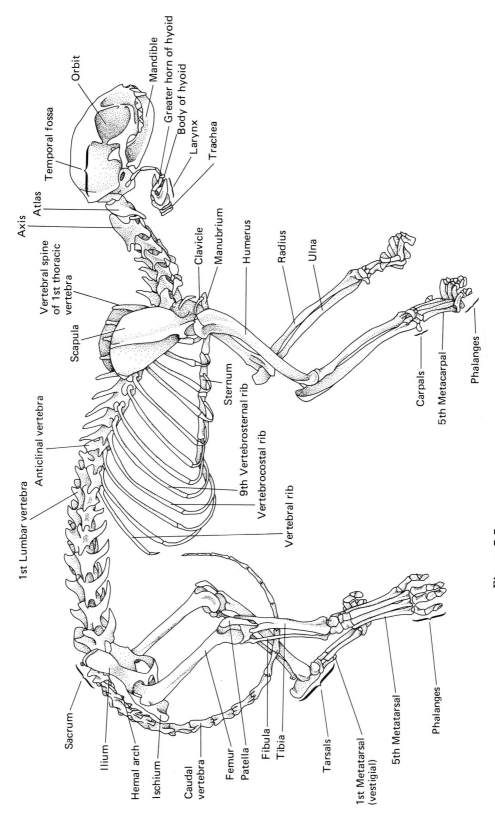

Figure 5-5
Lateral view of the skeleton of the cat. (*From Walker, A Study of the Cat.*)

Orbit

Mandible

Greater horn of hyoid

Body of hyoid

Larynx

Trachea

Temporal fossa

Axis

Atlas

Vertebral spine
of 1st thoracic
vertebra

Clavicle

Manubrium

Humerus

Radius

Ulna

Scapula

Carpals

5th Metacarpal

Phalanges

Anticlinal vertebra

1st Lumbar vertebra

Sternum

9th Vertebrosternal rib

Vertebrocostal rib

Vertebral rib

Sacrum

Ilium

Hemal arch

Ischium

Caudal
vertebra

Femur

Patella

Fibula

Tibia

Tarsals

1st Metatarsal
(vestigial)

5th Metatarsal

Phalanges

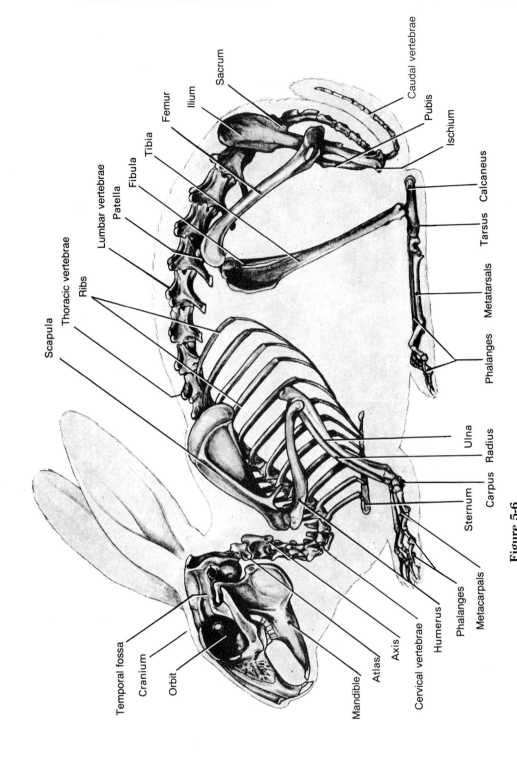

Figure 5-6
Lateral view of the skeleton of a rabbit. *(After Barone, Pavaux, Bolin, and Cuq.)*

◆ **(B) THORACIC VERTEBRAE**

The thoracic vertebrae should be studied first, as they are more similar to the vertebrae of *Necturus* than are the vertebrae in other regions. Examine one from near the middle of the series. Identify the **vertebral arch** with its long **spinous process,** the **vertebral canal,** and the **vertebral body,** or centrum (Fig. 5-7). Each end of the vertebral body is flat, a shape termed **acoelous.** In life, small fibrocartilaginous **intervertebral discs** are located between successive bodies. Articulate two vertebrae and notice how the back of one vertebral arch overlaps the front of the one behind it. Disarticulate the vertebrae and look for smooth articular facets in the area where the vertebral arches came together. These are the zygapophyses (*articular processes*). The facets of the pair of **cranial**

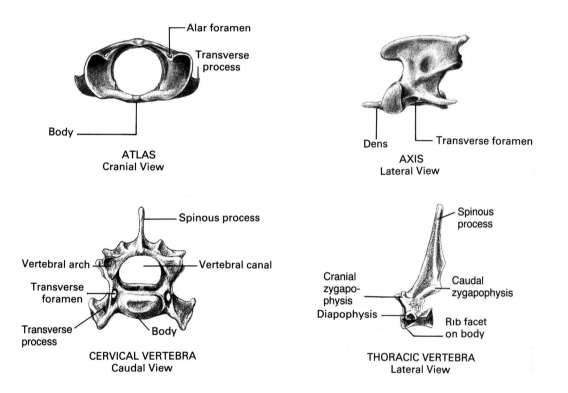

ATLAS
Cranial View

AXIS
Lateral View

CERVICAL VERTEBRA
Caudal View

THORACIC VERTEBRA
Lateral View

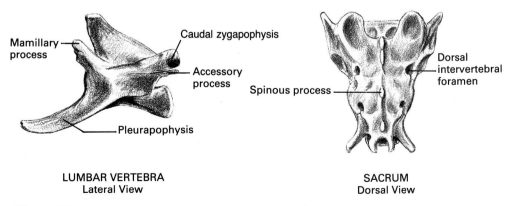

LUMBAR VERTEBRA
Lateral View

SACRUM
Dorsal View

Figure 5-7
Drawings of selected vertebrae of the cat. In those shown in lateral view, cranial is toward the left.

zygapophyses are located on the cranial surface of the vertebral arch and face dorsally; those of the pair of **caudal zygapophyses** are on the caudal surface of a vertebral arch and face ventrally. A lateral transverse process **(diapophysis)** projects from each side of the vertebral arch. Notice the smooth articular facet for the tuberculum of a rib on its end. In the cranial and middle portion of the thoracic series, the head of a rib articulates between vertebrae, so part of the facet for a head is located on the front of one vertebral body and part on the back of the next cranial one. Why does the head have an intervertebral articulation? The somewhat constricted portion of the vertebral arch between the diapophysis and vertebral body is called its **pedicle.** When several vertebrae are articulated, you will see openings for the spinal nerves, the **intervertebral foramina,** between successive pedicles.

Compare one of the thoracic vertebrae from the middle of the series with those near the cranial and caudal ends of the thoracic region and observe gradual changes in the size and inclination of the spinous processes, the plane of the zygapophyseal articulations, the size and direction of the transverse processes, and the positions of the facets for articulations of the ribs, changes that result from changes in the forces acting on them (Figs. 5-5 and 5-6). For example, the plane in which the articular facets of the zygapophyses lie changes as the major plane of vertebral movement changes. The most caudal thoracic vertebrae lack transverse processes, for their ribs lack a tuberculum, and the facets for their rib heads are entirely on one vertebral body. Articulate the last two thoracic vertebrae and notice how a process of the pedicle that extends caudad lateral to the zygapophyses reinforces the articulation of the zygapophyses. This is an **accessory process.** A skilled observer can find differences among the vertebrae sufficient to identify each one precisely—as the third or eighth thoracic, for example. However, it will be sufficient for you to be able to distinguish the major group to which a vertebra belongs. All thoracic vertebrae have at least one articular facet for a rib; no other vertebrae have such facets.

♦   **(C) LUMBAR VERTEBRAE**

Lumbar vertebrae are characterized by their large size and by prominent, bladelike transverse processes (Fig. 5-7). The transverse process is primarily a **pleurapophysis,** for a rib has united with a small diapophysis. Lumbar vertebrae also have a small bump for the attachment of ligaments and tendons dorsal to the articular surface of each cranial zygapophysis. This is called a **mamillary process.** You can also see traces of mamillary processes on the more caudal thoracic vertebrae. Most of the lumbar vertebrae also have accessory processes.

♦   **(D) SACRAL VERTEBRAE**

You can easily distinguish the sacral vertebrae because they fuse into a single piece (the sacrum), and because they present a broad surface for the articulation with the pelvic girdle. Examine the sacrum closely and you will be able to detect the spinous processes, zygapophyses, pleurapophyses, and so on, of the individual vertebrae that compose it. Notice how the distal ends of the pleurapophyses have fanned out and united with each other lateral to the intervertebral foramina. This has produced separate dorsal and ventral foramina for the respective rami of spinal nerves.

◆ **(E) CAUDAL VERTEBRAE**

Caudal vertebrae are characterized by their small size and progressive incompleteness. The more cranial vertebrae have the typical vertebral parts, but soon there is little left but an elongated vertebral body. Some mammals have V-shaped **hemal arches,** which protect the caudal artery and vein. The cat has traces of such bones on the anterior caudal vertebrae, but they are usually lost in a mounted skeleton. However, the points of articulation of a hemal arch will appear as a pair of tubercles, **hemal processes,** at the front end of the ventral surface of a vertebral body.

◆ **(F) CERVICAL VERTEBRAE**

Most of the cervical vertebrae can be recognized by their characteristic transverse processes, each of which (except for the most caudal) is perforated by a **transverse foramen** through which the vertebral blood vessels pass. The last cervical vertebra normally lacks this foramen and, aside from the absence of rib facets, is much like the first thoracic vertebra. Most of the cervical vertebrae also have low spinous processes and wide vertebral arches.

> The portion of the transverse process situated lateral and ventral to the center of the transverse foramen has developed from an embryonic rib and hence is a pleurapophysis in the strict sense of the term (p. 102); the basal part of the transverse process represents a diapophysis dorsally and a parapophysis ventrally (Cave, 1975).

The first two cervical vertebrae, the **atlas** and the **axis,** respectively, are very distinctive. The atlas is ring-shaped, with winglike transverse processes perforated by the transverse foramina. Its vertebral arch lacks a spinous process and is perforated on each side by an **alar foramen,** through which the vertebral artery enters the skull and the first spinal nerve leaves the spinal cord. Cranially the vertebral arch has facets that articulate with the two occipital condyles of the skull; caudally it has facets that articulate with the vertebral body of the axis. The body of the atlas is reduced to a thin transverse rod.

The axis is characterized by an elongated spinous process that extends over the vertebral arch of the atlas; very small transverse processes; rounded articular surfaces at the cranial end of its body; and a median, tooth-shaped process—the **dens**—that projects from the front of the body of the axis into the atlas. In life, a strong **transverse ligament** within the atlas crosses the dorsal surface of the dens.

Up-and-down movements at the head and the neck occur primarily at the occipital-atlas joint; rotational movements occur at the atlas-axis joint. Rotational movements were made possible in the reptile-to-mammal transition by the loss of zygapophyses between the atlas and axis and their replacement by new articular surfaces between the body of the axis and the vertebral arch of the atlas. Although the dens lies in the axis of rotation, Jenkins (1969) has shown that it does not function primarily as a pivot; rather, together with the transverse ligament that crosses it dorsally, it prevents undue flexion at the atlas-axis joint and drooping of the head, which would otherwise occur with the loss of the zygapophyses between these vertebrae. The dens evolves as an outgrowth of the body of the axis.

♦   **(G) RIBS**

Study the **ribs** (*costae*) of a cat from disarticulated specimens and on a mounted skeleton. Most of them have both heads characteristic of tetrapods—a proximal **head** articulating with the vertebral body and a more distal **tuberculum** articulating with the tranverse process (Fig. 5-8). But the last three ribs have only the head. The portion of the rib between its two heads is its **neck;** the long distal part is its **body,** or shaft. A **costal cartilage** extends from the end of the shaft. Those ribs whose costal cartilages attach directly to the sternum are called **vertebrosternal ribs;** those whose costal cartilages unite with other costal cartilages before reaching the sternum are **vertebrocostal ribs;** those whose costal cartilages have no distal attachment are **vertebral ribs** (Fig. 5-5). Sometimes vertebrosternal ribs are called **true ribs** and all the others **false ribs.**

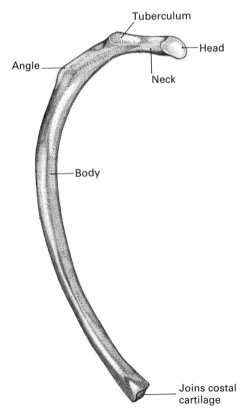

Figure 5-8
Left rib of a cat as viewed from the caudal surface.

Floating rib is an alternate name for a vertebral rib. The cat normally has nine vertebrosternal ribs, three vertebrocostal ribs, and one vertebral rib.

♦   **(H) STERNUM**

The **sternum** is composed of a number of ossified segments called the **sternebrae.** The first of these constitutes the **manubrium,** the last is the **xiphisternum,** and those between make up the **body** of the sternum. A cartilaginous **xiphoid process** extends caudad from the xiphisternum. Costal cartilages unite with the sternum between the sternebrae. In many mammals, but not in the cat, the clavicles of the pectoral girdle articulate with the cranial end of the manubrium.

The vertebral column of mammals is the primary girder of the skeleton. Not only does it carry the body weight and transfer this to the appendages; through its extension and flexion, it also participates in the locomotor movements of the body. To facilitate interpretation of its structure, some authors have compared the vertebral column to a cantilever bridge, but Slijper (1946) and others have taken a broader view and have compared the entire trunk skeleton and associated muscles to a bow-string bridge or to an archer's bow (Fig. 5-9). The thoracic and lumbar vertebrae form one arch, or bow; the cervical vertebrae form another that arches in the opposite direction. The vertebral bodies are the supporting elements of the arches. These arches are dynamic, and their curvature changes as the head is lowered or raised or as the back is flexed and extended. The thoracic-lumbar arch tends to straighten out because of the elasticity of certain dorsal ligaments and the tonus of the dorsal muscles. This is prevented by the "bow-string"—the sternum and ventral abdominal muscles such as the rectus abdominis. The "bow" and "string" are connected caudally by the pelvic

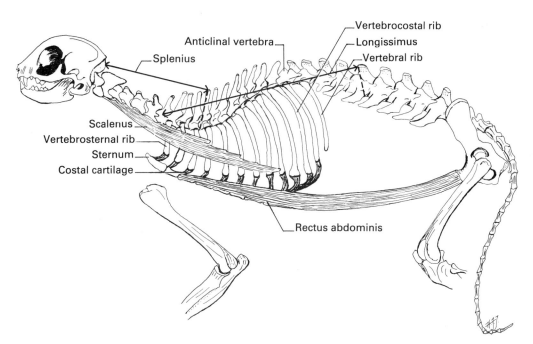

Figure 5-9
Diagram of a lateral view of a cat's skeleton illustrating Slijper's views on the biomechanics of the trunk. The scapula has been omitted. Heavy arrows indicate the line of action of certain muscles upon the spinous processes. *(Modified from Slijper.)*

girdle and cranially by the stout cranial ribs, which are held in place by the action of the scalenus and other muscles. The "string" of the cervical bow is the splenius and other dorsal cervical muscles, and the **ligamentum nuchae** that interconnects the back of the skull with the spinous processes of the cervical and anterior thoracic vertebrae. The ligamentum nuchae is very large and strong in horses and other species that have large and heavy heads. The vertebral column is a dynamic girder that can both support the body and participate in its movements.

The spinous processes are lever arms that transmit the pull of the muscles that are attached to them to a center or rotation (a fulcrum) between vertebral bodies. The direction of their inclination tends to be perpendicular to the major muscle forces acting upon them. Since several muscles may attach onto a single spinous process, the situation becomes quite complex, but in carnivores and many other mammalian quadrupeds, the tips of the spinous processes of the lumbar vertebrae point toward the head, partly in response to a very powerful longissimus muscle (see Figs. 7-41 and 7-42, pp. 203 and 204), whereas the spinous processes of the thoracic vertebrae slope in the opposite direction, partly in response to a powerful splenius muscle. The vertebra near the middle of the trunk, where the angle of inclination of the spinous processes reverses, is called the **anticlinal vertebra.**

Since the spinous processes are lever arms, an increase in their height increases the mechanical advantage of muscles acting upon them. The great length of the cranial thoracic spinous processes is related to the fact that the splenius and other muscles support the relatively heavy head.

The vertebral spines are blade-shaped because muscles pull upon them primarily in the sagittal plane of the body. These forces are resisted by an increase in the anteroposterior dimension of the spines. The spines need not be thick because few forces pull them to the side.

CHAPTER SIX

◆ ◆

THE APPENDICULAR SKELETON

The appendicular skeleton consists of the skeletal supports of the paired fins of fishes and the limbs of terrestrial vertebrates, and the girdles in the body wall to which the appendages attach. Most vertebrates have both a **pectoral** (shoulder) **girdle** and **pelvic** (hip) **girdle** and **appendage.** Nearly all of the appendicular skeleton is part of the endoskeleton because it is composed of cartilage or cartilage replacement bone. Several dermal bones contribute to the pectoral girdle of most species, so that this girdle consists of both endoskeletal and dermal components.

FISHES

A few extinct ostracoderms had finlike pectoral flaps of uncertain structure and function, but most agnathous vertebrates lack an appendicular skeleton. Recall the lamprey (p. 17). Both pectoral and pelvic fins are present in all jawed fishes, unless they have been secondarily lost. Paired fins aid in maneuvering and help to stabilize a fish against roll, pitch, and yaw as the animal moves through the water. Pectoral fins that are attached low on the body also have a hydroplaning effect and generate a lift force that elevates the front of the body.

Paired fins appear to have evolved when fishes became more active. They may have evolved as fleshy flaps behind spines that protruded from the side of the body. Many different types of paired fins were found in early jawed fishes. The earliest cartilaginous fishes had fan-shaped paired fins that were supported by many parallel rods and attached to the trunk by a broad base (Fig. 6-1**A**). In later species of cartilaginous fishes, the fin base narrowed to three basal elements (Fig. 6-1**B**). Such a tribasic fin continued to provide stability and had the advantage that it could twist and turn at the narrow base and increase maneuverability. Among bony fishes, the earliest actinopterygians had a fan-shaped tribasic fin. The fin was more elongated and lobate in sarcopterygians. In the crossopterygian group of sarcopterygians, the fin was supported by a single basal element, which was followed distally first by two and then by an increased number of radial elements (Fig. 6-1**C**). This type of fin, called a **crossopterygium,** may have evolved from the wider-based fin and, in turn, is the type of fin from which the tetrapod limb evolved (Fig. 6-1**D**).

◆ **(A) THE PECTORAL GIRDLE AND FIN**

Squalus   The appendicular skeleton of *Squalus,* which we will use as an example of the fish condition, is reasonably representative of the type found in many living fishes. But,

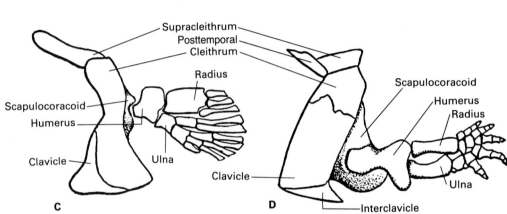

Figure 6-1
Types of paired appendages. **A,** The broad-based fin of an ancestral shark, *Cladoselache;* **B,** the tribasic pectoral fin of a contemporary shark; **C,** the pectoral girdle and appendage of a crossopterygian; **D,** the pectoral girdle and appendage of a labyrinthodont. *(A, After Bendix-Almgreen;* **B,** *after Mivart;* **C,** *after Gregory;* **D,** *from Romer and Parsons.)*

like other parts of the skeleton, it is atypical in being entirely cartilaginous. Also, the dermal bones that became associated with the pectoral girdle at an early point during vertebrate evolution are absent but can be seen in bony fishes such as *Amia.*

Examine the pectoral girdle and fin on a skeleton of the dogfish. The girdle is located just caudal to the gills (Fig. 6-2), for there is no neck region in fishes. It consists of a U-shaped bar of cartilage (Fig. 6-3). At the top of each limb of the U there is a separate **suprascapular cartilage,** 1 to 2 centimeters long. The rest of the girdle is formed of a single piece, the **scapulocoracoid.** In some sharks, but not in *Squalus,* it is obvious that the scapulocoracoid was formed from elements that have fused in the midventral line. The part of the scapulocoracoid that is located ventral to the point where the fin attaches—at the **glenoid surface**—may be called the **coracoid bar** since it has the same position as the coracoid of tetrapods. The rest of it may be called the **scapular process.** A small **coracoid foramen** for vessels and nerves can be found just cranial to the glenoid surface.

The pectoral fin is narrow at the base but widens distally. It is supported proximally by a series of cartilages, collectively called the **pterygiophores,** and distally by fibrous fin

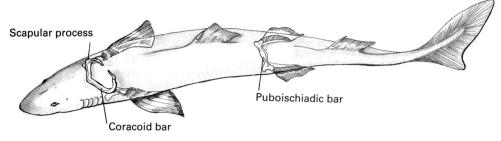

Scapular process

Puboischiadic bar

Coracoid bar

Figure 6-2
Dorsal view of *Squalus,* drawn as though the animal were transparent, to show the location and orientation of the pectoral and pelvic girdles.

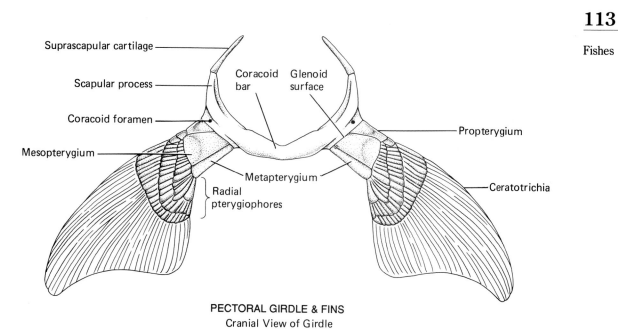

Suprascapular cartilage

Scapular process

Coracoid foramen

Mesopterygium

Coracoid bar Glenoid surface

Propterygium

Metapterygium

Ceratotrichia

Radial pterygiophores

PECTORAL GIRDLE & FINS
Cranial View of Girdle
Cranioventral View of Fin

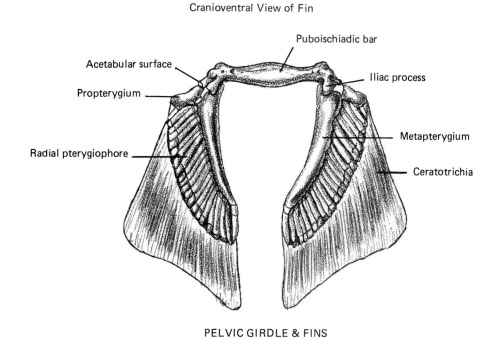

Puboischiadic bar

Acetabular surface

Propterygium

Iliac process

Metapterygium

Ceratotrichia

Radial pterygiophore

PELVIC GIRDLE & FINS
Dorsal View

Figure 6-3
Drawings of the girdles and fins of *Squalus.*

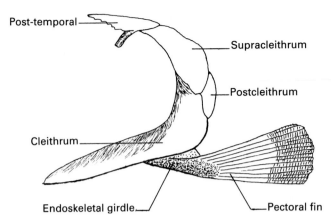

Post-temporal

Supracleithrum

Postcleithrum

Cleithrum

Endoskeletal girdle

Pectoral fin

Figure 6-4
A lateral view of the left pectoral girdle and fin of *Amia;* cartilage is stippled. Anterior is to the left.

rays, called the **ceratotrichia,** that develop in the dermis on each surface of the fin. The three large pterygiophores that articulate with the girdle are the **basal pterygiophores;** the rest are **radial pterygiophores.** The three basals are, from cranial to caudal, the **propterygium, mesopterygium,** and **metapterygium.** Note that the metapterygium is the longest.

> In many cartilaginous fishes and actinopterygians, the metapterygium forms the axis of the fin. The metapterygium is probably homologous to the single "basal" in the crossopterygium fin.

Amia   Examine a skeleton of the bowfin, *Amia.* Notice that the endoskeletal part of the pectoral girdle consists, on each side, of a small area of unossified cartilage that lies between the pectoral fin and a conspicuous arch of bone posterior to the **operculum** (dermal gill covering). This arch of bone, which is of dermal origin, forms most of the pectoral girdle. It consists of four elements (Fig. 6-4). A large ventral **cleithrum** begins ventral to the caudal visceral arches and continues dorsally a bit beyond the fin and endoskeletal girdle. A **supracleithrum** continues from the cleithrum nearly to the roof of the skull; a **posttemporal,** which is a part of the girdle even though it has surface sculpturings similar to the skull bones, articulates the supracleithrum to the back of the skull. Finally, a small **postcleithrum** is located caudal to the junction of cleithrum and supracleithrum.

The **clavicle,** a ventral dermal girdle element characteristic of many fishes and terrestrial vertebrates, has long been thought to be absent in *Amia.* However, Liem and Woods (1973) proposed that the clavicle is represented by a slender, rather superficial sliver of bone that overlaps the front of the ventral part of the cleithrum. It is not usually retained in commercial preparations of the skull and pectoral girdle of *Amia.*

♦   **(B) THE PELVIC GIRDLE AND FIN**

The pelvic girdle of *Squalus* consists of a simple transverse rod of cartilage, the **puboischiadic bar,** located in the ventral abdominal wall just cranial to the cloaca (Fig.

6-2). Each end of it extends dorsally as a very short **iliac process** (Fig. 6-3), but these processes do not reach the vertebral column. There are often two small nerve foramina near the base of each iliac process.

Like the pectoral fin, the pelvic fin consists of a series of proximal cartilaginous pterygiophores and distal ceratotrichia. There are only two **basal cartilages** in *Squalus*—a long **metapterygium** extending caudally from the girdle and clearly forming the main support of the fin, and a short **propterygium** projecting laterally from the girdle. Many **radials** extend into the fin from the two basals. In males, a **clasper,** the skeleton of which is formed by enlarged and modified radials, extends caudad from the end of the metapterygium. The fin attaches to the **acetabular surface** of the girdle.

AMPHIBIANS AND REPTILES

In the evolution of terrestrial vertebrates from crossopterygian fishes, the fin was transformed into a limb (see Fig. 6-1**C** and **D**) consisting of three segments. In the pectoral appendage, these segments are the **brachium,** containing the **humerus;** the **antebrachium,** containing the **radius** and **ulna;** and the **manus,** containing the **carpals, metacarpals,** and **phalanges.** Corresponding segments and bones in the pelvic appendage are the **femur,** with the **os femoris;**[10] the **shank,** with its **tibia** and **fibula;** and the **pes** with its **tarsals, metatarsals,** and **phalanges.** The number of toes present in the earliest tetrapods varied. Late Devonian labyrinthodonts in which the hands or feet have been fossilized had six, seven, or eight toes. Later the number stabilized at five (except in the hand of contemporary amphibians, in which the number is never more than four). The idealized five-fingered, or **pentadactyl,** limb is primitive for amniotes but not for all tetrapods. We do not know whether there was a selective advantage to having five fingers instead of more or fewer fingers, or whether this was a contingent event, an accident of evolution.

In early amphibians and reptiles, the limbs were not carried under the body as they are in a cat or dog; instead, the humerus and femur protruded laterally so that they were held horizontally when the animal moved (see Fig. 3-6). The distal part of the limb extended vertically to the ground. The front foot pointed forward and laterally, but maintenance of this placement as the humerus was retracted required a rotation of the antebrachial bones at the elbow. The hind foot pointed forward when first placed on the ground, but it rotated at the ankle joint as the femur was retracted and pointed caudally when the foot was removed from the ground. There was no rotation of lower leg bones. An increase in toe length from medial to lateral facilitated maintaining toe contact with the ground as the leg was retracted. Early amniotes had two phalanges in the first toe, three in the second, four in the third, five in the fourth, and then an abrupt decrease to three or four in the fifth toe. This is expressed as a **phalangeal formula** of 2-3-4-5-3 (or -4).

Originally, the limbs were used as a supplement to lateral undulations of the body in locomotion. But in the evolution through more advanced reptiles to birds and mammals, the limbs became increasingly important in support and locomotion. In this connection, the limbs of mammals, and the pelvic limbs of birds, have rotated closer to or beneath the body so that the humerus and femur move back and forth close to or in the vertical plane. The elbow is directed caudally, and the manus points forward, so the radius and ulna cross when the sole of the foot is on the ground. Both knee and hind foot point forward so there is no need for the lower leg bones to cross (review the changes in limb posture described in the section on external anatomy on p. 48).

Correlated with the increased importance of the appendages, the girdles became

---

[10]See footnote 4, p. 44.

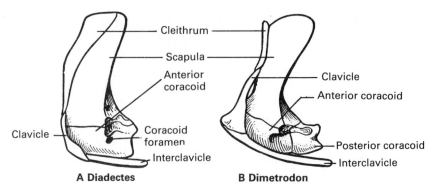

A Diadectes

B Dimetrodon

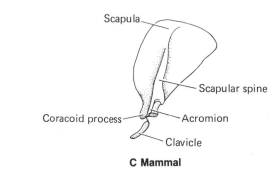

C Mammal

Figure 6-5
The evolution of the pectoral girdle. *A,* The girdle of an early reptile; *B,* the girdle of a mammal-like reptile; *C,* the girdle of an eutherian mammal. All are lateral views of the left side. The clavicle, interclavicle, and cleithrum are dermal elements. *(From Romer and Parsons.)*

more massive and stronger. In the endoskeletal part of the pectoral girdle of labyrinthodonts (see Fig. 6-1**D**) there was only one ossification on each side (a **scapulocoracoid**), but in early reptiles (Fig. 6-5**A**) a **scapula** ossified dorsal to the glenoid cavity, and an **anterior coracoid** ossified ventral to the cavity. Although these elements were large and platelike, the endoskeletal girdles of opposite sides neither united with each other ventrally nor connected directly with the vertebral column. Muscles transfer forces between the pectoral girdle and trunk skeleton in all vertebrates. Correlated with the evolution of a distinct neck, the dermal pectoral girdle lost its connection with the back of the skull. However, the cleithrum and clavicle persisted (Fig. 6-5**A**), and a new dermal element, the **interclavicle,** became associated with the girdle. The interclavicle is a median ventral element that connects the clavicles of opposite sides.

The endoskeletal part of the pectoral girdle of living amphibians, reptiles, and birds is derived from the type just described and is usually close to it in essential pattern. But in the evolution toward mammals, a third ossification, the **posterior coracoid** (or, simply, coracoid), appeared in the endoskeletal part of the girdle caudal to the anterior coracoid (Fig. 6-5**B**). In mammals, the scapular area is greatly expanded and the coracoid region is reduced (Fig. 6-5**C**). The anterior coracoid is completely lost in eutherian mammals, and the posterior coracoid is represented only by a small coracoid process of the scapula. Expansion of the dorsal part of the girdle and regression of the ventral part are correlated with the shift in limb position. Since much of the weight of the body is transferred to the ground through limb bones, ventral limb muscles become less important in raising the body from the ground and supporting it. Dorsal muscles, which are well situated to move the limb fore and aft, become more important. In all living tetrapods, the dermal portions of the girdle have become reduced, the cleithrum being completely lost except in certain frogs.

The pelvic girdle also enlarged during the transition from water to land, and in tetrapods it typically consists of three cartilage replacement bones—a **pubis** and an **ischium** ventral to the acetabulum and an **ilium** that extends from the acetabulum to

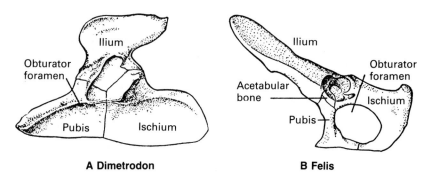

A Dimetrodon   **B Felis**

Figure 6-6
The evolution of the pelvic girdle. **A,** A mammal-like reptile; **B,** a young cat. *(From Romer and Parsons.)*

the sacral ribs and vertebrae (Fig. 6-6**A** and **B**). The ventral elements of opposite sides unite with each other. Primitively, the ventral elements formed a broad plate, and the ilium connected with only a single sacral vertebra. But in birds and mammals the ventral elements have become relatively smaller, and the ilium expands, turns forward, and unites with more sacral vertebrae. This, too, is correlated with the shift in limb position discussed previously.

Necturus

Since *Necturus* is neotenic, many parts of its girdles and appendages are unossified. *Necturus* does not use its limbs much, and they do not need to be strong, for this species does not leave the water. The position and shape of the girdles is representative of early terrestrial vertebrates.

♦   **(A) THE PECTORAL GIRDLE AND APPENDAGE**

Examine the pectoral girdle on a skeleton of *Necturus,* comparing it with Figure 6-7**A**. The two halves of the pectoral girdle overlap slightly ventrally but are not united. On each side an ossified **scapula** extends dorsally cranial to the **glenoid fossa** (a depression for the articulation of the girdle with the humerus). The scapula is capped by a **suprascapular cartilage.** The ventral part of the girdle remains unossified and is called the **coracoid plate.** A **procoracoid process,** not to be confused with the anterior coracoid bone of other tetrapods, extends forward from the coracoid plate. A **coracoid foramen,** for vessels and nerves, may be seen in the coracoid plate ventral to the scapula.

Study the pectoral limb skeleton, noting first the position of the different segments of the limb and the preaxial and postaxial surfaces (see p. 44). A single bone, the **humerus,** extends from the glenoid fossa to the elbow joint. Two bones of approximately equal size compose the forearm—a **radius** on the preaxial (medial) side and an **ulna** on the postaxial (lateral) side. The manus (hand) consists of a group of six cartilaginous **carpals** in the wrist, four ossified metacarpals in the palm of the hand, and the ossified **phalanges** of the digits, or fingers. Terms for the individual carpals are derived from their positions relative to other bones (see Fig. 6-11), but the individual carpals are usually not distinct in dried skeletons. Note that only four toes are present, a number characteristic of living amphibians. These probably are homologous to the second through the fifth toes of

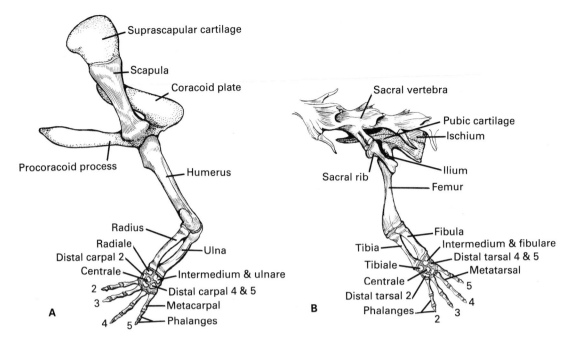

Figure 6-7
Dorsolateral views of the pectoral girdle and appendage (**A**) and pelvic girdle and appendage (**B**) of *Necturus*.

amniotes. What is the phalangeal formula? (The absent digit is represented as "0" in the formula.)

◆   **(B) THE PELVIC GIRDLE AND APPENDAGE**

Each half of the pelvic girdle has a narrow ossified **ilium** that extends dorsally from the socket for the leg articulation—the **acetabulum**—to attach on a single sacral rib and vertebra (Fig. 6-7**B**). Ventrally there is a broad **puboischiadic** plate, which contains a pair of ossified **ischia** caudally and a **pubic cartilage** cranially. An **obturator foramen,** for a nerve of the same name, may be seen in the pubic cartilage. Note that the pelvic girdle, together with the sacral rib and vertebra, forms a ring of bone around the caudal end of the trunk. The passage through this ring is called the **pelvic canal.** The intestine and urogenital ducts pass through this canal.

Study the pelvic limb, noting its position and its preaxial and postaxial surfaces. A **femur** forms the proximal segment of the limb; a **tibia** and **fibula** are contained within the shank (the former lying along the preaxial surface); and the pes consists of a group of six cartilaginous **tarsals** and ossified **metatarsals** and **phalanges.** The individual tarsals cannot be distinguished in dried skeletons. Many salamanders have five toes, but *Necturus* has only four, the most medial probably being homologous to the second toe of amniotes. What is the phalangeal formula?

The Turtle

Although the appendicular skeleton of the turtle is specialized in some respects, certain primitive features can be seen more clearly in it than in the appendicular skeleton of *Necturus*.

◆ (A) THE PECTORAL GIRDLE AND APPENDAGE

Examine a mounted skeleton and an isolated pectoral girdle of the turtle (Figs. 6-8**A** and 6-9). The endoskeletal part of the pectoral girdle, which has an unusual triradiate shape, is located between the **plastron** (the bottom shell) and the **carapace** (the top shell). The dorsal prong of the endoskeletal part of the girdle, which articulates with the carapace, represents the **scapula;** the cranioventral prong, a part of the scapula known as the **acromion;** and the expanded caudoventral prong, the **anterior coracoid.** In life, the

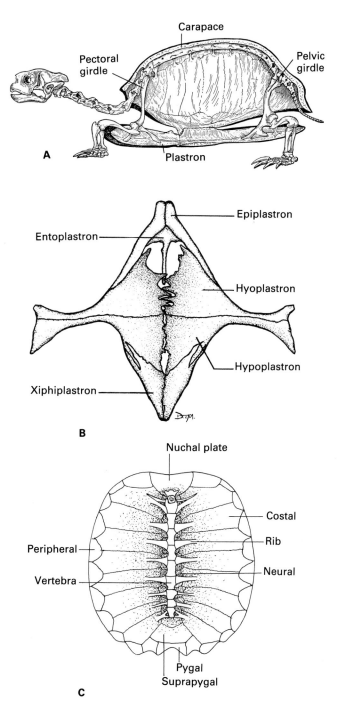

Figure 6-8
A, Lateral view of the skeleton of a turtle showing the orientation of the girdles and limbs; **B,** ventral view of the plastron of a snapping turtle (*Chelydra*) after removal of the horny laminae; **C,** ventral view of the carapace of a snapping turtle. *(A, From W. K. Gregory.)*

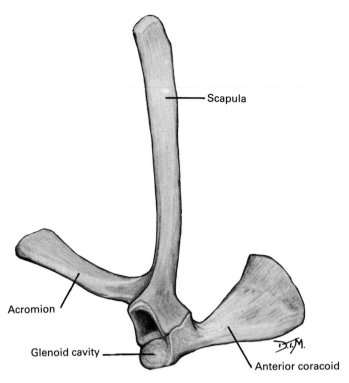

Figure 6-9
Lateral view of the left pectoral girdle of the snapping turtle, *Chelydra*. Anterior is toward the left.

acromion is connected by a ligament to the **entoplastral plate** of the plastron, and the scapula is connected by a movable joint to the carapace. An **acromiocoracoid ligament** may be seen extending between the tips of the acromion and coracoid. A **glenoid fossa** is present for the articulation of the humerus.

> The pectoral girdle of the turtle is highly specialized in relation to the presence of a rigid shell that prevents rib movements and change in size and volume of the thoracic cavity. Such movements play an important role in ventilating the lungs of most reptiles. The lungs of turtles are ventilated instead by the contraction and relaxation of the muscles lining the pockets into which the limbs can be withdrawn, and by the rotation of the pectoral girdle at its articulations with the shell at the tips of the scapular prong and the acromion. The girdle also rotates during locomotion and adds slightly to the length of the step.

Parts of the dermal girdle are present but are incorporated in the cranial plates of the plastron. Examine a plastron in which the epidermal scutes (**laminae**) have been removed and the **dermal plates** exposed (Fig. 6-8B). The front of the plastron is formed by a pair of **epiplastra,** caudal to which are a median **entoplastron** and three additional paired plates—the **hyoplastra,** the **hypoplastra,** and the **xiphiplastra.** All these plates represent, in part, an ossification in the dermis of the skin of the underside of the body. But during embryonic development, the originally separate primordia of the clavicles and interclavicle become incorporated in the first three plates, which are therefore compound plates. The epiplastra include the paired **clavicles;** the entoplastron includes the **interclavicle.** The remaining plastral plates may include the **gastralia** of early amphibians and reptiles. In these early tetrapods, the gastralia were riblike rods of dermal bone found on the ventral abdominal wall. They were remnants of the piscine dermal scales and may have had a protective function. They are retained in a few living reptiles—*Sphenodon,* crocodiles, and, possibly, turtles. Bones of the carapace (Fig. 6-8**C**) develop primarily from dermal ossifications, but ribs are incorporated in the costal plates.

Study the pectoral appendage (Fig. 6-10). How does the position of the limbs of a turtle compare with that of one of the early amphibians or reptiles (labyrinthodont or ancestral reptile)? Identify the preaxial and postaxial surfaces. The long bones of the appendage are a **humerus** in the upper arm and a **radius** and **ulna** in the forearm. The proximal end of the humerus has a round **head** that fits into the glenoid fossa and two prominent processes for the attachment of muscles. Of the forearm bones, the radius is the one on the preaxial (cranial or medial) surface. Both radius and ulna are about equal in size, but the ulna tends to extend over the distal end of the humerus, whereas the radius articulates on the underside of the end of the humerus. As in early tetrapods generally, both the radius and ulna articulate with the wrist bones, and there is no distal radioulnar joint.

The manus consists of a group of **carpals** in the wrist, a row of five **metacarpals** in the palm, and the **phalanges** in the free part of the toes. There are five toes, the first being the most medial. The phalangeal number has been reduced from the primitive number to 2-3-3-3-3 in the snapping turtle.

The carpus of the turtle is very similar to that of amphibians, and the individual components should be identified. The carpals can be grouped into a proximal and distal row. The proximal row consists of three bones—an **ulnare** adjacent to the ulna, an **intermedium** lying between the distal ends of the radius and ulna, and an elongated element distal to the radius and intermedium, which represents a fusion either of two centralia or of a **centrale** and a **radiale** (Fig. 6-11**A**). The distal row consists of five distal carpals that are numbered according to the digit to which they are related—distal carpal 1, distal carpal 2, and so on. In addition there may be a small **sesamoid** bone on the lateral edge of the carpus. Sesamoid bones develop in the tendons of muscles and are rather variable. However, the bone on the lateral edge of the wrist adjacent to the ulnar occurs in nearly all tetrapods and is called the **pisiform.** Sesamoid bones facilitate the movement of a tendon across a joint or alter slightly the direction of pull of a muscle.

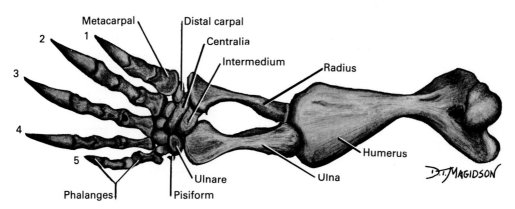

Figure 6-10
Dorsal view of the left pectoral appendage of the snapping turtle, *Chelydra.*

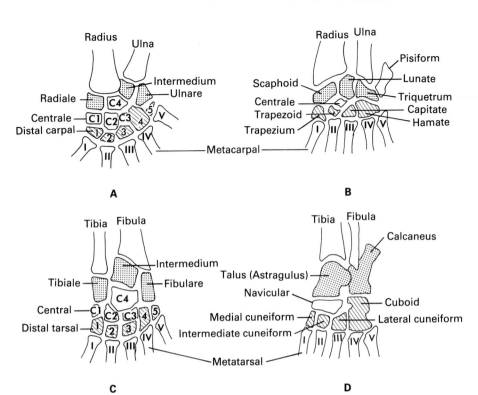

Figure 6-11
Evolution of the carpus (*top*) and tarsus (*bottom*). The left wrist and ankle are shown for an early
amphibian (**A** and **C**) and for a mammal (**B** and **D**). *(From Walker,* A Study of the Cat; *modified from Romer
and Parsons.)*

♦ **(B) THE PELVIC GIRDLE AND APPENDAGE**

Study the pelvic girdle of the turtle on a mounted skeleton and from an isolated specimen
(Figs. 6-8 and 6-12). Each half of the girdle consists of a dorsal **ilium** that inclines
caudodorsally and articulates with two sacral ribs and vertebrae, a cranioventral **pubis,**
and a caudoventral **ischium.** All three elements share in the formation of the
acetabulum, the socket for the leg articulation. The pubis and ischium of opposite sides
are united by a **symphysis.** An **epipubic cartilage,** which may be partly ossified,
extends forward in the midventral line from the pubic bones. Both the pubis and ischium
have a lateral process that is directed ventrally and, in life, rests on the plastron. A large
puboischiadic fenestra, which develops in association with the origin of a muscle, lies
between the pubis and ischium on each side. A separate obturator foramen seen in many
fishes and amphibians is not present in turtles since the obturator nerve also passes
through the puboischiadic fenestra.[11]

---

[11]The terminology and homology of the various pelvic openings are unfortunately confused. In most
amphibians and reptiles, an obturator foramen (pubic foramen) for an obturator nerve perforates the
pubis cranial to the acetabulum (Fig. 6-6*A*). The rest of the puboischiadic plate is solid. In some
reptiles (lizards) an additional opening, known as the puboischiadic fenestra (thyroid fenestra),
develops between the pubis and ischium in association with the origin of certain pelvic muscles. In
mammals the fenestrations of the puboischiadic plate includes both the primitive obturator foramen
and the puboischiadic fenestra. Such an opening is also termed an obturator foramen. The turtle
would seem to parallel this condition, but most authorities call the opening a puboischiadic fenestra.

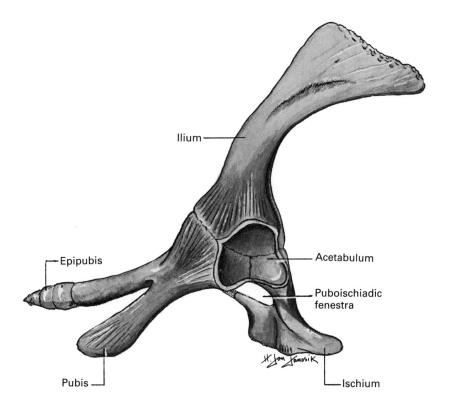

Figure 6-12
Lateral view of the left side of the pelvic girdle of the snapping turtle, *Chelydra.* Anterior is toward the left.

Examine the pelvic limb, noting its position and its preaxial and postaxial surfaces (Fig. 6-13). The long bone of the thigh is the **femur.** Its proximal end bears a round **head** that fits into the acetabulum, and two prominent processes for the attachment of muscles. The long bones of the shank are the **tibia** and **fibula,** the former being the larger and more cranial or medial.

The pes consists of a group of **tarsals** in the ankle region, a row of five **metatarsals** in the sole of the foot, and **phalanges** in the free part of the five digits. The phalangeal formula is reduced to 2-3-3-3-3 in the snapping turtle. The metatarsal of the fifth toe is flat and broad rather than round and elongate like the others.

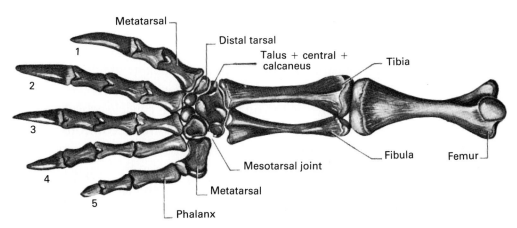

Figure 6-13
Dorsal view of the left pelvic appendage of the snapping turtle, *Chelydra.*

The individual tarsals should be studied and compared with Figures 6-11C and 6-13. There is a row of four **distal tarsals** next to the metatarsals. The fourth distal tarsal is larger than the others and is associated with the fourth and fifth toes. All the remaining elements of the tarsus tend to fuse into a single bone, but close inspection often reveals the lines of union between the three major elements: **talus** (fused tibiale, intermedium and proximal centrale), **distal centrale,** and **calcaneus** (fibulare). Sometimes the calcaneus remains distinct. The main ankle joint of the turtle, as in other reptiles, is a **mesotarsal joint,** for it lies between the large proximal element(s) and the distal tarsals. Rotation of the foot during locomotion occurs at this joint.

MAMMALS

The appendicular skeleton of mammals is well suited for terrestrial locomotion. The limbs have rotated so that they lie close to the trunk, with the elbow directed caudally and the knee cranially. In cats and many other species, they lie beneath the body, where they can provide good mechanical support and swing fore and aft through a long arc. Ancestral mammals were **plantigrade**—that is, they walked with the soles of their feet on the ground. But in **digitigrade** mammals such as the cat, a longer step and stride are made possible by walking on the digits with the soles of the feet off the ground. Ungulates, such as the horse and cow, carry the tendency further and walk on their toe tips; these animals are called **unguligrade.**

♦ **(A) THE PECTORAL GIRDLE AND APPENDAGE**

Study the appendages and girdles on mounted specimens and from disarticulated bones of the cat and other species available. Learn to distinguish the individual girdles and the long bones of the appendage and to recognize whether they are from the left or the right side. The pectoral girdle consists primarily of an expanded triangular **scapula** located on the side and back of the cranial end of the thorax (Figs. 5-5 and 6-14). In eutherian mammals the anterior coracoid has been lost, and the posterior coracoid is reduced to a small, hooklike **coracoid process** that can be seen medial to the cranial edge of the **glenoid fossa** (socket for the articulation with the humerus). Correlated with the shift of the limbs under the body, the glenoid fossa is directed ventrally rather than laterally.

If we picture the scapula as an inverted triangle, the glenoid fossa is at the apex, and the curved top of the scapula—its **dorsal border**—is at the base of the triangle. The cranial edge of the scapula is its **cranial border,** and the straight caudal edge, which is adjacent to the armpit, is its **caudal border.** A prominent ridge of bone, the **scapular spine,** extends from the dorsal border nearly to the glenoid fossa. The ventral tip of the spine continues as a process known as the **acromion.** The clavicle articulates with this process in those mammals that have a prominent clavicle. A **metacromion** (*suprahamate process*) extends caudally from the spine dorsal to the acromion. It is exceptionally long in the rabbit (see Fig. 5-6). That portion of the lateral surface of the scapula caudal to the spine is called the **infraspinous fossa;** that portion cranial to the spine is the **supraspinous fossa.** The spine represents the primitive anterior edge of the scapula, for it continues to the acromion. The acromion is on the anterior border of the scapula in early tetrapods. The portion of the scapula cranial to the spine is added during the course

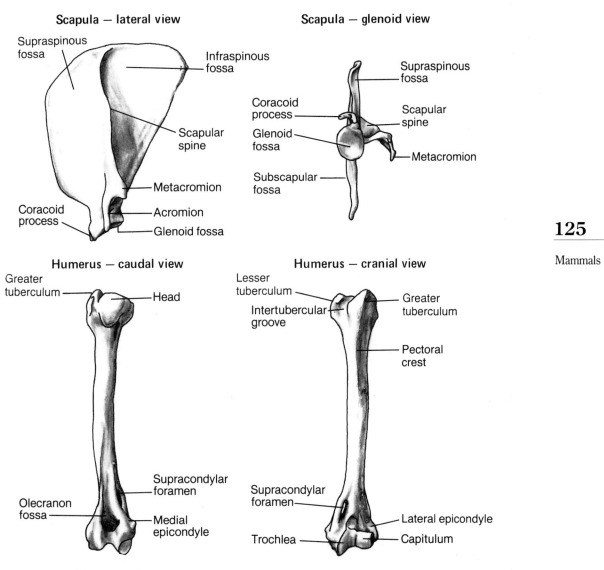

Scapula — lateral view

Supraspinous fossa

Infraspinous fossa

Scapular spine

Metacromion

Coracoid process

Acromion

Glenoid fossa

Scapula — glenoid view

Supraspinous fossa

Coracoid process

Scapular spine

Glenoid fossa

Metacromion

Subscapular fossa

Humerus — caudal view

Greater tuberculum

Head

Supracondylar foramen

Olecranon fossa

Medial epicondyle

Humerus — cranial view

Lesser tuberculum

Intertubercular groove

Greater tuberculum

Pectoral crest

Supracondylar foramen

Lateral epicondyle

Trochlea

Capitulum

Figure 6-14
The left scapula and humerus of the cat. *(From Walker,* A Study of the Cat.*)*

of evolution. The medial surface of the scapula is called the **subscapular fossa.** In human beings, from whom most of our anatomical terminology is derived, there is a prominent fossa here, but in the cat this surface is flat.

Of the elements of the dermal girdle, only the **clavicle** is present. In many mammals, human beings included, the clavicle extends from the acromion to the manubrium. But in some mammals, including the cat, the clavicle is connected to the rest of the skeleton only by ligaments and is reduced to a sliver of bone imbedded in the muscles cranial to the shoulder joint (see Fig. 5-5).

Jenkins (1974) has studied differences in locomotion between terrestrial mammals that retain the clavicle and those that have lost it or had it experimentally removed. During the recovery phase of the step of a front leg, the leg is moved forward and the foot is placed on the ground in advance of its previous position. The scapula also rotates forward to some extent. During the propulsive phase of a step, muscles extending from the trunk to the pectoral girdle and appendage pull the trunk forward relative to the girdle and appendage. If a well-developed clavicle is present, it acts as a "spoke" and fixes the distance between the acromion and manubrium. As the trunk moves forward, the shoulder is deflected laterally, for the clavicle resists the medial component of the pull of the muscles. Relative movement between the manubrium and acromium takes the form of an arc. Claviculate terrestrial mammals tend to be those in which the legs have not rotated completely beneath the body so that the shoulder and elbow joints are not in the same parasagittal plane. A clavicle is lost in cursorial mammals, in which shoulder and elbow joints lie in the same parasagittal plane. With the absence of the clavicle, the thorax and scapula show parallel movement. The shoulder is not deflected laterally during the propulsive stroke, with consequent loss of forward thrust. It has also been proposed that the loss of the clavicle permits a greater fore-and-aft excursion of the scapula and hence a longer step and stride, but it has not been demonstrated whether the scapula has a greater excursion in cursorial mammals than in others.

The bone of the brachium is the **humerus** (Fig. 6-14). Its expanded proximal end has a smooth rounded **head** that articulates with the glenoid fossa, and two processes for muscular attachment—a lateral **greater tuberculum** and a medial **lesser tuberculum.** An **intertubercular groove** for the long tendon of the biceps muscle lies between the two tubercles on the craniomedial surface of the humerus. The distal end of the bone is also expanded and bears a smooth articular surface known as a **condyle.** Although there appears to be but a single condyle, it can be divided into a medial pulley-shaped portion— the **trochlea**—for the ulna of the forearm (the bone that comes up behind the elbow), and a lateral rounded portion—the **capitulum**—for the radius. You may have to articulate the ulna and radius with the humerus to determine the extent of the trochlea and capitulum. An **olecranon fossa,** for the olecranon of the ulna, is situated proximal to the trochlea. The enlargements to the sides of the articular surfaces are the **medial** and **lateral epicondyles,** respectively. A **supracondylar foramen** for a nerve and blood vessel is located above the medial epicondyle. This foramen is a primitive amniote feature found in early reptiles but lost in most mammals, including human beings. That portion of the humerus, or of any long bone, lying between its extremities is its **body,** or shaft. The faint ridges and rugosities upon it mark the attachment of certain muscles. The most conspicuous of these is the **pectoral crest** for the insertion of part of the pectoral complex of muscles (p. 170). This ridge extends distally from the greater tubercle on the craniolateral surface of the shaft.

The **ulna** (Fig. 6-15) is the longer of the two forearm bones; a prominent, semilunar-shaped **trochlear notch,** for the articulation with the humerus, will be seen near its proximal end. The end of the ulna lying proximal to the notch is the **olecranon,** or "funny bone." A **coronoid process** forms the distal border of the notch, and a **radial notch,** for the head of the radius, merges with the trochlear notch lateral to the coronoid process. The bone terminates distally in a **lateral styloid process,** which articulates with the lateral surface of the wrist. Note that, unlike the situation in reptiles, the ulna and radius articulate distally and that the ulna plays a relatively insignificant role in the formation of the wrist joint. In some mammals, the distal half of the ulna is lost.

The other bone of the forearm is the **radius.** The articular surfaces on its **head** are of such a nature that the bone can rotate on the humerus and ulna. Slightly distal to the head is a prominent **radial tuberosity** for the insertion of the biceps muscle. The distal end of the radius is expanded; it has articular surfaces for the ulna and carpus and a short **medial**

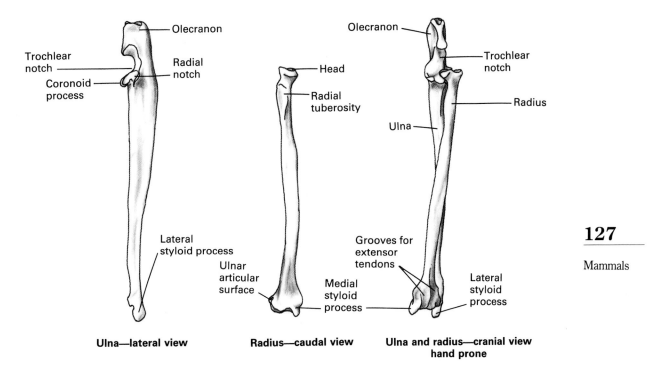

Figure 6-15
Left antebrachial bones of the cat. *(From Walker,* A Study of the Cat.*)*

Labels in figure:
Olecranon
Trochlear notch
Coronoid process
Radial notch
Olecranon
Head
Radial tuberosity
Trochlear notch
Radius
Ulna
Lateral styloid process
Ulnar articular surface
Grooves for extensor tendons
Medial styloid process
Lateral styloid process

Ulna—lateral view
Radius—caudal view
Ulna and radius—cranial view hand prone

styloid process. The articular surfaces between the radius and ulna in most mammals permit the radius to rotate on the ulna as the palm of the hand is turned toward the ground (pronation) or up (supination). Some mammals, including the rabbit (see Fig. 5-6), lose the ability to rotate the forearm. In rabbits, this appears to be an adaptation for hopping.

In amphibians and reptiles, the ulna extended straight down the lateral edge of the forearm and the radius straight down the medial edge. The articulation of the radius with the humerus, besides being medial, was slightly cranial to that of the ulna. The manus pointed forward. If the elbow rotated posteriorly only 90 degrees during the evolution of mammals, the manus would extend laterally. To bring the manus forward, the radius rotated at the elbow. The end result is that the radius continues to be the more medial bone at the wrist but is cranial and lateral to the ulna at the elbow (Fig. 6-15). The changes are best visualized by putting the arm of a mounted human skeleton into the various positions.

Study the **manus** (hand) of the cat. Its first portion, the **carpus** (wrist) (Figs. 6-11 and 6-16), consists of two rows of small **carpal bones.** The proximal row contains a large medial **scapholunate,** which represents the radiale, intermedium, and a centrale fused; a smaller **triquetrum,** which represents the ulnare; and, on the lateral edge, a large, caudally projecting **pisiform.** The pisiform is one of many small sesamoid bones found in the appendages. They are not supporting elements; rather, they are associated with the attachments of muscle tendons. Most are not named. The four elements of the distal row are, from medial to lateral, the **trapezium,** representing distal carpal 1; **trapezoid,** representing distal carpal 2; **capitate,** representing distal carpal 3; and **hamate,** representing distal carpal 4. Five **metacarpals** form the palm of the hand, and the free

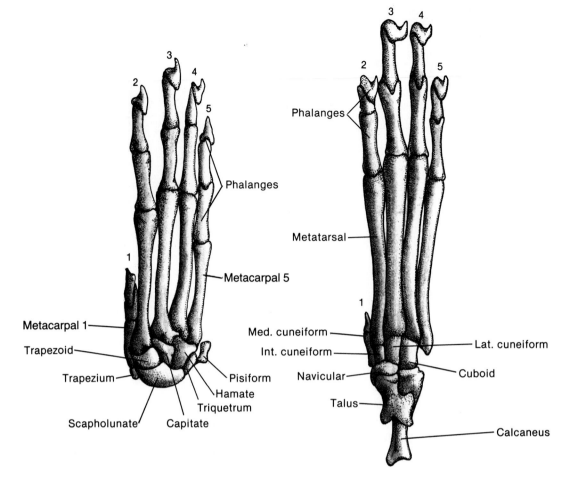

Figure 6-16
Dorsal views of the right manus (*left*) and pes (*right*) of the cat. *(From Walker,* A Study of the Cat.*)*

parts of the toes are composed of **phalanges.** The first toe is the most medial. Note that the number of phalanges has been reduced from that in ancestral reptiles. This correlates with the hand being directed forward throughout the step cycle. The terminal, or **ungual,** phalanx of the catlike carnivores is articulated in such a way that it, and the claw that it bears, can be either extended or pulled back over the penultimate phalanx.

♦ **(B) THE PELVIC GIRDLE AND APPENDAGE**

The ilium, ischium, and pubis on each side of the pelvic girdle have fused together in adult mammals to form an **os coxae,** or innominate bone, but they can be seen clearly in young specimens (see Fig. 6-6B). The **ilium** extends craniodorsally in the cat and most mammals from the **acetabulum,** or socket for the hip joint, to the sacrum (see Fig. 5-5). In a rabbit at rest on the ground, it extends dorsally (see Fig. 5-6). Its dorsalmost border is its **crest.** Notice that the ilium unites with three sacral vertebrae, more than does the ilium in amphibians and reptiles. The **ischium** surrounds all but the cranial portion, and some of the medial side, of the large opening—the **obturator foramen**—in the ventral portion of the girdle (Fig. 6-17). The enlarged caudolateral portion of the ischium is called its **tuberosity.** The **pubis** lies cranial to the foramen and completes its medial wall.

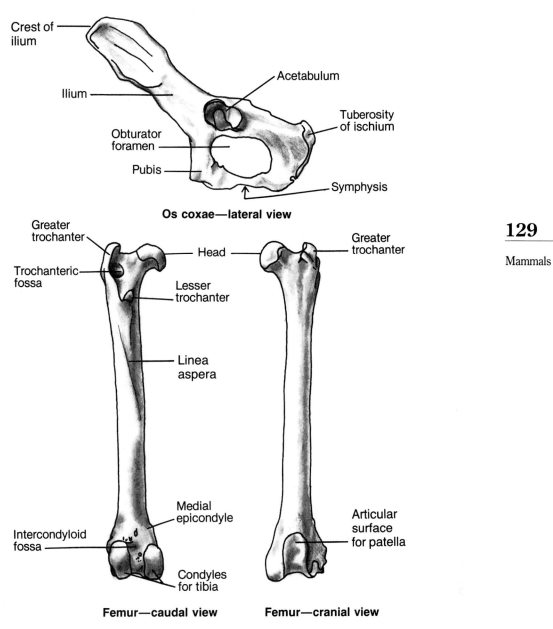

Crest of ilium

Acetabulum

Ilium

Tuberosity of ischium

Obturator foramen

Pubis

Symphysis

Os coxae—lateral view

Greater trochanter

Head

Greater trochanter

Trochanteric fossa

Lesser trochanter

Linea aspera

Medial epicondyle

Articular surface for patella

Intercondyloid fossa

Condyles for tibia

Femur—caudal view

Femur—cranial view

Figure 6-17
The left pelvis and femur of the cat. *(From Walker,* A Study of the Cat.*)*

Although the pubis enters the cranial portion of the acetabulum in most tetrapods, it is separated from it in the cat by a small **acetabular bone** (see Fig. 6-6**B**) of unknown phylogenetic significance. The pubes and ischia of opposite sides are united by **symphyses,** so that the pelvic girdle and sacrum form a complete ring, or **pelvic canal,** through which internal organs must pass to reach the anus and urogenital apertures.

Since the leg has rotated beneath the body, the thigh bone, or **femur,** articulates with the acetabulum by a **head** that projects from the medial side of the proximal end of the bone (Fig. 6-17). A large lateral **greater trochanter** and a small, caudal **lesser trochanter** can also be seen on the proximal end. These processes are for muscle attachments. A depression called the **trochanteric fossa** is situated medial to the greater trochanter. The most conspicuous of the ridges on the shaft on which muscles insert is the **linea aspera,** which extends diagonally across the caudal aspect of the shaft.

The distal end of the femur has a smooth articular surface over which the **patella,** or kneecap (a sesamoid bone), glides (see Fig. 5-5). Caudal to this are smooth **lateral** and **medial condyles** for articulation with the tibia. The rough areas above each condyle are **epicondyles,** and the depression between the two condyles is the **intercondyloid fossa.**

The **tibia** is the larger and more medial of the two shank bones (Fig. 6-18). Its proximal end has a pair of **condyles** for articulation with the femur, and a cranial, oblong bump—the **tuberosity**—for the attachment of the patellar ligament. Its shaft has a relatively sharp cranial margin, the **tibial crest,** which continues from the tuberosity. The distal end of the tibia, which articulates with the ankle, is prolonged on the medial side as a process called the **medial malleolus.** The **fibula** is a very slender bone. Notice that it articulates with the proximal end of the tibia and does not enter the knee joint. Distally, it also articulates with the tibia and serves to strengthen the ankle laterally. Its distal end has a small, pulley-like process known as the **lateral malleolus.** The tendon of the peroneus brevis muscle (p. 199) passes caudal to this process. Tibia and fibula are fused distally in some mammals, including the rabbit (see Fig. 5-6).

Examine the **pes** (foot). The ankle, or **tarsus,** of the cat is typical of that of mammals, for it consists of seven **tarsal bones** (Figs. 6-11 and 6-16). The one that articulates with the tibia and fibula is the **talus,** which appears to be homologous to the tibiale, intermedium, and one centrale. The main joint in the mammal ankle is between this bone and the tibia and fibula, rather than mesotarsal as in reptiles and birds. The large, caudally projecting heel bone is called the **calcaneus** (fibulare). A **navicular** (centrale) lies just distal to the talus, and a row of four bones lies distal to the navicular and calcaneus. These are, from medial to lateral, the **medial cuneiform** (distal tarsal 1), **intermediate cuneiform** (distal tarsal 2), **lateral cuneiform** (distal tarsal 3), and **cuboid** (distal tarsal

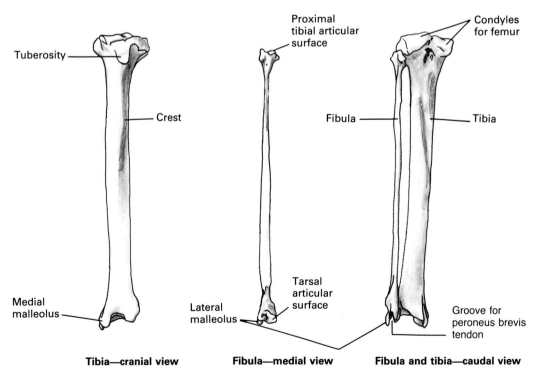

Figure 6-18
Left shank bones of the cat. *(From Walker,* A Study of the Cat.*)*

4). Five elongated **metatarsals,** which occupy the sole of the foot, normally follow the tarsals, but in the cat the first toe is lost, and its metatarsal is reduced to a small nubbin of bone articulated with the medial cuneiform. The bones in the free part of the digits are the **phalanges.** The phalangeal formula is reduced from that of ancestral amniotes to 2-3-3-3-3 in most mammals; in the cat it is 0-3-3-3-3.

Limb Joints in Tetrapods

Most of the joints between limb bones, and those between the limbs and girdles, are **synovial joints** that permit considerable movement between the bones. The basic anatomy of a synovial joint is shown in Figure 6-19. It can be seen in a demonstration of the shoulder joint, or by dissecting this joint after the muscles have been studied. **Articular cartilages** cover the ends of the bones that meet in the joint and provide an elastic, wear-resistant, and low-friction surface for movement. The joint is surrounded and supported by a fibrous **articular capsule** of dense connective tissue that is continuous with the periosteum covering the bones. Frequently, the capsule is strengthened by extracapsular **ligaments** extending between the bones, and by muscles and their tendons crossing the joint. A **synovial membrane** lines the joint capsule and secretes and reabsorbs a small amount of viscous **synovial fluid** that fills the joint cavity and lubricates the articular surfaces.

In some synovial joints, a fibrocartilaginous **articular disc** crosses the synovial cavity between the bones, or a fibrocartilaginous ring, or **meniscus,** may extend partway into the cavity from the articular capsule. There is a disc in the human sternoclavicular joint and a meniscus in the knee joint. A disc, or a meniscus, frequently improves the fit between the joint surfaces, acts as a shock absorber, or restricts certain movements.

The Age of Mammal Skeletons

Limb bones and other cartilage replacement bones grow in length in young individuals at plates of **epiphyseal cartilage** that extend across the bone between its ends **(epiphyses)** and its shaft. The joint between the epiphysis and the shaft is a synchondrosis (p. 88). The epiphyseal plate grows by mitosis of the cartilage cells and, as it grows, bone replaces the cartilage on the proximal and distal surfaces of the

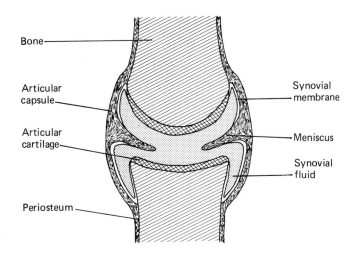

Figure 6-19
Diagram of a representative synovial joint. (*From Walker,* A Study of the Cat.*)*

Table 6-1   Aging a Skeleton

The approximate mean age is given for selected upper permanent teeth eruptions and epiphyseal fusions in the cat. (From E. Berman.)

Age (in months)	Tooth Eruptions, or Epiphyseal Fusions
3	Incisor 1
4	Incisor 3
5	Canine
12	Calcaneus
15	Distal ulnar
15	Distal femoral
18	Distal humeral
18	Proximal tibial
20	Distal radial

plate. Bone stops increasing in length when the epiphyseal plate stops growing and becomes completely ossified, and the epiphysis and shaft unite. This occurs at different ages for different epiphyses, so by observing which epiphyseal plates are present and which have disappeared, one can estimate the age of the skeleton (Table 6-1). The degree to which permanent teeth have replaced milk teeth and the degree of fusion of certain skull sutures also help in age determination.

CHAPTER SEVEN

◆ ◆

THE MUSCULAR SYSTEM

Continuing on the general theme of the organ systems whose activities support and move the body, we will now consider the muscular system. The study of this system is one of the more challenging parts of a comparative anatomy course because it makes special demands on your ability to learn and integrate a large number of different but interrelated data. Muscles, unlike most other structures of the vertebrate body, are primarily defined not by their shape and location but rather by their attachments to skeletal elements as well as their muscle fiber arrangement, histological structure, embryonic development, and their action. It is therefore important to review these various aspects of the muscular system before starting the dissections.

There are three basic types of muscle tissue, which are characterized by their histological structure, innervation, and association with organs and parts of the body. **Striated muscles** are usually associated with skeletal structures and hence are also called "skeletal muscles"; their innervation is under voluntary control. **Smooth muscles** are associated with internal organs, blood vessels, and glands; they are innervated by visceral nerves from the autonomic nervous system and are not under voluntary control. **Cardiac muscles** are unique to the heart; the modified muscle cells that act as an internal pacemaker and impulse-conducting system of the heart are regulated by visceral nerves of the autonomic nervous system. In this chapter we will study only the skeletal musculature; the smooth and cardiac musculatures will be studied as part of the organs in which they occur.

The Structure and Function of Skeletal Muscles

The **body** (or **belly**) of a muscle is enveloped by a connective tissue sheet, the **epimysium,** and consists of numerous **muscle fiber bundles** (or **fasciculi;** singular: **fasciculus**). These are surrounded by the **perimysia,** connective tissue envelopes that are connected with the epimysium (Fig. 7-1). Each muscle fiber bundle consists of several **muscle fibers** (muscle cells), which are individually surrounded by a network of connective tissue fibers, the **endomysium,** and by capillaries. The muscle fibers contain contractile **myofibrils.** These consist of actin and myosin filaments which in turn are arranged in **sarcomeres.** The naked eye can discern only muscle fiber bundles.

Within individual muscles, the muscle fiber bundles can be arranged in various ways (Fig. 7-2). They can be oriented in a generally parallel fashion as in spindle-shaped, straplike, or sheetlike muscles, or they can be arranged in a fanlike manner. In **pennate muscles,** muscle fibers assume a variety of orientations. Such muscles contain internal septa or external sheets of connective tissue to which the muscle fibers can attach. Some muscles, particularly those of the trunk region of fishes and amphibians, are **segmented.** In these muscles, the muscle fibers are arranged in a

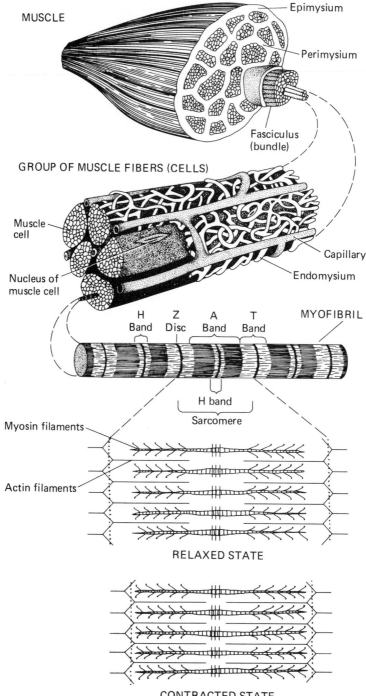

Figure 7-1
Structure of a striated muscle and its components at successive levels of magnification. *(Redrawn from Dorit, Walker, and Barnes, Zoology 1991; and Bloom and Fawcett.)*

parallel fashion, but the muscle itself is partitioned by transverse septa of connective tissue, called **myosepta,** which mark the borders between the embryonic myotomes (see p. 143) from which the muscle developed. Sometimes, a transverse connective tissue septum is found in an unsegmented muscle. In such cases, the septum is called a **tendinous intersection** and represents the line along which two originally separate muscles have become fused to each other.

The ends of muscles are attached to skeletal elements or connective tissue sheets that can move relative to one another. If a muscle attaches only to skeletal elements,

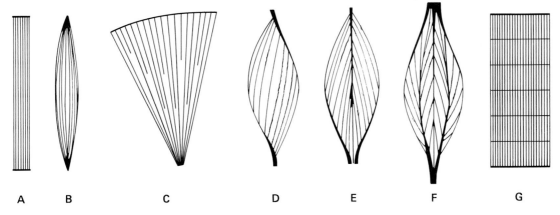

A   B   C   D   E   F   G

Figure 7-2
Arrangement of muscle fibers in muscles of different shape. **A**, Straplike muscle; **B**, spindle-shaped muscle; **C**, fan-shaped muscle; **D**, pennate muscle; **E**, bipennate muscle; **F**, multipennate muscle; **G**, segmented muscle. *(A, B, D, E, and* F, *redrawn from K. M. Dyce, W. O. Sack, and C. J. G. Wensing,* Textbook of Veterinary Anatomy, *Philadelphia, W. B. Saunders Company, 1987.)*

these are separated by at least one joint. Muscles that pass over two or more joints are called **biarticular muscles** or **polyarticular muscles,** respectively.

Muscles attach to skeletal elements always via **tendons,** which are made of densely and regularly arranged connective tissue fibers (mostly collagen). The connective tissue of these tendons is continuous with the connective tissue that surrounds and pervades the muscles. The ends of the muscle fibers are anchored to the connective tissue of the tendons by **myotendinous junctions.** The tendons in turn are anchored to skeletal elements by a blending of their connective tissue fibers with those of the periosteum or perichondrium and by penetration of the matrix of the skeletal elements. Tendons can assume a great variety of shapes, depending on the shape of the muscles to which they attach; if they are sheetlike, however, they are called **aponeuroses.** If a tendon is so short as to be barely visible with the naked eye, we speak of a **fleshy attachment** of the muscle, in contrast to a **tendinous** or **aponeurotic attachment.**

Besides muscle envelopes, tendons, and aponeuroses, other types of connective tissue structures are part of the muscle-bone system. A **fascia** is usually a dense sheet of connective tissue that invests muscles (e.g., as epimysium), muscle groups, and other organs; it can also serve as attachment site for certain skeletal muscles. A **ligament** is usually a band of connective tissue that connects skeletal elements to one another. **Loose connective tissue** (often incorrectly called "fascia") fills out spaces between muscles, nerves, and blood vessels. It is often associated with **adipose tissue** (or **fat**) and is usually picked away during the dissection process.

As we proceed now to consider the actions and functions of muscles, it is most important to keep in mind that as a muscle contracts and its myofibrils are shortening, the muscle as a whole generates tension that is equal at both of its ends. How this force affects the skeletal elements to which the muscle attaches is determined, however, by a multitude of factors in addition to the structure and physiology of the contracting muscle itself. One of these factors is the magnitude of the **load** (or the weight or external force counteracting the muscle force) placed on the skeletal elements to which the muscle attaches. If this load is smaller than the tension generated by the contracting muscle, the muscle will shorten and move one or both of its attachments; the process is called an **isotonic contraction.** If the load is as large as the tension generated by the contracting muscle, the muscle will still generate tension (i.e., it will become "tense"), but it will not shorten, and no movement will occur; the process is called an **isometric contraction.** And if the load is greater than the tension generated by the contracting muscle, the muscle will be extended as it generates tension; this process occurs, for example, in many mammalian limb muscles during locomotion and is called a **negative work contraction.**

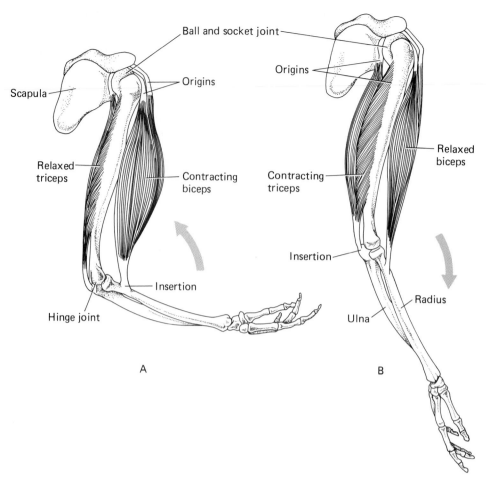

Figure 7-3
Lateral view of the biceps and triceps muscles in the human arm, as examples of muscular antagonists. **A**, Flexion; **B**, extension of the forearm through an angular rotation at the elbow joint. *(Redrawn from Villee et al., and Dorit, Walker, and Barnes, Zoology.)*

Although a contracting muscle exerts an equal force on both skeletal elements to which it attaches, during isotonic muscle contractions one element is usually more mobile than the other. The muscle attachment to the more mobile element, which is often also the more distal one, is called the **insertion** of a muscle; the muscle attachment to the more stationary element, which is often also the more proximal one, is called the **origin** or **head** of a muscle (see also Figs. 7-3 and 7-4). A single muscle may have more than one place of origin and more than one place of insertion (e.g., the mammalian biceps or triceps). Multiple fleshy attachments of a muscle are sometimes also called **slips.** It is important, however, to realize that particular skeletal elements may be stationary or mobile depending on the position of the animal. For example, in a walking cat, the serratus ventralis originates from the ribs of the stationary thorax and inserts on the moving scapula, but in a heavily breathing standing cat, the scapula can be more stable than the ribs of the thorax (see Fig. 7-24). Thus, the distinction between the origin and insertion of a muscle is often arbitrary. In doubtful cases, it is preferable to use the neutral term of **attachment** for all ends of a muscle.

A number of other factors determine or at least influence the action of contracting muscles. For example, while all muscles generate tension between their attachments as they contract, the direction of this tensile force is determined by the position and orientation of the muscles relative to the joints they span. Whether a muscle passes over the front or the back of a joint and whether it crosses a joint perpendicularly or diagonally fundamentally affects its actions on the skeletal elements to which it attaches (compare Figs. 7-3 and 7-4). As another example, the action of a contracting muscle is determined by the number of joints it spans, since a contracting muscle affects all the

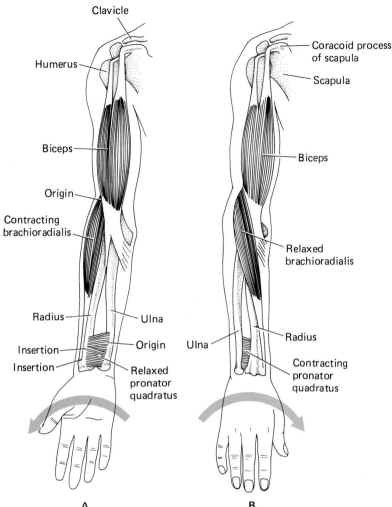

Figure 7-4
Frontal view of the brachioradialis and pronator quadratus muscles in the human arm, as examples of muscular antagonists. **A**, Supination; **B**, pronation of the hand through an axial rotation of the radius at the elbow joint. *(Redrawn from Young and Hobbs.)*

joints it crosses. As still another example, the shape and structure of the joints and their surrounding ligaments play a crucial role in controlling the direction of movements of skeletal elements. Whereas a **ball-and-socket joint** (e.g., the hip joint in mammals) allows a great versatility of movements, a **hinge joint** (e.g., the elbow joint in mammals) allows movement only in one plane. There are many other types of joints that allow different degrees of mobility.

The action of a particular muscle is usually also affected by the actions of muscles contracting at the same time. Movements of skeletal structures are rarely brought about by the contraction of a single muscle; in general, several muscles contract together or in overlapping sequences to move a skeletal element or to stabilize a joint. Muscles that work together for a particular movement, such as the flexion of the elbow joint, are called **synergists.** In a group of synergistic muscles, some muscles may stabilize a joint to allow movement only in one particular direction, while other muscles may generate the movement; some muscles may only initiate a movement, while other muscles may become active only as the skeletal element begins to move. Because muscles can only shorten to bring about a movement but cannot actively lengthen again, they must be extended by the contraction of other muscles, called **antagonists,** which cause the opposite movement (see Figs. 7-3 and 7-4). Antagonistic muscles do not necessarily need to have exactly opposite attachments relative to a particular joint (see, for example, Fig. 7-4). As with synergists, antagonists occur in groups of several muscles rather than as single muscles.

Although most natural movements of the skeletal system are quite complex, it is possible to identify several basic types of movements, which can combine to make up complicated motions. When using a term describing a basic type of movement, it is necessary to be very specific with respect to the identity of the moving structure and the axis and direction of the movement. A **flexion** is a movement that reduces the angle formed between skeletal elements, such as between the humerus and forearm (Fig. 7-3A); it is an angular rotation about a joint, in this case, about the elbow joint. Flexions can occur in different directions; for example, the head or the vertebral column can be flexed (or bent) forward (i.e., ventrally) or sideways (i.e., laterally). An **extension** is the opposite of a flexion and increases the angle formed between skeletal elements (Fig. 7-3B).

An **adduction** is a movement that brings a limb closer to a point of reference (e.g., the midventral line of the body), as during an angular rotation of the humerus about the shoulder joint towards the thorax with the help of the contracting pectoralis muscle (consult Fig. 7-24). An **abduction** is the opposite movement, rotating a limb away from a point of reference, such as the midventral line of the body. A **protraction** is a movement that brings a limb forward in a tetrapod, by rotating the humerus or femur anteriorly about the shoulder or hip joint, respectively. A **retraction** is the opposite, backward movement of a tetrapod limb. An **axial rotation** (in contrast to the angular rotations discussed previously) is a circular movement around a longitudinal axis, such as when the atlas, together with the head, rotates around the dens of the axis in a mammal (see also p. 108).

Supination and pronation are special cases of axial rotations. A **supination** is the lateral movement of the distal end of the radius around the long axis of the forearm, so that the palm of the hand (or the paw) faces forward or upward (Fig. 7-4A). A **pronation** is the opposite, medial movement of the radius, so that the palm of the hand (or paw) faces backward or downward (Fig. 7-4B).

The Classification of Muscles into Groups

The muscular system differs significantly among vertebrates and, in particular, between fishes, amphibians, and mammals (Table 7-1). In the course of evolution, muscles can (and did) undergo many modifications. For example, muscles may change their size and shape, divide into two or more muscles, fuse with other muscles, disappear altogether, or alter their internal architecture by modifying, increasing, or reducing intramuscular tendinous septa; they can also move their origins and insertions and thereby modify their actions. Such changes can usually be traced relatively easily for muscles in different representatives of the same vertebrate class. But the evolutionary processes responsible for the appearance of very different vertebrate life forms (e.g., fishlike versus tetrapod vertebrates) have affected the skeletal musculature often so deeply that the homology of muscles in representatives of different vertebrate classes is not always easily recognized, especially since the number of individually identifiable muscles varies greatly among classes (Table 7-1). In order to make meaningful comparisons between classes, we need to subdivide the muscular system into comparable parts in a manner that applies to all vertebrates.

One approach that has been very useful is based on the embryonic origin of muscles. Individual muscles develop always from the same, clearly identifiable embryonic tissues. Because the arrangement and organization of these embryonic tissues are very similar in all vertebrates at early stages of their embryonic development, we assume that muscles that have formed from the same particular embryonic structure can be clustered into muscle groups that are directly comparable in all vertebrates. Muscle groups that are defined by their embryonic origin are innervated usually by nerves that arise from the same or comparable divisions of the nervous system in all vertebrates (Table 7-1). Thus, the nerve supply of a muscle provides a useful clue to its embryonic

(*Text continues on p. 143*)

Table 7-1 Major Muscles of Vertebrates

Showing the major muscles of the vertebrates studied in the laboratory, the groups to which they belong, the main pattern of innervation, and the probable homologies.

Group	*Squalus*	*Necturus*	Mammal
		A. Somatic Muscles	
		I. **Axial Muscles**	
		1. Extrinsic Ocular Muscles	
From first somite (oculomotor nerve)	Dorsal rectus	Dorsal rectus	Levator palpebrae superioris / Dorsal rectus
	Ventral rectus / Medial rectus / Ventral oblique	Ventral rectus / Medial rectus / Ventral oblique	Ventral rectus / Medial rectus / Ventral oblique
Second somite (trochlear nerve)	Dorsal oblique	Dorsal oblique	Dorsal oblique
Third somite (abducens nerve)	Lateral rectus	Lateral rectus / Retractor bulbi	Lateral rectus / Retractor bulbi

2. Epibranchial Muscles (Dorsal Rami of Occipital and Anterior Spinal Nerves)

Group	*Squalus*	*Necturus*	Mammal
	Epaxial and hypaxial portions of the myomeres dorsal to gill region	Anterior part of the dorsalis trunci	Anterior part of the epaxial muscles

3. Hypobranchial Muscles (Ventral Rami of Spino-occipital Nerves in Anamniotes or of Hypoglossal Nerve and Cervical Plexus in Amniotes)

Group	*Squalus*	*Necturus*	Mammal
Prehyoid muscles	Coracomandibular	Genioglossus / Geniohyoid	Lingualis proprius / Genioglossus / Hyoglossus / Styloglossus / Geniohyoid
Posthyoid muscles	Rectus cervicis (or coracoarcual + coracohyoid) / Coracobranchials	Rectus cervicis / Omoarcuals / Pectoriscapularis	Sternohyoid / Sternothyroid / Thyrohyoid / _____ / _____

4. Axial Muscles of the Trunk (Spinal Nerves)

Group	*Squalus*	*Necturus*	Mammal	
Epaxial (dorsal rami)	Epaxial portion of myomeres	Interspinalis / Dorsalis trunci	Interspinalis / Intertransversarii / Occipitals / Multifidi / Spinalis / Semispinalis	Transverso-spinalis
			Longissimus / Splenius	Longissimus
			Iliocostalis	Iliocostalis

Table 7-1   Major Muscles of Vertebrates (Continued)

Group	*Squalus*	*Necturus*	Mammal	
Hypaxial (ventral rami)	Hypaxial portion of myomeres	Subvertebralis	Longus colli / Psoas minor / Quadratus lumborum	Subvertebral
		Levator scapulae (opercularis)	Omotransversarius	
		Thoraciscapularis	Serratus ventralis (part)	
		External oblique	Serratus ventralis (part) / Rhomboideus cervicis et thoracis / Rhomboideus capitis / Serratus dorsalis / Scalenus / Rectus thoracis / External oblique / External intercostals	Lateral
		Internal oblique	Internal oblique / Internal intercostals	
		Transversus	Transversus abdominis / Transversus thoracis	
		Rectus abdominis	Diaphragm muscles / Rectus abdominis	Ventral

II. **Appendicular Muscles** *(Ventral Rami of Spinal Nerves)*

1. **Pectoral Muscles**

Group	*Squalus*	*Necturus*	Mammal
Dorsal group	Abductor (extensor)	Latissimus dorsi	Cutaneus trunci (part) / Latissimus dorsi / Teres major
		Subcoracoscapularis	Subscapularis
		Scapular deltoid	Scapulodeltoid / Acromiodeltoid
		Procoracohumeralis	Cleidobrachialis / Teres minor
		Triceps	Triceps brachii / Tensor fasciae antebrachii
		Forearm extensors	Forearm extensors
Ventral group	Adductor (flexor)	Pectoralis	Cutaneus trunci (part) / Pectoralis complex
		Supracoracoideus	Supraspinatus / Infraspinatus
		Coracoradialis	Biceps brachii (part)
		Humeroantebrachialis	Biceps brachii (part) / Brachialis
		Coracobrachialis	Coracobrachialis
		Forearm flexors	Forearm flexors

2. **Pelvic Muscles**

Group	*Squalus*	*Necturus*	Mammal
Dorsal group	Abductor (extensor)	Iliotibialis	Sartorius
		Puboischiofemoralis internus	Iliacus / Psoas major / Pectineus / Vasti?
		Ilioextensorius	Rectus femoris
		Iliofibularis	Gluteus superficialis
		Iliofemoralis	Tensor fasciae latae? / Gluteus medius / Gluteus profundus
		Shank extensors	Shank extensors

Table 7-1 Major Muscles of Vertebrates (Continued)

Group	*Squalus*	*Necturus*	Mammal
Ventral group	Adductor (flexor)	Puboischiofemoralis externus	{ Obturator externus / Quadratus femoris
		(Adductor femoris. In *Necturus* this is not clearly separated from the preceding)	Adductor brevis et longus
		Pubotibialis	Adductor magnus
		Ischiofemoralis	{ Obturator internus / Gemelli
		Caudofemoralis	{ Crurococcygeus (absent in some mammals) / Piriformis
		Puboischiotibialis	Gracilis
		Ischioflexorius	{ Semimembranosus / Semitendinosus / Biceps femoris? / Abductor cruris caudalis? (absent in some mammals)
		Shank flexors	Shank flexors

B. Visceral Muscles
I. Branchiomeric Muscles

Group	*Squalus*	*Necturus*	Mammal
Mandibular muscles (trigeminal nerve)	{ Adductor mandibulae / Levator palatoquadrati / Spiracularis / Preorbitalis	Levator mandibulae (3 parts)	{ Masseter / Temporalis / Pterygoids / Tensor veli palati / Tensor tympani
	Intermandibularis	Intermandibularis	{ Mylohyoid / Anterior digastric
Hyoid muscles (facial nerve)	{ Levator hyomandibulae / Dorsal constrictor	Depressor mandibulae Branchiohyoideus	{ Stapedius / Platysma and facial muscles (part)
	{ Ventral constrictor / Interhyoideus	Interhyoideus Sphincter colli	{ Platysma and facial muscles (part) / Posterior digastric / Stylohyoid
Branchiomeric muscles of remaining arches (glossopharyngeal, vagus, and, in amniotes, spinal accessory nerves)	Cucullaris	{ Cucullaris / Levatores arcuum	{ Trapezius complex / Sternocleidomastoid complex
	Interarcuals	————	————
	Branchial adductors	————	————
	Superficial constrictors and interbranchials	{ Dilatator laryngis / Subarcuals / Transversi ventrales / Depressores arcuum	Intrinsic muscles of the larynx and certain muscles of the pharynx

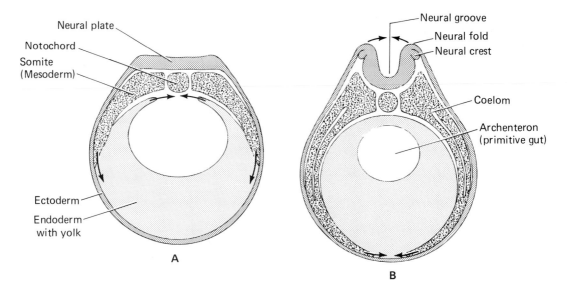

Figure 7-5

Diagrams illustrating the embryonic differentiation of the mesoderm in the trunk region. **A** through **F**, Cross sections through embryos with a moderate amount of yolk, such as in amphibians; the arrows indicate the direction of growth of embryonic tissues. **A**, Early stage of mesoderm formation; **B**, early stage of coelom formation; **C**, differentiation of the mesoderm into the epimere, mesomere, and hypomere; **D**, differentiation of the epimere into the dermatome, myotome, and sclerotome; **E**, differentiation of the dermatome into the dermis of the integument, of the myotome into the somatic musculature, and of the sclerotome into the axial skeleton; **F**, final differentiation of the mesodermal organs and division of the somatic musculature into epaxial and hypaxial muscles. (A through D, *Redrawn with permission from M. Hildebrand,* Analysis of Vertebrate Structure. *Copyright © 1974 and 1982 by John Wiley & Sons, Inc.; E and F, modified from Hyman.*)

origin and group affiliation, especially in those animals the embryonic development of which is not known.

In order to understand the rationale behind the system of muscle groups almost universally used in comparative anatomy, it is necessary to review the embryonic development of the musculature, which is intimately tied to the development and differentiation of the mesoderm (Figs. 7-5 and 7-6).

In the early stages of embryonic development, mesodermal tissue appears as **somites** along both sides of the notochord under the neural plate (Fig. 7-5A). The somites define the segmental organization of the vertebrate body. As the mesoderm grows ventrally between the ectoderm and endoderm, the mesoderm is split into an external and internal layer by the appearance of a space, the **coelom** (Fig. 7-5B). At the same time, the mesoderm differentiates into a large dorsal portion, the **epimere** (or somite), a small middle portion, the **mesomere** (or nephrogenic ridge), and a long ventral portion, the **hypomere** (or lateral plate) (Fig. 7-5C). In the latter, the segmentation disappears. The epimere subdivides into the lateral **dermatome,** the intermediate **myotome**[12], and the medial **sclerotome;** the mesomere will give rise to the urogenital organs; and the two layers of the hypomere, the lateral **somatic layer** and the medial **visceral layer,** are separated by the enclosed coelom (Fig. 7-5D). The dermatome expands under the ectoderm, loses its segmental organization, and differentiates into the dermis of the integument, including the dermal muscles (e.g., the arrector pili muscles of mammalian hair) and dermal skeletal structures (Fig. 7-5E and **F**). The sclerotome surrounds the notochord and neural tube and differentiates into the somatic part of the skeleton (p. 52). The myotome grows ventrally between the dermatome and the somatic layer of the hypomere and is subdivided by the horizontal skeletogenous septum into the epaxial and hypaxial musculature. The hypaxial musculature also gives rise to the appendicular musculature. The visceral layer of the hypomere differentiates into the smooth visceral musculature of the gut and inner

---

[12]See footnote 2, p. 7.

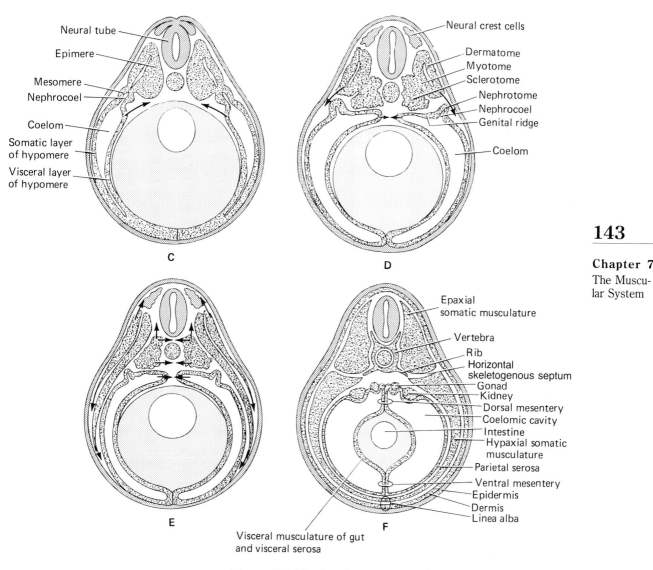

Figure 7-5 (Continued)

organs, the cardiac musculature of the heart, and the visceral serosa that covers the
inner organs and forms the mesenteries suspending them in the coelomic cavities. The
somatic layer of the hypomere becomes the parietal serosa that lines the coelomic
cavities, and provides the connective tissue associated with the appendicular
musculature.

In the head, or cephalic, region of the embryo, the developmental differentiation of
the skeletal musculature is somewhat different from that observed in the trunk region
described previously. The process is best illustrated by the embryonic development in
a shark (Fig. 7-6**A** and **B**).[13] In the cephalic region, the embryonic mesomeres are not
formed, and the unsegmented hypomere does not develop a coelom. The innermost
layer of the hypomere differentiates into the smooth pharyngeal constrictor
musculature, but the bulk of the cephalic hypomere develops into the striated
branchiomeric musculature as it becomes subdivided by visceral arches and

---

[13]It is important to realize that this description is based on interpretations of often incomplete, older
studies that were performed without the benefit of contemporary technical facilities. Some recent
studies (see, for example, footnotes 14 and 15) have raised doubts about the accuracy of some of
these interpretations. These interpretations may have to be revised as more data on this subject
become available. The traditional description has been retained here for its didactic value as an
integral part of the comprehensive theory concerning the somatic and visceral divisions of the
vertebrate body.

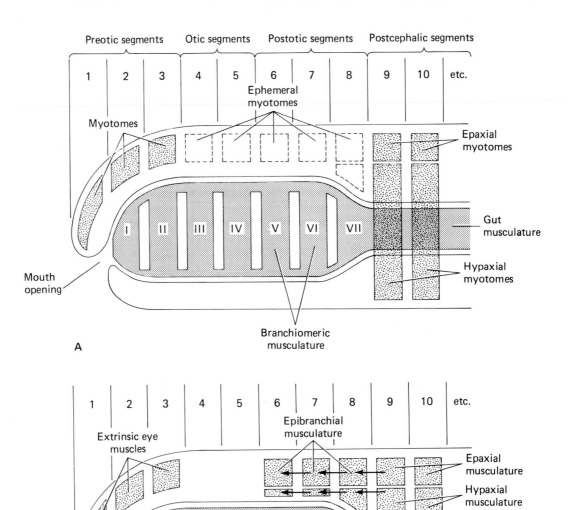

Figure 7-6
Diagrams illustrating the embryonic differentiation of the musculature from mesoderm in the cephalic region. **A** and **B**, Transparent lateral views of a shark embryo. **A**, Early embryonic stage; **B**, adult stage. *1* through *10*, Somatic segments; *1*, premandibular segment; *2*, mandibular segment; *3*, hyoid segment; *I* through *VII*, divisions of the branchiomeric musculature. Note that the visceral division *I* corresponds to the somatic segment *2*, and so on. *(Modified from Smith.)*

differentiates into the jaw, hyoid, facial, cucullaris, and laryngeal musculature.[14] Most original embryonic myotomes in the head region are ephemeral, except for the first three, which differentiate into the extrinsic eye muscles.[15] As the embryo grows, the

---

[14]Noden (1983, 1984) showed that in birds the branchiomeric musculature develops from cephalic myotomes, called somitomeres, and receives its connective tissue component from neural crest cells.

[15]As Starck (1979) explains, it is not quite clear whether the three embryonic tissue structures giving rise to the extrinsic eye muscles represent true myotomes, but see Meier and Tam (1982) for a new interpretation.

expanding branchial region pushes caudally against the proliferating postcephalic myotomes and divides these into a dorsal **epibranchial** portion and a ventral **hypobranchial** portion. Because the horizontal skeletogenous septum is situated slightly above the branchial region, the epibranchial musculature consists not only of epaxial muscle tissue but also includes a small part of the hypaxial muscle tissue (Fig. 7-6B). In chondrichthyan fishes, the epibranchial musculature grows forward into the cranial region, but in the other gnathostome vertebrates it does not develop further and is lost. The hypobranchial musculature projects forward below the branchial region and differentiates into lingual and cervical muscles.

The head of amniotes comprises more segments than that of anamniotes; the postcephalic segments of anamniotes become incorporated as occipital segments into the head of amniotes.

The Study of Muscles

As methods of body support changed during the transition from fish to tetrapod, and movements of the body and its parts became more complex during the evolution of vertebrates, the muscles became more numerous. It is not possible, in a course of this scope, to study all of them. Details of the muscles confined to the hand and foot have been omitted, as have those of the tail and perineum, but other groups are described with reasonable completeness. Insofar as possible, the muscles are described by the groups defined earlier. This will permit a further selection by the instructor of the muscles to be studied if time is short.

Muscle names are often confusing to the student, but they can be very helpful, for they describe one or more distinguishing features of a muscle: its shape (trapezius); location (intermandibularis); location and shape (dorsal rectus); location and fiber direction (external oblique); primary action (adductor mandibulae); or attachments in certain species (coracohyoid).

Confine your dissection of the muscles to one side of the body and, when cutting open the body, do so on the opposite side. Most of the figures show the muscles on the left side. Exercise care so as not to injure the underlying muscles when removing the skin. It is best to try to peel the skin off by tearing underlying connective tissue with your fingers or with forceps. If you must cut with a scalpel, cut toward the underside of the skin, not toward the muscles. After the skin has been removed, the muscles must be carefully separated from one another. This involves cleaning off the overlying connective tissue with forceps until you can see the direction of the muscle fibers. Ordinarily, the fibers of one particular muscle are held together by a sheath of connective tissue (epimysium), and all run in the same general direction between common attachments. The fibers of an adjacent muscle will be bound together by a different sheath and will have different muscle fiber orientations and attachments. This will give you a clue as to where to separate one from the other. Separate the muscles by picking away, or tearing, the connective tissue between them with forceps, watching the fiber direction as you do so. Do not try to cut muscles apart. If the muscles separate as units, you are dissecting correctly, but if you are exposing small bundles of muscle fibers, you are probably separating the parts of a single muscle.

It is best to expose and separate many of the muscles of a given region before attempting to identify them. It will be necessary sometimes to cut through a superficial muscle to expose deeper muscles. In such cases, be sure to expose first the entire muscle by identifying and freeing its borders, which run parallel to the muscle fiber direction. Then cut across the middle of the muscle (not across one of its tendons or aponeuroses!) at a right angle to the direction of its fibers; depending on the arrangement of the muscle, you may have to curve the cut in order to keep it at the desired angle relative to the muscle fibers. By doing this, you can reflect the cut ends to expose deep muscles; the cut ends can easily be put together again when necessary. If you have to cut through several layers of muscles, make the transversal cuts at different levels, so that the various cut ends cannot be confused.

The dissection will be more meaningful if you have a skeleton in front of you on which to visualize the points of attachment of the muscles as you proceed with the dissection. You may also be able to infer some basic muscle actions by observing the attachments of muscles and visualizing what structures would be brought together if the muscle shortened.

FISHES

The muscles of *Squalus* are a good example of the condition of the musculature in ancestral piscine vertebrates. The groups of muscles can easily be recognized and studied in the adult, for the groups have not lost their identity through the migration of muscles, as they have to some extent in tetrapod vertebrates.

Axial Muscles

◆   (A) TYPICAL MYOMERES OF THE TRUNK AND TAIL

The bulk of the musculature of fishes belongs to the axial group of somatic muscles, and the most conspicuous of these are the muscles of the trunk and tail. Remove a wide strip of skin from the posterior portion of the tail, and another strip from the front of the trunk between the pectoral and anterior dorsal fins (Fig. 7-7). These strips should extend from the middorsal to the midventral lines of the body. It will be difficult and time-consuming to

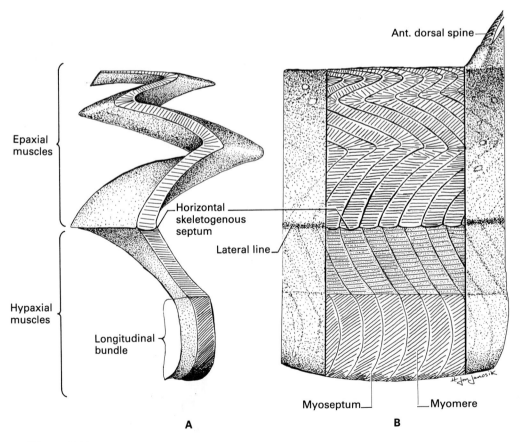

A   **B**

Figure 7-7
Lateral views of trunk muscles between the pectoral and anterior dorsal fins of a dogfish; cranial is towards the left. **A**, Diagram of an individual myomere; **B**, surface of a group of myomeres in situ.

separate the skin from the underlying muscles because many muscle fibers attach directly to the dermis (see p. 148). This is especially true for the muscle fibers of the ventralmost longitudinal bundle (Fig. 7-7). The best technique is to reflect the skin as far as possible and scrape the muscle fiber bundles carefully away from the internal skin surface with a scalpel.

Notice that the trunk musculature consists of muscle segments, or **myomeres,** which develop from the embryonic myotomes, and that the segments are separated from each other by connective tissue sheets, the **myosepta.** The myomeres are bent in a complex zigzag fashion and each is divided into dorsal and ventral portions by a longitudinal connective tissue sheet, the **horizontal skeletogenous septum** (see Fig. 5-1, p. 96), which lies deep to the lateral line. The dorsal portions of the myomeres constitute the **epaxial musculature;** the ventral portions constitute the **hypaxial musculature.** This division of the trunk musculature is found in all gnathostomes but is absent in the Agnatha. In addition to their divisions into epaxial and hypaxial masses, the myomeres can be divided, at the apices of the zigzags, into **longitudinal bundles.** Notice that the muscle fibers within the two somewhat darker longitudinal bundles above and below the horizontal skeletogenous septum are oriented longitudinally, whereas the muscle fibers in the other longitudinal bundles are oriented obliquely.

Each myomere is a complex entity, as can be seen in Figure 7-7. The internal part of each V-like fold of a myomere is cone-shaped and extends further cranially or caudally than the fold does at the body surface. An analogous arrangement can be made by cutting a V-shaped notch out of the top half of several conical paper cups, and then fitting them together. The cone-within-a-cone nature of the folds can be seen if you make a cross section of the tail. In such a section, the overlapping V-shaped folds of several myomeres appear as a series of nearly concentric rings (see Fig. 5-1, p. 96). As described on page 95, the relationships of the myomeres to the vertebral column and to the skeletogenous septa also show in this view. In the abdominal region, the ventral band of connective tissue that separates the myomeres of opposite sides is called the **linea alba.**

The darker, longitudinally oriented muscle fibers near the horizontal skeletogenous septum are **red fibers.** They contain considerable amounts of myoglobin, which facilitates the transfer of oxygen from the blood, and their metabolism is aerobic. They are used at normal cruising speeds, and their rate of contraction is only as fast (and the duration of action only as long) as their oxygen supply allows. They do not fatigue. When sudden bursts of speed are needed, the **white fibers** become active. They can contract very rapidly for short periods of time because their metabolism is anaerobic. Eventually they do fatigue and go into oxygen debt. Lactic acid then accumulates, but this is metabolized after the burst of speed.

To understand myomere structure, it is necessary to know how a fish swims. The dogfish has an **anguilliform** (eel-like) mode of locomotion in that the waves of curvature are pronounced and affect most of the trunk and tail (Fig. 7-8). The waves increase in amplitude as they approach the tail, which therefore moves to a greater extent than do the more anterior parts of the body and thus generates the most thrust. As the tail pushes laterally and caudally against the water, the water, in accordance with Newton's third law of motion, pushes back with an equal but opposite reaction force against the tail (Fig. 7-8). This reaction force can be resolved into a lateral and a forward component. As the shark moves its trunk and tail from side to side, the lateral components cancel one another, but the forward components add up and propel the shark forward.

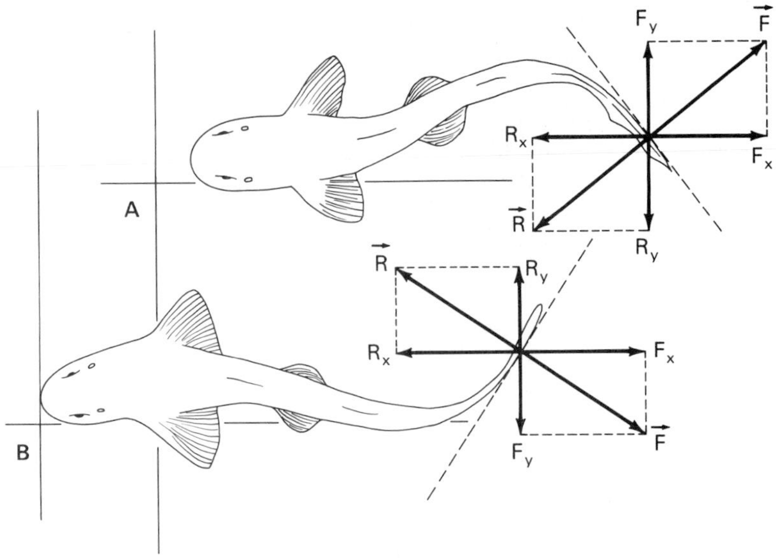

Figure 7-8
Dorsal views of a swimming shark drawn from individual pictures of a film. The interval between
A and **B** is 0.6 seconds. Note that the side-to-side movements are greater for the tail than for the
anterior part of the trunk. F, Force of the tail on the water; F_x, backward component; F_y, lateral
component; R, reaction force of the water on the tail; R_x, forward component; R_y, lateral
component. *(Modified from Gray, 1968; and Gans, 1974.)*

Segmented myomeres are well suited to generate the curvatures that sweep down
the body. Since the vertebral column prevents shortening of the body, contraction of
the myomeres of one side pulls myosepta together and bends the body; on the
opposite side, the myomeres are simultaneously extended, and the body is bent and
pushed against the water. The contraction of the myomeres is a highly synchronized
process. Myomeres on one side of the body contract sequentially in a wavelike
manner, starting anteriorly and continuing backwards. This contraction wave is soon
followed by another one on the opposite side of the body, so that the contraction
waves on the opposite sides of the body are out of phase with one another. At any one
time, a given intervertebral joint is affected by several myomeres; the most caudal
myomere may be just starting to contract, the middle myomeres may be reaching
maximal tension, and the cranial myomeres may be starting to relax. There is a
smooth flow of tension generation.

As we have seen, many of the superficial fibers of the myomeres attach onto the
dermis, which consists of layers of collagen fibers that follow helical paths along the
body (Wainwright, Vosburgh, and Hebrank, 1978). The skin acts as an exotendon;
some of the force generated by cranial myomeres is transferred by this tendon to the
tail. This, together with the more extensive folding of myomeres near the tail, causes
the more extensive movements of the tail. The exotendon also helps conserve energy.
The tendon is stretched on the convex side of a body curve, and this stored elastic
energy is released when this side of the body becomes concave.

In order to be able to swim horizontally, or even upwards, instead of sinking
continuously because of their weight, sharks take advantage of buoyancy, of which we
can distinguish two types. **Static buoyancy** is based on Archimedes' principle. A
shark, like any other immersed body, pushes downwards against the water with a
force that is equal to the shark's weight (i.e., the volume of the shark multiplied by the
specific weight of its body) (Fig. 7-9). (Specific weight is equal to the weight of a mass
divided by its volume.) The water pushes back against the shark with an upward
reaction force, the **buoyant force.** (Buoyant force is equal to the volume of the water
displaced by the shark multiplied by the specific weight of the water.) If the specific
weight of the shark were equal to that of the surrounding water, the weight of the
shark would be cancelled by the buoyant force, and the shark could simply float at any

level of the water. Animal tissues, however, are generally denser than water, but most fishes have evolved a way to reduce the difference between their weight and the buoyant force. Sharks, in particular, have compensated for the greater density of most of their tissues by not calcifying their skeleton and by storing in the liver large amounts of oil that has a lower density than water. (Density is equal to mass divided by volume.)

Although some sharks do achieve neutral buoyancy in this way (or, in other words, they equalize the average specific weight of their tissues with that of the surrounding water), most species do not do so, and their weight remains about 1% to 4% greater than the buoyant force of the surrounding sea water. This remaining surplus weight can be compensated for by **hydrodynamic buoyancy**, or **lift** (Fig. 7-9). This phenomenon is based on Bernoulli's theorem, according to which the total energy per unit volume at any two points of a fluid in smooth flow is constant. In the case of a streamlined swimming shark, this means that at any point of the surrounding water, the sum of the water's pressure and kinetic energy must be equal. (Kinetic energy is equal to one-half the mass times the square of the velocity.) In a shark, the dorsal surfaces of the head and pectoral fins are more curved and hence larger than those of the ventral surfaces. As a result, the velocity of the water flowing above the shark must be greater than that of the water flowing below the shark so that both water flows reach the end of the body or fin at the same time. As a consequence, the water pressure acting on the shark from below must be greater than the water pressure acting from above, and lift is created on the shark. This phenomenon is also known as the **hydrofoil effect.** As both upward forces generated at the head and pectoral fins act in front of the center of gravity, they tend to generate **torques,** or turning moments, about the center of gravity. (A torque is equal to force multiplied by the perpendicular distance from the center of rotation.) For the shark to remain horizontal, these torques must be compensated for by a torque in the opposite direction, such as the torque generated by a lift at the tail. [The exact mechanism of how the tail generates lift is not known at present; the asymmetry of the dorsal and ventral lobes of the tail fin, which is correlated with the degree of dorsoventral asymmetry of the body, does not seem to be a crucial factor by itself (Aleev 1969, Thomson 1990)].

In order for a shark to swim horizontally, the sum of the buoyant force and all lifts must be equal, but opposite, to the weight of the shark. This means that a shark may have to maintain a certain minimal speed to achieve this, since the magnitude of a lift is proportional to the swimming speed of the shark. A shark can change its swimming direction upward or downward by pointing its head slightly up or down and by adjusting the pitch of its pectoral fins as hydrofoils; or it can change its direction sideways by orienting its fins differently on each side.

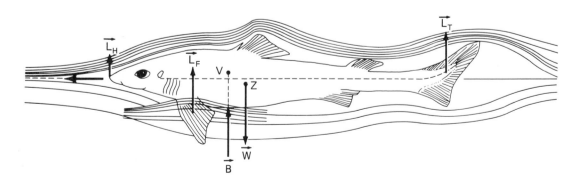

Figure 7-9
The interaction of the forces of hydrodynamic and static buoyancy preventing a swimming shark with a slightly higher density than the surrounding saltwater from sinking. B, Buoyant force of the water; W, weight of the shark; L_F, lift of the pectoral fin; L_H, lift of the head; L_T, lift of the tail fin; V, center of buoyancy; Z, center of gravity.

◆ **(B) EPIBRANCHIAL MUSCLES**

The axial musculature is interrupted in the head by the gill region but extends forward above and below the gills as **epibranchial** muscles and **hypobranchial** muscles, respectively. Remove a strip of skin dorsal to the gill region almost down to the gill slits, and you will see the epibranchial muscles as an extension of the epaxial musculature to the chondrocranium (see Fig. 7-11). Notice that the segmentation persists in this region. Certain of the deeper parts of this musculature, which are not easily seen, attach onto the tops of the branchial arches and on the ventral surface of the chondrocranium.

◆ **(C) HYPOBRANCHIAL MUSCLES**

◆ prehyoid muscles
    coracomandibular
◆ posthyoid muscles
    rectus cervicis
        coracoarcuals
        coracohyoids
        coracobranchials → ON DEMO

To see the hypobranchial muscles, which represent a forward extension of the hypaxial musculature, remove the skin on the entire underside region of the head from the pectoral region forward to the tip of the lower jaw, and laterally up to the gill slits and mandibular cartilages, so that the ventral surfaces of the latter are clearly exposed. A broad, thin, double-layered sheet of transverse muscle fibers covers the triangular area between the mandibular cartilages (Figs. 7-10 and 7-11). This sheet is part of the branchiomeric musculature (see p. 153), but will be studied now because you need to cut it to expose the underlying hypobranchial muscles.

Identify the midventral line by the presence of a longitudinal tendinous intersection, called a **raphe.** With a sharp scalpel, make a longitudinal cut 1 to 2 millimeters to the side of this raphe and about 2 millimeters deep. You will have cut deeply enough when you can see, upon separating the cut edges, an underlying muscle with longitudinally oriented fibers. The cut sheet of transverse musculature consists actually of two different muscles, namely the superficial **intermandibularis** and the deeper **interhyoideus.** Lift one cut edge of the muscle sheet to look at its internal surface; you will discover that the interhyoideus does not reach as far rostrally as the intermandibularis, and that its muscle fibers are more transversely oriented than the slightly obliquely oriented muscle fibers of the intermandibularis. Separate the two muscles, starting at the rostral edge of the interhyoideus and proceeding caudally and laterally until you reach the muscle attach-ments. The intermandibularis originates from the mandibular cartilage and the fascia covering the adductor mandibulae (Figs. 7-10 and 7-11) and inserts on the midventral raphe. Its caudal border can be lifted from the underlying musculature, because it is the most superficial muscle in this region. The interhyoideus originates from the ceratohyal; caudally it is continuous with the ventral hyoid constrictor (Fig. 7-11).

We can now proceed to study the hypobranchial muscles. The narrow midventral muscle (embryonically paired, but now fused to form an unpaired muscle), which is

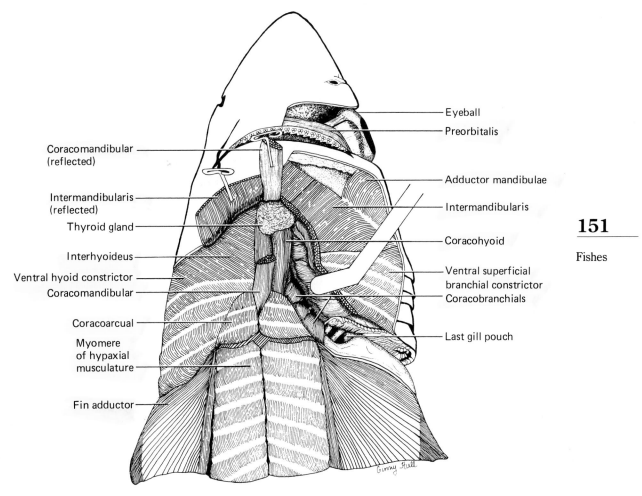

Figure 7-10
Ventral view of the head of *Squalus,* showing the hypobranchial and branchiomeric muscles.

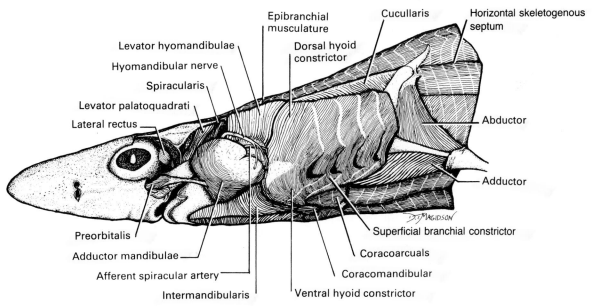

Figure 7-11
Lateroventral view of the superficial muscles of the pectoral appendages and head of *Squalus.*

exposed after the intermandibularis has been reflected, is called the **coracomandibular** (Fig. 7-10). It arises caudally from the surface of other muscles (see the following paragraph) and inserts on the tip of the lower jaw. It represents the **prehyoid** portion of the hypobranchial musculature, but in fishes this muscle commonly extends caudad to the hyoid, often as far as the coracoid. Cut through the caudal attachment of the coracomandibular and pull it forward. The dark or pink mass beneath its front end is the **thyroid gland.**

The paired muscles now exposed belong to the **posthyoid** portion of the hypobranchial musculature. The entire complex, which arises on the pectoral girdle and inserts on the hyoid arch, may simply be called the **rectus cervicis;** however, it is sometimes considered to be two pairs of muscles, for the fibers of the caudal half converge toward the midline, while those of the cranial half extend longitudinally to the hyoid arch. The caudal pair is called the **coracoarcuals;** the cranial pair is called the **coracohyoids.**

With scissors, make a cut from the ventral corner of the last gill slit to the caudolateral corner of the coracoarcual muscle, and then dissect forward, with two forceps, gradually separating the branchial region from the rectus cervicis. Be careful not to injure blood vessels unduly, but separate deeply enough to expose a deep group of muscles, the **coracobranchials,** which arise from the coracoid bar, from the wall of the pericardial cavity beneath the coracoarcuals, and from the dorsal surface of the rectus cervicis itself. A sheet of connective tissue covers the coracobranchials and must be pulled off to reveal the muscle fiber direction. The five coracobranchial muscles insert onto the ventral ends of the branchial arches (Fig. 7-10). They can also be seen in sections of the head (Fig. 10-8, p. 302, and Fig. 11-13, p. 356).

All the hypobranchial muscles are involved in opening the mouth (the coracomandibular) or in expanding the oropharyngeal cavity (the mouth cavity and pharynx) during feeding and inspiration of the respiratory current of water (see p. 305).

♦   **(D) EXTRINSIC MUSCLES OF THE EYE**

The segments rostral to the epibranchial muscles disappear during embryonic development except for the most rostral three, and these give rise to the extrinsic muscles of the eye. These muscles extend from the wall of the orbit to the surface of the eyeball, and are responsible for the movements of the eyeball. We will consider them when we study the eye (p. 220).

Appendicular Muscles

♦   **(A) MUSCLES OF THE PECTORAL FIN**

   ♦   adductor
   ♦   abductor

The appendicular group of somatic muscles is small and simple in most fishes, for the paired fins are used primarily to give the body lift, stability, and maneuverability, not to generate propulsive thrust. Remove the skin on the dorsal, ventral, and cranial surfaces of the base of the pectoral fin. The appendicular muscles of *Squalus* consist of a single dorsal and a single ventral muscle mass, each of which arises from the pectoral girdle and inserts by slips onto either the dorsal or ventral surface of the pterygiophores of the fin as far distal as the radial cartilages. The dorsal muscle mass—the **abductor** or extensor—elevates the fin and pulls it caudally (Fig. 7-11). The ventral muscle mass—the **adductor** or flexor, which also spreads onto the cranial surface of the fin—depresses the fin and pulls it forward.

♦ **(B) MUSCLES OF THE PELVIC FIN**

♦ adductor
♦ abductor

Remove the skin from the base of the pelvic fin. In a female, the dorsal muscle mass, the **abductor,** arises from the surface of the caudal trunk myomeres, from the iliac process, and from the metapterygium. It inserts by slips onto the radial cartilages. Most of the ventral mass—the **adductor**—is divided into proximal and distal portions. The former arises from the puboischiadic bar and inserts on the metapterygium; the latter arises from the metapterygium and inserts on the radials.

The appendicular muscles of the male are fundamentally the same, but portions of both the dorsal and ventral mass extend onto the clasper as distinct small muscles. Also, the adductor cannot be seen until you reflect a long muscular sac (the siphon, Fig. 12-7, p. 406) that lies between it and the skin. Do not remove the siphon completely, as it will be studied with the reproductive system.

Branchiomeric Muscles

♦ mandibular muscles
    adductor mandibulae
    levator palatoquadrati
    spiracularis
    preorbitalis
    intermandibularis
♦ hyoid muscles
    interhyoideus
    hyoid constrictor
    levator hyomandibulae

♦ branchial muscles
    superficial constrictors
    interbranchial
    branchial adductors
    interarcuals
    cucullaris

Very carefully remove the skin from the side of the gill region, jaws, and spiracle, and from the back and underside of the eye. Be especially careful when removing the skin from the areas between the gill slits; the superficial musculature is very thin and is easily overlooked, but try not to damage it. In the spiracular region, remove the skin, but do not otherwise injure the rostral surface of the spiracular valve (p. 42). After the skin is removed, carefully clean the area by cutting and picking away connective tissue.

Study the muscles, beginning with those of the mandibular arch. The large mass of muscle at the angle of the jaws is the **adductor mandibulae** (Fig. 7-11). It arises from the caudal part of the palatoquadrate (note the large adductor mandibulae process in this region on a skeleton, Fig. 4-4, p. 59) and inserts on the mandibular cartilage. Dorsal to this, and in front of the spiracle, is another muscular mass that arises from the side of the otic capsule and inserts on the palatoquadrate (Fig. 7-11). Careful dissection reveals that this mass is divided into a rostral **levator palatoquadrati** and a caudal **spiracularis.** The latter is on the cranial wall of the spiracular valve. Lift the eye, and a **preorbitalis** will be seen ventral to it. This muscle arises from the underside of the chondrocranium and inserts by a tendon on the mandibular cartilage. The last of the mandibular muscles, the **intermandibularis,** was cut through during the dissection of the hypobranchial muscles. Find it again.

Find also the **interhyoideus** again (p. 150). The **ventral hyoid constrictor** is located caudal to the interhyoideus and is continuous with it. The **dorsal hyoid constrictor** is partially separated from its ventral counterpart by a connective tissue plate; it lies above and behind the angle of the jaw. The **levator hyomandibulae** lies between the spiracle and the dorsal hyoid constrictor. It arises from the surface of the epibranchial musculature and otic capsule and inserts mainly on the hyomandibular cartilage; a few muscle fibers, however, extend onto the palatoquadrate. The dorsal and ventral hyoid constrictors are followed caudally by four **dorsal** and **ventral superficial branchial constrictors,** which are separated from one another by connective tissue intersections.

The levator muscles of the five branchial arches, which correspond to the levator hyomandibulae for the hyoid arch, have united with one another to form a triangular muscle, the **cucullaris** (partly homologous to the trapezius of mammals) (see Table 7-1), which lies dorsal to the superficial branchial constrictors. Observe that the cucullaris arises from the surface of the epibranchial musculature and passes caudally to insert on the pectoral girdle and last branchial arch. Completely separate the cucullaris and levator hyomandibulae from the epibranchial muscles, and push the branchial region ventrally. The large cavity exposed is the anterior cardinal sinus, a part of the venous system. Clean out the coagulated blood, if necessary, and remove the membrane that covers the branchial arches and their muscles. Push apart the epibranchial musculature from the branchial region as far as necessary to expose the dorsal surfaces of the branchial arches. Several nerves cross this surface; push them away to clear the view, but do not cut them, if possible. Medially, you will see small, ribbon-like muscles extending between the pharyngobranchial cartilages of the various branchial arches; these are the **dorsal interarcuals.** The **lateral interarcuals** are shorter, broader muscles extending between the epibranchial and pharyngobranchial elements of the individual branchial arches.

Open all the branchial pouches by cutting through the superficial branchial constrictors along a line parallel to the connective tissue intersections. Open them as wide as you can without cutting through the margins of the internal gill slits, which are communicating with the pharynx. In this way, you can mobilize and inspect the individual interbranchial septa (Fig. 7-12). Interbranchial septa are plates of mostly soft tissue. They are attached to the individual branchial arches and supported by the gill rays radiating from the branchial arches. (The last branchial arch lacks an interbranchial septum and the associated muscles.) The surfaces of the septa are covered with a thin epithelium and a mass of capillaries forming the respiratory gill lamellae (see also Fig. 10-11). Scrape away the gill

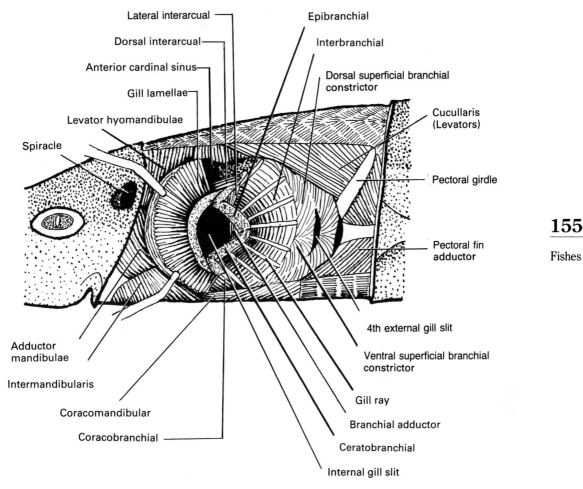

Lateral interarcual
Dorsal interarcual
Anterior cardinal sinus
Gill lamellae
Levator hyomandibulae
Spiracle
Epibranchial
Interbranchial
Dorsal superficial branchial constrictor
Cucullaris (Levators)
Pectoral girdle
Pectoral fin adductor
Adductor mandibulae
Intermandibularis
Coracomandibular
Coracobranchial
4th external gill slit
Ventral superficial branchial constrictor
Gill ray
Branchial adductor
Ceratobranchial
Internal gill slit

Figure 7-12
Lateral view of the head of *Squalus,* in which the third definitive gill pouch has been cut and spread open and the epithelial gill lamellae stripped from its caudal wall, in order to show the musculature associated with the fifth visceral (or third branchial) arch.

lamellae from the rostral surface of one interbranchial septum, so that you can see circularly arranged muscle fibers that interconnect the gill rays. These muscle fibers form the **interbranchial** muscle, which represents a deep part of the constrictor musculature. With scissors, cut through one mobilized interbranchial septum at the level of the gill slit; the cut must extend from the surface of the body all the way through the joint between the epibranchial and ceratobranchial cartilages into the pharyngeal cavity. Examine the cut surface medial to the branchial arch, and you will see a cross section of the **branchial adductor,** a short muscle that extends from the epibranchial to the ceratobranchial.

As we have seen, the basic pattern of the branchiomeric musculature is much the same for all but the last of the branchial arches and, to a certain degree, even for the hyoid arch. Most of the embryonic muscle plate of one of these visceral arches forms a deep interbranchial and a superficial constrictor, but parts of that muscle plate separate as a levator, an interarcual group, and an adductor. The interbranchials, superficial constrictors, interarcuals, and adductors pull the parts of a branchial arch together and compress the branchial pouches, thus aiding in expelling the respiratory current of water. The branchial pouches are expanded and take in water through the elastic recoil of the branchial skeleton aided by the contraction of the levators and the coracobranchials of the hypobranchial muscle group. This basic muscle pattern is modified for the hyoid and, especially, for the mandibular arches in accordance with their modified functions.

AMPHIBIANS

Many changes occur in the muscular system during the evolution from fish through tetrapods. These are correlated primarily with changes in the mode of support and locomotion, with the development of head movements independent of those of the trunk, with changes in the method of respiration, and with the development of a mobile tongue. As regards locomotion, fishlike lateral undulations of the trunk help to advance and retract the appendages of many amphibians and reptiles, including salamanders, and trunk muscles remain conspicuous in these species; but appendicular muscles play an increasingly important role as tetrapods become more terrestrial. The primitive dorsal and ventral appendicular muscle masses differentiate into a number of separate muscles as the appendages become the organs of propulsive thrust, and their movements become more complex. The appendicular muscles, however, retain their primitive dorsal and ventral groupings (see Table 7-1).

Although trunk muscles become less important in locomotor movements, the epaxial trunk muscles brace and tie the elements of the vertebral column together and hence help to support the body. They remain prominent. Superficial portions tend to fuse into long muscle bundles crossing many body segments, but the deeper parts remain segmented. Deeper parts of the hypaxial trunk musculature also become associated with the vertebral column as a **subvertebralis.** Most of the rest tends to thin out, to lose its segmentation in regions where ribs are absent, and to differentiate into broad, antagonistic layers that support the trunk wall and play a role in breathing movements. Parts of the hypaxial musculature attach onto the pectoral girdle, where they act as a muscular sling to transfer weight from the trunk to the pectoral girdle and appendage. The development of this feature is first seen in *Necturus,* where the muscular sling is represented by the **thoraciscapularis.**

Hypobranchial musculature remains very fishlike in *Necturus,* but in most tetrapods it differentiates as a muscular tongue evolves and as hyoid and swallowing movements become more complex. This musculature is also very important in the breathing movements of terrestrial amphibians; it is less important in opening the jaws than in a fish.

With the shift from gill to pulmonary respiration, the more caudal visceral arches and most of their muscles are reduced. Some of these muscles become associated with the larynx. An exception is the levators of the posterior arches, which form the cucullaris in a fish and become attached to the pectoral girdle. The cucullaris is still fishlike in *Necturus,* but, as head and girdle movements become more complex, it enlarges and differentiates.

The cranial visceral arches contribute to the jaws, auditory ossicles, and hyoid apparatus. Their muscles have a comparable history to a large extent. Mandibular muscles close the jaws. Most of the hyoid musculature loses its connection with the hyoid arch, which is now moved primarily by hypobranchial muscles. A derivative of the hyoid musculature, the **depressor mandibulae,** attaches onto a retroarticular process of the lower jaw and acts to open the jaws in amphibians, reptiles, and birds. Most of the rest of the hyoid muscles become rather superficial, forming in *Necturus* a **sphincter colli** over the neck.

Necturus exemplifies the early tetrapod condition very well, for the changes just described are in an early stage. Remember, however, that *Necturus* is a permanently aquatic larva and hence retains many fishlike attributes, among them external gills and more of the branchial apparatus than would be found in an adult amphibian.

Axial Muscles

◆ **(A) MUSCLES OF THE TRUNK**

◆ epaxial muscles
dorsalis trunci
interspinalis

◆ hypaxial muscles
external oblique
internal oblique
transversus
subvertebralis

Confine your study of the muscles of *Necturus* to one side of the body, preferably the side that has not been cut open to inject the circulatory system. From the middle of the trunk, remove a strip of skin that is about 5 centimeters long and extends from the middorsal to the midventral line. Clean off any extraneous connective tissue, and note that the musculature still consists of **myomeres,** separated by **myosepta,** and divided into **epaxial** and **hypaxial** portions. The myomeres, however, are not as complexly folded as they are in fishes.

Although myosepta remain in the epaxial and epibranchial regions, most of the muscle fibers of adjacent myomeres fuse to form one functional longitudinal muscle, the **dorsalis trunci** (Figs. 7-13C and 7-14). Deeper muscle fibers passing between the vertebrae and ribs remain segmented. You can see some of these by making a deep transverse cut through the dorsalis trunci. Take a firm grip with a pair of forceps on a bundle of muscle fibers beside the middorsal line, and pull the fibers caudad. This will expose the tops of the vertebral arches and a series of short **interspinalis** muscles. Each interspinalis arises on the edge of a caudal vertebral zygapophysis of one vertebra, and inserts on the arch of the vertebra behind it.

Examine the hypaxial musculature and note that most of its superficial fibers incline posteriorly and ventrally. These constitute the **external oblique** muscle. The fibers next to the midventral line, however, extend longitudinally and form an incipient **rectus**

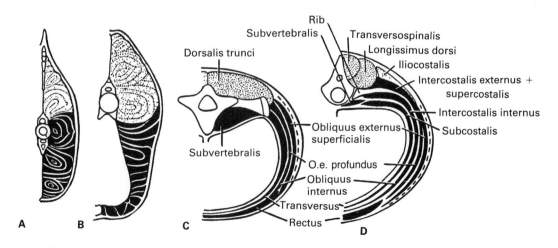

Figure 7-13
Diagrammatic cross sections to show the divisions of the trunk musculature in **A**, a shark tail; **B**, the shark trunk; **C**, a urodele; **D**, a lizard. The epaxial muscles are stippled; the hypaxial muscles are black. In **D**, the dorsal labels of the hypaxial muscles pertain to a region in which ribs are present; the more ventral labels are those of the corresponding abdominal muscles. *(From Romer and Parsons, mainly after Nishi.)*

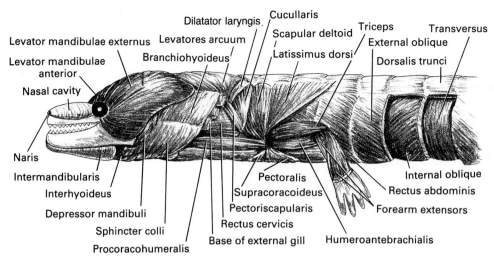

Figure 7-14
Lateral view of the trunk, pectoral, and head musculature of *Necturus.*

abdominis. Since *Necturus* is an incompletely metamorphosed species, the rectus abdominis is not as completely differentiated as it is in adult amphibians. Carefully cut into a myomere, reflecting the fibers of the external oblique as you do so. Presently, you will see a second layer, the **internal oblique,** whose fibers extend cranially and ventrally at nearly right angles to those of the external oblique. Uncover the internal oblique in several myomeres. Its ventral fibers form a part of the rectus. Now cut into the internal oblique in one segment and reflect its fibers. You will soon see a third layer, the **transversus,** whose fibers are more vertically oriented than those of the overlying muscle layers. Cut through this muscle, and you reach the parietal serosa and body cavity. Another hypaxial muscle is the **subvertebralis,** which lies on the sides of the vertebral bodies ventral to the transverse processes (Fig. 7-13**C**). It extends from vertebra to vertebra, attaching to the centra and transverse processes. You may see it now by making a small incision through the body wall slightly to one side of the midventral line, and pushing the viscera aside; or you may postpone this observation until you study the abdominal viscera. Several other hypaxial muscles attach onto the girdles and will be seen later.

♦ **(B) HYPOBRANCHIAL MUSCLES**

♦ prehyoid muscles
   geniohyoid
   genioglossus

♦ posthyoid muscles
   rectus cervicis
   omoarcuals
   pectoriscapularis

To approach the hypobranchial muscles, remove the skin from the entire underside of the head and neck as far caudally as the pectoral region. Carefully cut through the thin transverse sheet of branchiomeric muscles that lies under the head, and reflect it to the sides (Fig. 7-15). Cut slightly to one side of the midventral line. A pair of midventral, longitudinal muscles will be seen extending caudally from the symphysis of the jaw. Their caudal end inserts on the caudal portion of the hyoid apparatus (basibranchial 2). These muscles, the **geniohyoids,** represent the bulk of the prehyoid portion of the hypobranchial musculature. In amphibians, deep fibers derived from the geniohyoid begin to spread into the newly evolved tongue. *Necturus* has one such deep layer, **genioglossus,** which can be seen most clearly after cutting through the caudal attachment of the geniohyoid and pulling it forward. The genioglossus appears as a thin layer of more or less longitudinal fibers that arise on the chin, extend caudally in the floor of the mouth, and insert on a transverse fold of mucous membrane at the base of the tongue. Its lateral portions are best developed.

Most of the posthyoid hypobranchial musculature is represented by a **rectus cervicis,** a wide band of muscle extending caudad in the neck from the hyoid apparatus toward the pectoral girdle. It is continuous caudally with the rectus abdominis. Note that it retains traces of segmentation in the form of tendinous intersections. When you dissect the pectoral region, you will see slips of the rectus cervicis that are called **omoarcuals** attaching onto the procoracoid process of the girdle. At this time you will also see a **pectoriscapularis,** which is a derivative of the rectus cervicis.

Appendicular Muscles

♦   **(A) PECTORAL MUSCLES**

> ♦   ventral appendicular muscles
> > pectoralis
> > supracoracoideus
> > coracoradialis
> > humeroantebrachialis
> > coracobrachialis
> > forearm flexors

> ♦   dorsal appendicular muscles
> > latissimus dorsi
> > procoracohumeralis
> > subcoracoscapularis
> > triceps
> > forearm extensors
> > scapular deltoid
> ♦   other muscles acting on girdle
> > omoarcuals
> > cucullaris
> > pectoriscapularis
> > thoraciscapularis
> > levator scapulae

Cut off the external gills on one side and remove the skin from the pectoral appendage and the lateral and ventral surfaces of the trunk in its vicinity. Be careful, for some muscles may adhere to the skin. A large fan-shaped muscle, the **pectoralis,** will be seen on the ventral surface (Fig. 7-15). It arises from the linea alba and hypaxial musculature and inserts on the proximal end of the humerus. It lies caudal to the middle of the coracoid region. The ventral surface of the coracoid itself is covered primarily by a smaller fan-shaped muscle, the **supracoracoideus,** which also inserts proximally on the humerus. The posterior part of the supracoracoideus lies beneath the anterior part of the pectoralis. By carefully picking away connective tissue and watching the direction of the muscle fibers, you will be able to separate the two. To see the supracoracoideus more clearly, cut across the pectoralis and reflect its ends. Still another fan-shaped muscle, the **coracoradialis,** lies deep to the supracoracoideus, but do not look for it until you find its tendon of insertion passing along the brachium (p. 161). A **procoracohumeralis** will be found cranial and lateral to the supracoracoideus. It lies on the ventrolateral surface of the procoracoid process, from which it arises, and inserts proximally on the humerus. Some slips of the rectus cervicis (a hypobranchial muscle) attach onto the medial surface of the procoracoid process. These slips are sometimes called the **omoarcuals.**

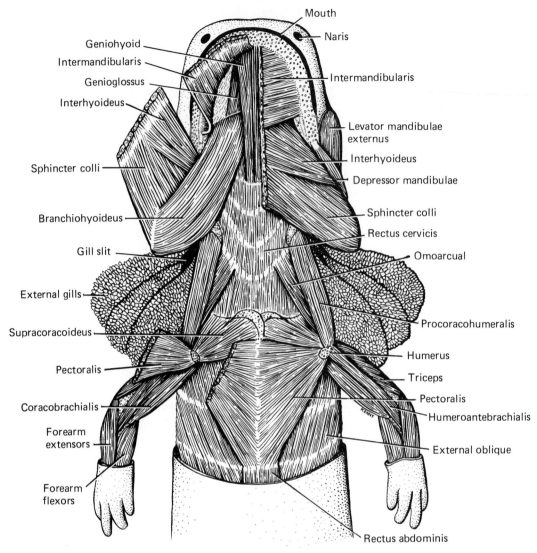

Figure 7-15
Ventral view of the head and pectoral muscles of *Necturus*.

Examine the lateral surface of the pectoral girdle (Fig. 7-14). The most caudal of a series of muscles that converge toward the shoulder joint is a large, fan-shaped **latissimus dorsi.** This muscle arises from the surface of the dorsalis trunci. Next cranial is the **scapular deltoid,** which arises from the lateral surface of the scapula and suprascapular cartilage. Both of these insert on the proximal end of the humerus. The small muscle just cranial to the scapula is the **cucullaris,** a muscle belonging to the branchiomeric series but inserting on the base of the scapula. The still narrower muscle cranial to the cucullaris is the **pectoriscapularis,** a hypobranchial muscle attaching onto the scapula.

Reflect the origin of the latissimus dorsi and carefully pull the caudal border of the scapula forward. You will see a band of muscle extending from the hypaxial musculature forward to insert on the medial side of the dorsal border of the scapula. This muscle, the **thoraciscapularis,** belongs to the hypaxial group (Fig. 7-16). Another, and narrower, band of muscle may also be seen extending forward from the front of the dorsal border of the scapula. You will obtain a better view of it by cutting and reflecting the cucullaris. This muscle, which is also a hypaxial derivative, is called the **levator scapulae.** Its anterior

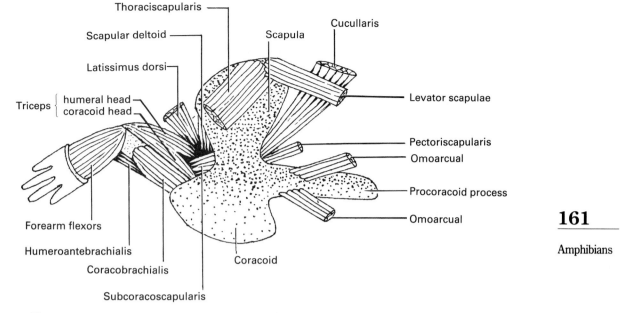

Figure 7-16
Medial view of the left pectoral musculature of *Necturus* after removal of the girdle from the trunk. Anterior is toward the right.

end attaches to the occipital region of the skull. It is this muscle that becomes the opercularis, which is associated with the ear in terrestrial amphibians (p. 233). Cut through the belly of the pectoralis, if this has not already been done, and also through the latissimus dorsi. Pull the arm forward so you can see the medial side of the girdle. (The sternal cartilages referred to on p. 102 can now be seen.) Clean away blood vessels and nerves dorsal to the glenoid region, and you will find a small muscle, the **subcoracoscapularis,** arising from the caudal edge of the scapula and coracoid and inserting on the humerus. It passes between two heads of the triceps (Fig. 7-16).

The upper arm is covered for the most part by three muscles. A **humeroantebrachialis,** arising from the humerus and inserting on the proximal end of the radius, is located on the ventrolateral surface (Figs. 7-14 and 7-15). A **coracobrachialis,** sometimes divided into a **brevis** portion and a **longus** portion, is located on the ventromedial surface. It arises from the coracoid and inserts on the shaft and distal end of the humerus. The tendon of the **coracoradialis,** mentioned earlier, can be found between the coracobrachialis and humeroantebrachialis. After you find it, trace it medially to the belly of the muscle, which lies on the ventral surface at the coracoid deep to the supracoracoideus. A **triceps** covers the dorsal surface of the upper arm. Although the triceps arises from the girdle and humerus by three heads, only two of these are readily distinguishable—a medial **coracoid head** located on the medial side of the arm dorsal to the coracobrachialis, and a dorsolateral **humeral head** separated from the coracoid head by the subcoracoscapularis (Fig. 7-16). All the heads converge and unite, forming a common tendon that passes over the elbow to insert on the proximal end of the ulna.

Clean the fascia from the surface of the forearm, and note that the muscles are aggregated into two groups—the **extensors** and the **flexors.** The extensors lie on the dorsal surface of the forearm and arise from the lateral surface of the distal end of the humerus. The flexors have the opposite relationships. They lie on the ventral surface and arise from the medial surface of the distal end of the humerus.

♦ ventral appendicular muscles
   pubotibialis
   puboischiofemoralis externus
   puboischiotibialis
   ischioflexorius
   caudofemoralis
   ischiofemoralis
   popliteus
   shank flexors

♦ dorsal appendicular muscles
   puboischiofemoralis internus
   iliotibialis
   ilioextensorius
   iliofibularis
   iliofemoralis
   shank extensors
♦ hypaxial muscles
   ischiocaudalis
   caudopuboischiotibialis

Remove the skin from one of the pelvic appendages and the adjacent trunk and tail. Also skin the cloacal region and remove, on one side, the large **cloacal gland,** which lies between the skin and muscles. Clean the area, and note the ilium on the lateral surface dorsal to the base of the appendage. The hypaxial musculature attaches to it. Other parts of the pelvic girdle are covered by muscle.

Study first the ventral surface of the girdle and thigh. One of the most distinct muscles, and hence one that will serve as a good landmark, is the **pubotibialis**—a narrow band on the ventral surface of the thigh that arises from the lateral edge of the pubic cartilage and inserts on the proximal end of the tibia (Fig. 7-17). The ventral surface of the girdle, and of the thigh medial to the pubotibialis, is covered by a large triangular mass of muscle. The cranial two thirds of this represents a **puboischiofemoralis externus;** the caudal third represents a **puboischiotibialis.** You can best detect the line of separation between them near the distal end of the thigh deep to the pubotibialis; then follow it toward the midventral line of the body. Note that the front of the puboischiotibialis overlaps the back of the puboischiofemoralis externus. The puboischiofemoralis externus, in turn, overlaps the attachment of the rectus abdominis on the front of the pubic cartilage. These two muscles arise from the ventral surface of the girdle. The puboischiofemoralis externus inserts on the femur; the puboischiotibialis, on the proximal end of the tibia. An **ischioflexorius** lies along the medioventral surface of the thigh caudal to the

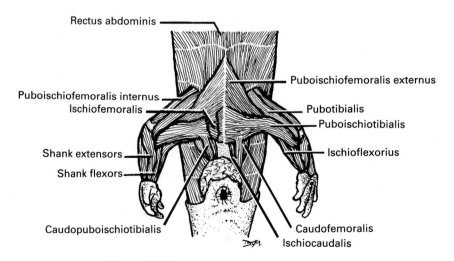

Figure 7-17
Ventral view of the pelvic muscles of *Necturus.*

puboischiotibialis and at first appears to be a part of this muscle. The two can be separated most easily at the proximal end of the shank, and from the dorsal side. Try to follow the muscles medially. The ischioflexorius inserts on the fascia of the distal end of the shank. It arises from the caudal end of the ischium. You may notice that the muscle is divided into distal and proximal bellies separated by a short tendon.

Dissect in the area lateral to the cloaca, and you will find two longitudinal muscles. The narrower one beside the cloaca is the **ischiocaudalis.** It inserts on the caudal vertebrae and arises from the caudal end of the ischium. The more lateral muscle, the **caudopuboischiotibialis,** has a similar insertion but arises from the puboischiotibialis. Both of these muscles are hypaxial muscles that attach in the pelvic region. They serve to flex the tail. Dissect dorsal to the caudopuboischiotibialis and you will find a third longitudinal muscle. This is the **caudofemoralis,** an appendicular muscle arising from the caudal vertebrae and inserting on the proximal end of the femur.

Cut across the puboischiotibialis and reflect its ends. The small triangular muscle lying caudal to the puboischiofemoralis externus is the **ischiofemoralis.** It arises from the ischium and inserts on the proximal end of the femur. The slender muscle, arising from the underside of the femur near the insertion of the puboischiofemoralis externus and continuing across the underside of the knee joint to the fibula, is the **popliteus.** It is really a shank muscle rather than a thigh muscle. (An **adductor femoris,** another deep muscle found cranial to the origin of the puboischiofemoralis externus in some urodeles, is absent as such in *Necturus.*)

The large muscle lying dorsal to the pubotibialis along the preaxial edge of the thigh is the **puboischiofemoralis internus.** Its origin is from the internal surface of the pubic cartilage and ischium. To expose it, cut through the pubic attachment of the rectus abdominis, and dissect deeply. The muscle inserts along most of the femur.

The dorsal surface of the thigh is occupied by a pair of muscles that arise from the base of the ilium, extend over the knee as a tendon, and insert on the tibia (Fig. 7-18). The cranial muscle is the **iliotibialis;** the caudal muscle is the **ilioextensorius.** Often these two appear as one muscle, for the separation between them may be obscure. Caudal to these, along the postaxial surface of the thigh, is a very slender **iliofibularis.** It too arises

163

Amphibians

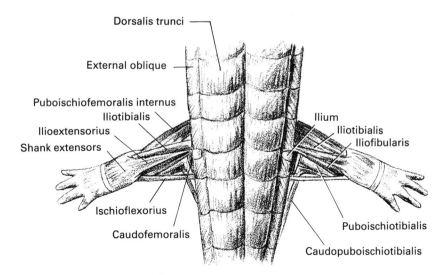

Figure 7-18
Dorsal view of the pelvic muscles of *Necturus.*

from the base of the ilium, but it diverges from the others and passes to insert on the fibula. Dissect away blood vessels and nerves caudal to the origin of the iliofibularis, and you will find a small, deep, triangular muscle that arises from the base of the ilium and inserts on the caudal edge of the femur. This is the **iliofemoralis.**

As with the forearm muscles, the muscles of the shank fall into extensor and flexor groups. The **extensors** lie on the dorsal surface of the shank; the **flexors** lie on the ventral surface. The extensors arise for the most part from the dorsal surface of the distal end of the femur; the flexors arise from the ventral surface.

Branchiomeric Muscles

♦ mandibular muscles
    intermandibularis
    levator mandibulae anterior
    levator mandibulae externus
♦ hyoid muscles
    interhyoideus
    sphincter colli
    depressor mandibulae
    branchiohyoideus

♦ branchial muscles
    levatores arcuum
    dilatator laryngis
    cucullaris
    subarcuals
    transversi ventrales
    depressores arcuum

Remove the skin from the underside of the head and the top and one side of the head. Be particularly careful above and below the stumps of the external gills. Note again the transverse sheet of muscle on the underside of the head that was cut and reflected during the study of the hypobranchial muscles. Approximately the rostral half of this represents an **intermandibularis,** a mandibular arch muscle; the caudal half an **interhyoideus,** a hyoid muscle (Fig. 7-15). Both insert on a median tendinous intersection, but they are distinct in their origins. The intermandibularis arises from the mandible; the interhyoideus arises from the hyoid arch medial to the angle of the jaw and from the surface of a large muscle posterior to the angle of the jaw. The latter portion of the interhyoideus is sometimes called the **sphincter colli.**

The rest of the mandibular arch musculature is represented by a levator mandibulae complex that arises from the top and sides of the skull, inserts on the mandible, and serves to close the jaws (Fig. 7-14). A **levator mandibulae anterior** lies on the top of the head lateral to the middorsal line; a **levator mandibulae externus** lies posterior to the eye.

The rest of the hyoid musculature is represented by the large muscle mass lying between the levator mandibulae complex and the gills. Close examination will reveal that this mass is divided into two muscles that arise in common from the first branchial arch in front of the gills. The more cranial muscle, the **depressor mandibulae,** inserts on the caudal end, or retroarticular process, of the mandible and is the major muscle involved in opening the jaws. The larger and more caudal muscle, the **branchiohyoideus,** continues forward beneath the interhyoideus and intermandibularis to insert along the ventral portion of the hyoid arch (the ceratohyal).

The remaining branchiomeric muscles belong to the branchial arches, although some have acquired other attachments. A series of **levatores arcuum** will be found dorsal to the bases of the external gills. Caudal to these, and crossing the origin of the cucullaris, you will see a **dilatator laryngis.** The dilatator laryngis looks like a levator, but note that it curves in toward the base of the neck (on its way to the larynx) rather than attaching onto the gills. The **cucullaris,** observed during the dissection of the shoulder region, completes the dorsal branchial muscles.

The ventral branchial muscles can be found by dissecting deep between the branchiohyoid and rectus cervicis. Pull the branchiohyoid forward and the rectus cervicis caudally. A series of **subarcuals** will be seen—a longitudinal muscle extending from the ceratohyal to the first branchial arch; another longitudinal muscle extending from the first and second to the third branchial arch; and an oblique muscle medial to this and extending between the first and second branchial arches. The oblique muscle fibers posterior to the subarcuals, and deep to the rectus cervicis, constitute the **transversi ventrales** (Fig. 11-15, p. 363). These muscle fibers extend from the third branchial arch on one side to that on the other. The remaining ventral muscles are a series of three **depressores arcuum** that extend from the ventral portion of the three branchial arches to the bases of the three external gills. They are small and may have been removed with the skin.

MAMMALS

The changes that were seen beginning in the muscles of early tetrapods continue in the evolution toward mammals. (Review the general remarks on tetrapod muscle evolution on p. 137). Details are summarized in Table 7-1 and will be seen during the dissections. We will note a few general trends at this time. Ocular muscles remain essentially the same throughout the vertebrate series. Mammals have one additional muscle in this group: the **levator palpebrae superioris** raises the upper eyelid; some also have a **retractor bulbi.**

Epaxial trunk muscles and subvertebral hypaxial muscles are well developed and support and move the vertebral column and head. Most of the hypaxial musculature is in the form of thin, antagonistic layers that support the thoracic and abdominal walls and are important in breathing movements. The muscles of the **diaphragm** are derived from this group. An increased number of hypaxial muscles transfer body weight from the trunk to the pectoral girdle and appendage. They also are responsible for movements between the trunk and girdle. Of particular importance in this connection are the **serratus ventralis** and **rhomboideus** complexes.

The hypobranchial musculature becomes subdivided into many units that give a precise control over the tongue and hyoid movements needed in mastication and swallowing.

Appendicular muscles are very conspicuous and numerous, for the appendages are crucial for body support and locomotion. As the appendages are positioned closer to the axis of the body and move in or close to a parasagittal plane during locomotion (see Fig. 3-9), they provide better support than in amphibians and suspend the body above the ground. They can also move fore and aft without the need for lateral undulations of the trunk and tail as in urodele amphibians. Ventral appendicular muscles are less important in holding the body off the ground, and some shift to a dorsal position where they can assist the dorsal muscles in protracting and retracting the limb. The **supraspinatus** and **infraspinatus** are examples of ventral appendicular muscles that have shifted to a dorsal position. Some fibers derived from appendicular muscles form the **cutaneous trunci,** which spreads over the trunk and moves the skin.

Mastication and precise occlusion of the teeth require strong and complex jaw muscles; thus, mandibular muscles have increased in strength and number. A small part of the mandibular musculature follows the old jaw joint into the middle ear cavity to become the **tensor tympani.** A new jaw-opening muscle, the **digastric,** evolves partly from mandibular and partly from hyoid musculature and replaces the depressor mandibulae found in amphibians, reptiles, and birds. A small part of the hyoid musculature, the **stapedius,** attaches to the stapes in the middle ear, and another part, the **stylohyoid,** retains a connection with the hyoid, but most hyoid muscles are superficial and form the **platysma** and **facial muscles** that move the skin of the neck, face, and lips. The primitive cucullaris, which evolved from caudal branchial levators, enlarges and divides into the **trapezius** and **sternocleidomastoid** complexes as shoulder and head movements become more complex and independent of the trunk.

Many of the groups of muscles have lost their clear definition in adult mammals, for some of the muscles have migrated from their original positions. This makes it more difficult to study the muscles by natural groups, but it can still be done, if a few concessions are made to topography.

Cutaneous Muscles

◆ cutaneus trunci
◆ platysma
◆ facial muscles

As the first step of your dissection, it is necessary to remove the skin from your specimen. With a sharp scalpel, make incisions through the skin completely around the neck behind the skull; around the paws at the level of the wrists and ankles, respectively; and around the base of the tail, anus, and external genitals. If your specimen is a male, first locate the testicles before making the incision; do not remove the skin over the testicles and (in both sexes) around the anus and urogenital orifices. For the time being, the skin will also be left on the head, tail, and paws. Lay your specimen on its belly, find the middorsal line by feeling the spinous processes of the vertebral column, and make a longitudinal, middorsal incision through the skin that extends from the circular incision around the neck to that around the base of the tail.

Beginning on the back, gradually separate the skin from the underlying muscles by tearing through the fibrous connective tissue of the **superficial fascia** with a pair of blunt forceps. As you separate the skin from the trunk, notice the fine, parallel brownish lines that adhere to its undersurface. They represent the muscle fiber bundles of the **cutaneus trunci,** the largest of the cutaneous muscles that move the skin. (This muscle is not present in humans.) The cutaneus trunci arises from the surface of certain appendicular muscles of the shoulder (latissimus dorsi and pectoralis), and from the **linea alba** (a midventral band of connective tissue), fans out over most of the trunk, and inserts on the dermis of the skin. It should be removed with the skin, except for that portion attached to the shoulder muscles posterior to the armpit. Several smaller cutaneous muscles, derived from the caudal musculature, become associated with the caudal part of the cutaneus trunci. They may not be noticed.

Much of the top and the sides of the neck are covered by another cutaneous muscle, the **platysma,** which is derived from the hyoid musculature. As the platysma spreads over the face it breaks up into many, small cutaneous muscles associated with the lips, nose, eyes, and ears. They are collectively known as the **facial muscles.** You may see them later when the head is skinned. These muscles are also present in human beings and are responsible for facial expression.

As you continue to skin your specimen, you will come upon narrow tough cords passing to the skin. These are **cutaneous blood vessels** and **nerves** and must be cut. Note that they tend to be segmentally arranged along the trunk. If your specimen is a pregnant or lactating female, the **mammary glands** will appear as a pair of large, longitudinal, glandular masses along the ventral side of the abdomen and thorax. They should be removed with the skin.

As you reach the limbs, try to tear the connection between the skin and the underlying muscles while leaving the skin around the limbs intact as "sleeves." Do so by prying and tearing with blunt forceps and probes under the skin continuing from the skinned part of the trunk as well as starting from the circular cut around the wrists and ankles until you can pull out the limb from the sleevelike skin. In this way you will be able to preserve the skin as some sort of jumpsuit that can easily be pulled on and off the specimen. If your specimen is an older male cat, this may not be possible, as the connective tissue may be very tough; in this case, make a longitudinal incision on the dorsal or lateral surface of the limbs to facilitate skinning.

After the specimen is skinned, clean away the excess fat and connective tissue on the side that is to be studied, but do not clean an area thoroughly until it is being dissected. If your specimen is a male, be particularly careful in removing the wad of fat in the groin, for it contains on each side the proximal part of the **cremasteric pouch**—a part of the scrotum containing blood vessels and the sperm duct extending between the abdomen and scrotal skin. First find this pouch. It is large in the rabbit, but rather small in the cat. Clean away connective tissue deep in the groin, or inguinal region, so as to find the actual boundary between the thigh and abdomen. In the rabbit, a shiny, white **inguinal ligament** will be seen in this region, extending from the pubis to the ilium.

Caudal Trunk Muscles

All the axial muscles cannot be studied at the same time, for many of them are located deep to the shoulder muscles. Those located on the trunk between the pectoral and pelvic appendages will be examined now, and the more cranial muscles will be considered after the appendages have been studied (p. 199).

◆ **(A) HYPAXIAL MUSCLES**

◆ muscles of the abdominal wall
    external oblique
    internal oblique
    transversus abdominis
    rectus abdominis

◆ subvertebral muscles
    quadratus lumborum
    psoas minor

Continue to clean off the surface of the trunk between the pectoral and pelvic appendages. The wide sheet of tough, white fascia covering the lumbar region on the back is the **thoracolumbar fascia.** The wide sheet of muscle that runs cranially and ventrally from the anterior part of this fascia and disappears in the armpit is the latissimus dorsi (an appendicular muscle) (see Figs. 7-20 and 7-21). The large triangular muscle that covers the underside of the chest is the pectoralis (another appendicular muscle). The borders of the latissimus dorsi and pectoralis appear to run together caudal to the armpit. Separate the two in this region by removing the cutaneus trunci and carefully trace their edges forward. Remove the fat and loose connective tissue from beneath them. The hypaxial trunk muscles can now be studied.

As in other tetrapods (Fig. 7-13), the abdominal wall is composed of three layers of muscle, plus a paired longitudinal muscle along the midventral line. All serve to compress the abdomen. The **external oblique** forms the outermost layer (Fig. 7-19). This muscle arises by slips from the surface of a number of caudal ribs and from the thoracolumbar fascia. Part of its origin lies beneath the caudal edge of the latissimus dorsi. Its fibers then extend obliquely caudally and ventrally to insert by an aponeurosis along the length of the linea alba. In the rabbit, some of the fibers insert on the inguinal ligament. Remove fat and connective tissue from the surface of the external oblique and identify its borders. The cranial border is very short, thin, and hidden under the latissimus dorsi and pectoral muscles; the caudal border is almost longitudinally oriented. Bisect the external oblique perpendicularly to the muscle fiber direction. Start the cut at one of the borders by separating, as far as you can, first the sheetlike external oblique from the underlying muscle (the internal oblique); then cut with scissors through the part of the muscle that can be lifted. Continue the separation of the two muscle layers as far as you can, cut, and so on,

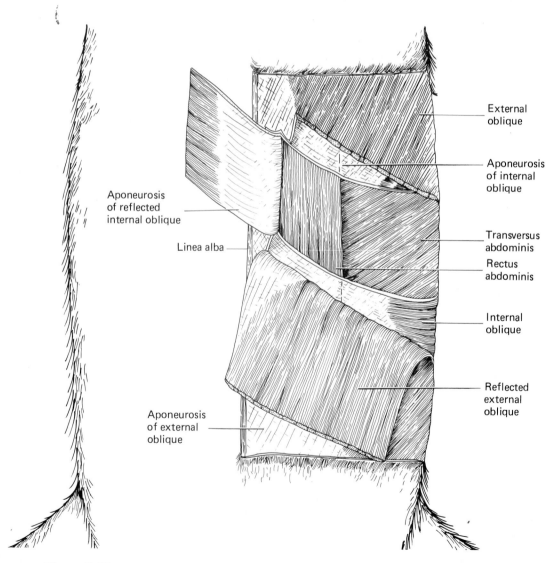

Figure 7-19
Ventral view of the abdominal muscle layers on the left side of a cat. Parts of the external oblique and internal oblique have been reflected to show the deeper layers.

until you reach the other border. Make sure to cut through the middle of the muscle fibers and not through the aponeurosis; the completed cut will be curved. Reflect the bisected muscle to expose the entire underlying muscle.

The **internal oblique** lies beneath the external oblique. It is most apparent high up on the side of the abdomen near its main origin from the thoracolumbar fascia. The fibers of the internal oblique extend ventrally and slightly cranially at right angles to the fibers of the external oblique and soon lead into a wide aponeurosis that inserts along the linea alba. Only the dorsal half of the muscle is fleshy; the ventral half is represented by an aponeurosis, but this may not be apparent at first, for one can see the third muscle layer, the transversus abdominis, through this aponeurosis.

To get at the **transversus abdominis,** make a longitudinal cut through the fleshy portion of the internal oblique in the same manner as described for the external oblique, and reflect the bisected muscle. The transversus abdominis arises primarily from the medial surface of the more caudal ribs, and from the transverse processes of the lumbar vertebrae. The latter portion of the origin lies in a furrow deep to the epaxial muscles. Note that this furrow has the same location as the horizontal skeletogenous septum of fishes. The fibers of the muscle extend ventrally, and slightly caudally, to insert along the linea alba by a narrow aponeurosis. Separate some of the fibers of the transversus abdominis, and you will expose the **parietal peritoneum** lining the abdominal cavity. (There is actually an internal fascia between the transversus abdominis and the parietal peritoneum, but it cannot be seen easily because it usually fuses with the transversus abdominis.)

The reflection of the internal oblique also exposes a longitudinal band of muscle lying lateral to the midventral line. This is the **rectus abdominis.** Transverse tendinous intersections can sometimes be seen in it, but these do not correspond to the primitive myosepta delimiting muscle segments. The rectus abdominis arises from the pubis and inserts on the sternum and cranial costal cartilages. In the cat, its cranial portion lies directly under the external oblique; its middle portion lies between two layers of the aponeurosis of the internal oblique, and its caudal portion lies between two layers of the aponeurosis of the transversus abdominis. In the rabbit, most of the rectus abdominis lies between the internal oblique and transversus abdominis.

In addition to the muscular layers of the abdominal wall, the caudal hypaxial musculature includes a subvertebral group (see Fig. 7-13) that lies ventral to the lumbar and caudal thoracic vertebrae. This group, which includes the **quadratus lumborum** and **psoas minor,** is associated with certain pelvic muscles and is described in connection with the hind leg (p. 195).

♦ **(B) EPAXIAL MUSCLES**

♦ multifidi
♦ erector spinae

Lift up the thoracolumbar fascia with a pair of forceps, and make a longitudinal incision through it about half a centimeter to one side of the middorsal line (identify its location by feeling for the spinous processes of the vertebral column). Extend the incision from the latissimus dorsi to the sacral region, and reflect the superficial layer of the fascia. A deeper layer of this fascia will now be seen encasing the epaxial muscles. Make a longitudinal cut through it about 1 centimeter from the middorsal line. The narrow band of muscle beside the spinous processes of the vertebrae is composed of a series of short **multifidi** that interlace the vertebrae and whose muscle fiber bundles run from caudolateral to craniomedial (see Figs. 7-41 and 7-42). The wider lateral band of muscle represents the **erector spinae.** In the cat, a part of the fascia covering the caudal part of the erector spinae dips into it and subdivides it, but these subdivisions do not correspond to the subdivisions that will be seen later on the cranial part of the trunk. Ignore them until the cranial part of the epaxial musculature is studied. The muscle fiber bundles of the erector spinae run from caudomedial to craniolateral, except in the most medial subdivision, which has longitudinally oriented muscle fibers.

Pectoral Muscles

Most of the muscles of the pectoral region are, of course, appendicular muscles, but a number of axial and several branchiomeric muscles have become associated with the pectoral girdle and appendage.

♦   **(A) PECTORALIS GROUP**

♦ pectoralis superficialis
   pectoralis descendens
   pectoralis transversus

♦ pectoralis profundus
   xiphihumeralis (cat)

The large triangular muscle complex covering the chest is the **pectoralis.** It arises from the sternum and passes to insert primarily along the humerus. Its major actions are to pull the humerus toward the chest (adduction) and caudally (retraction), but it also helps in transferring body weight from the trunk to the pectoral girdle and appendage (see Fig. 7-24). Clean the surface of the muscle enough so that you can see the direction of its fibers.

The pectoralis of quadrupeds is divided into a pectoralis superficialis and a pectoralis profundus, which are homologous to the pectoralis major and the pectoralis minor of human beings, respectively, but the profundus is the larger muscle in quadrupeds. Each part may be further subdivided.

The **pectoralis superficialis** arises from approximately the cranial one third of the sternum. Its fibers extend more or less laterally to insert on the humerus. The insertion in the cat (but not in the rabbit) extends to the distal end of the humerus and onto the antebrachium. The superficial pectoralis can be divided into two parts (Fig. 7-20). A narrow band of very superficial fibers, the **pectoralis descendens,** extends from the cranial end of the sternum to the middle of the humerus (rabbit) or to the fascia antebrachii (cat). The rest of the pectoralis superficialis is known as the **pectoralis transversus** (Fig. 7-20).

A cleidobrachialis muscle covers the front of the shoulder and part of the pectoralis transversus (Figs. 7-20 and 7-43). Separate the borders of the cleidobrachialis and

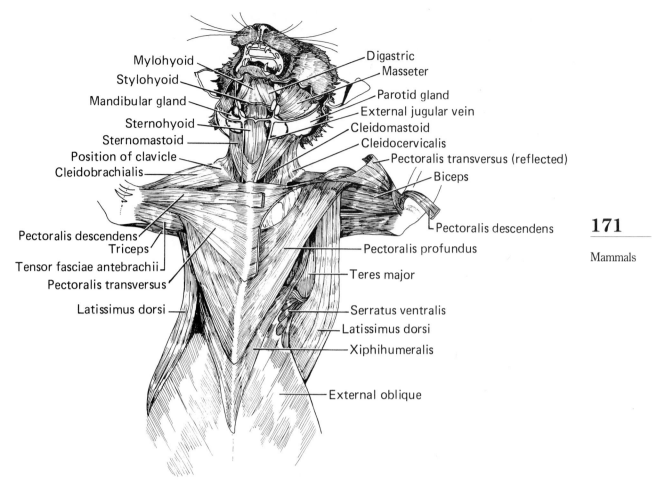

Mylohyoid
Stylohyoid
Mandibular gland
Sternohyoid
Sternomastoid
Position of clavicle
Cleidobrachialis

Digastric
Masseter
Parotid gland
External jugular vein
Cleidomastoid
Cleidocervicalis
Pectoralis transversus (reflected)
Biceps

Pectoralis descendens
Triceps
Tensor fasciae antebrachii
Pectoralis transversus
Latissimus dorsi

Pectoralis descendens
Pectoralis profundus
Teres major
Serratus ventralis
Latissimus dorsi
Xiphihumeralis

External oblique

Figure 7-20
Ventral view of the muscles in the pectoral region of the cat. Superficial muscles are shown on the left side of the drawing; deeper muscles, after removal of the pectoralis transversus, on the right side. *(From Walker,* A Study of the Cat.*)*

pectoralis transversus by breaking and picking away the connective tissue that binds them together. The clavicle is imbedded on the underside of the cleidobrachialis at the base of the neck; you can feel it if you palpate the muscle between your thumb and forefinger. Cut through the connective tissue that ties the medial end of the clavicle to the manubrium, so that you can push the cleidobrachialis forward and expose the entire pectoralis, but be careful not to cut through the jugular vein and cranial border of the pectoralis.

The **pectoralis profundus** originates from the caudal two thirds, or from the full length, of the sternum. Its fibers extend craniolaterally and disappear beneath those of the superficial pectoralis. Toward the armpit, the lateral border of the pectoralis profundus is strongly tied to the ventral border of the latissimus dorsi and cannot be separated easily; you may need to cut the two muscles apart with scissors. In the cat, the insertion is confined to the humerus, but in the rabbit some fibers also attach onto the clavicle (see Fig. 7-43), and a group of deep fibers pass dorsal to the clavicle and sweep over the front of the shoulder to insert on the scapular spine. This portion will be seen later. The scapular fibers probably help to pull the dorsal border of the scapula forward, an action that occurs in hopping. The most caudal fibers of the cat's pectoralis profundus form a distinct, narrow band known as the **xiphihumeralis,** which continues cranially on the dorsal side of the pectoralis profundus.

♦   **(B) TRAPEZIUS AND STERNOCLEIDOMASTOID GROUP**

    ♦   trapezius
            cleidocervicalis (cat)
            (cleidocervicalis + cleidobrachialis = brachiocephalicus)
            cervical trapezius
            thoracic trapezius
    ♦   sternocleidomastoid
            sternomastoid
            cleidomastoid
            cleido-occipitalis (rabbit)

The muscles belonging to the trapezius and sternocleidomastoid group are branchiomeric muscles that have become associated with the pectoral girdle, for they evolved from the cucullaris of fishes and amphibians. The human trapezius is a large, undivided muscle, but it is subdivided in the mammals being considered here. Most parts act on the scapula, pulling it toward the middorsal line (abduction), cranially (protraction), and caudally (retraction). In the cat, the most cranial part of the trapezius (cleidocervicalis) inserts on the clavicle and, together with the cleidobrachialis, protracts the arm. The sternocleidomastoid complex of muscles acts primarily upon the head, turning it to the side and flexing it downward. Clean off connective tissue from the ventral, lateral, and dorsal surfaces of the neck and from the dorsal part of the shoulder. You may have to remove more skin from the back of the head. Do not injure the large external jugular vein located superficially on the ventrolateral surface of the neck.

The **thoracic trapezius** is a thin sheet of muscle covering the cranial part of the latissimus dorsi, from which it should be separated (Figs. 7-21 and 7-22). From their origin on the middorsal line of the thorax, the muscle fibers of the thoracic trapezius converge to insert on the dorsal part of the scapular spine. A **cervical trapezius** lies cranial to the thoracic trapezius. It arises from the middorsal line of the front of the thorax and neck, and from an aponeurosis that interconnects the left and right cervical trapezius. Its origin extends as far forward as the nuchal crest of the skull in the rabbit, but only to the base of the neck in the cat. The muscle fibers of the cervical trapezius converge to insert on the ventral portion of the scapular spine and its metacromion.

The cat has a third component to the trapezius, the **cleidocervicalis** (clavotrapezius) (Fig. 7-21). From their origin on the nuchal crest and middorsal line of the cranial part of the neck, the muscle fibers of the cleidocervicalis extend caudally and ventrally to merge, at the level of the clavicle, with those of the cleidobrachialis, the muscle seen earlier covering the humeral insertion of the pectoralis.

The **cleidobrachialis,** which is a part of the deltoid group, continues caudally from the clavicle to insert on the humerus (rabbit) or ulna (cat). The cleidobrachialis together with the cleidocervicalis of the cat constitute the **brachiocephalicus.** The cleidobrachialis and cleidocervicalis tend to merge in the cat, since the clavicle is reduced, but careful dissection may reveal a tendinous intersection between them. In vertebrates with a well-developed clavicle, the cleidobrachialis and cleidocervicalis are distinct muscles.

In all of the mammals under consideration, a **sternomastoid** arises from the manubrium and extends cranially and dorsally to insert on the mastoid region of the skull (Figs. 7-21, 7-22, 7-40, and 7-43). As it extends forward, the muscle passes deep to the large external jugular vein. Muscle tissue superficial to the vein should be removed; it is either platysma or, in the rabbit, a specialized part of the platysma extending from the manubrium to the ear base, called the **depressor conchae posterior.** The sterno-mastoid of the rabbit is a narrow band, but it is wide in the cat and parallels the cranio-ventral border of the cleidocervicalis. Part of the insertion spreads onto the occipital region of the skull in the cat, and some authors distinguish these fibers as a sterno-occipitalis. Left and right sternomastoids of the cat merge near the sternum (Fig. 7-20) and should be cut apart.

The cat and rabbit have a **cleidomastoid** extending from the clavicle (where it joins other muscles attaching on the clavicle) to the mastoid region of the skull (Figs. 7-20 and 7-22). Much of it lies deep to the wide sternomastoid and cleidocervicalis in the cat, but it lies dorsal to the sternomastoid in the rabbit.

In the rabbit only (Fig. 7-22), a **cleido-occipitalis** arises from the clavicle lateral and dorsal to the origin of the cleidomastoid. It extends forward deep to both the cleidomastoid and sternomastoid to insert on the basioccipital region of the skull.

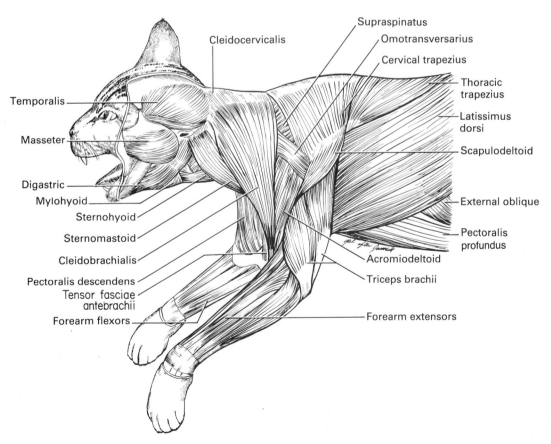

Figure 7-21
Lateral view of the pectoral, neck, and head muscles of the cat.

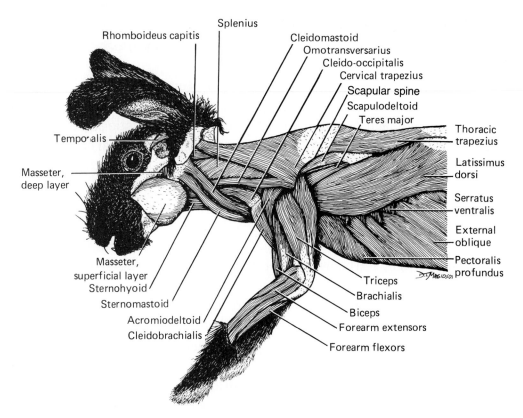

Figure 7-22
Lateral view of the pectoral, neck, and head muscles of the rabbit.

♦ **(C) REMAINING SUPERFICIAL MUSCLES OF
THE SHOULDER**

- ♦ omotransversarius
- ♦ deltoid
  - cleidobrachialis
  - acromiodeltoid
  - scapulodeltoid
- ♦ latissimus dorsi

The band of muscle whose caudal end can be seen on the side of the neck between the cleidocervicalis and cervical trapezius (cat) or between the sternocleidomastoid complex and trapezius (rabbit) is the **omotransversarius,** or levator scapulae ventralis (Figs. 7-21 and 7-22). It originates on the metacromion of the scapula ventral to the insertion of the cervical trapezius and extends forward to insert primarily on the transverse process of the atlas. In the cat, much of it lies deep to the cleidocervicalis. The omotransversarius is a hypaxial muscle that helps to pull the scapula forward (protraction).

The deltoid muscle lies caudal and ventral to the scapular attachment of the omotransversarius. In the mammals being considered, it is subdivided into three parts. The **cleidobrachialis** (clavodeltoid) has been observed arising from the clavicle (Figs. 7-21 and 7-22). It inserts on the humeral body in the rabbit and on the ulna in common with the brachialis (see the following section) in the cat. The **acromiodeltoid** lies dorsal and caudal to the cleidobrachialis. It arises from the acromion near the attachment of the omotransversarius and inserts on the proximal portion of the humeral body. The **scapulodeltoid** lies along the caudal border of the scapula. It arises from the scapular spine and, in the rabbit, from the surface of the muscle covering the lower half of the scapula (infraspinatus). It inserts on the proximal end of the humerus. The muscle is thinner in the rabbit and passes ventral to the large metacromion. But if the metacromion is broken and raised, the essential relationship will be seen to be the same as in the cat. The cleidobrachialis protracts the arm, but the caudal parts of the muscle complex are retractors and abductors of the humerus.

The **latissimus dorsi** has already been observed on the side of the trunk caudal to the arm (p. 167). It arises from the thoracolumbar fascia and from the spinous processes of the last thoracic vertebrae. From here it passes forward and ventrally to insert on the proximal end of the humerus in common with the teres major. In the cat, a small part of the muscle often forms a tendon that inserts with the pectoralis. The latissimus dorsi retracts the humerus.

♦ **(D) DEEPER MUSCLES OF THE SHOULDER**

- ♦ supraspinatus
- ♦ infraspinatus
- ♦ teres major
- ♦ teres minor
- ♦ rhomboideus cervicis et thoracis
    - rhomboideus cervicis (rabbit)
    - rhomboideus thoracis (rabbit)
- ♦ rhomboideus capitis

- ♦ serratus ventralis
    - serratus ventralis cervicis (rabbit)
    - serratus ventralis thoracis (rabbit)
- ♦ subscapularis

Cut across the center of the latissimus dorsi at right angles to its fibers, and also across the thoracic and cervical trapezius. Reflect the ends of these muscles, and clean out the fat and loose connective tissue from beneath them so as to expose the deeper muscles of the shoulder. In the rabbit, the pectoralis profundus can be seen sweeping over the front of the shoulder and scapula to insert on the scapular spine (see p. 171). It should be cut near its insertion and reflected.

The supraspinous fossa of the scapula is occupied by the **supraspinatus** (Fig. 7-23). The muscle inserts on the greater tuberculum of the humerus and protracts the humerus. The infraspinous fossa is occupied by the **infraspinatus.** Most of this muscle is covered by the scapulodeltoid in the rabbit. The infraspinatus inserts on the greater tuberculum of the humerus and rotates this bone outward.

A **teres major** arises from the caudal border of the scapula caudal and ventral to the infraspinatus. The teres major passes forward to insert on the proximal end of the humerus in common with the latissimus dorsi. It rotates the humerus inward and retracts it.

Lift up the cranioventral border of the scapulodeltoid and infraspinatus and dissect deeply between the infraspinatus and the long head of the triceps (the large muscle on the posterior surface of the brachium). You will eventually come upon a very small triangular muscle arising by a tendon from the middle of the caudal border of the scapula and

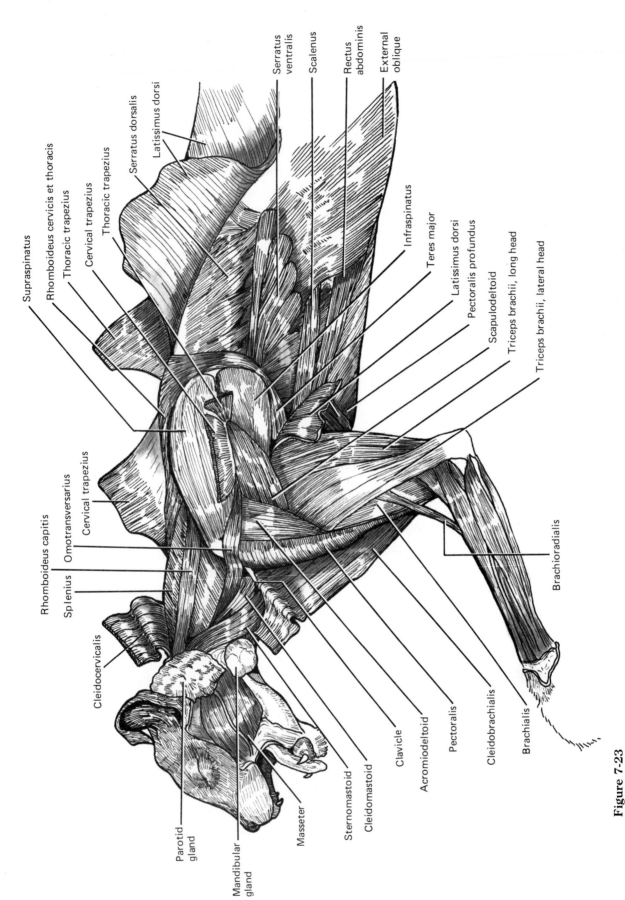

Figure 7-23
Lateral view of the pectoral and neck muscles of the cat after reflection of the trapezius muscle group and the latissimus dorsi. *(From Walker, A Study of the Cat.)*

Supraspinatus

Rhomboideus cervicis et thoracis

Thoracic trapezius

Cervical trapezius

Thoracic trapezius

Serratus dorsalis

Latissimus dorsi

Serratus ventralis

Scalenus

Rectus abdominis

External oblique

Infraspinatus

Teres major

Latissimus dorsi

Pectoralis profundus

Scapulodeltoid

Triceps brachii, long head

Triceps brachii, lateral head

Brachioradialis

Brachialis

Cleidobrachialis

Pectoralis

Acromiodeltoid

Clavicle

Cleidomastoid

Sternomastoid

Masseter

Mandibular gland

Parotid gland

Cleidocervicalis

Splenius

Rhomboideus capitis

Omotransversarius

Cervical trapezius

inserting on the greater tuberculum of the humerus. This is the **teres minor;** it helps the infraspinatus rotate the humerus outward.

Examine the scapular region from a dorsal view. The large muscle that arises from the tops of the posterior cervical and anterior thoracic vertebrae, and inserts along the dorsal border of the scapula, is the **rhomboideus cervicis et thoracis.** It forms a continuous sheet of muscle bundles in the cat, but in the rabbit it is clearly divided into a cranial **rhomboideus cervicis** and a caudal **rhomboideus thoracis** (see Fig. 7-26). The most cranial muscle bundle in all cases extends more rostrally than the others to its attachment on the back of the skull and is called the **rhomboideus capitis.** The rhomboideus is a hypaxial muscle that pulls the scapula toward the vertebrae, helps to hold it in place, and assists in its protraction and retraction.

Muscles that lie deep to the rhomboideus and extend from the dorsal border of the scapula to the ribs and to the transverse processes of the cervical vertebrae constitute the serratus ventralis complex. Cut across the rhomboideus, pull the top of the scapula

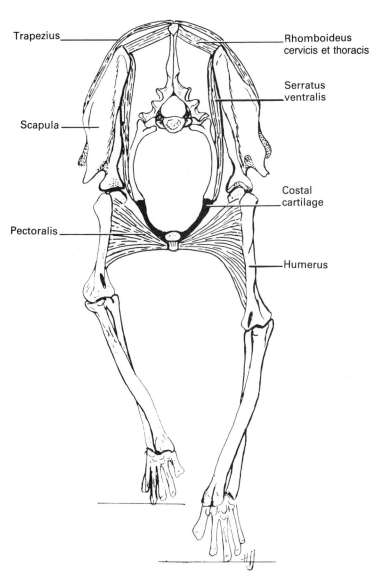

Figure 7-24
Diagrammatic cranial view of a section through the thorax of a cat at the level of the pectoral girdle, to show muscular connections between the trunk and appendages.

laterally, and clean away fat and loose connective tissue from the area exposed. The large fan-shaped muscle that you see is the **serratus ventralis** proper (Figs. 7-23 and 7-24). It is divided into a **serratus ventralis cervicis** and a **serratus ventralis thoracis** in the rabbit (Fig. 7-26). Approach the serratus ventralis also from the ventral side by bisecting and reflecting the pectoralis. (When doing so, cut the individual parts of the pectoralis separately in order to cut perpendicularly to the direction of the particular muscle fibers. Make sure to leave the underlying nerves and blood vessels intact for later study.) The serratus ventralis arises by a number of slips from the ribs just dorsal to the junction of ribs

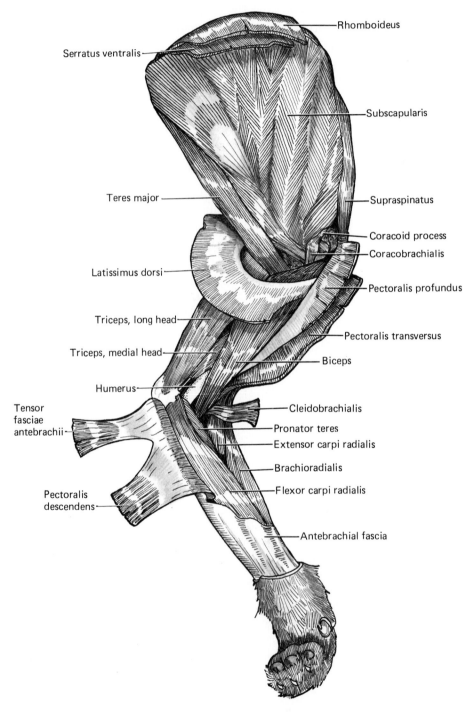

Figure 7-25
Medial view of the muscles of the left scapula and arm of the cat after removal of the appendage from the thorax. *(From Walker,* A Study of the Cat.*)*

and costal cartilages, and from the transverse processes of the posterior cervical vertebrae. It inserts on the dorsal border of the scapula ventral to the insertion of the rhomboideus. The serratus ventralis is a hypaxial muscle. Together with the pectoralis, it forms a muscular sling that transfers much of the weight of the body to the pectoral girdle and appendages (Fig. 7-24). The serratus ventralis is the major component in the sling.

Clean out the area between the serratus ventralis and the medial surface of the scapula. The subscapular fossa is occupied by a large **subscapularis,** which passes ventrally to insert on the lesser tuberculum of the humerus (Figs. 7-25 and 7-26). This muscle pulls the humerus medially (adduction).

♦ **(E) MUSCLES OF THE BRACHIUM**

♦ extensors
    tensor fasciae antebrachii
    triceps brachii
    anconeus (cat)

♦ flexors
    brachialis
    biceps brachii
    coracobrachialis

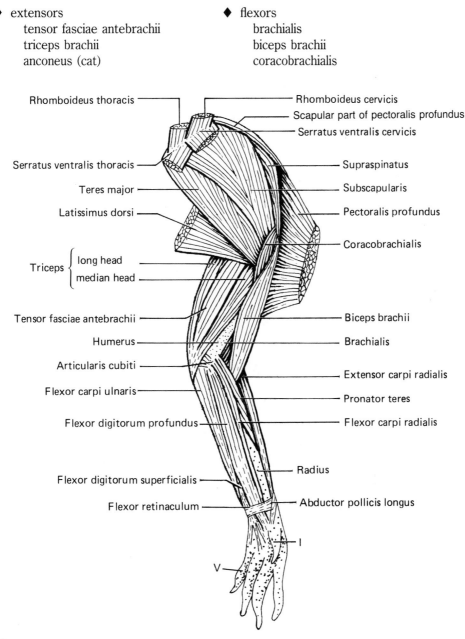

Figure 7-26
Medial view of the muscles of the scapula and arm of a rabbit after removal of the appendage from the thorax.

Clean the muscles of the brachium and separate them from one another. The large muscle that covers the caudal surface and most of the medial and lateral surface of the humerus is the triceps brachii. A **tensor fasciae antebrachii** (dorsoepitrochlearis) is closely associated with the medial surface of the triceps and should be studied first. The tensor fasciae antebrachii arises primarily from the lateral surface of the latissimus dorsi and extends distally along the median surface of the arm to insert on the tendon of the triceps and on the antebrachial fascia (Figs. 7-25 and 7-26).

Bisect the tensor fasciae antebrachii and reflect its halves. The **triceps brachii** is now exposed clearly (Figs. 7-21, 7-25, and 7-26). It has three main heads. The **long head,** located on the caudal surface of the humerus, is the largest. Note that it arises from the scapula caudal to the glenoid cavity. A large **lateral head** arises from the proximal end of the humerus and covers much of the lateral surface of this bone. It is quite distinct in the cat but is partly united with the long head in the rabbit. A small **medial head** can be found on the medial surface of the humerus deep to several nerves and blood vessels. It arises from most of the shaft of the humerus. All the heads of the triceps insert in common on the olecranon of the ulna.

In the cat, a small **anconeus** arises from the distal portion of the lateral surface of the humerus and inserts on the olecranon beside the insertion of the triceps. To see it, cut

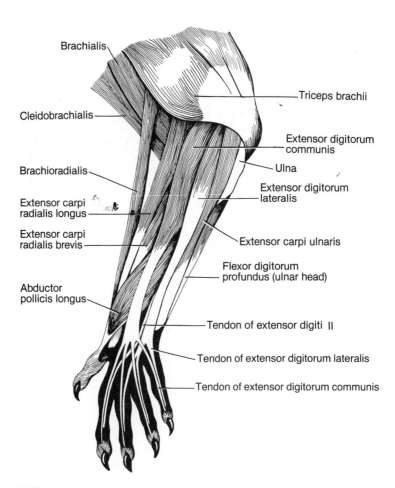

Figure 7-27
Lateral view of the extensor muscles of the left forearm of the cat after removal of the antebrachial fascia and extensor retinaculum. *(From Walker,* A Study of the Cat.*)*

through and reflect the distal end of the lateral head of the triceps. The triceps, tensor fasciae antebrachii, and anconeus extend the forearm.

The craniolateral surface of the humerus is covered by the **brachialis.** It arises from the shaft of the humerus and inserts on the proximal end of the ulna (cat, Fig. 7-23), or on the ulna and radius in common with the biceps (rabbit, Fig. 7-22).

The **biceps brachii** lies on the anteromedial surface of the humerus (Figs. 7-20, 7-25, and 7-26). To see it clearly, reflect the pectoralis near its insertion on the humerus. The biceps brachii has a single origin by a tendon from the dorsal border of the glenoid cavity on the scapula. This tendon lies in the intertubercular groove between the greater and lesser tubercles of the humerus. The biceps brachii derives its name from the presence, in human beings, of a second head arising from the coracoid process. The muscle forms a prominent belly beneath the humerus, and inserts by a tendon on the radial tuberosity (cat) or on both the radius and ulna (rabbit). Biceps brachii and brachialis are the major flexors of the forearm. The biceps brachii also assists in supination of the forearm.

Clean off connective tissue from the medial side of the shoulder joint, and you will find a short band of muscle arising from the coracoid process and inserting on the proximal end of the humerus. This is the **coracobrachialis** (Figs. 7-25 and 7-26). It helps to pull the humerus toward the body (adduction). In some individuals, a long portion of the coracobrachialis passes to the distal end of the humeral shaft.

♦   **(F) MUSCLES OF THE ANTEBRACHIUM**

- ♦ extensors
  - brachioradialis (cat)
  - extensor carpi radialis
  - extensor carpi ulnaris
  - extensor digitorum communis
  - extensor digitorum lateralis
  - extensor digiti I
  - extensor digiti II
  - abductor pollicis longus
  - supinator (cat)

- ♦ flexors
  - articularis cubiti (rabbit)
  - pronator teres
  - flexor carpi radialis
  - flexor carpi ulnaris
  - flexor digitorum superficialis
  - flexor digitorum profundus
  - pronator quadratus (cat)

Before you can study the muscles of the forearm and hand, you will have to remove the very extensive **antebrachial fascia.** (When removing this fascia, be careful not to damage the brachioradialis muscle; see Figs. 7-23 and 7-27). The deeper part of the antebrachial fascia is continuous with the tendons of the tensor fasciae antebrachii, triceps, and (in the cat) pectoralis descendens. It forms a dense fibrous sheet that dips down between many of the muscles and also attaches onto the ulna and radius. At the level of the wrist, part of this fascia forms ligaments that encircle the wrist and hold the muscle tendons in place. A band of dense fibers, the **extensor retinaculum,** bridges the tendon grooves on the extensor surface of the radius, and a comparable **flexor retinaculum** lies on the palmar side of the wrist (Fig. 7-26). The antebrachial fascia continues into the hand and, on the palmar side, helps to form fibrous sheaths, the **vaginal ligaments,** through which the flexor tendons of the fingers run. After the fascia has been removed, separate the major muscles before attempting to identify them.

The muscles of the forearm and hand can be sorted into an extensor group, located on the craniolateral surface of the forearm and the back of the hand, and a flexor group located on the caudomedial surface of the forearm and palm of the hand. In the elbow region, the insertions of the biceps, brachialis, and (in the cat) cleidobrachialis pass between these groups at the cranial border of the arm, and the ulna and insertion of the triceps separate them caudally.

Most of the extensors arise from or near the lateral epicondyle of the humerus. You can recognize the following superficial muscles at the level of the elbow and beginning from the medial (thumb) side of the forearm: a brachioradialis (absent in the rabbit), an extensor carpi radialis complex, an extensor digitorum communis, an extensor digitorum lateralis, and an extensor carpi ulnaris (Fig. 7-27). The **brachioradialis** arises more proximally from the humerus than the other muscles and inserts on the styloid process of the radius. The extensor carpi radialis complex can be divided into a more superficial and cranial **extensor carpi radialis longus** and a deeper and more caudal **extensor carpi radialis brevis.** The tendons of both muscles pass deep to the tendon of the abductor pollicis longus (see later; Fig. 7-27) to insert upon the bases of the second and third metacarpals, respectively. The **extensor carpi ulnaris** has a comparable position on the ulnar border of the forearm, and its broad tendon inserts upon the base of the fifth metacarpal. All these muscles act to extend the hand at the wrist, and the brachioradialis also assists the supinator (see later) in rotating the radius in such a way that the palm is turned to face dorsally (supination).

The long digital extensors are more complex. The **extensor digitorum communis** breaks up into four tendons that pass along the dorsal surface of digits 2 to 5. The tendons are bound to each phalanx of the digits by connective tissue, but the most conspicuous attachments are on the terminal phalanges. The tendon of the **extensor digitorum lateralis** passes through a groove on the ulna and divides at about the level of the wrist into two (rabbit), or three or four (cat), parts that pass down the dorsal surface of digits 2 to 5 or 3 to 5. These tendons at first lie somewhat on the ulnar side of those of the extensor digitorum communis but eventually unite with those of the latter muscle. These two muscles are extensors of digits 2 to 5.

Four muscles lie deep to the preceding ones on the extensor surface of the forearm (Fig. 7-28). Expose them by bisecting and reflecting the muscular portions of the two long digital extensors and the extensor carpi ulnaris. A narrow **extensor digiti I** (or extensor pollicis) and a narrower **extensor digiti II** (or extensor digiti secundi) lie deep to the extensor carpi ulnaris, where they arise, partly in common, from the proximal three fourths of the lateral surface of the ulna. They insert by tendons that extend to the middle phalanx of the first and second digits, respectively. These muscles assist in the extension of these digits. The muscle to the first digit is often missing.

A powerful **abductor pollicis longus** arises from much of the lateral surface of the ulna and adjacent parts of the radius. Its fibers converge, go deep to the tendon of the extensor digitorum communis, and form a tendon that goes superficial to the extensor carpi radialis tendons to insert upon the radial side of the first metacarpal. This muscle abducts the thumb.

Finally, in the cat but not in the rabbit, a **supinator** passes obliquely across the proximal half of the radius deep to the belly of the extensor digitorum communis and proximal to that of the abductor pollicis longus (Fig. 7-28). It arises from the lateral

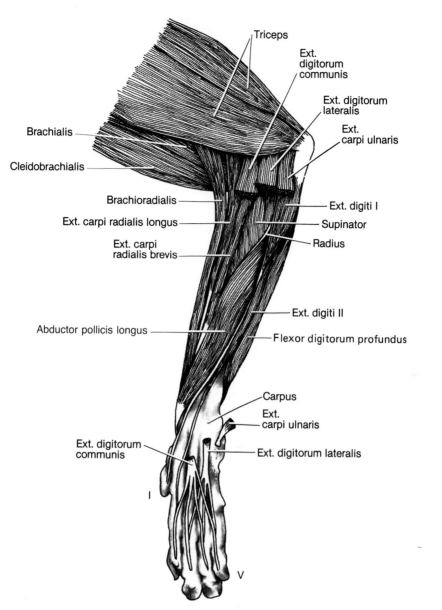

Figure 7-28
Lateral view of the deep extensor muscles of the left forearm of the cat after removal of the extensor digitorum communis, extensor digitorum lateralis, and extensor carpi ulnaris. *(From Walker,* A Study of the Cat.*)*

epicondyle of the humerus and elbow ligaments and inserts upon the radius. Its diagonal muscle fibers enable it to act as a powerful supinator of the hand.

Most of the flexor muscles arise from or near the medial epicondyle of the humerus. In the rabbit, but not in the cat, a small **articularis cubiti** strengthens the medial side of the humeroulnar joint. The remaining superficial muscles are, from cranial to caudal, the pronator teres, flexor carpi radialis, flexor digitorum superficialis, and the flexor carpi ulnaris (Figs. 7-26 and 7-29). The **pronator teres** passes diagonally from the medial epicondyle to the medial border of the radius, and its action (pronation) rotates the forearm in such a way that the palm faces the ground.

The **flexor carpi radialis** extends down the radial border of the forearm and forms a long tendon that passes deep to other tendons in the hand to insert on the proximal ends of the second and third metacarpals. The **flexor carpi ulnaris** has a comparable position on the ulnar border of the forearm. It arises by two heads, one from the medial epicondyle of the humerus and one from the surface of the ulna; it inserts upon the carpal pisiform bone. The chief action of these two muscles is to flex the hand at the wrist.

The **flexor digitorum superficialis** lies between the two carpal flexors in the cat. It arises partly from the medial epicondyle of the humerus and partly from the surface of a deeper muscle (flexor digitorum profundus).[16] (The individual heads will be identified during the dissection of the deeper flexor muscles.) At the level of the wrist it forms a tendon that passes deep to the flexor retinaculum. Part of this tendon has cutaneous attachments to the foot pads, but most of the tendon divides into four tendons that go to the bases of the second through fifth fingers. Near their insertion, each tendon splits and attaches onto the sides of the middle phalanx of each digit. The tendons of the flexor digitorum profundus pass through the arches formed by the splitting of the superficialis tendons and are held in place by these arches. Clean the muscular part of the flexor digitorum superficialis and remove as much of the connective tissue surrounding the tendons as possible, so that you can see the insertions on the phalanges. The flexor digitorum superficialis flexes the digits near their middle. Although present, the flexor digitorum superficialis is very small in the rabbit so that the flexor digitorum profundus is more evident.

In the cat, however, the flexor digitorum superficialis covers the flexor digitorum profundus and must be mobilized and pushed to the side to expose the underlying muscles. With small, pointed scissors, cut through the flexor retinaculum along the medial border of the superficial flexor digitorum, so that you can push the muscle and its tendons laterally and expose the single wide, thick tendon of the **flexor digitorum profundus.** Return to the superficial flexor for a moment and identify its superficial layer with the broad main muscle head and the five tendons leading to each digit. There is also a deep layer of the superficial flexor consisting of a small muscular head arising from the surface of the flexor digitorum profundus (see later) and sending three to four tendons to the digits (Fig. 7-29). Put a blunt probe under the tendon of the flexor digitorum profundus to isolate it from the tendons of the superficial flexor. The five muscle heads of the deep flexor all converge towards the common tendon; you can identify them by following them upward from the tendon (Fig. 7-30). If necessary, bisect the muscle heads of overlying muscles. Identify first the very large ulnar and radial heads. The ulnar head arises from most of the length of the outer border of the ulna, and part of it is visible from the extensor side of the forearm (Fig. 7-27). The radial head is the most medial and deepest of the five heads; it arises from the middle third of the radius, the interosseus ligament stretching between the radius and the ulna, and the adjacent parts of the ulna. The three heads lying between the radial and ulnar heads arise more or less in common from the medial epicondyle of the humerus. Two of these heads lie deep to the flexor digitorum superficialis and the flexor carpi radialis, respectively. The central humeral head is more superficial, and sometimes part of it can be

---

[16]Some authors call the superficial head of this muscle the palmaris longus, but it is not comparable to the human palmaris longus which passes superficially to the flexor retinaculum to end in the palmar aponeurosis. *Nomina Anatomica Veterinaria* does not list a palmaris longus for the mammals that it covers.

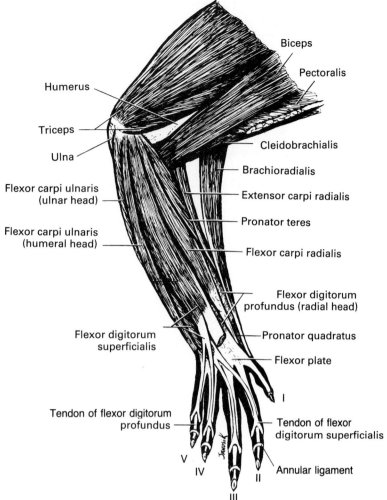

Biceps

Pectoralis

Humerus

Triceps

Ulna

Cleidobrachialis

Flexor carpi ulnaris
(ulnar head)

Brachioradialis

Extensor carpi radialis

Pronator teres

Flexor carpi ulnaris
(humeral head)

Flexor carpi radialis

Flexor digitorum
profundus (radial head)

Pronator quadratus

Flexor digitorum
superficialis

Flexor plate

I

Tendon of flexor digitorum
profundus

Tendon of flexor
digitorum superficialis

V

IV

II

Annular ligament

III

Figure 7-29
Medial view of the flexor muscles of the forearm of the cat after removal of the antebrachial fascia,
flexor retinaculum, palmar aponeurosis, and part of the flexor digitorum superficialis.

seen between the flexor carpi radialis and the flexor digitorum superficialis. Its distal
tendon serves as site of origin for the deep head of the flexor digitorum superficialis.

The common tendon of the flexor digitorum profundus extends into the palm as a
powerful flexor plate and then breaks up into five strong flexor tendons, which run through
ligamentous sheaths, beneath annular ligaments, and finally insert on the terminal
phalanges of the digits. The tendons on the second to fifth fingers perforate the tendons of
the flexor digitorum superficialis. The flexor digitorum profundus flexes all segments of
the digits. Certain small intrinsic muscles in the hand, called **lumbricales,** arise superficially from the flexor plate. Other intrinsic hand muscles are not described here.

Spread apart the radial head and common tendon of the flexor digitorum profundus
and the tendon of the flexor carpi radialis. In the cat, a bluish, shiny fascia will be revealed.
Make a longitudinal incision through this fascia and expose the **pronator quadratus** (Fig.
7-30). Its muscle fibers pass diagonally from their origin on the distal third of the ulna to
their insertion on the outer border of the radius. It assists in hand pronation. Rabbits lack
this muscle along with the brachioradialis and supinator on the extensor surface of the
forearm, for the configurations of their radius and ulna do not permit axial rotation of the
forearm (Fig. 5-6, p. 104).

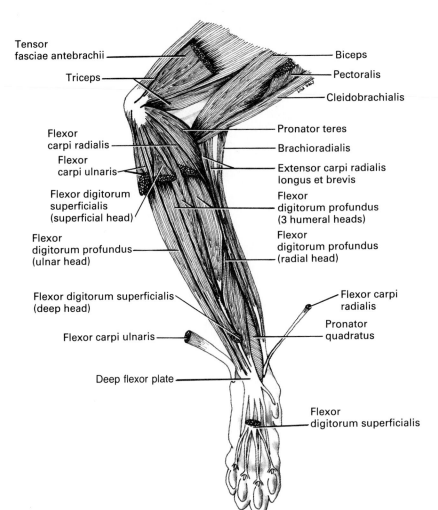

Tensor fasciae antebrachii

Triceps

Biceps

Pectoralis

Cleidobrachialis

Flexor carpi radialis

Flexor carpi ulnaris

Pronator teres

Brachioradialis

Extensor carpi radialis longus et brevis

Flexor digitorum superficialis (superficial head)

Flexor digitorum profundus (3 humeral heads)

Flexor digitorum profundus (ulnar head)

Flexor digitorum profundus (radial head)

Flexor digitorum superficialis (deep head)

Flexor carpi radialis

Pronator quadratus

Flexor carpi ulnaris

Deep flexor plate

Flexor digitorum superficialis

Figure 7-30
Medial view of the deep flexor muscles of the left forearm of the cat after removal of the flexor digitorum superficialis, flexor carpi ulnaris, and flexor carpi radialis. *(From Walker,* A Study of the Cat.*)*

Pelvic Muscles

Although the thigh muscles of all mammals are essentially the same, there are some distinct differences in relationships between certain muscles of cats and rabbits. You should refer frequently to the figures. Clear the fat and superficial fascia from the surfaces of the pelvic region and thigh as a preliminary to the dissection. When doing so, take care not to damage the aponeuroses of the gluteal muscles and of the tensor fasciae latae and biceps femoris (Fig. 7-31). Remove also the large pads of fat lateral to the tail base and in the depression—the **popliteal fossa**—behind the knee. Start to separate the more obvious muscles.

◆ **(A) LATERAL THIGH AND ADJACENT MUSCLES**

- ◆ sartorius
- ◆ tensor fasciae latae
- ◆ biceps femoris

- ◆ semitendinosus
- ◆ caudofemoralis
- ◆ abductor cruris caudalis

The most cranial muscle on the thigh of the cat is the **sartorius** (Fig. 7-31). It is a muscle band extending from its origin on the crest and ventral border of the ilium to its insertion on the patella and medial side of the thigh. Most of the muscle lies on the medial surface of the thigh, but part of it can be seen laterally. In the rabbit, the muscle is entirely on the medial surface (Fig. 7-36). It originates from the inguinal ligament and inserts on the medial tibial condyle. It is fused with the gracilis distally. The sartorius adducts the femur and rotates it forward. It helps to extend the shank.

A tough, white fascia, the **fascia lata,** lies on the lateral surface of the thigh caudal to the sartorius in the cat, or on the craniolateral surface of the thigh in the rabbit. The **tensor fasciae latae** arises from the ventral border of the ilium and from the surface of adjacent muscles, and inserts on the fascia lata. This muscle lies on the lateral surface of the thigh in the cat (Fig. 7-31), but it extends onto the medial side in the rabbit (Figs. 7-32 and 7-36).

The lateral surface of the thigh caudal to the fascia lata is covered by a very broad **biceps femoris** (Figs. 7-31 and 7-32). It has a narrow origin from the tuberosity of the ischium and then fans out to insert by a broad aponeurosis on the patella and much of the tibial shaft. The biceps femoris forms the lateral wall of the popliteal fossa. Its action is to flex the shank and abduct the thigh.

A more slender **semitendinosus** lies caudal to the origin of the biceps femoris and contributes to the medial wall of the popliteal fossa. It also arises from the ischial tuberosity, but in the rabbit a part of it arises from the fascia overlying the biceps femoris.

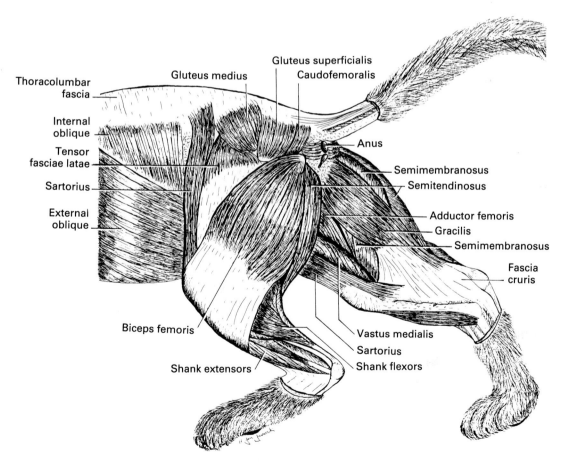

Figure 7-31
Superficial pelvic and thigh muscles of the cat. Lateral muscles can be seen on the left leg and medial muscles on the right leg.

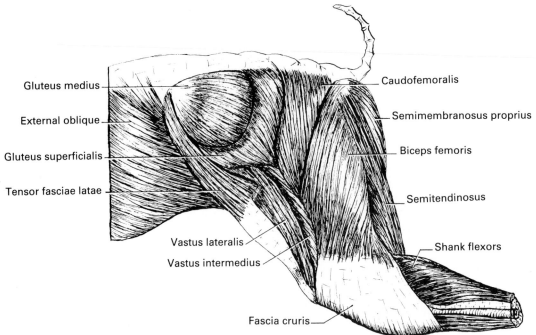

Figure 7-32
Lateral view of the pelvic and thigh muscles of the rabbit. Part of the fascia lata has been removed.

It inserts on the medial side of the proximal end of the tibia. In the rabbit, part of its insertion joins the calcaneus tendon of several of the shank flexors (see later). It acts to flex the shank and retract the thigh.

The **caudofemoralis** (coccygeofemoralis) forms a band of muscle cranial and dorsal to the origin of the biceps femoris. It arises from the more posterior sacral and anterior caudal vertebrae, passes beneath the cranial border of the biceps femoris, and forms a narrow tendon that inserts on the patella in common with part of the insertion of the biceps. Lift up the biceps to follow it. The caudofemoralis abducts and retracts the thigh and extends the shank. It is closely associated with the gluteal muscles (see p. 196) and is sometimes called the gluteobiceps.

Cut across the biceps femoris and caudofemoralis near, but not at, their origins, being careful not to cut a very slender muscle that lies beneath them. This muscle band, which will be seen on reflecting these muscles, is the **abductor cruris caudalis** (tenuissimus) (Fig. 7-33). It arises from a sacral or caudal vertebra, inserts on the tibia and adjacent crural fascia with the biceps femoris, and assists the biceps femoris in thigh abduction and shank flexion. The abductor cruris caudalis is closely associated with the biceps femoris and may have evolved from it.

◆ **(B) GLUTEAL COMPLEX AND DEEPER PELVIC MUSCLES**

◆ gluteus
    gluteus superficialis
    gluteus medius
    gluteus profundus
    gluteus accessorius (rabbit)
◆ piriformis
◆ articularis coxae (cat)

◆ gemellus cranialis
◆ gemellus caudalis
◆ obturator externus
◆ obturator internus
◆ quadratus femoris

The muscle mass covering the dorsolateral surface of the sacrum between the caudofemoralis and sartorius (cat), or between the caudofemoralis and tensor fasciae latae (rabbit), is the gluteus complex (Figs. 7-31 and 7-32). Carefully remove the overlying loose connective tissue. The most superficial part is the gluteus superficialis; the next deeper layer is the gluteus medius. The gluteus superficialis is also somewhat caudal and distal to the gluteus medius. (The gluteus superficialis is homologous to the human gluteus maximus, but it is usually not as large as the gluteus medius in quadrupeds.)

The **gluteus superficialis** arises from the sacral fascia and from the spinous processes of sacral and anterior caudal vertebrae. In the rabbit, part of it also arises from the ventral border of the ilium. Fibers converge to insert on the greater trochanter of the femur (cat), or, in the rabbit, on a third trochanter, which lies on the lateral surface of the femur slightly distal to the greater trochanter.

Bisect the sartorius. Transect the tensor fasciae latae through its muscular part by starting to cut at its cranial border. Recall that the tensor fasciae latae of the rabbit extends onto the medial side of the thigh and is partly united with the vastus medialis (see Fig. 7-36). In the cat, the tensor fasciae latae appears to be almost divided into two portions by a dorsal extension of the fascia lata. The caudal part of the tensor fasciae latae is relatively short and tends to fuse caudally with the gluteus superficialis; if this is the case, separate the two muscles on the basis of their different insertions. The cranial part of the tensor fasciae latae is longer and thicker and covers the anterior surface of the thigh. Dorsally, the cranial part of the tensor fasciae latae tends to fuse with the cranial border of the gluteus medius. You do not need to separate these two muscles now, because later they can be reflected together.

Bisect the belly of the gluteus superficialis and reflect its ends. The **gluteus medius,** which lies partly under the gluteus superficialis, arises from the crest and lateral surface of the ilium and adjacent vertebrae. It inserts upon the greater trochanter of the femur. Both the gluteus superficialis and gluteus medius act primarily as thigh abductors.

Bisect the gluteus medius with a scalpel by cutting slowly deeper and deeper. In the cat, the gluteus medius tends to fuse with the underlying **piriformis;** as soon as you have reached a muscle layer that is heavily interspersed with tendons, you have reached the piriformis. Separate now the gluteus medius from the piriformis by reflecting the gluteus medius; if this is not possible, proceed to cut through the piriformis. A prominent ischiadic nerve emerges from beneath the caudal border of the piriformis. For now, do not cut deeper than the level of the ischiadic nerve. The piriformis arises from the last sacral and first caudal vertebrae and inserts on the greater trochanter of the femur. It acts as an abductor of the femur. Reflect the two halves of the piriformis.

The **gluteus profundus** lies cranial to the piriformis in the rabbit and is associated with the **gluteus acccessorius** (Fig. 7-34). In the cat, the gluteus profundus consists of a cranial part, which is cylindrical and enveloped in a thick, shiny tendinous cover, and a caudal part, which extends caudally under the piriformis as a fan-shaped muscle. The gluteus profundus arises from the lateral surface of the ilium and inserts on the greater trochanter of the femur along the distal border of the insertions of the gluteus medius and piriformis. It acts as an abductor of the thigh.

If you are dissecting the cat, look deep to the cranioventral border of the gluteus profundus (Fig. 7-33). The short muscle observed is the **articularis coxae.** It arises from part of the lateral iliac surface, inserts upon the proximal end of the femur, and helps to flex the thigh. This muscle is absent in the rabbit.

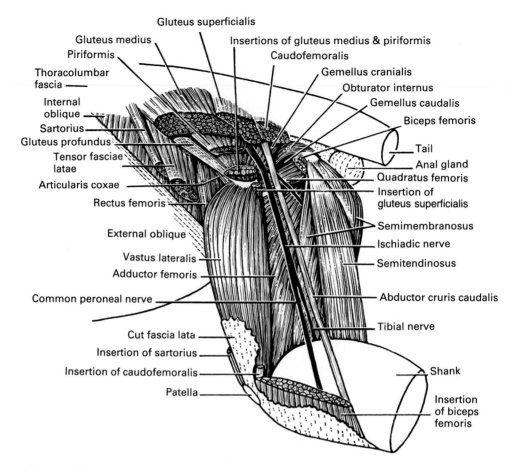

Figure 7-33
Lateral view of the deep muscles of the pelvis and thigh of the cat. The sartorius, tensor fasciae latae, biceps femoris, gluteus medius, gluteus superficialis, caudofemoralis, and piriformis have been largely cut away to expose the deeper muscle. However, the origins and insertions of these muscles are shown as points of reference.

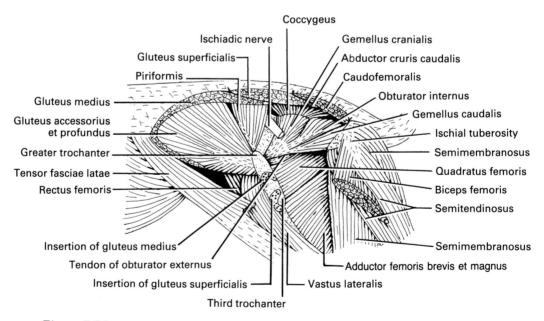

Figure 7-34
Lateral view of the deep pelvic muscles of the rabbit after removal of the caudofemoralis, abductor cruris caudalis, gluteus superficialis, and gluteus medius.

Bisect the ischiadic nerve and the abductor cruris caudalis (if it is not already torn). Three small muscles lie deep to the ischiadic nerve on the lateral surface of the ischium; in the rabbit, but not in the cat, they converge and insert together in the trochanteric fossa (Figs. 7-33 and 7-34). The most cranial of these muscles, the **gemellus cranialis,** arises from the dorsal rim of the ilium and ischium (feel for the bony ridge!). In the cat, it seems to merge cranially with the gluteus profundus. The two muscles can be distinguished on the basis of their different insertions; the feline gemellus cranialis inserts along the proximal border of the insertions of the gluteus medius and piriformis on the greater trochanter of the femur. The **obturator internus** follows caudally to the gemellus cranialis. It originates from the medial surface of the ischium down to the obturator foramen and passes over the dorsal rim of the pelvis, extending cranially and ventrally towards its insertion next to that of the gemellus cranialis. The last of the three muscles, the **gemellus caudalis,** arises from the lateral surface of the ischium just cranial to the ischial tuberosity and inserts caudal to the insertion of the obturator internus. Sometimes, the origin of the gemellus caudalis merges with that of the obturator internus if the latter extends its origin along the caudal rim of the ischium all the way to the ischial tuberosity.

The **coccygeus** is a deep axial muscle (Fig. 7-34). At first, it may look like a dorsal extension of the gemellus cranialis, but it originates from the dorsal edge of the ilium and ischium, extends dorsally to the origin of the gemellus cranialis, and inserts on the tail vertebrae. It helps to form the wall of the pelvic canal.

The **quadratus femoris** is the rather thick band of muscle distal to the gemellus caudalis. It arises from the ischial tuberosity deep to the origin of the biceps femoris and passes to insert on the femur at the bases of the greater and lesser trochanters. It is primarily a thigh retractor.

Separate the gemellus caudalis and quadratus femoris. The **obturator externus** can be seen deep between them, as a muscle in the cat and as a tendon in the rabbit. It arises on the lateral surface of the pelvis from the borders of the obturator foramen and inserts deep in the trochanteric fossa. Its action is thigh rotation and retraction.

◆   **(C) QUADRICEPS FEMORIS COMPLEX**

    ◆ quadriceps femoris
       rectus femoris
       vastus lateralis
       vastus medialis
       vastus intermedius

The front portion of the thigh of mammals is covered by a group of four muscles that insert in common on the patella and **patellar tendon,** which, in turn, attaches to the tuberosity of the tibia. The whole complex is often referred to as the **quadriceps femoris,** and it is the primary shank extensor. The patella is a sesamoid bone.[17] It helps form the cranial surface of the knee joint and permits the common tendon of these muscles to slide easily across the joint.

Reflect and separate the tensor fasciae latae from adjacent muscles and clean the area exposed. The large muscle on the craniolateral surface of the thigh, which was largely covered by the tensor fasciae latae, is the **vastus lateralis** (Figs. 7-32 and 7-33). It arises from the greater trochanter and adjacent parts of the femoral body. A **rectus femoris** lies on the cranial thigh surface medial to the vastus lateralis (Figs. 7-35 and 7-36). Since it arises from the ilium just cranial to the acetabulum, it also acts across the hip joint and helps to protract the thigh. A **vastus medialis** lies on the medial surface of the thigh caudal to the rectus femoris. It arises from the femoral body (Figs. 7-31 and 7-36). The **vastus intermedius** of the cat can be found by dissecting deeply between the vastus lateralis and rectus femoris. It arises from the femur lateral to the origin of the vastus medialis. It has a similar origin and similar relationships in the rabbit but is larger, so that part of the muscle can be seen on the lateral surface of the thigh between the vastus lateralis and caudofemoralis if these muscles are pulled apart (Fig. 7-32).

♦ (D) CAUDOMEDIAL THIGH MUSCLES

- ♦ gracilis
- ♦ semimembranosus
    semimembranosus proprius (rabbit)
    semimembranosus accessorius (rabbit)

- ♦ pectineus
- ♦ adductor femoris
    adductor longus
    adductor brevis et magnus

The medial surface of the thigh caudal to the sartorius and quadriceps femoris is largely covered by the **gracilis** (Figs. 7-31, 7-35, and 7-36), a broad, thin muscle that arises from the pubic and ischial symphyses. It inserts by an aponeurosis onto the tibia and crural fascia. The distal part of the rabbit's sartorius units with it. The gracilis adducts and retracts the thigh, and flexes the crus.

Cut through the gracilis and reflect its ends. Notice again the semitendinosus seen earlier from the lateral surface (see p. 187). It forms much of the medial wall of the popliteal fossa. The broad, thick muscle lying deep to the gracilis and cranial to the semitendinosus is the **semimembranosus.** It arises from the tuberosity and caudal border of the ischium and inserts upon the medial epicondyle of the femur and the adjacent part of the tibia. It is a retractor of the thigh and flexor of the shank. The semimembranosus of the rabbit can be divided into two parts: the **semimembranosus proprius,** which you have exposed, and a **semimembranosus accessorius** imbedded within the proprius. In order to see the latter, carefully cut through the belly of the semimembranosus proprius until you come to a darker, cylindrical portion. This is the semimembranosus accessorius. Its origin is limited to the ischial tuberosity and its insertion to the medial condyle of the tibia.

---

[17]Sesamoid bones are embedded in tendons (less frequently, in ligaments), where tendons are subjected to excessive pressure and friction. The sesamoid bones become part of the joints they overlie.

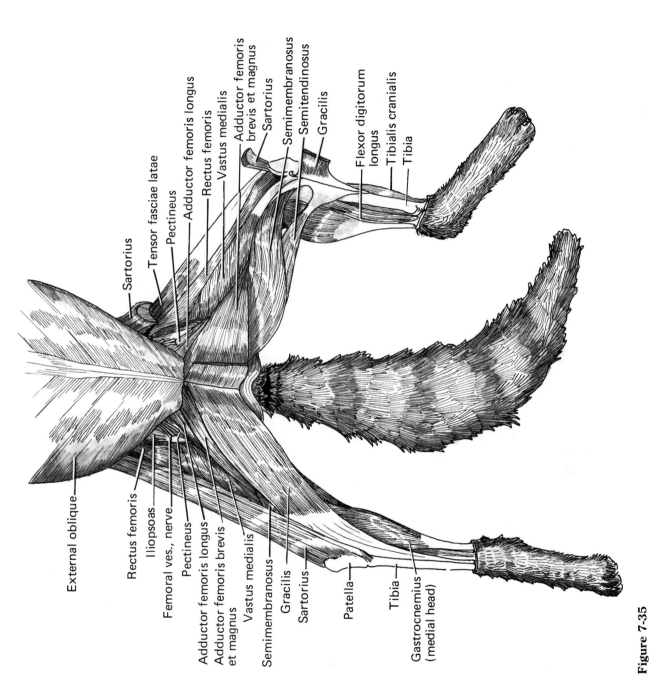

Figure 7-35
Medial view of the thigh muscles of the cat. Superficial muscles are shown on the left side of the drawing; deeper muscles are shown on the right side.

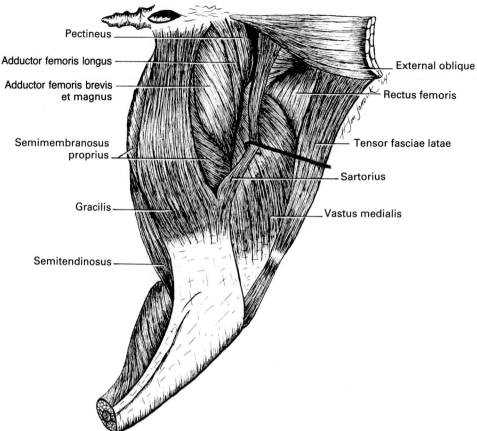

Pectineus

Adductor femoris longus

Adductor femoris brevis
et magnus

Semimembranosus
proprius

Gracilis

Semitendinosus

External oblique

Rectus femoris

Tensor fasciae latae

Sartorius

Vastus medialis

Figure 7-36
Medial view of the thigh muscles of the rabbit.

A small, triangular **pectineus** arises from the cranial border of the pubis just caudal to the point where the femoral blood vessels emerge from the body wall. It inserts on the femoral body beside the origin of the vastus medialis. It is a thigh adductor.

An **adductor femoris** complex of muscles lies between the pectineus and semimembranosus and takes its origin from the cranial border of the pubis and from the pubic and ischial symphyses. The adductor femoris inserts along the femoral body and adducts the thigh. The slender, cranial component of the muscle is an **adductor femoris longus,** and the rest constitutes an **adductor femoris brevis et magnus.** These subdivisions are less clear in the cat than in the rabbit.

♦ **(E) ILIOPSOAS COMPLEX AND ADJACENT
MUSCLES**

- ♦ iliopsoas
  - psoas major
  - iliacus

- ♦ subvertebral muscles
  - psoas minor
  - quadratus lumborum

Observe the thick bundle of muscle that emerges from the body wall medial to the origin of the rectus femoris. This is the iliopsoas complex. You may study it now or postpone it until you have dissected the abdominal viscera. If you choose to dissect it now, trace the bundle forward by cutting through the muscle layers of the abdominal wall. The muscle complex lies in a retroperitoneal position. The main part of the bundle represents the **psoas major** (Fig. 11-29, p. 384). This portion arises primarily from the bodies of the last two or three thoracic vertebrae and from the bodies of all the lumbar vertebrae. It inserts on the lesser trochanter of the femur and protracts and rotates the thigh.

Lateral and slightly dorsal to its extreme caudal portion, you will see a group of muscle fibers arising from the ventral border of the ilium and, in the rabbit, from the adjacent vertebrae. These muscle fibers represent the **iliacus.** They insert and act in common with the psoas major. The psoas major and iliacus are more intimately united in the cat than in the rabbit; together they may be called the **iliopsoas.**

The thin muscle medial to the psoas major is the **psoas minor** (Fig. 11-29). In the cat, it arises from the bodies of the caudal thoracic and cranial lumbar vertebrae; in the rabbit, it arises from the bodies of the caudal lumbar vertebrae. It inserts on the pelvic girdle near the origin of the pectineus. The psoas minor is one of the subvertebral hypaxial muscles referred to on page 169, rather than an appendicular muscle. Its action is flexion of the back.

Lift up the lateral border of the psoas major near its middle. The thin muscle that lies on the ventral surface of the transverse processes is the **quadratus lumborum,** another subvertebral hypaxial muscle. It arises primarily from the bodies and transverse processes of the lumbar and last several thoracic vertebrae. It extends farther forward in the rabbit, and some of its fibers also arise from the ribs. It inserts on the ilium cranial to the origin of the iliacus muscle. Its action is lateral flexion of the vertebral column.

♦ **(F) MUSCLES OF THE SHANK**

- ♦ shank flexors
  - triceps surae
    - gastrocnemius (two heads)
    - soleus
  - flexor digitorum superficialis
  - popliteus
  - flexor digitorum longus
  - flexor hallucis longus
  - tibialis caudalis

- ♦ shank extensors
  - tibialis cranialis
  - extensor digitorum longus
  - peroneus longus
  - peroneus brevis
  - peroneus tertius

The shank is covered by the tough **fascia cruris,** which is partly united with the tendons of certain thigh muscles, including the biceps femoris and gracilis. Remove the fascia cruris, bisect and reflect the gracilis and semitendinosus, and reflect the sartorius, which was bisected earlier. Try to leave the tendons of the reflected muscles intact. Separate the more obvious shank muscles. They are very similar in the cat and rabbit.

As was the case in the forearm and hand, the muscles of the shank and foot fall into extensor and flexor groups, but the groups are not quite as clearly separated as in the forearm. The extensors lie on the craniolateral surface of the shank; the flexors lie on the caudomedial surface. They are separated by an exposed strip of the tibia on the medial side (see Fig. 7-38); the position of the fibula indicates their separation laterally, although the fibula is not exposed at the surface (Fig. 7-37).

The large calf muscle on the caudal surface of the shank is a functional unit, the shank flexor, composed of the gastrocnemius, flexor digitorum superficialis, and soleus (Figs. 7-37 and 7-38). The lateral head of the **gastrocnemius** arises from the lateral epicondyle of the femur, lateral surface of the patella, and adjacent parts of the tibia and, in the cat, by a small slip from the crural fascia. Its medial head arises from the medial epicondyle of the femur. Although a part of the **soleus** is visible on the lateral surface of the shin (Fig. 7-37), most of it lies deep to the lateral head of the gastrocnemius, where it arises from the proximal one third of the fibula (cat), or by a narrow tendon from the proximal end of the fibula (rabbit). Because the tendons of the two heads of the gastrocnemius and the soleus converge, these muscles together are sometimes called the **triceps surae.** Dissecting from the caudal surface of the shank, separate the two heads of the gastrocnemius. The muscle lying between them is the **flexor digitorum superficialis** (plantaris of some authors). It takes its origin deep to the lateral head of the gastrocnemius from the lateral epicondyle of the femur and adjacent part of the patella. Its tendon joins that of the triceps surae to form a large common tendon, the **calcaneus tendon** (Achilles tendon), which inserts upon the calcaneus. As mentioned on p. 188, part of the tendon of the semitendinosus joins this one in the rabbit. Remove the connective tissue around the common calcaneus tendon, so that you can separate the individual tendons of the gastrocnemius, soleus, and flexor digitorum superficialis. You will see that the tendon of the flexor

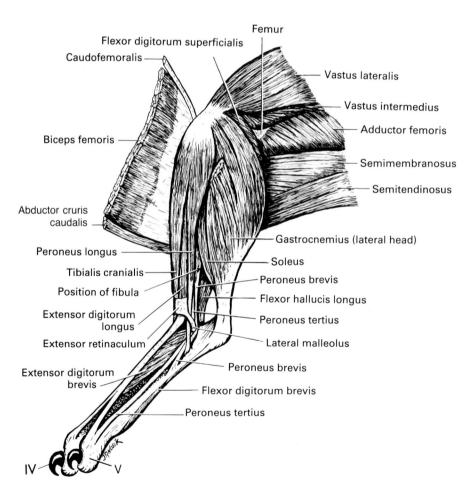

Figure 7-37
Lateral view of the shank muscles of the cat after reflection of the biceps femoris and removal of the crural fascia.

digitorum superficialis forms a cap over the calcaneus, the **calcaneal cap,** and extends distally to the tips of the toes (see later). From an evolutionary viewpoint, these muscles belong to the ventral limb musculature or flexors, and their action upon the foot is one of plantar flexion. (This action is often called "extension of the foot" in human anatomy.) These muscles are particularly important in thrusting the foot upon the ground and pushing the body upward during the propulsive phase of a step—hence their large size.

The remaining four flexor muscles of the shank can be seen from the medial side, where they lie between the tibia and the group just described (see Fig. 7-38), or from the lateral side if you cut through and reflect the triceps surae and flexor digitorum superficialis (Fig. 7-39). A triangular **popliteus** arises by a narrow tendon from the lateral epicondyle of the femur, extends toward the medial side of the shank, passing caudal to the knee joint, and fans out to insert upon the proximal one third of the tibia. It helps to flex the shank and also slightly rotates it medially, turning the foot toward the midventral line.

Return for a moment to the calcaneal cap of the flexor digitorum superficialis and follow the superficial flexor tendon distally. A thin, broad muscle sheet, the **flexor digitorum brevis,** attaches to and surrounds much of the superficial flexor tendon. Distally, the superficial flexor tendon splits into four tendons that attach to the four digits and give off branches to the foot pads. The **flexor digitorum longus** arises from the head of the fibula and the shaft of the tibia deep to the insertion of the popliteus. The **flexor**

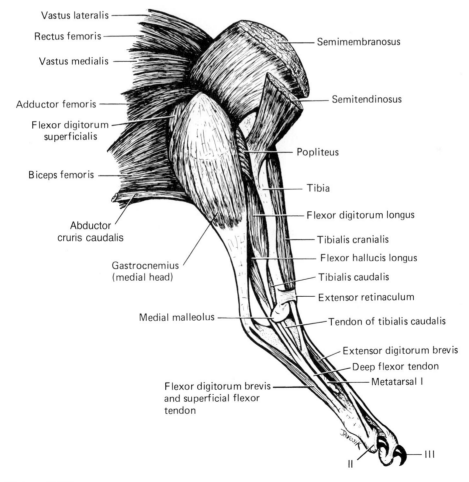

Figure 7-38
Medial view of the shank muscles of the cat after reflection of the tensor fasciae latae, gracilis, semitendinosus, and semimembranosus, and removal of the crural fascia.

hallucis longus arises from the caudal surface of much of the rest of the fibula and tibia lateral to the flexor digitorum longus. Each muscle forms a powerful tendon that passes caudal to the medial malleolus of the tibia before uniting to form a broad tendon, the deep flexor tendon, which covers much of the sole of the foot deep to the superficial flexor tendon and the flexor digitorum brevis. To expose the deep flexor tendon down to the tips of the toes, free the medial border of the flexor digitorum brevis from underlying connective tissue and push the muscle with the superficial flexor tendon to the side. At the level of the toes, the deep flexor tendon breaks up into four tendons that extend down the flexor side of the digits to the terminal phalanges. All these muscles flex the toes and assist in plantar flexion of the whole foot.

Last of the flexors is a small **tibialis caudalis,** which lies between the flexores digitorum longus and hallucis longus. Its origin from the fibula and tibia extends proximally to the insertion of the popliteus. Its tendon of insertion, too, passes caudal to the medial malleolus and inserts upon certain distal tarsals. It assists the other muscles in plantar flexion of the foot.

The extensor musculature of the shank is much less massive than the flexors, for it is involved primarily in the recovery movements of the limb. Most cranial of the extensors is

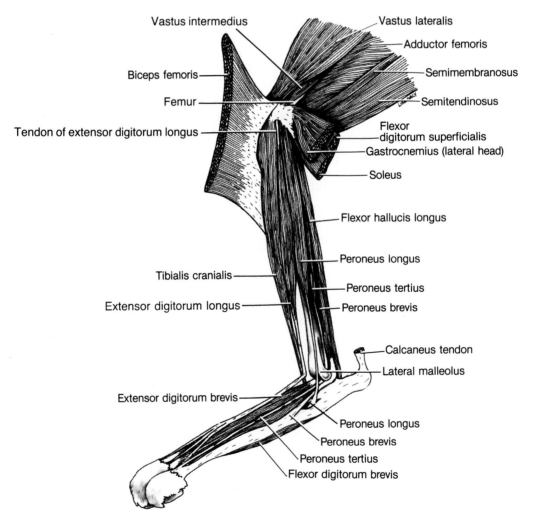

Figure 7-39
Lateral view of the left shank muscles of the cat after removal of the gastrocnemius, flexor digitorum longus, and soleus. *(From Walker, A Study of the Cat.)*

the **tibialis cranialis** (Figs. 7-38 and 7-39). It arises from about the proximal one third of the fibula and adjacent parts of the tibia, forms a long tendon that crosses the cranial surface of the tibia, goes beneath a band of connective tissue near the ankle joint, the **extensor retinaculum,** and finally inserts upon the first metatarsal. Together with other muscles in this group, it extends the foot—a motion often called dorsiflexion.

The remaining extensors all lie on the lateral side of the shank (Fig. 7-39). An **extensor digitorum longus** is located caudal and deep to the tibialis cranialis, which partly covers it. It arises from the lateral epicondyle of the femur by a tendon that traverses the knee joint capsule. After passing beneath the extensor retinaculum near the ankle joint, the muscle breaks up into four tendons that pass down the dorsum of digits 2 to 5, finally inserting on the terminal phalanges. Its attachments on the digits are closely associated with those of intrinsic foot muscles, including the **extensor digitorum brevis.** The extensor digitorum longus extends the digits and assists in dorsiflexion of the foot.

A peroneus complex lies caudal to the extensor digitorum longus and takes its origin from the full length of the fibula. The complex is subdivided into three components, which, although they arise from different parts of the fibula, are most distinct at their insertions (Fig. 7-39). The tendon of the **peroneus longus** passes through a groove on the surface of the lateral malleolus of the tibia and then runs through a diagonal groove deep in the sole of the foot attaching onto metatarsals 2 to 4. Its primary action is abduction and eversion of the foot, and it assists in plantar flexion. The tendons of the **peroneus tertius** and **peroneus brevis** pass through a groove on the caudal border of the lateral malleolus. That of the peroneus tertius continues down the dorsum of the fifth digit, finally uniting with the tendon of the extensor digitorum longus. The much stouter tendon of the peroneus brevis inserts onto the lateral side of the fifth metatarsal. These muscles help extend the foot and toes.

Cranial Trunk Muscles

The caudal trunk muscles were described before the appendages were dissected, and some other trunk muscles were seen during the dissection of the shoulder. Now that the appendages have been examined, it is possible to resume the study of the cranial trunk muscles.

◆ **(A) HYPAXIAL MUSCLES**

◆ acting on pectoral girdle
  omotransversarius
  rhomboideus cervicis et thoracis
  rhomboideus capitis
  serratus ventralis

◆ part of thoracic wall
  rectus abdominis
  external intercostals
  internal intercostals
  transversus thoracis
  scalenus
  serratus dorsalis

◆ subvertebral
  longus colli

All the trunk muscles that become associated with the pectoral girdle belong to the hypaxial group. They are the **omotransversarius, rhomboideus cervicis et thoracis, rhomboideus capitis,** and **serratus ventralis.** Find them again.

Lay your specimen on its back, reflect the pectoralis, and examine the muscles on the ventrolateral portion of the thoracic wall. The **rectus abdominis** will be seen passing forward to its insertion on the sternum and cranial costal cartilages (see Figs. 7-19, 7-23, and 7-40). The thoracic wall is composed of three layers of muscle comparable to those of the abdominal wall. Observe that the outermost layer, the **external intercostals,** consists of fibers that pass from one rib caudally and ventrally to the next caudal rib. This layer does not extend all the way to the midventral line. Cut through and reflect an external intercostal and you will expose an **internal intercostal.** Its fibers extend cranially and ventrally. The third layer, the **transversus thoracis,** is incomplete and is found only near the midventral line. To see it, lift up the rectus abdominis, and cut through and reflect the ventral portion of an internal intercostal. The transversus thoracis arises from the dorsal surface of the sternum and inserts by a number of slips onto the internal surfaces of the costal cartilages; its muscle fibers run parallel to the ribs. You will have a better view of the muscle when you open the thorax (p. 322).

In addition to these muscle layers, other muscles are associated with the thoracic wall (Fig. 7-40). The diagonal muscle that arises near the middle of the sternum and crosses the cranial end of the rectus abdominis to insert on the first rib is the **rectus thoracis.** Dorsal to it is a fan-shaped muscle complex that extends between the cervical vertebrae and the ribs. This is the **scalenus.** It arises from the transverse processes of most of the cervical vertebrae (cat), or the last four (rabbit), and has multiple insertions on various ribs. In the cat, one portion of its insertion extends as far caudad as the ninth or tenth rib. In the rabbit, much of the insertion passes between the cranial and caudal halves of the serratus ventralis.

Turn your specimen on its side, reflect the latissimus dorsi, and pull the top of the scapula away from the trunk. Medial to the scapula and serratus ventralis, you will see a number of short muscular slips that arise from the thoracolumbar fascia and insert on the dorsal portion of the ribs. These constitute the **serratus dorsalis** (Fig. 7-40). A cranial part with muscle fibers directed obliquely can be distinguished from the caudal part, whose muscle fibers are oriented more transversely. The last slip of the caudal part inserts on the last rib; you can see it if you reflect the latissimus dorsi.

All these thoracic muscles, together with the muscular diaphragm and the abdominal muscles, are involved with respiratory movements. The muscular movements of respiration are very complex. During inspiration, the volume of the thoracic cavity is increased by the contraction of the diaphragm, which compresses the abdominal viscera, and by the forward movement of the ribs caused by the actions of such muscles as the scalenus and the cranial portion of the serratus dorsalis. During expiration, thoracic volume is decreased because these muscle relax; the diaphragm is pushed cranially by the contraction of abdominal muscles and the pressure of the abdominal viscera; and the ribs move caudally by their elastic recoil and by the action of such muscles as the rectus thoracis and the caudal part of the serratus dorsalis. Electromyographic studies in humans have shown that the intercostal muscles do not play a crucial role in respiratory movements; they primarily maintain a constant distance between the ribs as other forces expand and contract the thoracic cage. It is not known whether this is also the case in quadrupeds.

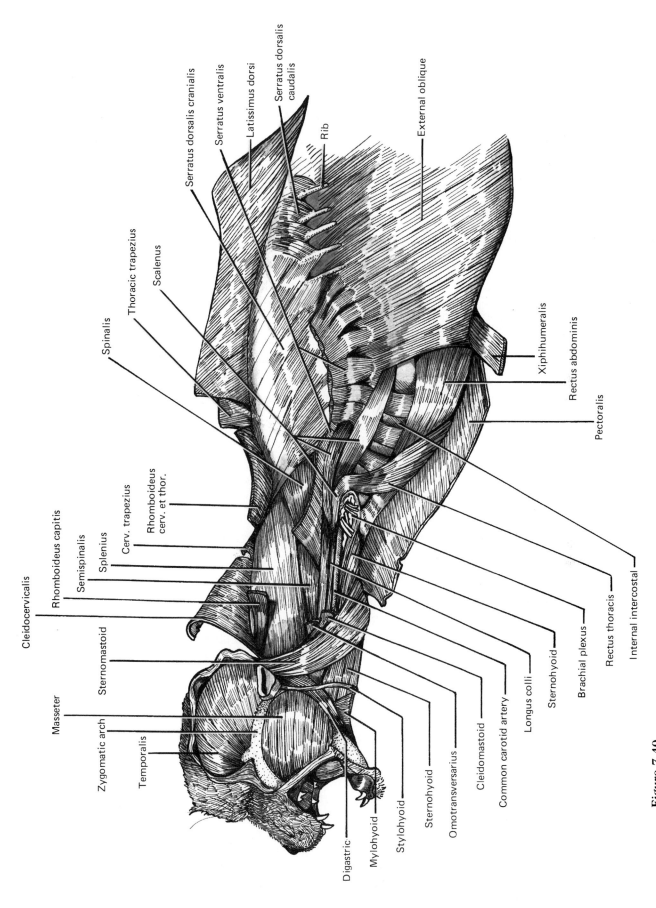

Figure 7-40
Lateral view of the muscles of the neck and thorax of the cat after removal of the shoulder and arm. *(From Walker, A Study of the Cat.)*

Masseter

Zygomatic arch

Temporalis

Cleidocervicalis

Rhomboideus capitis

Sternomastoid

Semispinalis

Splenius

Spinalis

Cerv. trapezius

Thoracic trapezius

Rhomboideus cerv. et thor.

Scalenus

Serratus dorsalis cranialis

Serratus ventralis

Latissimus dorsi

Serratus dorsalis caudalis

Rib

External oblique

Xiphihumeralis

Rectus abdominis

Pectoralis

Internal intercostal

Rectus thoracis

Brachial plexus

Sternohyoid

Common carotid artery

Longus colli

Cleidomastoid

Omotransversarius

Sternohyoid

Stylohyoid

Mylohyoid

Digastric

The subvertebral portion of the hypaxial musculature in the neck and thoracic regions is represented primarily by the **longus colli** (Fig. 7-40). This muscle appears as a band in the neck lying ventral and medial to the origin of the scalenus. It arises from the ventral surfaces of the bodies of the first six thoracic vertebrae, and as it passes forward it receives other slips of origin from the transverse processes and bodies of the cervical vertebrae. Portions of it insert on the bodies and transverse processes of each of the cervical vertebrae. Its action is ventral and lateral flexion of the neck.

♦ **(B) EPAXIAL MUSCLES**

- ♦ splenius
- ♦ multifidi
- ♦ spinalis

- ♦ erector spinae
  longissimus dorsi
  longissimus capitis
  iliocostalis
- ♦ semispinalis cervicis et capitis

Much of the thoracolumbar fascia was reflected during the study of the caudal epaxial muscles and the serratus dorsalis. Complete the reflection of this fascia. Identify and clean the surface of the serratus dorsalis (p. 202), and bisect the entire muscle. When doing this, make sure to cut through the muscular portion, not through the prominent aponeurosis. The epaxial mass should now be well exposed. Find the multifidi and erector spinae caudally (p. 170) and trace them foward.

The **multifidi** can be seen most clearly in the lumbar region. They consist of bundles of muscle fibers that extend from the mamillary processes, transverse processes, and zygapophyses of caudally lying vertebrae to the spinous processes of more cranial vertebrae. Most bundles cross two vertebrae between their origin and insertion. More cranial parts of the multifidi lie deep to the spinalis and will not be seen.

The **spinalis** lies lateral to the spinous processes of the thoracic and the more caudal cervical vertebrae. In the cat, most of it arises from the inner surface of the fascia covering the erector spinae, and for this reason the spinalis is sometimes considered a division of the erector spinae. Deeper parts arise from the dorsal surface of the vertebrae. The spinalis inserts on the spinous processes of the vertebrae. The spinalis of the rabbit is similar except that it is not as intimately associated with the erector spinae. The muscle fibers of the spinalis, like those of the multifidus, extend diagonally from caudolateral to craniomedial.

As the **erector spinae** continues forward from its origin on the iliac crest and dorsal surfaces of the more caudal trunk vertebrae, it splits into a **longissimus dorsi** lying lateral to the spinalis and a more lateral **iliocostalis.** You may have to flip the latissimus dorsi back and forth to see the transition between the caudal erector spinae and its cranial extensions, the longissimus dorsi and the iliocostalis. Muscle fibers of these muscles extend diagonally from caudomedial to craniolateral; those of the iliocostalis insert on the ribs, and those of the longissimus insert chiefly on the transverse processes of thoracic and cervical vertebrae.

Before we can study the rest of the epaxial muscles, we need to reflect some muscles

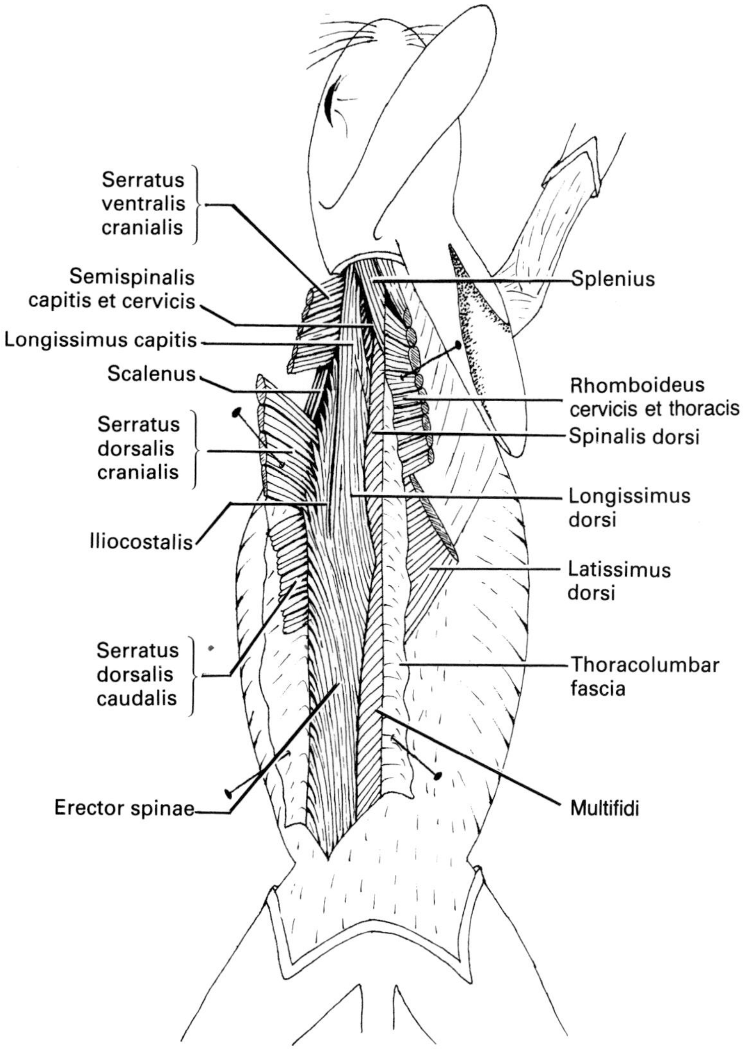

Figure 7-41
Dorsal view of the epaxial muscles on the left side of the rabbit after reflection of the superficial muscles and thoracolumbar fascia.

of the sternocleidomastoid group (p. 172). Find the cleidocervicalis and its insertions on the skull and neck; then find the cleidomastoid and its attachment on the clavicle, deep to that of the cleidocervicalis. Separate the two muscles. Bisect the cleidocervicalis between the clavicle and the attachment on the neck and skull. (Because the cleidobrachialis was bisected earlier, only the middle portion of the brachiocephalicus and the origin of the cleidomastoid remains attached to the clavicle.) Reflect the dorsal half of the cleido-cervicalis after having separated its cranial border from the caudal border of the sterno-mastoid (p. 173). Clean the surface of the sternomastoid and separate the salivary glands from its surface (without damaging or removing them), until you can see its attachment to the mastoid process of the skull. Bisect the sternomastoid and reflect its halves; then reflect the exposed cleidomastoid (with the clavicle). Now the entire splenius is revealed.

The most superficial of the cranial epaxial muscles is the **splenius** (Figs. 7-41 and 7-42). It is a thin but broad triangular muscular sheet that covers the back of the neck. The splenius arises from the middorsal line of the neck and passes forward and laterally to insert on the occipital region of the skull (nuchal crest) and transverse process of the atlas.

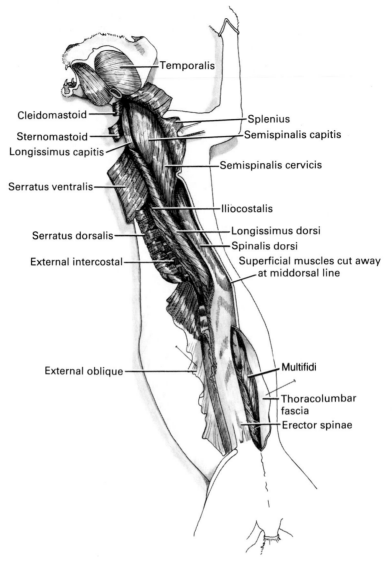

Figure 7-42
Dorsal view of the epaxial muscles of the cat after removal of the superficial muscles and reflection of the splenius. *(From Walker,* A Study of the Cat.*)*

Each splenius individually acts as a lateral flexor of the head; together they elevate the head (extension). Identify its ventral border and separate it from the adhering **longissimus capitis,** which extends cranially from the longissimus dorsi.

Bisect the splenius perpendicularly to its muscle fibers (there is no dorsal border to speak of). Reflect the two flaps of the splenius and expose the underlying **semispinalis cervicis et capitis.** In the cat, the semispinalis cervicis, which has a parallel arrangement of muscle fibers, can be distinguished from the semispinalis capitis, which is more fan-shaped. The semispinalis cervicis et capitis arises from the vertebrae between the cranial ends of the spinalis and longissimus capitis and inserts on the back of the skull.

Deeper epaxial muscles are not described here. Collectively, the epaxial muscles are extensors and lateral flexors of the back, neck, and head.

As shown in Figure 7-13, the epaxial muscles of reptiles fall into three groups—transversospinalis, longissimus, and iliocostalis. The epaxial group becomes further complicated in mammals, but the reptilian subdivisions can still be recognized. The transversospinalis system of reptiles is represented by several deep muscles (interspinalis, intertransversarii, occipitals) and by the multifidus spinae, spinalis, and semispinalis; the longissimus is represented by the longissimus and splenius and the iliocostalis by the iliocostalis.

Hypobranchial Muscles

The hypobranchial muscles are a hypaxial group located on the ventral side of the neck and throat (see p. 145). All move the larynx, hyoid apparatus, and tongue. The muscle group plays a crucial role in opening the mouth, in manipulating food in the mouth, and in swallowing. Complete the skinning of this region as far forward as the chin. Be careful not to damage the thin and fragile stylohyoid muscle (see Fig. 7-20), which lies directly under the platysma. Clean away the loose connective tissue and fat. You may turn back and push aside, but not destroy, prominent glands, ducts, blood vessels, and nerves.

♦ **(A) POSTHYOID MUSCLES**

- ♦ sternohyoid
- ♦ sternothyroid
- ♦ thyrohyoid

Find the sternomastoid muscles (p. 173) and reflect them. The **sternohyoid** is a thin, midventral band of muscle that covers the windpipe (trachea) and extends from its origin on the cranial end of the sternum to its insertion on the hyoid (Figs. 7-43 and 7-44). The sternohyoid occurs as a pair of muscles, but they are generally fused in the cat.

Carefully separate the sternohyoid from another band of muscle that lies dorsal and lateral to it. This band, the **sternothyroid,** has a similar origin but passes forward to insert on the thyroid cartilage of the larynx. The larynx, or "Adam's apple," lies caudal to the hyoid. Thus the sternothyroid is not as long a muscle as the sternohyoid. A short band of muscle, the **thyrohyoid,** lies on the lateral surface of the larynx. It arises at the point of insertion of the sternothyroid and passes forward to insert on the hyoid. Sternothyroid and thyrohyoid appear as one band unless their attachments on the larynx are carefully exposed.

♦ **(B) PREHYOID MUSCLES**

- ♦ geniohyoid
- ♦ hyoglossus
- ♦ genioglossus
- ♦ styloglossus
- ♦ lingualis proprius

Three branchiomeric muscles must be studied and reflected in order to expose the prehyoid muscles. Note the stout band of muscle that is attached to the ventral border of the mandible. This is the **digastric** (Fig. 7-43). In the cat, it has a fleshy origin from the

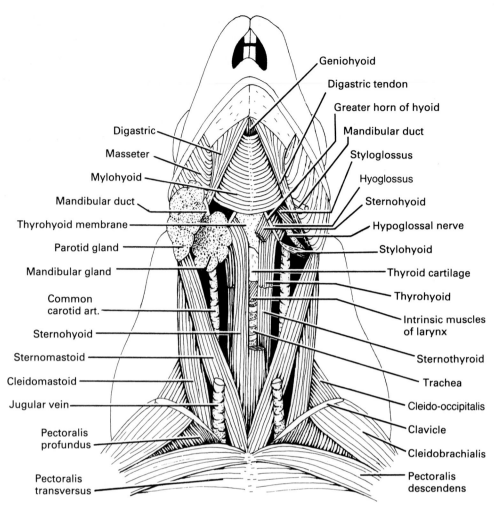

Digastric

Masseter

Mylohyoid

Mandibular duct

Thyrohyoid membrane

Parotid gland

Mandibular gland

Common
carotid art.

Sternohyoid

Sternomastoid

Cleidomastoid

Jugular vein

Pectoralis
profundus

Pectoralis
transversus

Geniohyoid

Digastric tendon

Greater horn of hyoid

Mandibular duct

Styloglossus

Hyoglossus

Sternohyoid

Hypoglossal nerve

Stylohyoid

Thyroid cartilage

Thyrohyoid

Intrinsic muscles
of larynx

Sternothyroid

Trachea

Cleido-occipitalis

Clavicle

Cleidobrachialis

Pectoralis
descendens

Figure 7-43
Ventral view of the muscles in the neck and floor of the mouth of a rabbit. Salivary glands and the sternohyoid muscle have been removed from the right side of the drawing.

paracondyloid and mastoid processes of the skull. In the rabbit, its origin is by a tendon from just the paracondyloid process. The digastric muscle derives its name from its condition in human beings in which it is divided by a central tendon into two distinct bellies. Sometimes cats have a tendinous intersection in the center of the muscle. It inserts along the ventral border of the lower jaw and is the primary jaw-opening muscle. A narrow, vulnerable ribbon of muscle, the **stylohyoid** (Figs. 7-20, 7-43, and 7-44), crosses the middle of the digastric. Cut and reflect it. Now disconnect the digastric from the mandible and reflect it. The sheet of more or less transverse fibers that lies between and deep to the insertions of the digastric muscles of opposite sides of the jaw is the **mylohyoid.** It arises from the mandible and inserts on a medial tendinous intersection, or raphe, and on the hyoid. It acts to raise the floor of the mouth. Make a longitudinal incision through the mylohyoid and reflect it on one side. It is not very thick.

Longitudinal muscles that lie deep to the mylohyoid constitute the prehyoid portion of the hypobranchial musculature. To see these muscles clearly, it may be necessary to cut through the intermandibular suture and spread the two halves of the lower jaw apart. The midventral band of muscle (really a pair of muscles that have united) is the **geniohyoid**

(Figs. 7-43 and 7-44). It arises from the front of the mandible and inserts on the hyoid. Cut across the geniohyoid and reflect its ends. The band of muscle that arises from the chin deep to the origin of the geniohyoid is the **genioglossus.** It passes caudally and dorsally into the tongue. The muscle that arises from the hyoid lateral and deep to the insertion of the geniohyoid is the **hyoglossus.** It passes forward into the tongue. Pull the tip of the tongue and you will note that the muscle is moved. The band of muscle that arises from the mastoid process at the base of the skull and passes forward into the tongue is the **styloglossus.** It lies lateral to the rostral portion of the hyoglossus. Hyoglossus and styloglossus are distinct at their origins, but they merge as they extend into the tongue. The glossus muscles, together with intrinsic muscle fibers within the tongue collectively called the **lingualis proprius,** form the substance of the tongue and manipulate this organ.

The body of the tongue in mammals consists mainly of muscles and connective tissue. The hyoid skeleton serves only as an anchor place for certain tongue muscles but does not enter the body of the tongue. Smith and Kier (1989) called such a structure a "muscular hydrostat." When the intrinsic tongue musculature contracts, the tissues inside the body of the tongue are compressed, and their internal pressure rises. The tissues become rigid under this compression and act as a hydrostatic skeleton.

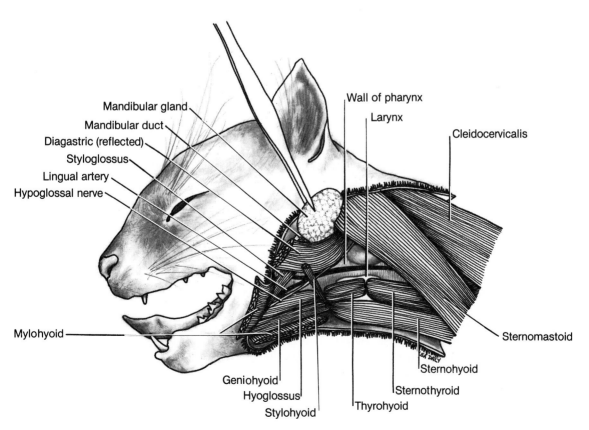

Figure 7-44
Lateral view of the hypobranchial and adjacent muscles of the cat. *(From Walker,* A Study of the Cat.*)*

Branchiomeric Muscles

♦ **(A) MANDIBULAR MUSCLES**

♦ mylohyoid
♦ digastric (rostral part)
♦ masseter
♦ temporalis

♦ pterygoid
♦ tensor tympani
♦ tensor veli palatini

The **mylohyoid** and the rostroventral half of the **digastric,** which were seen during the dissection of the hypobranchial muscles, are branchiomeric muscles of the mandibular arch. To see other mandibular arch muscles, remove the skin from the top of the head and the cheek region on one side. Also cut off the auricle. The platysma, facial muscles (both belonging to the hyoid group; see later), and loose connective tissue must be removed, but be careful not to injure glands, nerves, and vessels in this region. Exercise special care in skinning and cleaning the cheek, for the duct of one of the salivary glands crosses the cheek just beneath the facial muscles. Find the zygomatic arch. The powerful muscle that lies ventral to the arch is the **masseter.** It arises from the arch and inserts in the masseteric fossa and adjacent parts of the mandible (see Figs. 7-21 and 7-22). Deep and superficial layers can be recognized in most mammals, but they are particularly distinct in the rabbit, where the deep layer arises from the caudal part of the zygomatic arch and extends rostrally and ventrally nearly at right angles to the superficial layer.

Another mandibular muscle, the **temporalis,** lies dorsal to the zygomatic arch. In the cat, it is a sizeable muscle and fills the large temporal fossa. It arises primarily from the surface of the cranium, but some fibers spring from the dorsal edge of the zygomatic arch. It passes deep to the zygomatic arch and inserts on the coronoid process of the mandible.

In the rabbit, the temporal fossa and muscle are very small. To see the muscle, cut through a powerful facial muscle that passes from the top of the skull caudally to the base of the auricle. The **temporalis** extends from a point above the base of the auricle to the back of the orbit (see Fig. 7-22). Its insertion is by a tendon that passes down the caudal wall of the orbit to the coronoid process of the mandible.

A **pterygoid** muscle extends from the skull base to the medial side of the lower jaw. You will notice it when you open the mouth and pharynx. Other mandibular muscles are impractical to dissect at this time. A **tensor tympani** attaches to the malleus in the middle ear (p. 236), and a **tensor veli palatini** extends from the base of the skull into the soft palate.

Jaw mechanics are quite different in a cat and in a rabbit (Fig. 7-45). The jaw joint of a cat, as is characteristic of carnivores, is in line with the tooth row, so that the upper and lower jaws come together in the manner of a pair of scissors. The shape of the condyle and mandibular fossa are such that only a hinge action is permitted. However, the lower jaw as a whole can move slightly to the left or right side so that the carnassial teeth can engage more intimately on one side or the other. In the rabbit, as is characteristic of gnawing and herbivorous mammals, the jaw joint is situated well dorsal to the tooth row so that all of the teeth of the upper and lower jaw come together simultaneously. The shape of the condyle and mandibular fossa permits the lower jaw to move back and forth and from side to side, actions that occur in gnawing and grinding.

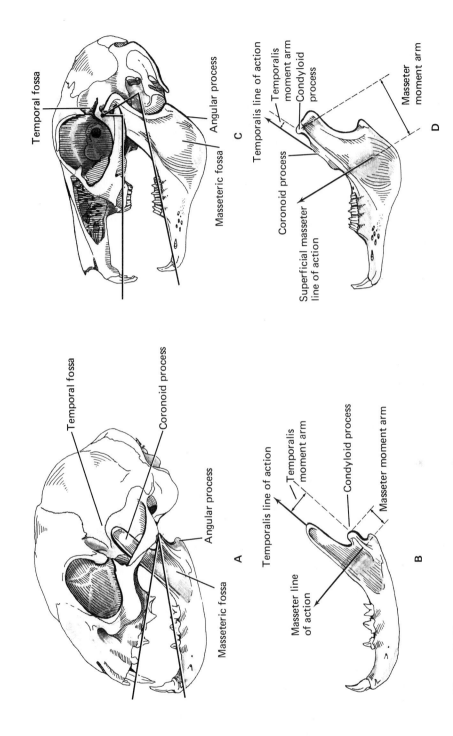

Figure 7-45
Jaw mechanics in a cat (**A** and **B**) and rabbit (**C** and **D**). **A** and **C**, Jaw closure; **B** and **D**, lines of action and moment arms of the masseter and temporalis muscles.

The adductor muscles of the jaws must exert a strong force for closing the jaws, and also balance forces at the jaw joint so that there is little tendency for the jaws to become disarticulated. Both masseter and temporal muscles are large in carnivores, but a low condyloid process and a high coronoid process give the temporal muscle a somewhat longer moment arm (the perpendicular distance between the line of action of the muscle and the jaw joint), and hence a greater mechanical advantage, than in the masseter. Both exert strong forces for closing the jaws. The direction of pull of the temporal muscle also resists any forward pull on the canine teeth that may occur when a carnivore is seizing prey, and the pull of the masseter helps to resist any tendency for the jaw condyle to slip ventrally and caudally out of the mandibular fossa, which could occur if the temporal muscle alone were used in cutting up the food. The pterygoids of carnivores are modest-sized muscles whose actions are similar to those of the masseters.

The masseter of herbivores is a very large muscle divided into superficial and deep layers that are oriented nearly at right angles to each other. The superficial layer provides a particularly strong force for closing the jaws because it has a very long moment arm. Its tendency to pull the lower jaws forward out of the mandibular fossa is opposed by the deep layer. These two layers, and a large pterygoid, provide the forces needed to close the jaws, move them back and forth and from side to side in a grinding action, and balance forces at the jaw joint. The temporal muscle is not an important muscle in herbivores, and it, together with its points of attachment on the skull and jaws, is small.

◆ (B) HYOID MUSCLES

- ◆ platysma
- ◆ facial muscles
- ◆ digastric (caudal part)

- ◆ stylohyoid
- ◆ stapedius

The major hyoid arch muscles are the **platysma, facial muscles,** caudal half of the **digastric,** and the **stylohyoid.** Find the stylohyoid again (p. 206). It is a small ribbon of muscle that lies lateral (cat) or caudal (rabbit) to the caudal portion of the digastric (Fig. 7-44). In the cat, it arises from the stylohyal of the hyoid bone and inserts on the body of the hyoid. In the rabbit, it arises from the base of the skull and inserts onto the greater horn of the hyoid. It acts on the hyoid. The **stapedius,** which has followed the stapes into the middle ear, can be seen later (p. 236).

◆ (C) CAUDAL BRANCHIOMERIC MUSCLES

- ◆ trapezius
- ◆ sternocleidomastoid

- ◆ intrinsic laryngeal muscles
  thyroarytenoid
  cricoarytenoid
  cricothyroid

Much of the caudal branchiomeric musculature is lost during the course of evolution, but some part becomes associated with the pectoral girdle as the **trapezius** and **sternocleidomastoid** complexes. These were described in connection with the shoulder region (p. 172). Some remaining part of the caudal branchiomeric musculature forms the **intrinsic muscles of the larynx,** such as the **thyroarytenoid,** the **cricoarytenoid,** and the **cricothyroid** (see Fig. 10-20, p. 321), and some is contributed to the wall of the pharynx. Certain of the intrinsic muscles of the larynx can be seen on the ventral surface of the larynx deep to the cranial end of the sternohyoid.

CHAPTER EIGHT

♦ ♦

THE SENSE ORGANS

Although the sense organs and nervous system integrate the activities of all parts of the body, they may appropriately be considered at this time, for the most conspicuous effector organs are the muscles described in the previous chapter. The sense organs, the central nervous system, and the basic pattern of distribution of the peripheral nerves will be the topic of this chapter and the one following. If separate heads cannot be provided for the study of *Squalus,* and separate brains for the mammal, instructors should postpone this unit of work until the end of the course.

Types of Receptors

To survive in a changing world, organisms must monitor changes in their external and internal environments and make appropriate responses. The individual cells of many simple invertebrates, such as hydras and jellyfish, can both receive and respond to stimuli. More specialized animals have **receptors** that receive stimuli, nerve cells, or **neurons,** that transmit and integrate information, and specialized muscle cells, gland cells, ciliated cells, and other **effectors** that mediate the response.

Receptors may be the ends of neurons or specialized cells that in turn activate sensory neurons. In vertebrates, free sensory neuron endings, which sometimes are encapsulated in specialized layers of connective tissue, detect many cutaneous sensations such as pain, touch, temperature, and pressure. Odoriferous particles also are detected by specialized neuronal endings in the nose. Most other sensations are detected by specialized receptor cells that are particularly sensitive to one type of environmental stimulus or modality, such as light or mechanical deformation. A sense organ is an aggregation of receptor cells or neuron endings together with other tissues that support and protect the cells and frequently direct or amplify the environmental stimulus—for example, the lens of the eye and the tympanic membrane of the ear.

Receptor cells and receptive neurons are often classified according to the sensory modality to which they are attuned. Mechanoreceptors are activated by a physical deformation of a part of the cell, photoreceptors by a change in light, thermoreceptors by temperature changes, and so forth. As with the skeleton and muscles, it is convenient to sort this array of receptors, and of the sensory neurons that lead from them, into somatic and visceral groups. **Somatic sensory receptors** lie in the "outer tube" of the body—the skin and the somatic muscles (proprioceptors). Occasionally, proprioceptive organs, which are the organs of muscle sense, are found in branchiomeric muscles. **Visceral sensory receptors** are associated with the "inner tube" of the body, that is, the viscera.

Before studying the nervous system, we will study the grossly visible sense organs, because they are more superficial. The deeper nervous system cannot be studied without considerable damage to the sense organs.

THE NOSE

The ability to detect chemical changes in the environment is known as **chemoreception;** it occurs in all animals. **Gustatory (taste) receptors** are chemoreceptors that respond to chemical changes in the immediate environment of the animal. Usually the chemicals are in contact with a body surface, such as the tongue. **Olfactory (smell) receptors** are chemoreceptors that respond to stimuli emanating from a greater distance. The distinction between taste and smell may be clear to us, but the difference becomes blurred in aquatic animals. The mechanisms for the detection of gustatory and olfactory stimuli, however, are quite distinct. Gustatory receptors are microscopic groups of specialized epithelial cells known as **taste buds.** Since taste buds develop from endodermal cells, they are considered visceral receptors. Most are confined to the lining of the mouth and pharynx, but in some fishes, including catfishes, many taste buds migrate onto the body surface. The tentacles, or barbels, around the mouth are particularly rich in taste buds. Taste buds are too small for us to consider further.

Olfactory receptors are specialized neurons, often called **neurosensory cells** because they receive environmental stimuli and transmit nerve impulses. They develop, with other parts of the nervous system, from ectoderm and are considered somatic receptors. The olfactory cells are too small for us to see, but we will consider the nasal sacs to which they are confined.

The sense of taste and smell are the most important senses in many animals, even in some animals that have eyes and ears. It is through these senses that many animals find food and mates and often gain important clues used in navigation. We may find this difficult to appreciate because our sense of smell is somewhat reduced.

In fishes, the nose typically consists of a pair of **olfactory sacs,** each of which connects to the surface by a pair of external nostrils **(nares),** through which water carrying odoriferous particles circulates. In sarcopterygian fishes and tetrapods, each olfactory sac, now usually called a **nasal cavity,** has but one naris, but each opens into the mouth through an internal nostril **(choana).** The nose of tetrapods serves as an air passage as well as retaining its original olfactory function. Olfactory and respiratory roles of the nasal cavities become segregated to some extent, the olfactory epithelium becoming restricted for the most part to the dorsal parts of the cavities. But in many tetrapods, a part of the olfactory epithelium remains in the ventral part of the cavity, where it forms the **vomeronasal organ** (Jacobson's organ). The vomeronasal organ frequently has a separate connection to the mouth cavity. The receptive cells of this organ resemble other olfactory cells. However, their fibers travel in a special bundle of the olfactory nerve and terminate in an accessory olfactory bulb of the brain. There is evidence that the vomeronasal organ plays a special role in social interactions, enabling organisms to follow prey trails and to recognize other individuals and potential sex partners.

In order for a substance to stimulate the olfactory epithelium, it must be in solution. Tetrapods have solved the problem by evolving glands whose secretions keep the epithelium moist. The secretions of these glands, and the mucosa of the nasal passages as a whole, also condition the air that passes to the lungs by moistening, cleaning, and, in birds and mammals, warming it. In some reptiles and in birds and mammals, the mucosal surface is increased through the evolution of **conchae,** or turbinate bones (p. 89). In mammals, the respiratory passages are prolonged through the evolution of a bony hard palate (p. 79) and a fleshy soft palate.

♦ **(A) FISHES**

Study the nose of *Squalus.* Note that each **naris** is divided into two openings (Fig. 8-1A) by a superficial flap of skin and a deeper ridge. The lateral opening, which also faces rostrally to a slight extent, is the incurrent aperture, through which a current of water

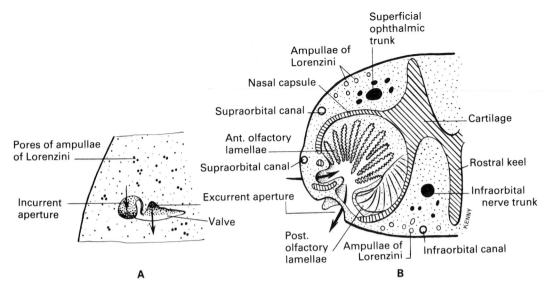

Figure 8-1
A, Ventral view of the right naris of *Squalus*; rostral is toward the top of the page. **B**, Diagram of the left olfactory sac as seen in a cross section taken just caudal to the naris. Viewed from behind. Arrows indicate the direction of the water current. The ampullae of Lorenzini and canals are parts of the lateral line system.

enters the olfactory sac; the medial one is the excurrent aperture. A thin, flaplike valve along its caudal margin prevents water from entering the excurrent aperture. Remove the skin and other tissue from around the **olfactory sac** on the side of the head to be used for dissecting the sense organs. The nasal capsule and antorbital process of the chondrocranium, which nearly surround the olfactory sacs, must also be cut away. You can then see that the olfactory organ is a round sac having no connection with the mouth. An **olfactory tract** of the brain extends to the caudomedial surface of the sac and there expands into an **olfactory bulb.** Cut open the sac and note how its internal surface is increased by many septa-like folds **(olfactory lamellae),** which in turn bear minute secondary folds (Fig. 8-1**B**). The olfactory cells are concentrated at the bases of the troughs formed by the lamellae.

> The sense of smell is highly developed and very important; indeed, sharks have been described as "swimming noses." Experiments have shown that smell is a major sense that guides a shark to the general location of prey, but other senses are important in directing the strike. Recordings from individual olfactory cells have shown that the odor of cutaneous mucus and blood of marine animals most frequently elicits responses. Smell also appears to play a role in reproductive behavior.

◆ (B) AMPHIBIANS

Note the pair of widely separated **nares** on your specimen of *Necturus*. Remove the skin between the naris and the eye on the side of the head used for the dissection of the muscles. You will see the long, tubular **nasal cavity** after picking away surrounding connective tissue (Fig. 7-14, p. 158). Open the mouth, cutting through the angle of the jaw on the side being studied, and find the slitlike **choana** through which the nasal cavity communicates with the mouth. It lies lateral to the most caudal, and shortest, row of teeth (pterygoid teeth). (See Fig. 10-15, p. 313.) Cut open the nasal cavity and find the **olfactory lamellae** within it. A vomeronasal organ is not developed in *Necturus* but is present in metamorphosed amphibians and most reptiles.

♦ **(C) MAMMALS**

The nose of mammals is to be studied on sagittal sections of the head cut in such a way that one half shows the nasal septum, and the other shows the inside of the nasal cavity. The nose should be studied from demonstrated preparations, unless this unit of work has been postponed to the end of the course, or unless heads from specimens of a previous year's class have been saved for the purpose. This work should also be correlated with the description of the sagittal section of the skull on page 88. The following account is based on the cat but is applicable to many other mammals.

The **nares,** which are close together in mammals, lead into paired **nasal cavities.** The nasal cavities occupy the area of the skull rostral to the cribriform plate of the ethmoid bone and dorsal to the hard palate. On the larger section, you can see that they are completely separated from each other by a **nasal septum** (Fig. 8-2). The ventral portion of the septum is formed by the **vomer bone;** the caudodorsal portion is formed by the **perpendicular plate of the ethmoid.** The rest is formed by **cartilage.**

On the smaller section, you can see that each nasal cavity is filled to a large extent by three folded **conchae,** or turbinate bones (see Figs. 4-26, 8-2, and 10-17). These are, of course, covered with the nasal mucosa. The **ventral concha** (maxilloturbinate) is represented by a simple fold that extends from the dorsal edge of the naris caudally and ventrally to about the middle of the hard palate. The nasolacrimal duct that drains tears from the eye enters lateral to the ventral concha. It is best seen on a skull in which this concha has been removed. The **dorsal concha** (nasoturbinate) is represented by a single longitudinal fold in the dorsal part of the nasal cavity lying deep to a median, perpendicular plate of the nasal bone. The area between and caudal to these two conchae is filled by the highly folded **middle concha** (ethmoturbinate). Most of the olfactory epithelium is limited to the more caudal parts of the conchae. With low magnification, you may see

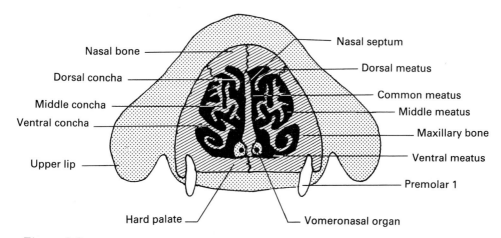

Figure 8-2
Semidiagrammatic cross section through the nasal cavity of a cat at the level of the first premolar tooth.

olfactory nerves passing through the cribriform foramina of the cribriform plate to the olfactory bulb of the brain.

Air passages lead from the naris between the conchae. The most prominent of these is the **ventral meatus,** which lies between the ventral and middle conchae on the one hand and the hard palate on the other. It opens by the **choana** into the nasopharynx. The choanae are located at the caudal border of the hard palate. The nasopharynx is separated from the oropharynx, into which the mouth cavity leads, by the fleshy soft palate. A **dorsal meatus** lies dorsal to the dorsal concha. The **middle meatus** is represented by the spaces formed by the complex folds of the middle concha. A **common meatus** lies between the nasal septum and the conchae. Certain of the air passages communicate with air sinuses in adjacent bones, including the frontal and sphenoid (see Fig. 10-17). They serve to lighten the skull.

Paired **vomeronasal organs** are present in the cat. The entrance to one can be seen on the roof of the mouth just caudal to the first incisor tooth. From here an **incisive duct** leads through the palatine fissure (p. 91) to the organ. You can find the vomeronasal organ by carefully dissecting away the rostroventral portion of the nasal septum. It appears as a cul-de-sac with a cartilaginous wall lying on the hard palate and against the nasal septum. It extends approximately 1 centimeter caudad to the incisor teeth.

THE OCTAVOLATERALIS SYSTEM I:
THE LATERAL LINE SYSTEM

Fishes, larval amphibians, and a few aquatic adult amphibians have a superficial sensory system known as the **lateral line system,** by which the animal can detect currents and other water movements. The lateral line system is closely related to the ear. The receptive cells in both the lateral line and ear are **hair cells.** In the lateral line they are clustered into groups called **neuromasts.** The surface of each hair cell bears a single modified cilium called a **kinocilium** and a cluster of microvilli known as **stereocilia** (Fig. 8-3). The stereocilia are graded in length, with the longest located

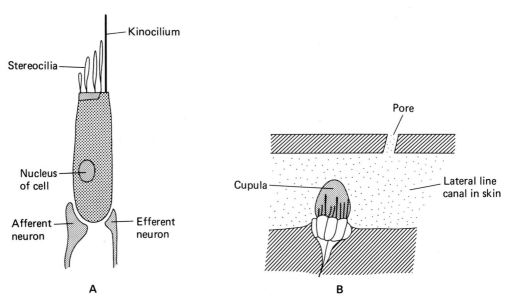

Figure 8-3
Receptor cells of the octavolateralis system. **A,** An individual hair cell; **B,** a group of hair cells forming a neuromast in a lateral line canal.

next to the kinocilium. The kinocilium and stereocilia are capped by a gelatinous matrix, called a **cupula** in the lateral line. The cupula is displaced by low-frequency vibrations and water movements, and this displacement bends the hairs.

The hair cells in the lateral line are tonic receptors; that is, they have a certain base level of activity at all times. This activity activates sensory (afferent) neurons that synapse with the cells. If the movement of the cupula is toward the kinocilium, the rate of activity of the cells increases. Movement of the cupula in the opposite direction decreases the rate of activity of the hair cells. Motor (efferent) neurons modulate the level of sensitivity of the hair cells.

The hair cells in both the lateral line and ear develop embryonically from adjacent ectodermal thickenings, or placodes. Both the lateral line and ear are supplied by neurons that terminate in adjacent and closely related centers in the brain. The neurons from the ear form the eighth cranial nerve; those from the lateral line system, called **lateralis neurons,** travel in the adjacent seventh, ninth, and tenth cranial nerves. Because of the similarities in the structure of the hair cells, their mode of embryonic development, and their connections within the brain, many investigators speak of the lateral line and ear together as the **octavolateralis system.** We will consider the lateral line and related electroreceptors at this time but will postpone study of the ear until the eye has been dissected. It is easier to dissect the ear of the dogfish after the eye has been removed.

The Lateral Line System and the Ampullae of Lorenzini

The lateral line system of the dogfish consists of some neuromasts that lie in pits beneath specialized scales. These are called **pit organs.** Most neuromasts lie within canals in the skin that open by many pores to the body surface. The canals are filled with sea water. Experiments on a variety of fishes have shown that the lateral line enables them to localize objects at a distance, even in turbid water in which some species live, either by the disturbance produced by a moving object, or by the reflected waves set up by the fish's own movements. Since the neuromasts in the canals have different polarities (for example, the kinocilium may be located toward the anterior or the posterior of the body), a distinction can be made in the direction of water movements and the source of the disturbance. The lateral line system has been called one of "distant touch." (See Boord and Cambell, 1977, for a review.)

All cartilaginous fishes, other nonteleostean fishes, and a few amphibians (caecilians) have groups of modified lateral line receptors known as ampullary organs or (in cartilaginous fishes) **ampullae of Lorenzini.** The receptor cells are neuromasts in which the hairlike processes are reduced on the hair cells and no cupula is present. Efferent neuron fibers are also absent. The receptors lie in small swellings (ampullae) at the base of jelly-filled tubes that open on the body surface. The tubes lie just beneath the skin and parallel to it. They vary in length and collectively extend in all directions; that is, some groups extend rostrally, some caudally, some medially, and so forth. Although the ampullae do respond to tactile stimuli and to changes in temperature and salinity, their primary function is electroreception. The skin and wall of the ampullary organs have such a high electrical resistance that a small voltage charge in the external environment does not affect deeper tissues directly through them. The jelly in the ampullary organs, however, acts as an electrical capacitor. It can hold a voltage charge for a moment and transmit it without a voltage drop to the receptive cells at the base of the organ. The length and direction of a tube determine the stimulus intensity at the receptive cells. Tubes directly in line with the voltage gradient would be most affected. This would enable the fish to detect the source of the voltage. The ampullae are sensitive enough to detect the direct current and low-frequency bioelectric fields generated by mucus secretions and the activity of respiratory and other muscles of prey. The electroreceptive system can lead a shark to its prey in the absence of other sensory clues, but field experiments have shown that sharks are attracted into an area by olfactory clues. They use the electroreceptive system to direct their final attack. Attacks can be provoked by a voltage gradient as low as 5 nanovolts per centimeter.

Interactions between the earth's magnetic field and a swimming shark, or one passively drifting in a current, also induce electrical fields. Experiments have shown that cartilaginous fishes can detect such fields. It has been suggested that they may use this information in orientation and navigation.

The lateral line system occurs in all fishes and larval amphibians; the electroreceptive system occurs in nonteleostean fishes and certain amphibians. These sensory systems were lost during the evolution of reptiles and were never reacquired, even in such aquatic tetrapods as the Cetacea. Parts of the systems have been noted during the study of the external features of *Petromyzon* (p. 18), *Squalus* (p. 42), and *Necturus* (p. 45). We will consider them in detail only in *Squalus*.

◆   **(A) THE AMPULLAE OF LORENZINI**

Examine your specimen of a large head of *Squalus*. The pores on the snout through which a jelly-like substance extrudes when the area is squeezed are the openings of the ampullae of Lorenzini. Note their distribution. They are very abundant on the ventral surface of the snout (Fig. 8-4). Other smaller groups are located caudal to the jaws. Make a deep V-shaped cut through the skin in one of the dorsal snout patches. The apex of the V should be directed caudally. Pull the skin flap and underlying tissue forward and examine its underside. Note that each pore leads into a long, jelly-filled tube, which in this region extends rostrally to a round swelling, the ampulla proper. Small nerve twigs attach onto the ampullae.

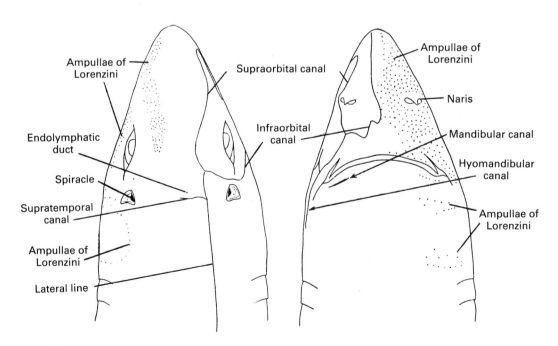

Figure 8-4
Dorsal and ventral views of the lateral line canals, and ampullae of Lorenzini, of the dogfish, *Squalus suckleyi*. *(From Daniel.)*

◆ **(B) THE LATERAL LINE CANALS AND PIT ORGANS**

Make a fairly long transverse cut through the skin, caudal and medial to the spiracle. Examine the cut surface. The hole that you can see in the deeper layer of the skin is a cross section of the **lateral line canal** proper. This canal extends the length of the trunk and tail. Examine a demonstration dissection of the ramifications of the system over the head, or, if time permits, trace it forward by carefully slicing off the superficial layers of the skin with a sharp scalpel. The objective is to make a horizontal section of the canal, taking the top half off and leaving the bottom half on the specimen. Keep your eye on the canal as you expose it to be sure you are cutting at the right level. Pores leading into the canal can be seen on the underside of the skin you remove. Farther forward on the head, the lateral line canal leads into others (Fig. 8-4), which should be uncovered in the same manner.

A **supratemporal canal** crosses the top of the head caudal to the endolymphatic pores. An **infraorbital canal** passes ventrally caudal to the eye, zigzags beneath the eye, and then extends forward to the tip of the snout in a somewhat meandering fashion, passing medial to the nostril. A **hyomandibular canal** extends caudally from the bottom of the zigzag of the infraorbital canal. A **supraorbital canal** passes forward dorsal to the eye and onto the snout. It then turns on itself and extends caudally to connect with the infraorbital canal. The latter portion of the supraorbital canal passes just dorsal to the nostril. A short **mandibular canal** is not connected with the others. It appears as a row of pores overlying the mandible caudal to the labial pocket. Make a cut through the skin at right angles to the row of pores, examine the cut surface, and verify that a canal is present.

Pit organs are difficult to discern grossly. There are a pair rostral to each endolymphatic pore, a mandibular row on the lower jaw, and scattered pit organs on the flank dorsal to the lateral line canal. Dermal denticles adjacent to them usually are specialized and overlap the organs.

THE EYEBALL AND ASSOCIATED STRUCTURES

The eyes are somatic sensory organs whose photoreceptive cells develop embryonically as outgrowths from the diencephalon of the brain. All vertebrates have a pair of lateral **image-forming eyes,** unless they have been secondarily lost. In addition, many fishes, some amphibians, and some reptiles have a median **pineal eye,** a **parietal eye,** or both (Fig. 8-5). Recall the pineal eye of the lamprey (p. 18). Experiments on larval lampreys and other vertebrates have shown that the pineal and/ or parietal eye detects the level of ambient light and affects the level of activity and behavior of the animal. In some species, the median eye complex is a neuroendocrine transducer that translates light signals into chemical messengers. These messengers have been shown to affect metamorphosis and the development of sexual maturity in some species. The median eye complex is transformed into the endocrine **pineal gland** in birds and mammals.

We will study the lateral, image-forming eyes at this time. Although the structure of the eyeball is essentially the same in all vertebrates, there are differences in the role of the cornea and lens in light refraction, the ways and speed with which the eye adapts to changes in light intensity, methods of accommodation, or focusing, and so forth. Many of these are associated with the different properties of water and air and problems associated with vision in these media. The need for protecting and cleansing the surface of the eyeball is also different. The surface of the fish eye is bathed in

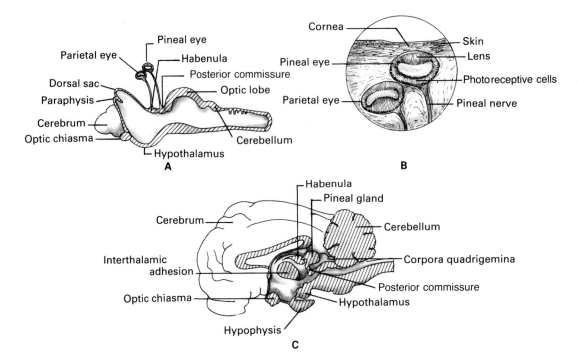

Figure 8-5
Diagrams of the pineal-parietal complex. **A,** Relationship of the median eyes to the brain in a lamprey; **B,** enlargement of the median eyes of a lamprey; **C,** the pineal gland of a sheep and its relation to the brain. *(From Walker,* A Study of the Cat.*)*

water, which keeps it moist and clean. Tetrapods have evolved movable eyelids of various types, as well as tear glands and ducts that protect, cleanse, and moisten the eye surface.

♦   **(A) FISHES**

Study the eye of *Squalus* on the same side of the head on which you dissected the nose and lateral line. *Squalus,* like other cartilaginous fishes, has upper and lower **eyelids** formed of skin folds, but they are immovable. (Some sharks have movable eyelids, but most fishes lack eyelids altogether.) Note that the epidermis on the inner surface of the eyelids reflects onto and over the surface of the eyeball as a transparent layer called the **conjunctiva.**

Remove the upper eyelid and free the eye from the lower lid by cutting through the conjunctiva. The **eyeball** (*bulbus oculi*) lies in a socket, the **orbit,** on the side of the chondrocranium. Cut away the cartilage that forms the roof of the orbit (supraorbital crest, antorbital process, postorbital process), but do not cut into the otic capsule. A mass of gelatinous connective tissue surrounds and helps to support the eye. It must be picked away.

A group of ribbon-shaped muscles passes from the medial wall of the orbit to the eyeball. These are the extrinsic muscles of the eye, and they are responsible for the various movements of the eyeball. In elasmobranchs, these muscles contain both white muscle fibers, which presumably are used in rapid eye movements, and red fibers, used in slower, more sustained actions. As explained on page 152, the extrinsic ocular muscles belong to the axial subdivision of the somatic muscles. Although they are derived from three myotomes (see Table 7-1), they form two groups in the adult: an oblique group and a rectus group. Push the eyeball caudally and note the two muscles that arise from the rostromedial corner of the orbit and insert on the eye (Fig. 8-6). The one that inserts on the dorsal surface is the **dorsal** (superior) **oblique;** the one that inserts on the ventral surface is the **ventral** (inferior) **oblique.** The muscles that arise from the caudomedial corner are all recti. Three can be seen in a dorsal view. The one that inserts on the top of the eyeball adjacent to the insertion of the dorsal oblique is the **dorsal** (superior) **rectus;** the one that lies medial to the eye is the **medial rectus;** the one that lies caudal to the eye is the **lateral rectus.** Lift up the eye and look at its ventral surface. A **ventral** (inferior) **rectus** passes to insert on the ventral surface of the eyeball beside the insertion of the ventral oblique.

Most of the other strands that are seen passing to the muscles and through the orbit are nerves and will be considered later; at this time, find the **optic nerve.** It is the large nerve passing from the eyeball caudal to the ventral oblique (Fig. 8-7).

Cut across the extrinsic muscles of the eye near their insertions on the eyeball, trying not to injure the nerves going to them. The stalk of cartilage that will be seen passing to the back of the eyeball between the four rectus muscles is the **optic pedicle.** It is shaped like a golf tee and anchors the eyeball to the skull. Disconnect it from the eyeball and also free the small nerve that crosses the medial surface of the eye. Cut the optic nerve and remove the eye. Some **conjunctiva** will cling to the front of the eye.

The outermost layer of the eyeball is a tough **fibrous tunic,** the anterior portion of which is modified to form the transparent **cornea,** through which you can see an opening, the **pupil,** surrounded by the pigmented iris. Conjunctiva and cornea are fused.

Submerge the eyeball in a dish of water and cut off its dorsal third. Note the large, spherical **lens.** Try not to pull it away from surrounding tissues. You can see the three layers or coats that make up the eyeball at the back of the eye. As mentioned, the outermost layer is the fibrous tunic. Its front portion is modified as the cornea; the rest of it constitutes the **sclera** (Fig. 8-7). Much of the dogfish's sclera is cartilaginous, and this provides a great deal of support for the eyeball. The pigmented layer internal to the sclera

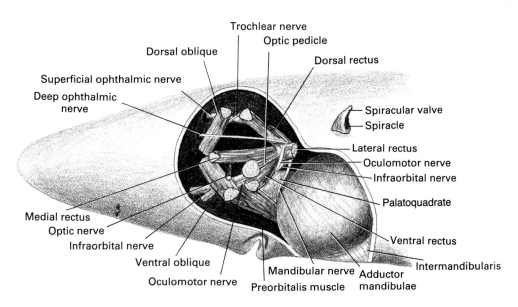

Figure 8-6
Lateral view of orbit of *Squalus* after removal of the eyeball.

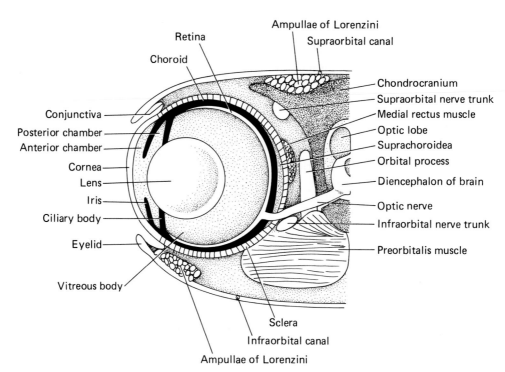

Figure 8-7
A cross section of the head of the dogfish at the level of the eye showing the structure of the eyeball and its relationship to surrounding structures.

is the **vascular tunic,** most of which forms a **choroid.** The **iris** is the modified anterior portion of the vascular tunic, and the **pupil** is a hole through the iris. The whitish layer internal to the choroid is the **retina.** It is an incomplete layer that disappears near the lens. The point at which the optic nerve connects with the retina is the **optic disc,** sometimes called the "blind spot" because there are no receptive cells here.

Carefully pull the lens away from the iris. A ring of black material will probably adhere to it. This black material is in the form of small radiate folds and represents the **ciliary body**—a modification of the vascular tunic located near the base of the iris. A gelatinous material (the **zonule**), which will not be seen, extends between the ciliary body and the lens. The lens is supported by the ciliary body, the zonule, a middorsal suspensory ligament (which cannot be seen grossly), and the **vitreous body.** The vitreous body is the gelatinous material that fills the large cavity **(chamber of the vitreous body)** between the lens and the retina. Other cavities, which are filled with a watery **aqueous humor,** are located in front of the lens. The very small cavity that lies between the lens and the iris is the **posterior chamber;** the larger one between the iris and the cornea is the **anterior chamber.**

Cut across the back of the eyeball where the optic pedicle attaches. The relatively thick layer of material between the sclera and choroid is known as the **suprachoroidea.** It is a vascular connective tissue that is found only in species with an optic pedicle.

Cut into the lens and observe that it is composed of layers of modified epithelial cells arranged concentrically like the skins of an onion.

The sclera is the supporting layer of the eyeball, and the presence of cartilage within it is not surprising, since the sclera develops in part from the optic capsule of the embryonic chondrocranium (p. 54). The **sclerotic bones** present in some large-eyed vertebrates ossify in this cartilage. The choroid performs several functions. It is vascular and helps to nourish the light-sensitive, avascular retina. Its pigment, along with pigment of the retina, prevents internal reflections of light. In addition, the choroid of many vertebrates, including the dogfish, is so constructed that it can reflect some light back onto the retina. Such a reflecting device, known as a **tapetum lucidum,** is found in vertebrates that live under conditions of varying light intensity—certain fishes and nocturnal tetrapods. The elasmobranch tapetum depends on silvery **guanine crystals** within certain choroid cells. It is among the most remarkable of vertebrate tapeta. Pigment is absent from the normally pigmented layer of the retina, so light passes through the retina to the choroid. The guanine plates in the choroid are set at such an angle that light is reflected directly back to the receptive cells of the retina and does not scatter and blur the image. Adjacent **choroid chromatophores** can extend or retract their pigment across the guanine layer, thus adapting the eye to light or dark conditions (Fig. 8-8). These chromatophores are called **independent effectors** because they detect changes in illumination and respond directly. Of course, the circular and radial muscle fibers within the iris, which respectively narrow and widen the pupil, further adjust the eye to light and dark. Although in elasmobranchs these muscles are under some degree of neural control, they also respond directly to the level of ambient light.

Since the refractive index of the cornea and humors of the eye is nearly the same as that of water, they do not take part in the refraction of light rays. Nearly all the refraction in fishes occurs at the thick, spherical lens. In elasmobranchs, the lens is held in such a position that moderately distant objects are in focus on the retina. In bright light, images are automatically brought into sharp focus, regardless of the distance of the object, by the contraction of the pupil. This is the principle used in an old-fashioned pinhole camera. To accommodate for very close objects, the lens can be pulled forward by the contraction of a small **protractor lentis** muscle that is located ventrally in the ciliary body. The pressure of the aqueous humor pushes the lens back to its resting position when this muscle relaxes.

The retina contains the photoreceptive cells. Both rods and cones are present in elasmobranchs, but the ratio of rods to cones varies with the species, the habitat it occupies, and its mode of life. Bottom-dwelling species have few cones, but pelagic species have more. The ratio of rods to cones is 50 to 1 in *Squalus*. The rods are particularly sensitive in low-intensity light; cones require higher light intensity. It is unclear to what extent elasmobranchs have color vision. As divers know, water quickly

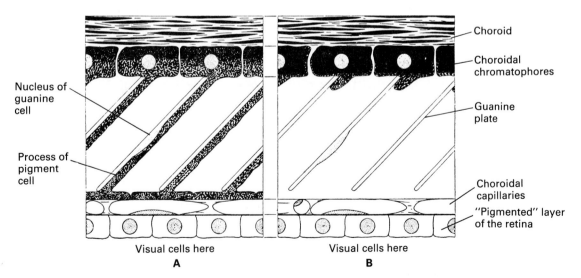

Figure 8-8

Diagrams of section through the choroid of the dogfish *Mustelus mustelus*, to show the tapetum lucidum adapted for (**A**) light, and (**B**) dark conditions. In the light-adapted eye, processes of the pigment cells cover the guanine plates; in the dark-adapted eye they are withdrawn and allow the guanine to reflect light back to the visual cells of the retina. *(From Walls, after Franz.)*

absorbs red, orange, and other colors that have longer wavelengths so the underwater world is mostly blue. In summary, the eyes of elasmobranchs vary with the environment in which they live. They are well developed anatomically and physiologically, and vision is probably an important sense at intermediate distances. (See Gruber and Cohen, 1978, for a review.)

♦ **(B) MAMMALS**

We may pass directly to the mammalian eye, for that of the aquatic *Necturus* contributes little to an understanding of the evolution of this organ. The basic structure of the mammalian eyeball is much the same as in *Squalus,* but details differ. The iris muscles are under the control of the autonomic nervous system and respond quickly to the rapid changes in light intensity that occur on land. Since the mammalian cornea has a much higher refractive index than air, it is important in bending light rays. The lens is less important in refraction and is more oval in shape. Lens accommodation occurs through changes in its curvature.

Tear Glands and Extrinsic Muscles   Examine the eye of your specimen on the side that you used to study the muscles. Movable **upper** and **lower eyelids** (*palpebrae*) are present. The slitlike opening between them is called the **palpebral fissure.** The corners of the eye where the lids unite are the **ocular angles.** Cut through the lateral ocular angle and pull the lids apart. You can now clearly see the **nictitating membrane** (*semilunar fold*). It is attached at the medial ocular angle, but its lateral edge can spread over the surface of the eye if the eyeball is retracted slightly. Our nictitating membrane is reduced to a small, semilunar fold that can be seen covering the medial corner of the eye. With a hand lens, examine the edge of each lid of your specimen 3 or 4 millimeters from the medial ocular angle. If you are fortunate, you will see on each a minute opening **(lacrimal punctum)** that leads into a lacrimal canaliculus. If you cannot find them in your specimen, look for one on the human eye by pulling down the lower lid and examining its edge near the most medial eyelash.

Cut off the upper and lower lids, leaving a bit of skin around the medial ocular angle. As you remove the lids, note that the **conjunctiva** on the underside of the lids reflects over the **cornea.** If the cornea is not too opaque, the **pupil** and **iris** can be seen. A facial muscle, the **orbicularis oculi,** encircles the eyelids and will be cut off with them. It closes the lids.

Free the eyeball and associated glands from the bony rims of the orbit by picking away connective tissue. Do not dissect deeply in the region of the medial ocular angle, and try not to destroy a loop of connective tissue attached to the rostrodorsal wall of the orbit. One of the ocular muscles passes through this loop (see Fig. 8-10). Using bone scissors, cut away the zygomatic arch beneath the eye; also cut the postorbital processes and the crest of bone above the orbit (supraorbital arch). Push the eye rostrally. The dark glandular mass that lies on the caudodorsal surface of the eyeball is the **lacrimal gland.** It is larger in the rabbit than in the cat and extends ventral to the eyeball (Fig. 8-9). Its secretions, the tears, enter near the lateral ocular angle, bathe the surface of the cornea, and pass into two lacrimal canaliculi through the **lacrimal puncta.** These canaliculi unite to form a **nasolacrimal duct** that enters the nose through the **lacrimal canal** in the lacrimal bone

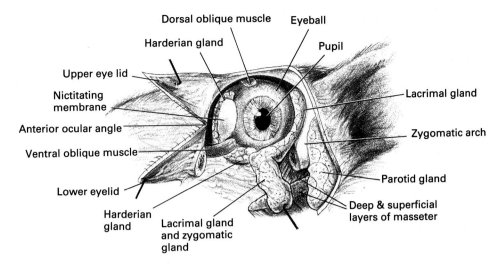

Figure 8-9
Dissection of the lacrimal apparatus of the rabbit's eye. Rostral is toward the left.

(p. 91). You may be able to find the nasolacrimal duct by dissecting ventral and rostral to the medial ocular angle. First find the rostral border of the orbit and the position of the lacrimal canal. Do not injure a muscle that is attached to the orbital wall near the lacrimal bone.

A second tear gland, the **gland of the nictitating membrane** (harderian gland), lies just rostral to the nictitating membrane. It is not large in carnivores but is quite large in the rabbit, and part of the rabbit's gland extends ventrally, going deep to the ventral oblique muscle, to appear on the underside of the eye.

The cat and rabbit have a **zygomatic gland** located in the floor of the orbit ventral to the eye. In the rabbit, it is difficult to separate this gland from the ventral portion of the

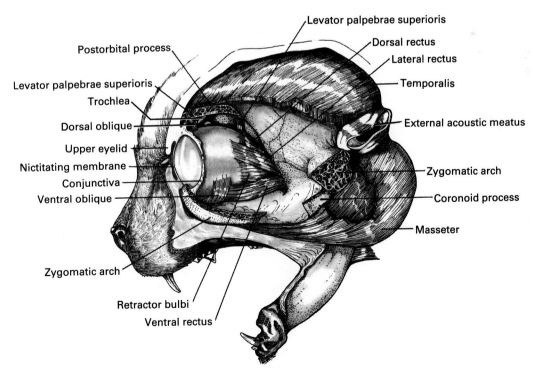

Figure 8-10
Lateral view of a dissection of the orbit of a cat to show the extrinsic ocular muscles.
(From Walker, *A Study of the Cat.*)

lacrimal gland. The zygomatic gland is not a part of the lacrimal apparatus; rather, it secretes saliva that is discharged into the mouth at a point near the last upper tooth.

Remove the glands and connective tissue from around the eye of your cat or rabbit and expose the extrinsic muscles of the eyeball (Fig. 8-10). The pattern of the muscles is much the same as in fishes, except for two additional muscles. A **levator palpebrae superioris** arises from the medial wall of the orbit dorsal to the optic foramen and inserts on the upper eyelid, which it raises. Its lateral end may be found unattached, for its insertion was probably cut in removing the eyelids. The rest of these muscles act on the eyeball. Two oblique muscles pass to the rostral wall of the eyeball. The **dorsal** (superior) **oblique** arises from the wall of the orbit slightly rostral to the optic foramen and goes through a connective tissue pulley, the **trochlea,** which is attached to the rostrodorsal wall of the orbit, before inserting on the eye. The **ventral** (inferior) **oblique** arises from the maxillary or lacrimal bone. Four recti arise from the margins of the optic foramen and pass to the caudal portion of the eyeball. A **dorsal** (superior) **rectus** inserts on the dorsal surface of the eyeball, a **ventral** (inferior) **rectus** on the ventral surface, a **medial rectus** on the medial surface, and a **lateral rectus** on the lateral or posterior surface. A **retractor bulbi,** which can be divided into four parts, passes to the eye deep to the recti. The derivations of the extrinsic ocular muscles are shown in Table 7-1.

The Eyeball   The mammalian **eyeball** *(bulbus oculi)* is similar in basic structure to that of a fish. Directions for its dissection are included here for those who wish to make a comparison or prefer to study the mammal's eye instead of the fish's. It can be seen most clearly in a large eye such as that of a cow or sheep, although your specimen's eye may be used. Carefully clean up the surface of the eyeball by removing the extrinsic ocular muscles and associated fat. Notice that the optic nerve attaches somewhat eccentrically, toward the rostroventral portion of the eyeball. Leave a stump of it attached. Some conjunctiva will adhere to the surface of the cornea.

Open the eyeball by carefully cutting a small window through its dorsal surface. Observe that its wall consists of the same three layers as in a fish's eye: an outer **fibrous tunic;** a middle dark layer, the **vascular tunic;** and the inner, whitish **retina** (Fig. 8-11). The fibrous tunic is a dense, supporting connective tissue. Approximately the medial two thirds of it form the opaque **sclera;** the lateral one third forms the transparent **cornea,** through which light enters the eye. The vascular tunic is rich in blood vessels; most of this layer is a **choroid** that lies behind the retina and helps to nourish it. The portion of the retina that you see is the nervous layer, which contains the receptive rods and cones on its choroid surface. In the embryo, there is a pigmented layer to the retina, but in the adult this becomes associated with the choroid. The pigment reduces light reflections within the eye.

With a pair of fine scissors, extend a cut from the window that you made, completely around the equator of the eyeball, thereby separating the eyeball into an anterior half containing the lens and cornea and a posterior half containing most of the retina. Cut through all the layers. The jelly-like mass filling the eyeball between the lens and retina is the **vitreous body,** a medium that helps to support the lens and also helps to refract light entering the eye. The space in which it lies is the **chamber of the vitreous body.** Keep the vitreous body with the anterior half of the eyeball. Submerge both halves in a dish of water.

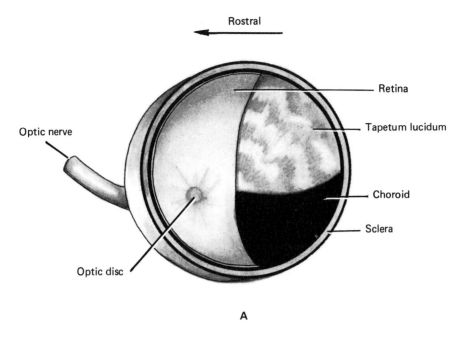

Rostral

Retina

Tapetum lucidum

Optic nerve

Choroid

Sclera

Optic disc

A

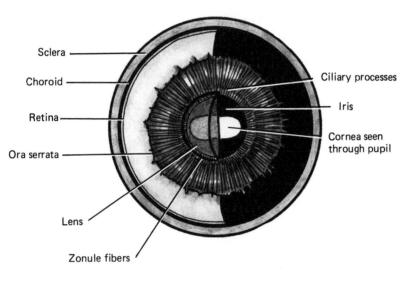

Sclera

Choroid

Retina

Ora serrata

Ciliary processes

Iris

Cornea seen
through pupil

Lens

Zonule fibers

B

Figure 8-11
Dissection of a cow eye. **A,** Posterior half; **B,** anterior half. A portion of the retina has been removed in each half to show the vascular tunic.

Examine the posterior half (Fig. 8-11A). The retina has probably become partly detached from the choroid but can be floated back into its normal position. Note the round spot **(optic disc),** at which the optic nerve attaches to the retina. This region is devoid of rods and cones and hence is often called the blind spot. Remove the retina and observe that an extensive section of the choroid dorsal to the optic disc is quite iridescent. This is the **tapetum lucidum,** an area that reflects some of the light passing through the retina back onto the rods and cones and hence facilitates the animal's ability to see in dim light. This tapetum depends on a compact layer of endothelial cells in the choroid, not on guanine granules as in a dogfish. Many mammals, including human beings, lack a tapetum.

Carefully remove the vitreous body from the anterior half of the eyeball; notice that the white, nervous layer of the retina does not extend far into this half of the eyeball. The line of demarcation between it and the dark choroid (plus embryonic pigment layer of retina) is the **ora serrata** (Fig. 8-11B). The portion of the vascular tunic that you see extending from the ora serrata toward the **lens** is the **ciliary body.** The portion of it next to the lens has a pleated appearance; the individual folds are the **ciliary processes.** While observing the area with a dissecting microscope, carefully stretch the region between the lens and the ciliary processes. You will see many delicate **zonule fibers** passing from the ciliary body to the equator of the lens.

Refraction and accommodation differ in mammals and fishes. Intraocular pressures cause the wall of the eyeball to bulge outward, and this force is transmitted via the zonule fibers to the elastic lens, which is consequently under tension and somewhat flattened. Under these circumstances, the lens has its minimum thickness and refractive powers, so the eye is focused on distant objects. In a terrestrial vertebrate, the cornea produces the greatest refraction of light because of the sharp contrast between its index of refraction and that of air; the role of the lens is more analogous to the fine adjustment of a microscope. Accommodation for a close object requires a greater bending of light rays. Muscle fibers in the base of the ciliary body contract and bring the base of the ciliary body a bit closer to the lens. This releases the tension on the zonule fibers and permits the lens to bulge and increase its thickness.

Carefully remove the lens and notice that it is not spherical, as in a fish, but somewhat flattened. The vascular tunic continues in front of the lens to form the **iris.** The **pupil,** of course, is the opening through the iris. Its diameter, and the amount of light it permits to pass, is regulated by circular and radial muscle fibers within the iris. The space between the lens and iris is the **posterior chamber;** that between the iris and cornea is the **anterior chamber.** Both are filled with a watery **aqueous humor** produced by the ciliary processes. This liquid maintains the intraocular pressure. Excess liquid is drained off by a microscopic **scleral venous sinus** (canal of Schlemm), which encircles the eye between the base of the cornea and the iris. If you make a vertical cut through the anterior half of the eyeball and examine it under a dissecting microscope, you may be able to see this canal and the ciliary muscles.

THE OCTAVOLATERALIS SYSTEM II: THE EAR

The ear of vertebrates, as we have discussed (p. 217), is closely related to the lateral line system. Some investigators have proposed that the inner, receptive part of the ear evolved by the invagination of a portion of the lateral line system. The inner ear, now isolated from surface disturbances, has become specialized to detect internal disturbances caused by changes in the orientation and movements of the body. The inner ear consists of a series of thin-walled ducts and sacs filled with a fluid known as the **endolymph.** These ducts and sacs are collectively called the **membranous labyrinth** (Figs. 8-12 and 8-13). They may lie within a single chamber in the otic capsule, but often they are imbedded within a series of parallel canals and chambers within the capsule known as the **cartilaginous** or **osseous labyrinth.** The membranous labyrinth and osseous labyrinth are separated from each other by spaces filled with fluid and crisscrossed by minute strands of connective tissue. This fluid is the **perilymph.**

Only an internal ear is present in fishes. As in terrestrial vertebrates, it is an organ of equilibrium, and in many fish, parts of it are sensitive to sound waves. A sound

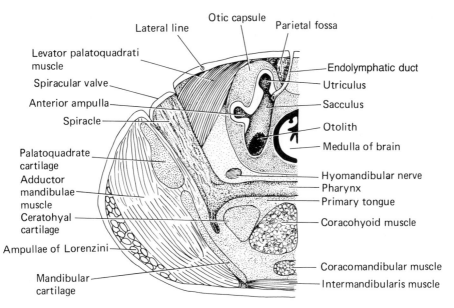

Figure 8-12
A cross section through the head of a dogfish at the level of the spiracle showing the relationship of the otic capsule and ear to surrounding structures.

generated under water has two components. First, there is physical disturbance of water particles analogous to the ripples generated when one drops a stone into a pond. Water particles are displaced with a certain velocity, and this **displacement wave** travels outward from the sound source as vibrations of low frequency and high amplitude. These "near field" waves do not travel far or fast. Second, **pressure** or **sound waves** are generated that have a higher frequency and lower amplitude. They propagate very rapidly and travel a considerable distance into a "far field." The displacement wave can be detected by the lateral line and by parts of the inner ear in some species. The sound waves, already in water, easily travel through the head and inner ear of a fish because the body tissues are mostly water. In a sense, a fish is transparent to sound waves. Fishes cannot detect these pressure waves unless they have a mechanism to transduce them into displacements that can affect certain hair cells of the inner ear. Connections between a swim bladder and the inner ear do this in some species of teleost fishes.

♦   **(A) FISHES**

You may study the ear of *Squalus* on demonstration preparations, or, if time permits, you may dissect it on the side of the head used for the study of other sense organs. The basic structure of the membranous labyrinth can be seen better in the cartilaginous fishes than in any other vertebrates, for the labyrinth is relatively large and can be freed more easily from cartilage than from bone. First uncover the otic capsule by removing the skin and muscles from its dorsal, lateral, and caudal surfaces. You may also cut away the spiracle and adjacent parts of the mandibular and hyoid arch (Fig. 8-12). Note the **endolymphatic duct** leading from one of the endolymphatic pores in the skin to the parietal fossa on the top of the chondrocranium. It ultimately connects with the sacculus of the membranous labyrinth.

To expose the **membranous labyrinth** of the inner ear, you must carefully shave away the surrounding cartilage of the otic capsule, beginning on the dorsal and lateral surfaces and gradually working ventrally and medially. You can generally see the various ducts and chambers through the cartilage shortly before you reach them. Use special care as you dissect the cartilage from around the parts of the membranous labyrinth, and try not to break them. You will first see the three **semicircular canals—anterior vertical, posterior vertical,** and **lateral**—of the cartilaginous labyrinth. Comparable **semicircular ducts** of the membranous labyrinth lie within the canals (Fig. 8-13). As you continue to dissect, you will come upon the **sacculus** lying in a large cavity medial to the lateral duct. You can dissect away the dorsolateral wall of the cavity without injuring the sacculus, for the sacculus lies deep within the cavity. Continue to remove cartilage from around the sacculus and semicircular ducts, tracing the latter to their points of attachment on chambers called utriculi. Each end of each duct attaches to one of two utriculi. The anterior and lateral ducts attach onto the **anterior utriculus;** the posterior duct attaches onto the **posterior utriculus.** The utriculi connect with the sacculus by inconspicuous openings. The ventral end of each duct bears a round swelling, the **ampulla,** containing a sensory patch of hair cells and a cupula known as a **crista.** Branches of the **vestibulocochlear** (statoacoustic) **nerve** can be seen leaving each ampulla, but the cristae may not be apparent. Dissect away as much of the cartilage as possible from the ventral side of the sacculus and observe the short extension or bulge from its caudoventral portion. This is the **lagena.** The sacculus and lagena contain large sensory patches of hair cells called **maculae.** Each sensory patch is overlaid by a mass of minute mineral concretions imbedded in a gelatinous matrix to form an **otolith.** In many sharks, the otolith is

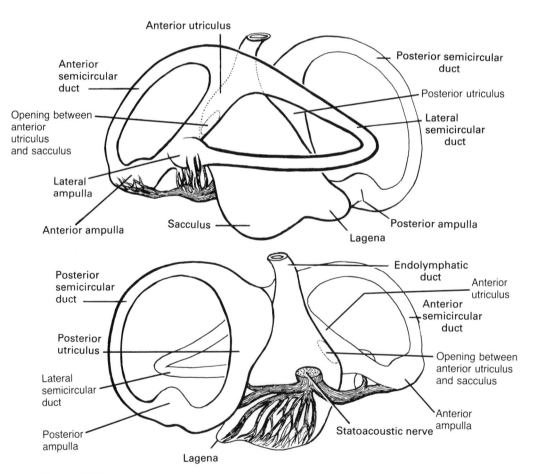

Figure 8-13
Left ear (membranous labyrinth) of the selachian, *Heptanchus maculatus.* Lateral view is above, medial view below. *(Redrawn from Daniel.)*

composed of secreted crystals of calcium carbonate, but in *Squalus,* the otolith is formed by sand grains that enter through the endolymphatic ducts. In preserved specimens, the matrix that holds the sand grains tends to disintegrate. Remove as much cartilage as possible from the medial side of the membranous labyrinth, cut the nerves coming from the sensory patches, and gently lift out the ear and float it in a dish of water. It is well to leave a bit of cartilage on the medial and ventral surfaces. Review the parts of the ear mentioned and compare your dissection with Figure 8-13.

No studies have been done on the acoustic properties of the ear of *Squalus,* but some have been done in larger reef and lemon sharks, which have ears structurally similar to those of the dogfish. (See Corwin, 1981, for a review.) Behavioral studies and recordings of the action potentials made on the vestibulocochlear nerve have demonstrated that the larger sharks are aroused by low-frequency vibrations of the type that would be generated by the movements of a wounded and struggling fish. Their highest sensitivity is for frequencies between 40 and 160 Hz. Moreover, they can localize the source of a sound coming from as far away as 250 meters.

The primary area of sensitivity in the ear is the previously little-studied **macula neglecta,** which is located in the **posterior vertical duct** (Fig. 8-14). This duct is the equivalent of the dorsal part of the posterior utriculus of *Squalus,* but instead of connecting ventrally with the posterior semicircular duct, the posterior duct terminates ventrally directly in the sacculus and on a **lateral membrane** adjacent to the large perilymphatic space. Low-frequency vibrations approaching the fish from above and to the side cause particle displacements of the type shown by arrows in Figure 8-14. Because of their low frequency, few of those vibrations impinging on the cartilage of the otic capsule enter the ear. But those hitting the skin over the **parietal fossa** travel easily through the loose, watery connective tissue filling the fossa and impinge upon the posterior duct canal, which reaches the fossa through the **perilymphatic foramen.** This foramen is analogous to the fenestra ovalis of tetrapods. Corwin gives

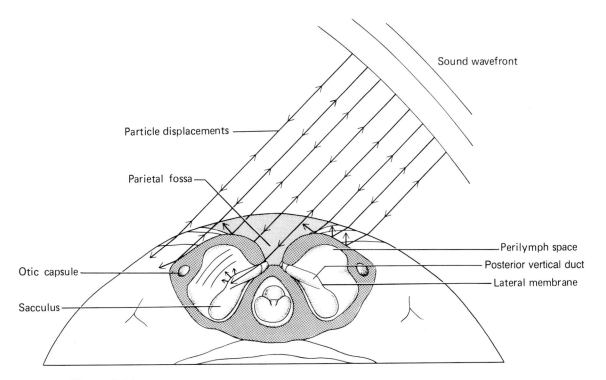

Figure 8-14
Diagram of a cross section through a shark's head indicating the path by which particle displacements induced by a low-frequency wave front reach the inner ear. *(From J. T. Corwin.)*

it this name, but, of course, it is not homologous to the oval window. Particle displacements travel directly through the posterior duct canal, displace the cupula overlying the very dense population of hair cells in the macula neglecta, and finally are released through the lateral membrane into the perilymphatic space. Notice from Figure 8-14 that vibrations coming from above and to the side affect primarily the contralateral ear. When the shark turns toward the sound source, both ears will be equally affected.

Electrical recordings from parts of the vestibulocochlear nerve have shown that the sacculus also responds to low-frequency vibrations but to a lesser extent than does the macula neglecta. Low-frequency vibrations, with their large amplitude, cause the entire fish to vibrate to some extent. Since the heavy otolith particles have more inertia than the rest of the fish, their movement lags, and this stimulates the underlying hair cells of the macula sacculus. Because the distribution of cilia-like processes on the hair cells determines the direction of the displacements to which they will respond, groups of cells within the macula with different patterns of distribution of the cilia-like processes will detect displacements coming from different directions. Sharks presumably used this system to detect sounds coming from behind or beneath them.

♦ **(B) AMPHIBIANS AND REPTILES**

Some terrestrial vertebrates lack an eardrum and respond primarily to low-frequency vibrations that are transmitted to the inner ear through certain skull bones. In those with ears sensitive to air-borne sounds, higher-frequency pressure waves impinge upon an eardrum, or **tympanum,** and are transmitted across an air-filled **tympanic cavity** (homologous to the spiracular pouch) by one or more auditory ossicles (see Fig. 8-16A and **B**). The foot plate of the **stapes,** which may be the only ossicle or the innermost of three, fits into a **fenestra vestibuli,** or oval window, on the side of the otic capsule. The stapes often also has a connection to the quadrate, so its relationships are essentially the same as those of the hyomandibular, to which it is homologous. The difference in size between a large tympanum and a small foot plate increases the pressure amplitude sufficiently to overcome the inertia of the liquids in the inner ear. The pressure wave is transduced into a displacement wave that travels through parts of the perilymph to the receptive part of the membranous labyrinth. Vibrations finally are released through the **fenestra cochleae,** or round window, back into the middle ear cavity. The location of the tympanic membrane, the relationship between the stapes and certain nerves, the number of auditory ossicles, the location of the receptive cells within the membranous labyrinth, and other aspects of the tetrapod ear vary so much among groups that investigators now believe that ears sensitive to air-borne pressure waves evolved independently of one another several times in the course of tetrapod evolution. (See Lombard and Bolt, 1979, for a review.)

You should study the structure of the ear of an amphibian on demonstration dissections of a bullfrog, since the urodele ear lacks the **tympanum** and middle ear cavity found in most terrestrial vertebrates. The large external tympanum is easily seen (Fig. 8-15). It is located high on the skull. Remove it on one side, and you will expose the **tympanic cavity,** or middle ear cavity. Open the mouth and note that the tympanic cavity communicates with the back of the buccopharyngeal cavity via the **auditory** (Eustachian) **tube.** A long, rodlike **stapes** crosses the middle ear cavity.

Expose the **otic capsule,** which contains the inner ear, by removing the overlying skin and muscle. Also remove the musculature lying caudal to the middle ear. Cut through the back of the middle ear cavity and trace the stapes to the otic capsule. Its inner end is associated with the knoblike, specialized **operculum** (homologous to the urodele

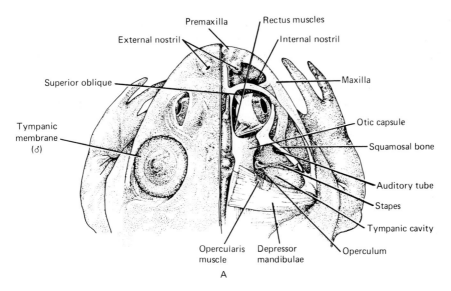

Figure 8-15
The tympanic membrane (tympanum) and middle ear of a bullfrog in a dorsal surface view (*left*) and dissected (*right*). (*After Walker.*)

structure of the same name, p. 69). Remove the operculum and the inner ear of the stapes and you will see the **oval window** into which they fit. Vibrations enter the inner ear at this point and are released through a fenestra, which, in living amphibians, opens into the cranial cavity dorsal to the vestibulocochlear nerve. You may find it by opening the cranial cavity and removing the brain. This fenestra is analogous to the **round window** that opens into the tympanic cavity of certain reptiles and mammals and is usually given this name.

Contemporary amphibians have evolved a unique auditory system. (See Capranica, 1976, for a review.) The inner ear has two areas sensitive to sound waves. A **basilar papilla** is sensitive to the high-frequency sounds associated with the mating and release calls of male frogs. The mating call identifies species, and the release call identifies sex because it signals a male that has inadvertently grasped another male to let go. An **amphibian papilla** is sensitive to low-frequency vibrations of the type that are generated and probably transmitted through the body tissues when a male grasps a female. It probably is sensitive to other low-frequency environmental sounds as well. Sounds reach the inner ear by way of the tympanic membrane, stapes, and operculum. The stapes and operculum are notched in such a way that tension exerted on a slender **opercularis muscle,** which extends from the operculum to the scapula, locks the two ossicles together. They then form a functional unit of considerable size and inertia that responds only to low-frequency sounds. When the opercularis muscle relaxes, the two ossicles uncouple, and then the smaller size and lower mass of the stapes allow it to respond to high-frequency sounds. The frog ear can be tuned to either low- or high-frequency sounds. Adult salamanders have no mating calls, and they detect only low-frequency environmental noises and ground vibrations. They lack a tympanic membrane and cavity but have the stapes, operculum, and opercularis muscle.

◆ (C) MAMMALS

Although many reptiles have a tympanic membrane and a middle ear located caudal to the quadrate bone (Fig. 8-16**A** and **B**), early mammal-like reptiles appear to have had a different type of sound-transmitting system (Allin, 1975). Their skulls did not have enough space for a tympanum large enough to amplify sound waves, and the stapes was a rather heavy bone that retained a connection with the quadrate. These reptiles probably detected low-frequency ground-borne vibrations via their lower jaw, and these were most likely transmitted to the inner ear by way of the jaw joint bones (articular and quadrate) and stapes. Advanced mammal-like reptiles had an unusual flange from the angular bone of the lower jaw (the reflected lamina, Fig. 8-16C), which extended ventrally and caudally around an air-filled space that probably lodged a relatively large "mandibular" tympanic membrane. Such a mechanism may have detected higher-frequency air-borne vibrations and thus supplemented the transmitting mechanism of the bone. The quadrate, articular, and angular were small, and, although involved with jaw mechanics, they were not bound tightly to adjacent bones, so they could vibrate to a limited extent. With the further reduction in the size of the jaw joint bones and the shift in jaw joint that occurred during the reptile-to-mammal transition (p. 83), the ear could become specialized for detecting air-borne sound waves. The quadrate became the **incus**; the articular became the **malleus** (Fig. 8-16E). The retroarticular process of the articular, no longer needed for the attachment of jaw-opening muscles, was in a strategic position to become a lever arm or handle for the malleus. The angular remains as the **tympanic bone** supporting the ear drum. An **auricle** develops about the external acoustic meatus. The sound-detecting portion of the mammal ear is a long, spiral **cochlea** (Fig. 8-16D). The cochlea consists of the **cochlear duct,** which is the part of the membranous labyrinth containing the receptive cells, and two perilymphatic ducts, the **scala vestibuli** and **scala tympani,** by which pressure waves enter and are released from the cochlea.

According to this view, ears sensitive to air-borne vibrations evolved independently in certain living amphibians and reptiles and in the late mammal-like reptiles. In the line of evolution toward mammals the stapes, quadrate, and articular were at all times part of a sound-transmitting system. They were used first to transmit ground vibrations, then for a combination of ground- and air-borne vibrations, and finally as specialized bones to transmit air-borne vibrations. The quadrate and articular were not fitted into an ear already specialized for detecting air-borne sound waves.

The external ear can be seen easily on your mammal specimen. It consists of the external ear flap, called the **auricle,** and a canal with a cartilaginous wall leading inward to the skull **(external acoustic meatus).** The **tympanum** lies at the base of the meatus between the external and middle ears. You can find it on the side of the head on which you removed the auricle by cutting away as much of the external acoustic meatus as possible and shining a light into the remainder. Note that the tympanum is set at an angle; its rostral portion extends more medially than its caudal portion. The opaque line seen through the membrane is the handle of the malleus.

The rest of the ear is difficult to dissect and should be observed on demonstration preparations, or studied from a diagram (Fig. 8-16D and E). Dissections can be prepared from the sagittal sections of the head used for the study of the nose. (The following account is based on the cat but is applicable to many other mammals.) Remove the muscles and other tissue from around the tympanic bulla except at its rostromedial corner. The middle ear, or **tympanic cavity,** lies within the bulla and opens into the nasopharynx by the **auditory tube** (Eustachian tube). The opening of the tube appears as a slit in the lateral wall of the nasopharynx (Fig. 10-17, p. 317). Careful dissection between this slit and the rostromedial corner of the bulla will reveal the tube; part of its wall is cartilaginous and part is bony.

The rest of the dissection will be much easier to do if you first decalcify the specimen by placing it in a weak solution (0.06%) of nitric acid for a few days. Break away the

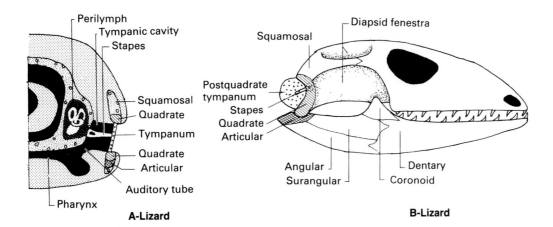

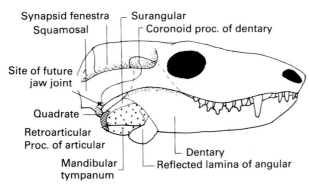

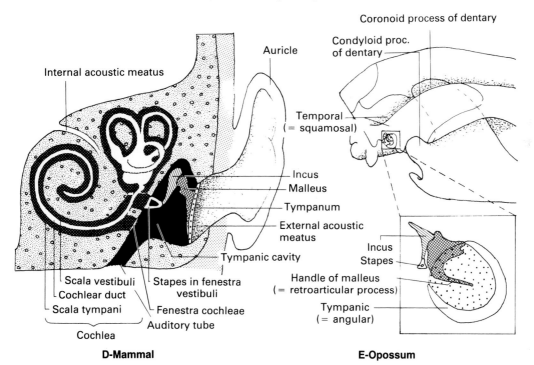

Figure 8-16
Diagrams of some types of ears found in terrestrial vertebrates. *A*, Transverse section through the otic region of a lizard; *B*, lateral view of the skull and ear of a generalized lizard; *C*, lateral view of the skull and ear of a late mammal-like reptile; *D*, transverse section through the otic region of a mammal skull; *E*, lateral view of the posterior portion of an opossum skull with the otic region enlarged below. *(B and C, Redrawn from Allin; E, redrawn from Crompton and Parker.)*

caudomedial portion of the bulla and also the mastoid and paracondyloid processes and adjacent parts of the nuchal crest, exposing the caudomedial chamber of the tympanic cavity. Note that it is largely separated from a smaller rostrolateral chamber by a more or less vertical plate of bone. A hole through the dorsolateral portion of this plate passes between the two chambers of the middle ear cavity. You can see the **fenestra cochleae,** or round window, through this hole (see Fig. 4-27). Carefully break away all this plate of bone and open up the rostrolateral chamber. You can see the handle of the **malleus** on the inside of the tympanum, and you will note that the fenestra cochleae is situated on a round promontory of bone. A finger-like process of cartilage extends from the caudolateral wall of the tympanic cavity between these two structures. The tiny nerve that runs along it and leaves its tip is the **chorda tympani,** a branch of the facial nerve going to the taste buds of the tongue and certain salivary glands. Break away bone from the rostromedial corner of the bulla and find the entrance of the auditory tube. The other auditory ossicles **(incus** and **stapes),** and the **fenestra vestibuli,** or oval window, in which the stapes fits, lie dorsal to the fenestra cochleae. To see them, you must remove a piece of bone caudal and dorsal to the external acoustic meatus without injuring the plate of bone supporting the tympanic membrane. You will also notice two small muscles passing to certain of the ossicles. A **stapedius** arises from the medial wall of the tympanic cavity caudad to the fenestra cochleae and inserts on the stapes. A **tensor tympani** arises from the medial wall rostral to the fenestra vestibuli and inserts on the malleus. These muscles are parts of the hyoid and mandibular musculature, respectively, that have followed parts of the hyoid and mandibular arches into the middle ear. They adjust the auditory ossicles and tympanum to the intensity of the sound waves. Their contraction, for example, reduces the amplitude of the vibration of the ossicles and protects the delicate structures of the inner ear from injurious movements resulting from loud noises.

The inner ear lies within the petrosal portion (otic capsule) of the temporal bone. Portions of it may be noted by removing the brain and chipping away pieces of the petrosal, but it cannot be dissected satisfactorily by this method. The **internal acoustic meatus** for the vestibulocochlear and facial nerves lies in the cranial cavity on the caudomedial surface of the petrosal.

◆ ◆

THE NERVOUS SYSTEM

Receptors alert an animal to changes in its internal and external environments. Often, the animal combines and sorts out sensory signals from different sources and draws upon memories of past experiences before making a motor response appropriate to the changed conditions. The motor response often involves the activation (or inhibition) of a score of muscles, glands, and other effectors. We refer to the totality of these processes as **integration** or **coordination.** Most rapid integration, such as maintaining balance when the body's position changes, is accomplished by nerve impulses travelling along nerve cells, or **neurons,** to a combination of specific effectors. Most slower responses that regulate continuing processes affecting many parts of the body, such as metabolism, growth, and reproduction, are mediated by chemical messengers, called **hormones,** secreted by **endocrine glands** and distributed through the blood to their targets. A hormone reaches all parts of the body, but only target cells with matching biochemical receptors on their surfaces will respond. Although there is a distinction between nervous and endocrine integration, the dichotomy is not absolute. Neurons usually secrete chemical messengers **(neurotransmitters)** that transfer a nerve impulse to another neuron or to an effector. A few neurons secrete hormones, and some physiological processes, including color changes in many fishes, are affected by both nerve impulses and hormones.

Since the endocrine glands are widely scattered, we will not discuss them in one place but as they are observed. We will consider the nervous system at this time, as it is the last of the group of organ systems that deals directly with the general function of body support and movement.

Neurons and Supportive Cells

The nervous system is composed of two cell types, both of ectodermal origin. The **neurons** are the functional units in the sense that they process and transmit nerve impulses, but they are supported, protected, and partly nourished by other cells. **Glial cells** perform these functions for neurons in the brain and spinal cord, and **neurilemma** cells of neural crest origin enwrap peripheral neurons. Certain glial and neurilemma cells may wrap themselves around the processes of neurons many times and surround them with a fatty **myelin sheath,** which increases the speed of nerve impulse transmission. A neuron can be divided into four regions. (1) The cell body, or **trophic region,** contains the nucleus, synthesizes materials the cell needs, and is responsible for the maintenance of the cell. (2) A **receptive region** receives stimuli from receptors or other neurons and initiates the nerve impulse. This region consists of relatively short processes known as **dendrites** that connect with the cell body or may consist of just the cell body itself. (3) The **conductive region** is composed of a long process known as the **axon,** or nerve fiber, that rapidly transmits the nerve impulse to the end of the cell. (4) A **transmitting region** at the end of the axon consists of branching **telodendria** that transmit the impulses across synapses to other neurons or to effector cells. Usually the cell body is located near the receptive end of the neuron, in which case the dendrites are short processes leading toward the cell body and the axon is a long one leading away from it, but the cell body may be located at any point along the neuron. In sensory neurons, it is set off to the side of a long axon. Parts of neurons are now defined in terms of the functions that they perform.

Aggregates of nerve cell bodies within the vertebrate brain are known as **nuclei;** aggregates of cell bodies outside the brain and spinal cord are called **ganglia. Tracts** are groups of fibers running together within the brain and spinal cord. **Nerves** are groups of fibers outside the brain and spinal cord.

Divisions and Components of the Nervous System

Grossly, the nervous system can be divided into a central portion and a peripheral portion. The **central nervous system** consists of the **brain** and **spinal cord** (Fig. 9-1). Both the spinal cord and brain are hollow, for they contain a **central canal,** which expands to form **ventricles** in certain regions of the brain. Recall that a single, dorsal, tubular nerve cord is one of the diagnostic characteristics of a chordate. Neurons are distributed within the central nervous system in such a way that we can speak of **gray matter** and **white matter.** The gray matter contains the cell bodies of neurons and unmyelinated fibers; the white matter contains myelinated fiber tracts. Most of the gray matter is centrally located. In the cord it forms a continuous column whose cross section usually resembles the letter H or a butterfly. **Sensory,** or **afferent, neurons** enter the dorsal column of the gray matter at the level at which they attach to the spinal cord, or they may ascend or descend in the white matter of the cord before entering the dorsal column. Their cell bodies are usually located in a peripheral ganglion. On entering the gray matter, some sensory neurons may extend directly to motor neurons, but most terminate on **interneurons** whose cell bodies comprise most of the gray matter of the dorsal column. The interneurons transmit the impulse to the motor neurons, or they may enter the white matter and ascend to the

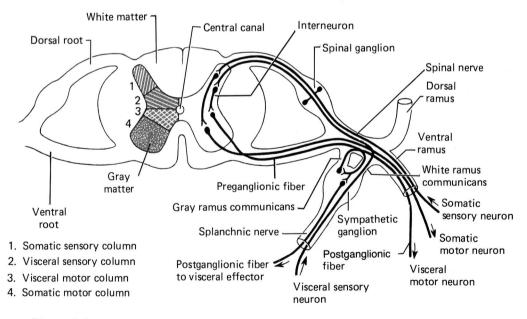

1. Somatic sensory column
2. Visceral sensory column
3. Visceral motor column
4. Somatic motor column

Figure 9-1

Diagram of a cross section through the spinal cord and a pair of spinal nerves to show the organization and functional components of the nervous system. Two interneurons are shown in the gray matter between the sensory and motor neurons. *(From Villee, Walker, and Barnes.)*

brain. During their ascent, they normally cross, or **decussate,** to the opposite side of the central nervous system. The cell bodies of **motor,** or **efferent, neurons** lie in the ventral column, and their axons extend to the effectors. They may receive impulses directly from sensory neurons, but usually they receive impulses from interneurons coming from the dorsal column or the brain. Interneurons from the brain decussate on the way caudad.

Sensory and motor neurons, and the gray columns where they end or begin, are divided into **somatic** and **visceral groups** (Fig. 9-1). Somatic sensory neurons are stimulated by somatic receptors and enter the somatic sensory portion of the dorsal column. Somatic motor neurons begin in the somatic motor column and transmit impulses to somatic muscles. Visceral neurons and columns serve the visceral receptors and effectors. Thus, we can speak of a somatic motor neuron, a somatic motor column, and so on. Further subdivisions are sometimes made. For example, visceral motor fibers that serve most of the gut muscles (general visceral motor fibers or autonomic fibers) are distinguished from those that serve branchiomeric muscles (special visceral motor fibers).

As the gray columns continue rostrally into the brain, they tend to break up into distinct islands of gray matter, or nuclei. The relationship among nuclei is similar to the relationship among the parts of the columns. The most dorsal nuclei are somatic sensory nuclei; the most ventral nuclei are somatic motor nuclei. Nuclei are closely associated with the terminations and origins of neurons that travel in the cranial nerves. In mammals, some gray matter migrates to parts of the brain surface and forms a conspicuous **cortex.**

The **peripheral nervous system** consists of all the neural structures—ganglia and nerves—lying outside the spinal cord and brain. **Spinal nerves** are segmentally arranged, and each connects to the central nervous system by a **dorsal root** and a **ventral root** (Fig. 9-1). The dorsal root bears the **spinal ganglion** containing the cell bodies of sensory neurons. More distally, the spinal nerve breaks up into branches, or rami, going to various parts of the body—a **dorsal ramus** to the epaxial region, a **ventral ramus** to the hypaxial region, and often one or more **communicating rami** with visceral connections. Most of the rami contain a mixture of sensory and motor neurons, but the neurons segregate in the roots. Sensory neurons always enter through the dorsal roots. In reptiles, birds, mammals, and a few fishes both visceral and somatic motor neurons leave through the ventral roots. In most anamniotes, however, some (or all) of the visceral motor neurons leave through the dorsal root, so that the ventral root primarily (or exclusively) contains somatic motor neurons.

Cranial nerves from the nose, eye, and ear evolved with these special sense organs. Many of the other cranial nerves, as we shall see, contain the combination of neurons that we find in either the dorsal root or ventral root of the spinal nerves of early vertebrates. This suggests that these cranial nerves may be serially homologous to either a ventral or dorsal root of the spinal nerves of ancestral vertebrates.

The **autonomic** portion of the peripheral nervous system consists of general visceral motor neurons going to the visceral organs, glands, and smooth muscle. These leave the spinal cord or brain in certain of the spinal and cranial nerves. Some of the fibers of the autonomic system remain in the spinal and cranial nerves, but some leave to travel in special branches to the organs in question. Thus, the autonomic system, while clearly definable in terms of function, is not completely separated from the rest of the peripheral nervous system morphologically.

Somatic muscles and branchiomeric muscles are innervated by somatic motor neurons or special visceral motor neurons, respectively, that extend all the way from the central nervous system to their effectors. In contrast, the general visceral motor fibers of the autonomic nervous system always relay in some peripheral ganglion on their way to the effectors (Fig. 9-1). Thus, we can distinguish between **preganglionic** and **postganglionic neurons.**

Another unique feature of the autonomic nervous system is its subdivision (in terrestrial vertebrates, at least) into **sympathetic** and **parasympathetic divisions.**

Most visceral organs receive fibers of both types. The sympathetic fibers of the mammalian autonomic nervous system leave the central nervous system through the thoracic and anterior lumbar spinal nerves; the parasympathetic fibers leave through certain cranial and sacral nerves. The peripheral relay of the sympathetic fibers is in a ganglion at some distance from the organ being supplied, so the postganglionic fiber is quite long. The parasympathetic relay, on the other hand, is in or very near the organ being supplied; thus, its postganglionic fibers are relatively short. Many of the sympathetic ganglia lie against the back of the body cavity, lateral and ventral to the vertebral column. Those on each side of the body may be interconnected by visceral fibers to form a chain known as the **sympathetic cord.**

Because sympathetic and parasympathetic postganglionic fibers release different neurotransmitters at their junctions with effectors, they generally have antagonistic effects. One stimulates an organ; the other inhibits it. The effect of sympathetic stimulation is similar to the effect of the hormone produced by the medullary cells of the adrenal gland. Both help a vertebrate adjust to stress by mobilizing resources that expend energy. Cardiac muscles contract with greater force and speed, blood pressure

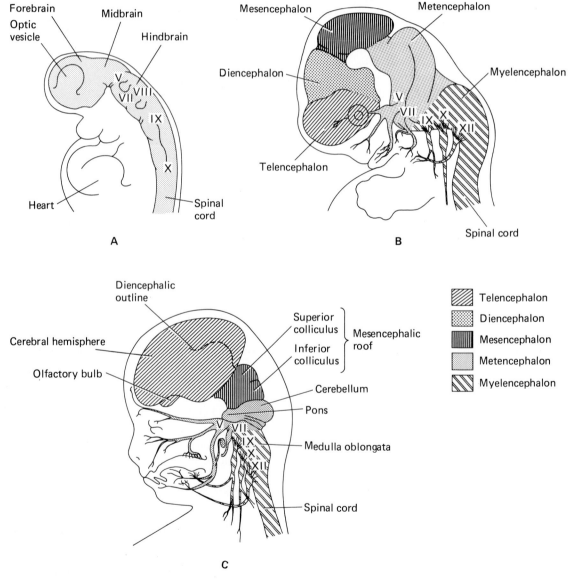

Figure 9-2
Diagrams of three stages in the development of the human brain to show the differentiation of the principal brain regions. Certain cranial nerves are identified by Roman numerals. *(Modified after Patten.)*

rises, blood flow to cardiac and skeletal muscles increases, blood sugar levels rise. Gut peristalsis and digestive functions are inhibited. Parasympathetic stimulation activates processes that produce and store energy. More blood is directed to the gut, digestive activities increase, and glucose is stored.

Development of the Brain

The brain is the most complex part of the nervous system. We can best gain an understanding of the various regions of which it is formed by considering its development. As shown in Figure 9-2, the brain arises as an enlargement of the cranial end of the neural tube. As it develops, it differentiates into three vesicles, or regions: an anterior forebrain, or **prosencephalon;** a middle midbrain, or **mesencephalon;** and a posterior hindbrain, or **rhombencephalon.**

The mesencephalon does not divide further, but the other two regions do. In most vertebrates, a pair of vesicles that will become the **telencephalon** evaginate and grow forward and laterally from the rostral end of the prosencephalon. The part of the prosencephalon remaining in the midline becomes the **diencephalon.** The rhombencephalon divides into a rostral **metencephalon** and a caudal **myelencephalon.** Each of these five regions further differentiates. The telencephalon gives rise to the cerebral hemispheres and olfactory bulbs; the diencephalon gives rise to the thalamus, hypothalamus, and epithalamus. The dorsal part of the mesencephalon forms a roof, or tectum, that differentiates in mammals into the paired superior colliculi (optic lobes) and inferior colliculi. The metencephalon gives rise to the cerebellum and, in mammals, the pons; the myelencephalon gives rise to the medulla oblongata.

FISHES

The nervous system of the dogfish should be studied carefully, for it shows the basic structure of the nervous system exceptionally well. Not only can you easily expose the system by removing cartilage, but it is in a morphologically primitive and generalized stage. The nervous system of *Squalus* is a good prototype for that of all vertebrates.

The Dorsal Surface of the Brain

You should perform the dissection of the nervous system on the large head of *Squalus* that you used for the study of the sense organs. The cranial nerves should be studied primarily on the intact side of the head, for certain of them may have been destroyed during the dissection of the sense organs. Remove the skin and underlying tissue from the dorsal surface of the chondrocranium and from around the eye. Be careful not to cut a large dorsal nerve (the **superficial ophthalmic nerve**) that lies on the dorsal side of the orbit and lateral to the rostrum (see Fig. 9-6). Cut away the cartilaginous roof of the cranial cavity by starting between the otic capsules and working slowly forward. As you do so, look into the rostral part of the cavity, and you may see a delicate, threadlike stalk extending from a depressed area of the top of the brain (diencephalon) to the epiphyseal foramen in the roof of the cranial cavity. This is the **epiphysis,** a homologue of the pineal eye of earlier vertebrates (Fig. 9-3). The epiphyseal foramen (p. 57) permits more light to impinge on this organ than on adjacent parts of the brain. The epiphysis contains photoreceptors and has been shown to be very sensitive to light (Gruber, 1977). Its biological role in sharks is not well understood, but it has been implicated in color changes in one species.

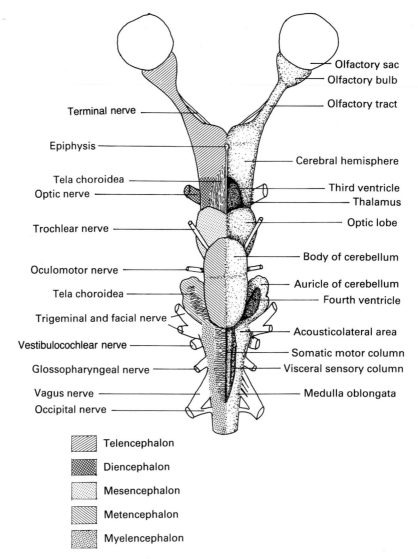

Terminal nerve

Epiphysis

Tela choroidea
Optic nerve

Trochlear nerve

Oculomotor nerve

Tela choroidea

Trigeminal and facial nerve

Vestibulocochlear nerve

Glossopharyngeal nerve

Vagus nerve

Occipital nerve

Olfactory sac
Olfactory bulb
Olfactory tract

Cerebral hemisphere

Third ventricle
Thalamus

Optic lobe

Body of cerebellum

Auricle of cerebellum
Fourth ventricle

Acousticolateral area
Somatic motor column
Visceral sensory column

Medulla oblongata

Telencephalon

Diencephalon

Mesencephalon

Metencephalon

Myelencephalon

Figure 9-3
Dorsal view of the dogfish brain; the tela choroidea has been removed from the right side.

Next, cut away the supraorbital crest and as much of the lateral walls of the cranial cavity as is possible without injuring nerves. Much of the ear on the intact side will have to be cut away. Be particularly careful not to break the small **trochlear nerve** that leaves the brain dorsally and passes to the dorsal oblique muscle of the eyeball (see Fig. 9-6).

The brain should now be well exposed. Its surface is covered with a delicate, vascular connective tissue, the **primitive meninx.** Strands of connective tissue pass from the meninx to the connective tissue lining the cranial cavity, or **endochondrium.** In life, a protective, mucoid perimeningeal tissue lies between the brain and endochondrium. It sometimes coagulates with preservation and must be washed out carefully.

The paired **olfactory bulbs** form the most rostral part of the brain (Figs. 9-3 and 9-6). They are the lateral enlargements in contact with the olfactory sacs, and they receive the primary olfactory neurons coming from the olfactory epithelium. Secondary olfactory neurons originate in the bulbs and form the **olfactory tracts** that extend caudally to the **cerebral hemispheres.** Most of these neurons terminate in a ventrolateral portion of the hemisphere that is essentially homologous to the mammalian piriform lobe. Olfactory bulbs and cerebral hemispheres constitute the **telencephalon.**

The **diencephalon** is the depressed area, often with a dark roof, situated caudal to the cerebral hemispheres. The roof of the diencephalon is the **epithalamus,** its lateral walls form the **thalamus,** and its floor (which will be seen later) is the **hypothalamus.** The **epiphysis** may be seen attaching to the caudal part of the epithalamus. Most of the roof is very thin, vascular, and membrane-like, forming a **tela choroidea.** The tela choroidea consists of only the ependymal epithelium that lines the central nervous system and the meninx that covers it. Carefully remove the tela choroidea on one side of the diencephalon and, while doing so, note that part of it extends onto the caudal surface of the cerebral hemispheres (Fig. 9-4). This part of the roof actually does not belong to the diencephalon but forms a thin-walled sac, the **paraphysis,** which is considered part of the telencephalon. A prominent fold extends from the tela choroidea down into the large cavity **(third ventricle)** within the diencephalon. This fold, the **velum transversum,** represents the rostral end of the diencephalon. The fold can often be discerned as an opaque transverse line visible through the intact tela choroidea of the third ventricle. A vascular tuft from the tela choroidea extends forward from the velum transversum into a **lateral ventricle** within each cerebral hemisphere. Each vascular tuft constitutes a **choroid plexus of the lateral ventricle.** Choroid plexuses secrete and absorb a lymphlike **cerebrospinal fluid** that slowly circulates through the ventricles of the brain. Cerebrospinal fluid helps maintain a proper nutritive and ionic environment for the brain. Only a small amount of cerebrospinal fluid escapes from the ventricles of fishes to circulate in meningeal spaces, but in mammals the central nervous system is bathed in cerebrospinal fluid. The small transverse bulge of nervous tissue at the very caudal part of the epithalamus under the tala choroidea is the **habenular region.**

The habenular region lies just rostral, and slightly ventral, to the large pair of **optic lobes.** The optic lobes develop in the roof, or **tectum,** of the **mesencephalon** (Fig. 9-3). The floor of the mesencephalon, which cannot be seen at this time, is known as the **tegmentum.**

The **metencephalon** lies caudal to the mesencephalon and consists dorsally of the **cerebellum.** The **body of the cerebellum** is the large, median, oval mass whose rostral end overhangs the optic lobes. Note that a longitudinal groove and a transverse groove partially subdivide it into four parts. The pair of earlike flaps that lie on either side of the caudal part of the body of the cerebellum are the **auricles of the cerebellum.** The ventral part of the metencephalon contributes to the medulla oblongata in fishes.

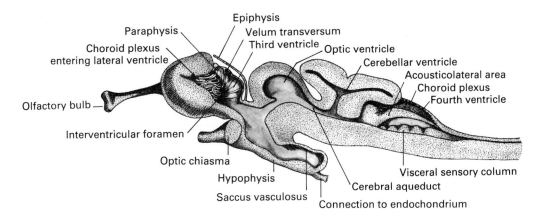

Figure 9-4
Sagittal section of the brain of *Squalus*.

The **myelencephalon** lies caudal to the metencephalon and forms the greater part of the **medulla oblongata**—the elongated region of the brain that is continuous caudally with the spinal cord. Most of the roof of the medulla consists of a thin, vascular tela choroidea, which covers a large brain cavity, the **fourth ventricle.** Carefully remove the tela choroidea and also its forward extension onto the dorsal surface of the auricles of the cerebellum. Vascular folds, the **choroid plexus of the fourth ventricle,** can be seen protruding into the ventricle. Lift up the caudal end of the body of the cerebellum and notice that the auricles are continuous with each other across the dorsal midline of the brain.

The columns of gray matter that continue from the spinal cord into the brain can be seen in the ventral and lateral walls of the fourth ventricle. The pair of midventral, longitudinal folds that lie on the floor of the ventricle are the **somatic motor columns** (Figs. 9-3 and 9-5). They contain the cell bodies of somatic motor neurons. There is a deep, longitudinal groove lateral to each somatic motor column. The **visceral motor column,** a column containing the cell bodies of visceral motor neurons, is situated laterally deep in this groove. A longitudinal row of small bumps lies dorsal to the visceral motor column. This is the **visceral sensory column,** and it receives and relays impulses from the visceral sensory neurons. The longitudinal groove between the visceral motor and visceral sensory columns is the **sulcus limitans,** a landmark separating the ventral motor portion from the dorsal sensory portion of the cord and brainstem. (The brainstem is the brain minus the cerebrum and cerebellum.) The dorsolateral rim of the fourth ventricle constitutes the **somatic sensory column,** and it receives and relays impulses from the somatic sensory neurons. You can visualize the relationship between these columns particularly well in a cross section, which you may make after the brain and nerves have been studied. The rostral part of the somatic sensory column is enlarged. This portion, known as the **acousticolateral area** (Fig. 9-3), receives the neurons from the ear and lateral line organs, but neurons from the ear, lateral line organs, and ampullae of Lorenzini terminate in their own nuclei within this area. These are not grossly distinct. The acousticolateral area is continuous rostrally with the auricles of the cerebellum. Although it cannot be seen in a gross dissection, these sensory and motor columns continue forward through the metencephalon and into the mesencephalon. Their rostral ends, however, become discontinuous, forming discrete patches, or nuclei, of gray matter.

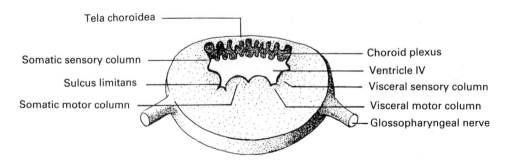

Figure 9-5
Cross section through the medulla oblongata of *Squalus* made at the level of the glossopharyngeal nerve.

Cranial and Occipital Nerves

We must consider the cranial and occipital nerves before examining the ventral and internal parts of the brain. You will have to remove more of the lateral wall of the cranium as you study the nerves.

♦ **(A) THE TERMINAL NERVE**

Fishes are usually described as having ten pairs of cranial nerves, and these are both named and numbered. However, an additional rostral nerve has been left out of this numbering system: the **terminal nerve** (Fig. 9-3), a tiny nerve that lies along the medial surface of the olfactory tract and extends between the olfactory sac and cerebral hemisphere. It is seen best at the medial angle formed by the junction of the olfactory tract with the cerebral hemisphere, for it separates slightly from the olfactory tract in this region. Although the terminal nerve is found in all vertebrates, except cyclostomes and birds, its function is uncertain. Autonomic and undefined sensory functions have been attributed to it, but more recent studies suggest that it has a chemosensory function. Studies in goldfish and some terrestrial vertebrates indicate that the terminal nerve is involved in helping to regulate reproductive functions by way of the nasal detection of pheromones (Demski, 1984). Experimental stimulation of the nerve in goldfish causes sperm release.

♦ **(B) THE OLFACTORY NERVE**

The **olfactory nerve (I)** carries olfactory impulses from the olfactory sac to the olfactory bulb. Since the sac and bulb are adjacent to each other in the dogfish, the olfactory nerve is not a compact structure but consists of a number of minute groups of neurons passing between these structures. You may see these by making a section through the olfactory sac and bulb. The olfactory neurons both receive stimuli and transmit nerve impulses.

♦ **(C) THE OPTIC NERVE**

The **optic nerve (II)** brings in optic impulses (somatic sensory) from the eye. Find it in the orbit and trace it medially (Fig. 9-6). It is a thick nerve. Push the brain away from the cranial wall and note that the optic nerve attaches to the ventral surface of the diencephalon. Since the retina of the eye develops embryonically from an outgrowth of the brain, the optic nerve is really a brain tract rather than a true peripheral nerve. Fibers in the optic nerve are the axons of ganglion cells whose cell bodies lie in the retina. The ganglion cells receive impulses from the photoreceptive rods and cones via short bipolar cells. In addition, cells within the retina make many horizontal interconnections, and the retina does considerable processing of visual information before sending signals to the brain.

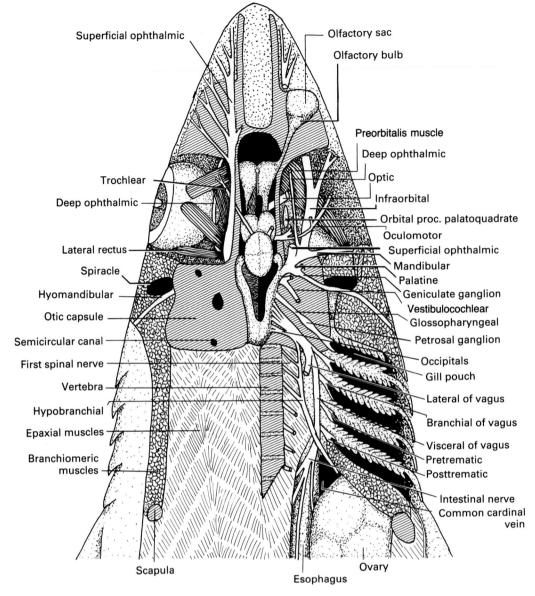

Figure 9-6
Semidiagrammatic dorsal dissection of the brain and cranial nerves of the dogfish. Deeper structures
are shown on the right side.

♦ **(D) THE OCULOMOTOR NERVE**

The **oculomotor nerve (III)** carries somatic motor impulses to most of the extrinsic
ocular muscles, receives proprioceptive impulses from these muscles (somatic sensory),
and carries autonomic fibers (visceral motor) to the ciliary body of the eye. To see it,
mobilize the eye on the intact side of the head in the manner described in connection with
the dissection of the eye (p. 220), remove the gelatinous connective tissue lying in the
orbit, and look on the ventral surface of the eyeball. The branch of the oculomotor going to
the **ventral oblique muscle** will be apparent (Fig. 8-6, p. 221). Follow it caudally and
medially. It passes ventral to the ventral rectus, and at the caudal margin of this muscle it
crosses a small and often whitish blood vessel. This vessel, an artery, follows the margin
of the ventral rectus and enters the eyeball. The autonomic fibers of the oculomotor nerve

form a small **ciliary nerve** that travels along this vessel, but these fibers can seldom be seen grossly. The branch of the oculomotor nerve that goes to the **ventral rectus muscle** lies between the ventral rectus muscle and the small vessel. Turn your specimen over and pick up the oculomotor nerve from the dorsal side. Cut the lateral rectus muscle near its insertion on the eyeball, and also the superficial ophthalmic nerve, and reflect them to see the oculomotor nerve clearly. It extends dorsal to the origin of the dorsal rectus muscle and enters the cranial cavity. Just before it enters, it gives off one branch to the **dorsal rectus muscle** and another to the **medial rectus muscle.** (Do not confuse the oculomotor nerve with another nerve of the same size, the deep ophthalmic nerve, which crosses the base of the oculomotor nerve and extends along the medial surface of the eyeball.) Push the brain away from the cranial wall and note the attachment of the oculomotor nerve on the ventral surface of the mesencephalon.

♦   **(E) THE TROCHLEAR NERVE**

The **trochlear nerve (IV)** has been noted crossing an optic lobe (p. 242). Lift up the rostral end of the body of the cerebellum and note where it attaches on the brain. The trochlear nerve passes through the cranial wall, goes ventral to, or perforates, the large superficial ophthalmic nerve, and extends to the **dorsal oblique muscle.** Like the oculomotor nerve, it is primarily a somatic motor nerve, but it carries a few proprioceptive (somatic sensory) fibers.

♦   **(F) THE ABDUCENS NERVE**

Skip the fifth nerve for a moment and consider the **abducens nerve (VI),** which carries somatic motor fibers to the **lateral rectus muscle** and returns proprioceptive (somatic sensory) impulses from this muscle. You can see it on the ventral surface of the lateral rectus. You will see its attachment on the ventral surface of the medulla later when you remove the brain (see Fig. 9-8).

♦   **(G) THE TRIGEMINAL NERVE**

Return to the **trigeminal nerve (V),** which is the nerve of the mandibular arch and the general cutaneous sensory nerve of the head (Figs. 9-3 and 9-6). The trigeminal nerve attaches in common with the facial (VII) and vestibulocochlear (VIII) nerves on the dorsolateral surface of the medulla just caudal to the auricles of the cerebellum. It is difficult to separate these nerves grossly at their attachments on the brain, but you can sort them out to some extent by their peripheral branches. In fishes the trigeminal nerve has four branches. A **superficial ophthalmic branch,** together with a comparable branch of the facial nerve, forms the large **superficial ophthalmic nerve** that has been noted passing through the dorsal region of the orbit and along the lateral surface of the rostrum. A **deep ophthalmic branch** (or **profundus nerve**) enters the orbit dorsal to the oculomotor nerve, adheres to the connective tissue on the medial surface (back) of the eyeball, and leaves the front of the orbit through a small foramen to join the superficial ophthalmic nerve. Trace the deep ophthalmic nerve on the side of the head in which the eyeball has been left. Both ophthalmic branches of the trigeminal nerve return somatic sensory impulses from general cutaneous sense organs (not lateral line organs) in the skin on the top and side of the head. In addition, the deep ophthalmic nerve has several minute and inconspicuous branches to the eyeball. These are considered to be homologous to the long ciliary nerve of mammals and, for the most part, return sensory fibers from parts of the eye other than the retina. In some selachians, there is evidence that they also carry a few autonomic fibers to the eye (Norris and Hughes, 1920).

You can find a **mandibular branch** of the trigeminal nerve by dissecting away the connective tissue on the caudal wall of the orbit. It is a fairly thick nerve that lies caudal to the lateral rectus muscle. The mandibular branch carries special visceral motor fibers to the branchiomeric muscles of the first visceral arch and returns some somatic sensory fibers from general cutaneous sense organs in the skin overlying the lower jaw.

The **maxillary branch** of the trigeminal nerve, together with the buccal branch of the facial nerve, forms the large **infraorbital nerve** that extends rostrally across the floor of the orbit. The infraorbital nerve is as wide as any of the ocular muscles and is easily confused with a muscle. It divides near the rostral border of the orbit and is distributed to the skin overlying the upper jaw and the underside of the rostrum. The maxillary portion of this nerve returns somatic sensory fibers from general cutaneous sense organs in this region.

Cut away enough cartilage and connective tissue from the caudomedial corner of the orbit to be able to see where all of the branches of the trigeminal nerve come together, and again note the attachment of the trigeminal nerve to the medulla. The main part of the trigeminal nerve bears a slight enlargement, the **semilunar ganglion,** that contains the cell bodies of the sensory neurons; however, it is unlikely that you can distinguish this ganglion.

♦   **(H) THE FACIAL NERVE**

The **facial nerve (VII)** is the nerve of the hyoid arch, spiracle, and the rostral lateral line organs. As mentioned previously, its **superficial ophthalmic** and **buccal branches** contribute to the superficial ophthalmic and infraorbital nerves, respectively. They return somatic sensory impulses from the lateral line canals, pit organs, and ampullae of Lorenzini. You can see the attachment of certain fibers from these trunks to the ampullae of Lorenzini. Cut away the skin adjacent to the caudoventral corner of the spiracle on the intact side of the head, and pick away the underlying connective tissue. The large nerve that will be seen is the **hyomandibular nerve,** a branch of the facial nerve (see also Fig. 7-11, p. 151). Follow it peripherally, noting that it is distributed to the hyoid muscles (special visceral motor fibers), skin (somatic sensory fibers from lateral line organs), and lining of the mouth cavity (visceral sensory fibers of both general nature and taste). Follow it medially to its union with the other branches and its attachment on the brain, which, as stated, is in common with that of the trigeminal and vestibulocochlear nerves. You will have to cut away some of the spiracle, surrounding muscles, otic capsule, and ear as you go, because it passes ventral to the ear. About 1 centimeter from the brain, the hyomandibular nerve bears a slight enlargement, the **geniculate ganglion,** that contains the cell bodies of sensory neurons (Fig. 9-6). Another, and smaller, branch of the facial nerve leaves from the rostroventral surface of this ganglion. This is the **palatine nerve,** and it returns visceral sensory neurons from the mouth lining.

◆   **(I) THE VESTIBULOCOCHLEAR NERVE**

You may have noted a part of the **vestibulocochlear nerve (VIII)**[18] coming from the ampullae of the anterior vertical and lateral semicircular ducts during the dissection of the hyomandibular nerve. Continue to remove the cartilage of the otic capsule and note another, and longer, part of this nerve coming from the ampulla of the posterior vertical semicircular duct, the sacculus, and parts of the utriculus. The vestibulocochlear nerve contains somatic sensory fibers from various parts of the inner ear.

◆   **(J) THE GLOSSOPHARYNGEAL NERVE**

The **glossopharyngeal nerve (IX)** is the nerve of the third visceral arch and the first of the five definitive gill pouches. It can be seen crossing the floor of the otic capsule caudal to the sacculus. It passes ventral to the caudal branch of the vestibulocochlear nerve and at first may be confused with this nerve. Cut away this part of the vestibulocochlear nerve and find the attachment of the glossopharyngeal nerve on the side of the medulla. Trace the glossopharyngeal nerve laterally; this will be easier if you open the first gill pouch by cutting through the skin and muscle dorsal and ventral to the first external gill slit. Cut all the way to, but not through, the internal gill slit (the opening between the gill pouch and pharynx). As the nerve leaves the otic capsule, it bears an oval-shaped swelling, the **petrosal ganglion,** that contains the cell bodies of sensory neurons. Four branches leave from the petrosal ganglion. Follow them distally and expose them to see at least the three larger branches. A large **posttrematic branch** passes down the caudal face of the first gill pouch to carry special visceral motor fibers to the branchial muscles and return visceral sensory fibers from this region. A smaller **pretrematic branch,** which consists entirely of visceral sensory fibers, passes down the cranial face of the first pouch. A still smaller **pharyngeal branch** follows the pretrematic branch a short distance and then curves around a tendon near the dorsal edge of the internal gill slit and is distributed to the wall of the pharynx. It, too, is entirely visceral sensory fibers. There is finally a small **dorsal branch** distributed to lateral line organs, and often to the skin in the supratemporal region, but it is impractical to find.

---

[18]This nerve is called the vestibulocochlear nerve in mammals, for it carries impulses from the vestibular apparatus (receptors for balance) and the cochlea. Although nonmammalian vertebrates lack a cochlea, we are applying this term to all vertebrates. The term *statoacoustic* is also widely used for this nerve.

♦ (K) THE VAGUS NERVE

The **vagus nerve (X)** is the nerve of the remaining visceral arches. Find its connection on the dorsolateral surface of the caudal end of the medulla and follow it caudally out of the otic capsule. To see the rest of the nerve, you must cut open the remaining gill pouches in the manner described for the first. If you cut as far as you should, you will cut into a large blood space, the **anterior cardinal sinus,** lying dorsal to the internal gill slits. If you did not expose this sinus during the dissection of branchial muscles (p. 154), open it now by a longitudinal incision. A large branch of the vagus nerve (the **visceral nerve**) lies beneath the connective tissue on the dorsomedial wall of the anterior cardinal sinus. It gives off four **branchial branches** that cross the floor of the sinus and are distributed to the remaining four visceral arches and pouches. Each branchial branch follows the pattern of the glossopharyngeal nerve, having a sensory ganglion from which **posttrematic, pretrematic,** and **pharyngeal branches** arise. Visceral sensory fibers in the posttrematic branch of the last branchial branch are distributed to the caudal surface of the last gill pouch, but there are no branchiomeric muscles here. Motor fibers in the last branchial branch form a small **accessory nerve** that goes to the cucullaris muscle. This nerve is difficult to find. After giving off the last branchial branch, the visceral nerve continues as the **intestinal nerve** along the wall of the esophagus to the viscera. It also sends a branch to the pericardial cavity. The intestinal nerve contains visceral motor fibers of the autonomic system and visceral sensory fibers.

Just before the visceral nerve enters the rostral end of the anterior cardinal sinus, the vagus gives off a **lateral** (or dorsal) **nerve.** This branch lies medial to the visceral nerve and extends caudally between the epaxial and hypaxial musculature. It receives somatic sensory fibers from the lateral line canal proper and, in some elasmobranchs, a few general cutaneous fibers from the skin in the gill region. Cell bodies of these neurons lie in a ganglion near the proximal end of this branch.

♦ (L) THE OCCIPITAL AND HYPOBRANCHIAL
    NERVES

Free the visceral branch of the vagus nerve from the wall of the anterior cardinal sinus. A **hypobranchial nerve** emerges from the epibranchial musculature, crosses the visceral nerve at about the level of its last branchial branch, and curves ventrally in the wall of the common cardinal vein (Fig. 9-6). It carries somatic motor fibers to the hypobranchial musculature and returns a few proprioceptive and cutaneous somatic sensory fibers.

Trace the hypobranchial nerve medially and cranially toward the vertebral column and chondrocranium. It becomes progressively narrower and more difficult to trace through the musculature, but if you are successful you will see that it is formed by the confluence of several **spinal nerves** and two or three **occipital nerves.** The first spinal nerve emerges between the chondrocranium and the first vertebra. The last occipital nerve can be seen between this point and the large root of the vagus nerve. More rostral occipital nerves lie deep to the root of the vagus. The occipital nerves resemble accessory rootlets of the vagus nerve, for they appear to join it. Actually, they travel with the vagus nerve only a short distance as they leave the chondrocranium; then they separate and, together with several spinal nerves, form the hypobranchial nerve.

The hypobranchial nerve, and its occipital and spinal components, are homologous to the amniote twelfth cranial nerve, the hypoglossal nerve. By convention, the hypobranchial nerve is not considered a cranial nerve in fishes because its origin relative to the caudal end of the skull varies in different groups. Frequently, as in *Squalus,* the origin is partly from the back of the brain (the occipital nerves) and partly from the front of the spinal cord (the spinal nerves). The occipital nerves are serially homologous to the ventral roots of spinal nerves. The dorsal roots of these nerves are believed to form the vagus nerve.

A summary of the distribution and types of neurons found in the cranial and occipital nerves is presented in Table 9-1 and shown diagrammatically in Figure 9-7. At first, this appears to be a confusing array of nerves, but you can sort them into three groups by examining the types of neurons they contain.

Group I includes the olfactory nerve (I), possibly the terminal nerve, the optic nerve (II), and the vestibulocochlear nerve (VIII). These are composed of somatic sensory neurons. These nerves undoubtedly evolved in conjunction with the special sense organs. Motor neurons that may modulate the sensitivity of the receptive cells have been found in the optic and vestibulocochlear nerves of some vertebrates.

Group II contains the oculomotor nerve (III), the trochlear nerve (IV), the abducens nerve (VI), and the occipital nerve. These nerves are composed of somatic motor neurons supplying the somatic musculature of the eye (extrinsic ocular muscles) and the epibranchial and hypobranchial musculature. As you might expect, they also contain a few somatic sensory proprioceptive fibers that provide the sensory feedback needed to regulate muscle activity. All connect ventrally on the brain except for the trochlear nerve, but on entering the dorsal surface of the brain, the trochlear fibers terminate in a ventral nucleus. The ventral roots of the spinal nerves of the earliest vertebrates also are composed of somatic motor neurons, so some investigators consider this group of cranial nerves to be the serial homologues of ventral roots. (We might add that the ventral and dorsal roots of spinal nerves do not unite to form a common spinal nerve in cyclostomes.)

Group III contains the remaining cranial nerves. They are composed of sensory (somatic sensory and/or visceral sensory) fibers and visceral motor fibers. They supply the skin of the head and lateral line organs (somatic sensory fibers) and the branchial pouches and musculature of the visceral arches (visceral sensory and visceral motor fibers). They contain the sorts of fibers we find in the dorsal roots of primitive spinal nerves and attach more dorsally on the brain. Some investigators regard them as the serial homologues of dorsal roots. A few proprioceptive fibers return from some of the visceral muscles.

The pattern of branching seen in the glossopharyngeal nerve (IX) may be the prototype for this group of branchiomeric nerves (Fig. 9-7 and Table 9-1). It has a small dorsal branch (somatic sensory fibers) to the adjacent skin and lateral line, and three branches—pharyngeal, pretrematic, and posttrematic—to the first gill pouch. All contain visceral sensory fibers, but the visceral motor fibers to the musculature of the third arch, which lies behind the pouch, are limited to the posttrematic branch. The vagus nerve (X) follows the same pattern as the glossopharyngeal nerve, except that it supplies four pouches and arches (the fourth through the seventh), and continues to the internal visceral organs. It may have evolved by the fusion of four branchiomeric nerves, each of which resembled the glossopharyngeal nerve. The first and second visceral arches lie anterior to the glossopharyngeal nerve, and the pouch that lay between them is reduced to a small opening called the spiracle or lost. There is no pouch anterior to the first visceral arch. It is not surprising, therefore, that the branchiomeric nerves—facial (VII) and trigeminal (V)—supplying this area do not have a full complement of branches. Each has a posttrematic branch, but the pretrematic branch and sometimes the pharyngeal branch are lost, and the dorsal branch supplying

(*text continues on p. 254*)

Table 9-1 Distribution and Components of the Cranial and Occipital Nerves of Selachians

The cranial and occipital nerves of selachians, together with their general distribution and functional components, are shown in this table. An X indicates the presence of the various components, and an (X), that the component is found in some but not all selachians. A few efferent fibers that probably modulate the sensitivity of the receptors travel in the optic and vestibulocochlear nerves.

Nerve	Branches	Distribution	Somatic Sensory General (Cutaneous)	Somatic Sensory Special including Lateral Line (L)	Somatic Sensory Proprioceptive	Visceral Sensory General	Visceral Sensory Special (Taste)	Visceral Motor General (Autonomic)	Visceral Motor Special (Branchiomeric)	Somatic Motor
Terminal		Olfactory sac		X?						
I. Olfactory		Olfactory epithelium		X						
II. Optic		Retina		X						
III. Oculomotor	4 muscular branches	Ventral oblique; ventral, dorsal, and medial rectus			X					X
	Ciliary	Ciliary body of eye						X		
IV. Trochlear		Dorsal oblique			X					X
V. Trigeminal	Superficial ophthalmic (dorsal)	Skin over top and side of head	X							
	Maxillary (dorsal)	Skin over upper jaw and underside of rostrum	X							
	Deep ophthalmic (profundus)	Skin over top and side of rostrum; sensory fibers from the eye and in some cases a few motor fibers to the eye	X					(X)		
	Mandibular (posttrematic)	Mandibular muscles and skin over lower jaw	X						X	
VI. Abducens		Lateral rectus			X					X

Nerve	Branch	Distribution									
VII. Facial	Superficial ophthalmic (dorsal)	Lateral line organs over top and side of head							X(L)		
	Buccal (dorsal)	Lateral line organs over upper jaw and underside of rostrum							X(L)		
	Hyomandibular (posttrematic)	Hyoid muscles; mouth lining; and lateral line organs near lower jaw		X		X	X		X(L)		
	Palatine (pharyngeal)	Mouth lining				X	X				
VIII. Vestibulocochlear (statoacoustic)		Sensory patches of inner ear									X
IX. Glossopharyngeal	Dorsal	Supratemporal lateral line organs; adjacent skin							X(L)	(X)	
	Pretrematic	Anterior wall of first typical gill pouch				X	X				
	Posttrematic	Posterior wall of first typical gill pouch		X		X	X				
	Pharyngeal	Pharyngeal lining				X	X				
X. Vagus	Lateral (dorsal)	Most of lateral line canal; skin in dorsolateral gill region							X(L)	(X)	
	Visceral	4 branchial branches to remaining pouches and then continues as the intestinal branch to the heart and abdominal viscera. Each branchial branch has pretrematic, posttrematic, and pharyngeal branches. At least one posttrematic supplies ventral pit organs.		X	X	X			X(L)		
Occipital (2-3 in *Squalus*)		Anterior epibranchial and hypobranchial musculature (the first 2-3 spinal nerves also supply the hypobranchial musculature in *Squalus*)	X					X			

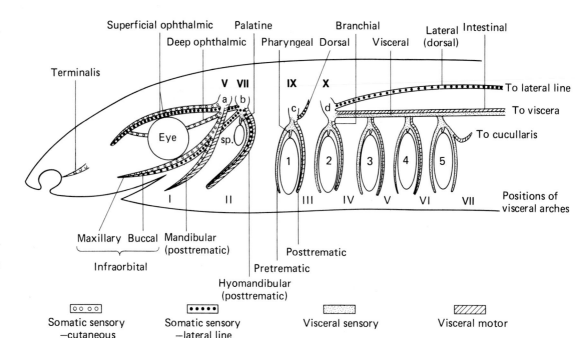

Superficial ophthalmic Palatine   Branchial   Lateral Intestinal
  Deep ophthalmic   Pharyngeal Dorsal   Visceral   (dorsal)

Terminalis

V VII   IX   X

a b   c   d   To lateral line

Eye   To viscera

sp.   To cucullaris

1   2   3   4   5

I   II   III   IV   V   VI   VII   Positions of
                                                    visceral arches

Maxillary Buccal Mandibular
              (posttrematic)   Posttrematic

Infraorbital   Pretrematic

Hyomandibular
(posttrematic)

□ o o o □   ■ • • • •   ▨   ▨
Somatic sensory   Somatic sensory   Visceral sensory   Visceral motor
—cutaneous   —lateral line

Figure 9-7

A diagram of the major branches and components of the branchiomeric nerves of a dogfish. Boldface Roman numerals indicate cranial nerves; lightface Roman numerals indicate the position of visceral arches. Ganglia: *a*, semilunar; *b*, geniculate; *c*, petrosal; *d*, jugal and nodose. Gill slits are indicated in Arabic numerals.

the skin and complex pattern of lateral line organs is divided. Interestingly, there is also a division of labor in the sensory functions of these two nerves: The facial nerve supplies the lateral line organs, and the trigeminal nerve serves the other cutaneous receptors.

The resemblance of certain cranial nerves to dorsal and ventral roots of spinal nerves led investigators early in this century to propose that the head of ancestral vertebrates was segmented in a manner similar to the trunk. There is no doubt that the caudal part of the head is of segmental origin, but there is some doubt as to how far forward this pattern can be projected. Zoologists now realize that the vertebrate head has a complex origin, part coming from body segments, but much of the front of the head developing embryonically from the neural crest, and this part apparently is not segmented.

The Ventral Surface of the Brain

It is now possible to return to the brain, remove it, and study its ventral surface. Before lifting the brain out of the braincase, cut across the caudal end of the medulla, the olfactory tracts, and the roots of the cranial nerves. Try to leave stumps of the nerves attached to the brain. After you have cut across the trigeminal, facial, and vestibulocochlear nerves, push the brain to one side and note the small abducens nerve leaving. It too must be cut. Lift up the caudal end of the brain and carefully pull it rostrally. You will soon see a part of the brain extending into a recess (the **sella turcica**) in the floor of the cranial cavity. It will probably be necessary to cut away some of the floor of the chondrocranium caudal to this recess to get the brain out intact.

After the brain has been removed, examine its ventral surface (Fig. 9-8). Identify the connections of the cranial nerves and the major regions of the brain. Several new

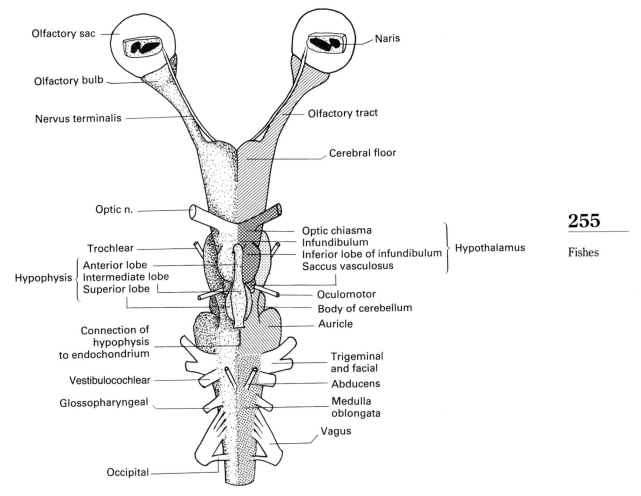

Figure 9-8
Ventral view of the dogfish brain. See Figure 9-3 for the key to brain regions.

structures can be seen in the region of the **hypothalamus** (floor of the diencephalon). The optic nerves attach at the rostral end of the hypothalamus, and their fibers decussate at this point, forming an X-shaped structure known as the **optic chiasma.** In most nonmammalian vertebrates, the decussation is complete, so that the left eye projects to the right side of the brain and vice versa. In some amphibians and reptiles, a few fibers do not decussate and remain ipsilateral. In many mammals, especially those species in which the visual fields of the two eyes overlap to a large extent and stereoscopic vision is present, many fibers remain ipsilateral and the rest decussate. The visual field of each eye thus projects to each side of the brain. After passing through the optic chiasma, the optic fibers form a band, the **optic tract,** that leads caudally and dorsally to the optic lobes. You will have to remove the primitive meninx to distinguish the tract. Most of the rest of the hypothalamus consists of a large, caudally projecting **infundibulum.** Just caudal to the optic chiasma, the infundibulum bears a pair of prominent lateral lobes known as the **inferior lobes** of the infundibulum. The **hypophysis,** or pituitary gland, lies partly between and partly caudal to the inferior lobes. The infundibulum finally forms a thin-walled, dark **saccus vasculosus,** which lies caudal to the inferior lobes and dorsal to the hypophysis. The saccus vasculosus of fishes is a specialized structure composed of neural epithelium and containing an extensive vascular plexus. Its function is unknown, but one hypothesis suggests that the saccus vasculosus detects water pressure and, hence, depth; another suggests that it helps control the composition of cerebrospinal fluid.

If the brain was carefully removed, you can recognize the various lobes of the hypophysis (Fig. 9-8). Two conspicuous lobes lie in the midventral line—an **anterior lobe** between the inferior lobes of the infundibulum and an **intermediate lobe** caudal to this. A pair of **superior lobes** lies between the intermediate lobe and the saccus vasculosus.

The hypophysis is an endocrine gland of dual embryonic origin. Most of its anterior lobe, which attaches to the infundibulum, develops as an outgrowth of the diencephalon and, therefore, is homologous to the mammalian **neurohypophysis.** The rest of the hypophysis develops from an embryonic hypophyseal pouch, which is an invagination from the stomodeum, and hence is comparable to the mammalian **adenohypophysis.** Functions of the fish hypophysis are not fully known, but they are probably similar to those of a mammal. The neurohypophysis stores and releases hormones involved in osmoregulation. These hormones are actually secreted by neuron cell bodies in the hypothalamus and reach the neurohypophysis by traveling along the axons of those cells. The adenohypophysis produces a hormone that helps regulate growth, and others that interact with the thyroid, adrenal, and gonadal hormones in the regulation of the level of body metabolism, dispersion of pigment in chromatophores, mineral metabolism, and reproduction.

The Ventricles of the Brain

The central cavity that characterizes the neural tube of chordates expands to form large chambers, or ventricles, in the brain (Fig. 9-4). You have already observed some. To see the others, make a sagittal section of the brain and also cut into one of the cerebral hemispheres. **Lateral ventricles** lie in the paired cerebral hemispheres. Each connects with the **third ventricle** of the diencephalon by a narrow passage known as the **interventricular foramen** (foramen of Monro). The third ventricle connects with the **fourth ventricle** of the hindbrain by a narrow, ventral passage known as the **cerebral aqueduct** (aqueduct of Sylvius). **Optic ventricles** in the optic lobe and a **cerebellar ventricle** in the body of the cerebellum lie dorsal to the cerebral aqueduct. In life, all are filled with cerebrospinal fluid.

The Spinal Cord and Spinal Nerves

The spinal cord lies in the vertebral canal of the vertebral column. To see it and the spinal nerves, remove the muscles overlying several centimeters of the vertebral column behind the chondrocranium and carefully shave away the dorsal part of the vertebral arch. You will soon see the **spinal cord** and the **dorsal roots** of the spinal nerves through the cartilage. A **spinal ganglion** is present on each but is rather small. Each dorsal root lies slightly caudal to its corresponding ventral root, so to find the **ventral root** you must cut away the lateral wall of the vertebral arch cranial to the dorsal root. The ventral root arises from the cord by a fan-shaped group of rootlets and passes out of the vertebral canal through a foramen in the neural plate. The two roots unite in the dogfish in the musculature lateral to the vertebral column, but this union, and the subsequent splitting of the spinal nerve into dorsal, ventral, and communicating rami, is difficult to find.

The components in the roots of the dogfish spinal nerves appear to be similar to those in the nerves of amniotes—sensory neurons in the dorsal roots, motor neurons in the ventral roots. Autonomic fibers present in the dorsal roots of many fishes and amphibians have not been positively identified in selachians, but some may be present. Certainly most of the preganglionic visceral motor fibers of the autonomic system extend through the ventral roots to synapse with postganglionic fibers in a series of ganglia lying dorsal to the posterior cardinal sinus and kidneys. The autonomic system of the dogfish thus has cranial and spinal contributions, but the visceral organs apparently do not have the double innervation found in mammals. Also, autonomic fibers do not go to the skin in the dogfish.

Functions of the Spinal Cord and Brain

The gross structure of the central nervous system of elasmobranchs is well known, and many of the neuronal pathways have been established, but less is known about the functions of the various regions. Experiments in which the connection between the spinal cord and brain has been severed have shown that the spinal cord of a dogfish is more than a center for reflex action. Such a "spinal" dogfish still displays normal, rhythmic swimming movements. Further experiments suggest the presence of a **central pattern generator** within the spinal cord that controls swimming because electrical recordings from individual motor neurons show a spontaneous rhythmic activity even when curare poisoning has blocked all muscular activity. Other studies have demonstrated that sensory feedback from the skin and active muscles also has a role and does at least modify any inherent rhythmic activity. It is also clear that tracts descending from the brain modify locomotor patterns. (See Williamson and Roberts, 1986, for a review.)

As in other vertebrates, a dense network of short interconnecting neurons, known as the **reticular formation,** extends through the floor of the medulla oblongata, metencephalon, and mesencephalon. The reticular formation receives sensory inputs from the cranial nerves attaching to these regions, as well as from many other parts of the central nervous system: spinal cord, cerebellum, optic tectum, thalamus, and hypothalamus. Its efferent fibers lead to the motor neurons of the cranial nerves, and some project into the cord. The reticular formation of elasmobranchs has been shown to integrate respiratory and cardiac movements, and it probably has a role in integrating other functions as well, including feeding and swallowing movements.

Sensory neurons from the inner ear, lateral line, and electroreceptors terminate in adjacent but distinct nuclei in the acousticolateral area of the medulla. Fibers from these nuclei extend into the spinal cord and cerebellum, particularly the auricles of the cerebellum. The cerebellum also receives visual, tactile, and proprioceptive inputs. As in other vertebrates, the cerebellum of elasmobranchs doubtless monitors the body's position in space and the degree of muscle contraction and sends corrective signals to motor neurons. Experiments have shown that the cerebellum sends signals to muscles of the pectoral fins that affect up-and-down movements of the fins. These movements help to maintain balance. Destruction of the cerebellum of fishes, however, results in fewer motor defects than its destruction in mammals.

The optic tectum has six layers of neuron cell bodies and fibers, making it anatomically the most complex region of the elasmobranch brain. Although it receives its major sensory input from the eyes, the tectum also receives projections that bring in most other sensory modalities. Most of its efferent fibers lead to the reticular formation, from where the tectum can affect motor activity at many levels. The coordination of eye movements with swimming movements is one example. Some investigators believe that the tectum is the primary integration center of the elasmobranch brain, analogous to the mammalian cerebral hemispheres.

The habenula in the roof of the diencephalon is an olfactory center. The thalamus, located in the lateral walls of the diencephalon, receives sensory inputs relayed from the reticular formation, cerebellum, and optic tectum. A few fibers from the eyes go directly to the thalamus. Neuronal connections exist that can send some of this information forward to parts of the cerebrum. Efferent fibers lead from the thalamus back to the reticular formation and motor neurons. The hypothalamus in the floor of the diencephalon receives projections from olfactory, gustatory, and general visceral senses. Most of its efferent fibers lead by way of the reticular formation to visceral motor neurons, but some of its neurons produce neurosecretions that affect the hypophysis. The hypothalamus of elasmobranchs appears to affect the level of body activity (rest versus wakefulness) and helps regulate water balance, blood sugar levels, and other aspects of homeostasis. In conjunction with parts of the cerebrum, it also controls aspects of behavior related to feeding, aggression, and sex.

The olfactory bulbs are the first olfactory integration centers. Additional olfactory integration occurs in limited parts of the cerebral hemispheres to which the olfactory bulbs project. Much of the cerebrum, however, is not olfactory. This region receives projections via the thalamus from visual and other sensory centers. Its efferent fibers lead to the hypothalamus and the thalamus, from where impulses can go to the reticular formation. The cerebrum has neuronal connections that imply an important role in integration, but little is known about its functions in elasmobranchs.

AMPHIBIANS

The nervous system of such early tetrapods as the amphibians has progressed little beyond the condition seen in fishes. As a matter of fact, the cerebellum is not as well developed in living amphibians as in active fishes like *Squalus*. *Necturus* may be studied as an example of the amphibian condition, but it need not be studied in detail, as the changes are not significant.

The Dorsal Surface of the Brain

Expose the dorsal surface of the cranium by removing the skin and muscles overlying it; then carefully chip away the bone that forms its roof. The primitive meninx of fishes is represented by two layers in amphibians—a tough outer **dura mater** and an inner, vascular **pia-arachnoid mater** applied to the surface of the brain. If you did not remove the dura mater with the skull bones, you will have to cut it off in order to see the brain.

The rostral region of the brain is formed by the paired **cerebral hemispheres** and **olfactory bulbs** (Fig. 9-9). The olfactory bulbs lie rostral to the hemispheres but are not clearly separated from them. At most, only a slight lateral indentation may be seen between bulbs and hemispheres. Since the olfactory bulbs are adjacent to the rest of the telencephalon rather than to the olfactory sacs, there is no long, narrow olfactory tract as there was in the dogfish. The bands of nervous tissue that extend between the olfactory sacs and the olfactory bulbs are not the tracts but the **olfactory nerves**. The small, dark body that lies between the caudal ends of the cerebral hemispheres is the **paraphysis**. This is an evagination from the telencephalon.

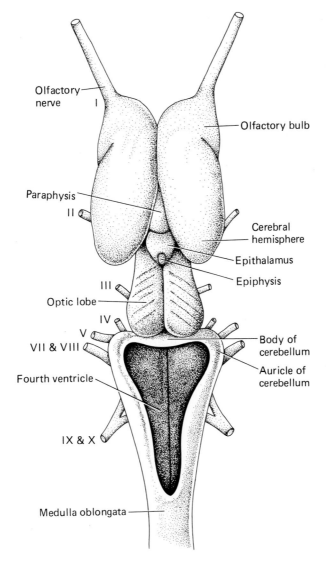

Olfactory
nerve I

Olfactory bulb

Paraphysis

II

Cerebral
hemisphere

Epithalamus

Epiphysis

III

Optic lobe

IV

V

VII & VIII

Body of
cerebellum

Auricle of
cerebellum

Fourth ventricle

IX & X

Medulla oblongata

Figure 9-9
Dorsal view of the brain and cranial nerves of *Necturus*.

The **epithalamic** region of the diencephalon appears as a small, triangular, light area posterior to the paraphysis. As usual, the **epiphysis** and **habenula** are present in this region, but they are difficult to distinguish. From the tela choroidea in this region, a **choroid plexus** has invaginated into the third ventricle and has extended caudad into the cavity of the midbrain. You can see it by cutting open the epithalamus.

The **optic lobes** lie caudal to the epithalamus. They are relatively small in *Necturus* and do not bulge dorsally to the extent seen in *Squalus*. The line of separation between optic lobes and epithalamus is not sharp.

The **medulla oblongata** lies caudal to the optic lobes; you can recognize it by the very dark tela choroidea that forms most of its roof. Remove the tela with its **choroid plexus** and note the heart-shaped **fourth ventricle**.

As mentioned at the outset, the **cerebellum** is very small in amphibians. Its **body** consists of only a narrow, transverse band of nervous tissue lying between the optic lobes and the medulla. This fold may be difficult to distinguish from the optic lobes. The **auricles** of the cerebellum are represented by the tissue that surrounds the anterolateral corners of the fourth ventricle.

The Cranial and Occipital Nerves

The **olfactory nerve (I)** was observed coming into the brain from the olfactory sac. The **optic nerve (II)** from the eye crosses the cranial cavity and attaches to the brain at the rostral end of the diencephalic floor.

The nerves to the eye muscles—the **oculomotor nerve (III)**, the **trochlear nerve (IV)**, and the **abducens nerve (VI)**—are present but can be seen only under magnification. They have the same relationships as in *Squalus* and supply the same muscles. The abducens, however, also goes to the newly evolved **rectractor** bulbi muscle.

The large trunk arising from the lateral surface of the rostral end of the medulla is the **trigeminal nerve (V);** the trunk caudal to it is the common attachment of the **facial (VII)** and **vestibulocochlear nerves (VIII)**. The roots caudal to this trunk represent the origin of the **glossopharyngeal nerve (IX)** and the **vagus nerve (X)**. The glossopharyngeal and vagus nerves run together until they leave the cranium; then they separate. The distribution and composition of these nerves is substantially the same as in *Squalus* but, of course, there would be some modification in the gill region. Lateral line fibers are present in *Necturus* but are lost in most metamorphosed amphibians.

Like other salamanders, *Necturus* has a **hypobranchial nerve** to the hypobranchial musculature. This nerve is formed primarily of fibers from the first spinal nerve, but it also receives contributions from the second spinal nerve and, in some salamanders, a small twig from the glossopharyngeal-vagus trunk. Certain of these contributions would obviously be homologous to the occipital nerves of *Squalus,* but just which ones is uncertain. It is of interest in this connection that the first, and sometimes the second, spinal nerve consists of only the ventral root.

The Ventral Surface of the Brain

Cut across the caudal end of the medulla and across the cranial nerves and remove the brain. Try to lift the hypophysis from the sella turcica without breaking it off. Note the major regions of the brain and the stumps of the cranial nerves, and study the floor of the diencephalon (hypothalamus) in more detail. The optic nerves decussate to form an **optic chiasma** at the rostral end of the hypothalamus, but the chiasma is small and inconspicuous in *Necturus*. As in *Squalus,* the greater part of the hypothalamus consists of a large, posteriorly projecting **infundibulum** to which the hypophysis is attached. A small **saccus vasculosus** is present in the roof of the infundibulum but is difficult to see grossly.

MAMMALS

In the evolution through reptiles to mammals, numerous changes occur in the nervous system that are related in large measure to the increased activity and flexibility of response of mammals. A major change is the tremendous expansion of a region of gray matter of the cerebrum, known in mammals as the **neopallium,** or **neocortex,** and its migration to the surface to form the gray cortex of the cerebral hemispheres. The neopallium becomes the dominant integrating region of the brain. To understand the evolution of the neopallium and its effects on other parts of the cerebrum, we must first examine the areas of gray matter in the cerebral hemispheres of amphibians. The gray matter in amphibians, and in elasmobranch fishes as well, lies deep within each cerebral hemisphere next to the lateral ventricle (Fig. 9-10A). The dorsolateral and dorsoventral portions of the gray matter form the **paleopallium** and **archipallium,** respectively. Both are olfactory centers. Ventrally, the gray matter forms a lateral **corpus striatum** (which is an integration center and a relay center for outgoing motor impulses) and a medial **septal region.** The septal region interacts with the hypothalamus in the control of emotional responses. Dorsally, a small **dorsal pallium** lies between the paleopallium and the archipallium. It is this seemingly insignificant area that expands to become the neopallium.

The cerebral hemispheres of modern reptiles have expanded in comparison with those of amphibians and have grown caudad to cover part of the diencephalon.

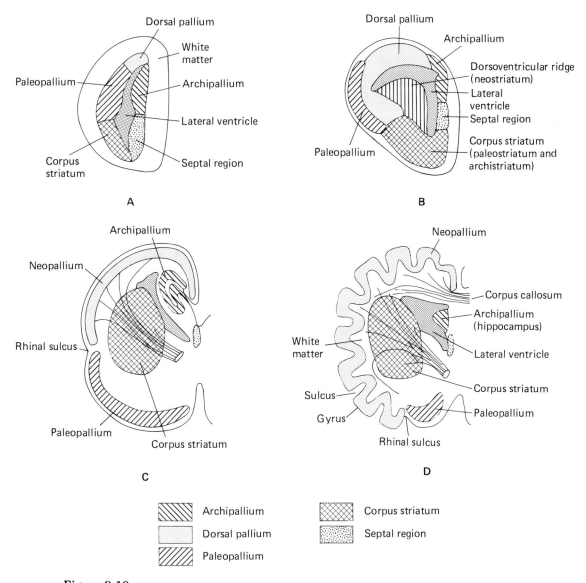

Figure 9-10
Diagrammatic cross sections of the left cerebral hemisphere of a modern amphibian (**A**), a modern reptile (**B**), an early mammal (**C**), and a eutherian mammal (**D**). Note the disposition of the areas of gray matter in the cerebrum, and in particular the changes in the dorsal pallium, which becomes the neopallium. (*Modified from Walker,* Functional Anatomy of the Vertebrates.)

Enlargement of the cerebral hemispheres derives partly from an expansion of the corpus striatum and partly from the formation of a dorsoventricular ridge in the floor of the lateral ventricle (Fig. 9-10**B**). The latter appears to have evolved from an inward growth and proliferation of the dorsal pallium. The cerebral hemispheres also enlarge on the line of evolution to mammals and grow caudad to cover the diencephalon and mesencephalon. Again, the dorsal pallium is involved in the enlargement, but it moves to the surface to form a gray cortex, and as it expands it pushes the archipallium and paleopallium farther and farther apart (Fig. 9-10**C** and **D**). The paleopallium, which remains an olfactory center, comes to lie on the ventral surface of a cerebral hemisphere. The archipallium is pushed into the medial wall of the lateral ventricle to become a structure known as the **hippocampus.** The hippocampus continues to receive some olfactory input, but it also acquires other functions, as we will see. In many mammals, especially the larger species, the cortical area is increased further by the evolution of a complex pattern of folds (gyri) and grooves (sulci). The neopallium also has an intricate architecture involving connections between six layers of neurons, whereas the older pallial areas have a simpler structure, with only three layers.

The evolution of the neopallium is coupled with a great increase in the size and importance of the thalamus because all sensory information projected to the neopallium is processed and relayed in dorsal thalamic nuclei, and many motor impulses coming from the neopallium are relayed in ventral thalamic nuclei. As integrating functions are transferred to the neopallium, the optic tectum becomes less important in integration but remains an important center for visual and auditory reflexes.

Other important changes occur in the cerebellum that are correlated with the increased complexity of muscular movement and the increased role of the cerebrum in initiating motor activity. A **neocerebellum,** represented primarily by the cerebellar hemispheres, is added to the primitive archicerebellum (auricles) and paleocerebellum (body). The neocerebellum, with the motor cortex of the neopallium, is involved in complex feedback mechanisms enabling the cerebellum to monitor motor activity, to promote even and steady contractions, and to synchronize the precise timing and duration of contraction of muscles.

Other changes in the brain tend to be correlated with the increased importance of the neopallium and the cerebellum. Important centers develop in the ventrolateral portions of the midbrain and metencephalon for the interconnection of the cerebellum and cerebrum. More and larger fiber tracts, both afferent and efferent, evolve in the mesencephalon and hindbrain because the cerebrum, the cerebellum, and the brain in general exert more influence over the body than they did in nonmammalian vertebrates.

The basic pattern of the cranial nerves of mammals remains much the same as in fishes (Table 9-1), but a few changes have been superimposed upon this plan during the evolution to mammals. The major changes affect the branchiomeric (dorsal root) nerves that supply the branchial apparatus: (1) The loss of the lateral line system in most adult amphibians is accompanied by the loss of lateral line neurons from nerves that previously carried them (VII, IX, X). (2) Except for the limited trigeminal outflow, mammals retain the autonomic fibers present in the cranial nerves of fishes and, in addition, acquire autonomic fibers in the facial and glossopharyngeal nerves. Those in the facial nerve go to the lacrimal and salivary glands; those in the glossopharyngeal nerve go to the parotid gland. (3) With the elaboration of the cucullaris muscle to form the trapezius and sternocleidomastoid complex of muscles, we find that the special visceral motor fibers that supply these muscles separate from the vagus nerve to form a new cranial nerve, the **accessory nerve (XI).** This change occurs first in reptiles. (4) The major change in ventral root nerves occurs when the caudal limit of the cranium becomes fixed, and the occipital nerves (plus a rostral spinal nerve or two) form a definite cranial nerve, the **hypoglossal nerve (XII).** Among living vertebrates, this change is first seen in the reptiles, but it probably occurred earlier, for there is a foramen for such a nerve in the skulls of crossopterygians and labyrinthodonts. In modern amphibians, the hypobranchial musculature is supplied by fibers that travel in the first spinal nerve.

Meninges

You should study the mammalian brain and the stumps of the cranial nerves on isolated sheep brains. As explained, the peripheral distribution and composition of the nerves is, with a few exceptions, essentially the same as in fishes. The foramina through which they leave the cranium are listed in Table 4-3 (p. 91) and you should review them as you study the nerves. During later dissections, you will see peripheral parts of certain of the nerves. If it is not possible to study the brain from isolated specimens, you may remove it from your own specimen, but if this is to be done, postpone this study until the end of the course. To remove the brain, first make a sagittal section of the head and then carefully loosen the halves of the brain and pull them out. Leave on the tough membrane (dura mater) covering the brain as you take it out. The cranial nerves will, of course, have to be cut, but leave stumps as long as possible attached to the brain.

The primitive meninx of fishes has differentiated in mammals into three layers: dura mater, arachnoid, and pia mater (Fig. 9-11). Since the brain fills the cranial cavity in mammals, the tough, outer **dura mater** has been pushed against the periosteum lining the cranial cavity and has fused with it. If the dura is still on your specimen, carefully remove it. As you do so, note that it sends one extension down between the cerebrum and cerebellum and another between the two cerebral hemispheres. The former extension is called the **tentorium;** the latter is the **falx cerebri** (Fig. 10-17, p. 317). These membranes help to stabilize the brain within the cranial cavity and prevent it from distorting during sudden rotational movements of the head. Cut across the falx cerebri and notice that it contains the **superior sagittal venous sinus.** Irregular pits evident on the cerebral surface of the falx lead into this venous sinus. In life, granule-shaped projections of the next deeper meninx (the arachnoid) protrude through these into the sinus.

The arachnoid and pia mater are difficult to distinguish grossly. The **pia mater** is the vascular layer that closely invests the surface of the brain. The **arachnoid** lies between the pia and dura mater. On the cerebrum it is most easily distinguished from the pia in the region overlying the grooves on the brain surface, for the arachnoid does not dip into them, whereas the pia mater does. In life, a narrow **subarachnoid space,** which is criss-crossed by weblike strands of connective tissue (hence the name *arachnoid*), lies between the pia mater and arachnoid (Fig. 9-11). The subarachnoid space expands into a large **cerebellomedullary cistern,** where the arachnoid leaves the caudal border of the cerebellum and extends over to the medulla. **Cerebrospinal fluid** circulates in the subarachnoid space of mammals as well as in the ventricular system.

Most cerebrospinal fluid is secreted into the ventricles of the brain. Sympathetic neurons that go to the blood vessels and to the epithelium of the choroid plexuses regulate the secretion. Stimulating the neurons reduces the production of cerebrospinal fluid, whereas cutting them enhances production (Lindvall, Edvinsson, and Owman, 1978). Cerebrospinal fluid leaves the cavities of the brain through minute pores in the roof of the fourth ventricle and slowly circulates in the subarachnoid space (Fig. 9-11). It re-enters the blood vessels through the **arachnoid granulations** that protrude into the superior sagittal and certain other cranial venous sinuses. The cerebrospinal fluid forms a liquid cushion around the central nervous system that protects and supports the exceedingly soft and delicate nervous tissue. Cerebrospinal fluid gives the brain a great deal of buoyancy. It has been calculated that a human brain weighing 1500 grams outside the body has an effective weight of only 50 grams in situ.

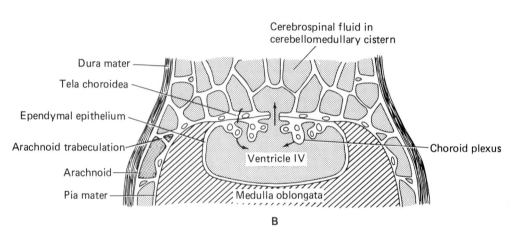

Figure 9-11
Diagram of the meninges and cerebrospinal fluid circulation in a mammal. **A,** Cross section through
the dorsal surface of the cerebral hemisphere showing the entrance of cerebrospinal fluid into the
venous system. **B,** Cross section through the medulla oblongata showing secretion of the cerebrospinal fluid into the fourth ventricle and its escape into an enlargement of the subarachnoid space
(cerebellomedullary cistern).

   The cerebrospinal fluid also helps to provide the brain with carefully selected
nutrients and other substances. The immediate environment of brain cells must be
more carefully regulated than that of other cells because the activity of brain cells is
adversely affected by the accumulation of waste products and by fluctuating levels of
sugars, other nutrients, hormones, and certain ions. Tight junctions between brain
capillary cells and the cells of the choroid plexuses form a **blood-brain barrier** to the
free passage of many substances. Gases and some other materials pass easily,
whereas other substances are held back, and some are actively transported by glial
cells and the cells of the choroid plexuses.

External Features of the Brain and the Stumps of the
Cranial Nerves

♦   **(A) THE TELENCEPHALON**

The paired cerebral hemispheres and the cerebellum are so large that little else is at first
apparent in a lateral or dorsal view. Notice that the **cerebral hemispheres** are

separated dorsally from each other by a deep longitudinal furrow known as the **longitudinal cerebral fissure.** Spread the dorsal parts of the cerebral hemispheres apart and note the thick transverse band of fibers that connects them. This is the **corpus callosum,** a neopallial commissure characteristic of eutherian mammals. The surface of each hemisphere is thrown into many folds (the **gyri**), which are separated from each other by grooves, called **sulci** (Figs. 9-12 and 9-13). A pair of **olfactory bulbs** project from the rostroventral portion of the cerebrum. They can be seen best in a lateral or ventral view. The olfactory bulbs lie over the cribriform plate of the ethmoid and receive the groups of olfactory neurons from the nose through the **cribriform foramina** (p. 89). These neurons, which will have been pulled off the bulb during brain removal, constitute the **olfactory nerve (I).** A whitish band, the **lateral olfactory tract,** extends at an angle caudally and laterally from each bulb. The ventral portion of the cerebrum, to which each olfactory tract leads, is known as the **piriform lobe.** It is separated laterally from the rest of the cerebrum by the **rhinal sulcus.**

The piriform lobe represents the paleopallium of other vertebrates. The remaining olfactory portion of the primitive cerebrum, the archipallium, has been pushed internally and does not show on the surface. Thus, all the superficial part of the cerebrum that lies lateral and dorsal to the rhinal sulcus is neopallium—the major integrating region of the brain. Sensory information of all types is projected here, memories of past experience may be drawn upon, and appropriate motor impulses are initiated. Cerebral pathways through which sensory impulses enter and motor impulses leave will be seen later.

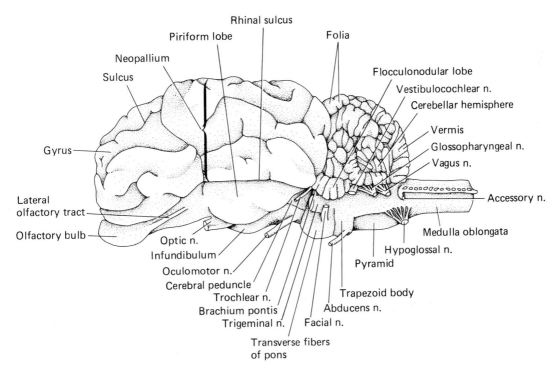

Figure 9-12
Lateral view of the sheep brain. *(From Walker,* A Study of the Cat.*)*

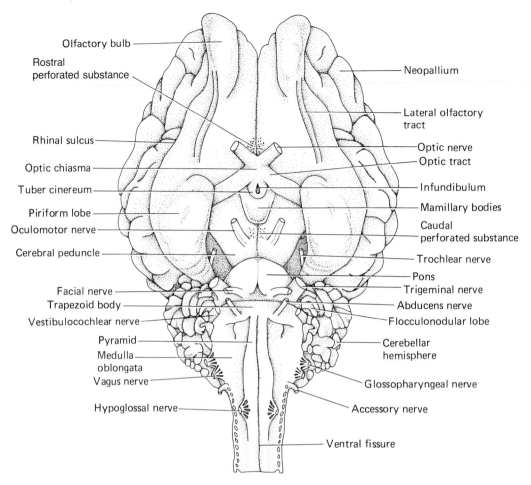

Figure 9-13
Ventral view of the sheep brain. *(From Walker,* A Study of the Cat.)

♦ **(B) THE DIENCEPHALON**

The telencephalon has enlarged to such an extent that it has grown back over and covers the diencephalon and much of the mesencephalon. To see the dorsal portion of the diencephalon **(epithalamus),** it is necessary to spread the cerebral hemispheres apart, cutting the corpus callosum. Pick away the tela choroidea and its choroid plexus forming part of the roof of the diencephalon. The longitudinal slit that is then exposed is the **third ventricle** (Fig. 9-14). The knoblike **pineal gland** lies caudal to the ventricle. The narrow transverse band of tissue between the pineal body and the ventricle is the **habenular commissure,** and the tissue forming the posterolateral rim of the ventricle is the **habenula.**

Turn the brain over and examine the ventral surface of the diencephalon—the **hypothalamus** (Fig. 9-13). The **optic nerves (II),** which enter the skull through the **optic canal** (p. 89), undergo a partial decussation at the rostral border of the hypothalamus, forming the prominent **optic chiasma.** The rest of the hypothalamus is the oval area lying caudal to the optic chiasma. The **hypophysis** (pituitary gland) may still be suspended by its narrow stalk, the **infundibulum,** from the hypothalamus. If so, remove it in order to get a clearer view of the region. The cavity in the infundibulum represents an

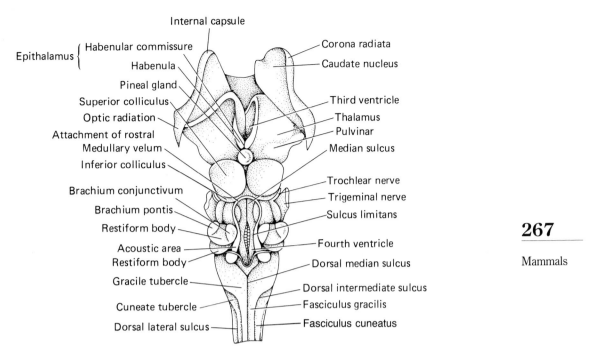

Internal capsule

Epithalamus { Habenular commissure — Corona radiata
Habenula — Caudate nucleus
Pineal gland
Superior colliculus — Third ventricle
Optic radiation — Thalamus
Attachment of rostral — Pulvinar
Medullary velum — Median sulcus
Inferior colliculus
Brachium conjunctivum — Trochlear nerve
Brachium pontis — Trigeminal nerve
Restiform body — Sulcus limitans
Acoustic area — Fourth ventricle
Restiform body — Dorsal median sulcus
Gracile tubercle
Cuneate tubercle — Dorsal intermediate sulcus
Dorsal lateral sulcus — Fasciculus gracilis
— Fasciculus cuneatus

Figure 9-14
Dorsal view of the sheep brainstem. The telae choroideae have been removed. *(From S. W. Ranson and S. L. Clark.)*

extension of the third ventricle. That portion of the hypothalamus adjacent to the attachment of the infundibulum is known as the **tuber cinereum.** A pair of rounded **mamillary bodies** forms the caudal end of the hypothalamus. Grossly they are not as obviously paired in the sheep brain as in the human, but nuclei within them are paired.

In order to glimpse the **thalamus,** or lateral wall of the diencephalon, you must carefully pull one of the cerebral hemispheres forward and look beneath it. You will get a better view after you have removed the cerebrum (p. 279), so postpone further study of the thalamus until then.

The pineal gland has been implicated in regulating sexual development, because secretions of the gland inhibit gonadal development in some species. In some mammals, exposure to prolonged light decreases activity of the gland and enhances gonadal development.

The habenula continues to be an important olfactory center.

As in other vertebrates, the hypothalamus of mammals is an important integrating center for many autonomic and visceral functions. Gustatory, olfactory, and general visceral sensory impulses are projected here. Rest or wakefulness, feeding behavior and digestion, blood sugar level, water and salt balances, sexual activity, and body temperature are among the activities regulated here. Efferent pathways lead to the reticular formation in the medulla and to visceral motor nuclei of the autonomic nervous system. The hypothalamus also has important connections to the hypophysis. Certain hormones are synthesized in the hypothalamus and travel along neurons to the neurohypophysis, where they are stored and released. Releasing factors for the numerous hormones secreted by the adenohypophysis are also produced in the hypothalamus and reach the adenohypophysis by way of a minute hypophyseal portal system of blood vessels.

We will consider functions of the thalamus when we study that structure.

♦   **(C) THE MESENCEPHALON**

You can see the roof, or **tectum,** of the mesencephalon by spreading apart the cerebrum and cerebellum. Four prominent, round swellings—the **corpora quadrigemina**—characterize this region (Fig. 9-14). The larger, rostral pair are the **superior colliculi;** the smaller, caudal pair are the **inferior colliculi.** Note that the **trochlear nerves (IV)** arise slightly caudal to the inferior colliculi.

A pair of **cerebral peduncles** lie along the ventrolateral surface of the mesencephalon (Figs. 9-12 and 9-13). Each emerges from beneath the tract of optic fibers extending dorsally and caudally from the optic chiasma. An **oculomotor nerve (III)** arises from the surface of each peduncle. The third, fourth, and sixth nerves (described later) converge and leave the skull together through the **orbital fissure** (p. 89) to enter the base of the orbit, where they are distributed to the extrinsic ocular muscles. The depression between the two peduncles is the **interpeduncular fossa.** If you strip the meninges from this region, you may be able to see small holes through which blood vessels enter the brain. This region constitutes the **caudal perforated substance.** A comparable **rostral perforated substance** lies rostral to the optic chiasma.

The evolution of the neopallium has robbed the optic tectum of its original importance as the major integrating area, but optic and auditory fibers are still projected to the superior and inferior colliculi, which remain as significant visual and auditory reflex centers, respectively. The peduncles are large bundles of fibers that extend caudally from the cerebral hemispheres. Most efferent impulses from the cerebrum pass back through them.

♦   **(D) THE METENCEPHALON**

The dorsal portion of the metencephalon forms the **cerebellum.** Its surface area is increased by numerous platelike folds, or **folia,** separated from each other by **sulci** (Fig. 9-12). The median part of the cerebellum, which has the appearance of a segmented worm bent nearly in a circle, is called the **vermis;** the lateral parts are the **hemispheres.** The lobe of each hemisphere that lies ventral to the main part of the hemisphere, and lateral to the region where the cerebellum attaches to the rest of the brain, is known as the **flocculonodular lobe.** These lobes are homologous to the cerebellar auricles, or archicerebellum, of the dogfish; most of the vermis is homologous to the body, or paleocerebellum; and most of the hemispheres are new additions (the neocerebellum) with cerebral connections.

The cerebellum is connected with other parts of the brain by three prominent fiber tracts, or peduncles (Figs. 9-14 and 9-15). Most of the **brachium pontis,** or middle peduncle, lies medial to the rostral half of the flocculonodular lobe. You will have to dissect off this lobe on one side to see the brachium clearly. Note that the brachium pontis connects ventrally with a transverse band of fibers known as the **transverse fibers of the pons** (Fig. 9-12). Trace it dorsally into the white matter of the cerebellum, the **arbor vitae.** The arbor vitae will be seen more clearly in the sagittal section (p. 275). The tissue caudal or slightly medial to the brachium pontis constitutes the **restiform body,** or caudal

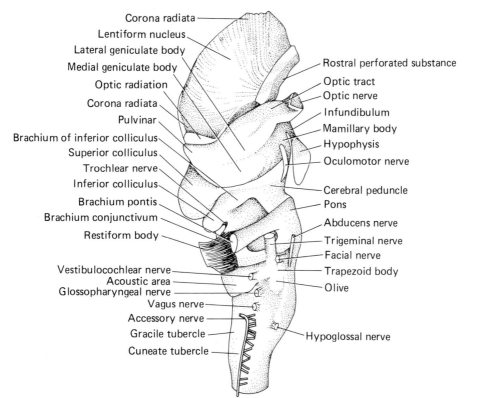

Corona radiata
Lentiform nucleus
Lateral geniculate body
Medial geniculate body
Optic radiation
Corona radiata
Pulvinar
Brachium of inferior colliculus
Superior colliculus
Trochlear nerve
Inferior colliculus
Brachium pontis
Brachium conjunctivum
Restiform body
Vestibulocochlear nerve
Acoustic area
Glossopharyngeal nerve
Vagus nerve
Accessory nerve
Gracile tubercle
Cuneate tubercle

Rostral perforated substance
Optic tract
Optic nerve
Infundibulum
Mamillary body
Hypophysis
Oculomotor nerve
Cerebral peduncle
Pons
Abducens nerve
Trigeminal nerve
Facial nerve
Trapezoid body
Olive
Hypoglossal nerve

Figure 9-15
Lateral view of the sheep brainstem. *(From Ranson and Clark.)*

peduncle. It also continues along the dorsolateral margin of the medulla. The **brachium conjunctivum,** or rostral peduncle, lies medial to the brachium pontis; you can see it by looking in the area between the cerebellum and caudal colliculi. An important auditory pathway, the **lateral lemniscus,** forms a bundle of fibers that ascends from between the brachium conjunctivum and brachium pontis.

In mammals, the ventral portion of the metencephalon has differentiated sufficiently from the medulla oblongata to be considered a distinct region—the **pons** (Fig. 9-13). Grossly, the pons includes the region of transverse fibers bordered rostrally by the interpeduncular fossa, and caudally by the trapezoid body (part of the myelencephalon). The large **trigeminal nerve (V)** arises from the lateral portion of the pons and extends rostrally across the base of the brachium pontis. As in fishes, the trigeminal nerve of mammals consists of three primary divisions: (1) the ophthalmic, returning sensory fibers from much of the head above the level of the orbit; (2) the maxillary, returning sensory fibers from the skin over the upper jaw; and (3) the mandibular, returning sensory fibers from the lower jaw area and carrying motor fibers to the mandibular muscles. These branches leave the cranial cavity through the three foramina just caudal to the optic canal: the **orbital fissure** (ophthalmic + III, IV, and VI), the **foramen rotundum** (maxillary), and the **foramen ovale** (mandibular) (p. 90).

The cerebellum is a major center for equilibrium and motor coordination (Fig. 9-16). Vestibular inputs from the ear are projected via the restiform bodies to the archicerebellum (flocculonodular lobe). Proprioceptive and tactile impulses also pass through the restiform bodies to reach the paleocerebellum (all but a small central

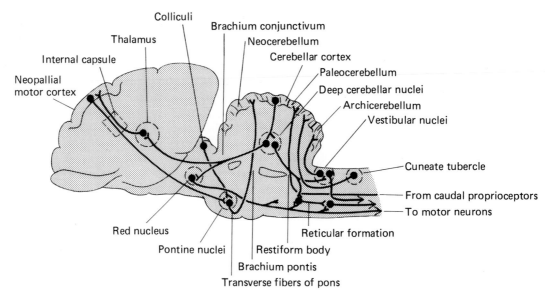

Labels in figure:
Colliculi
Thalamus
Brachium conjunctivum
Internal capsule
Neocerebellum
Cerebellar cortex
Neopallial motor cortex
Paleocerebellum
Deep cerebellar nuclei
Archicerebellum
Vestibular nuclei
Cuneate tubercle
From caudal proprioceptors
To motor neurons
Red nucleus
Reticular formation
Pontine nuclei
Restiform body
Brachium pontis
Transverse fibers of pons

Figure 9-16
Diagram on a lateral projection of the sheep brain of major cerebellar connections.

region of the vermis). The neocerebellum (the cerebellar hemispheres and the central region of the vermis) receives impulses from the motor cortex of the cerebrum. These are relayed in **pontine nuclei,** decussate in the transverse fibers of the pons, and enter the cerebellum through the brachia pontis. Afferent fibers ascend to their particular cortical areas through the arbor vitae. Integration occurs in the complex layered cerebellar cortex. Efferent fibers leave on the axons of large **Purkinje cells,** relay in several **deep cerebellar nuclei,** and leave the cerebellum through the brachia conjunctiva and restiform bodies. The Purkinje cells have an inhibitory effect upon their targets, but the degree of inhibition is altered by connections within the cerebellar cortex. Some impulses strengthen the degree of inhibition; others reduce it (disinhibition).

The effect of the archicerebellum is to adjust body and eye movements to changed positions of the head. The paleocerebellum regulates postural and locomotor movements relative to the current positions and activity of the limbs. The neocerebellum projects back to the motor cortex of the cerebrum via the thalamus and helps to regulate the timing and duration of muscle contractions, especially those related to skilled, voluntary movements.

♦   **(E) THE MYELENCEPHALON**

All the rest of the brain belongs to the myelencephalon and forms the **medulla oblongata.** In order to see the parts of the medulla clearly, you must strip off the meninges on at least one side, but before doing this you should identify the remaining cranial nerves (Figs. 9-12 and 9-13). At the border between the pons and the medulla oblongata are the **abducens nerve (VI)** and the **facial nerve (VII).** The abducens nerve lies more medial and extends rostrally across the pons. The facial nerve extends laterally and dorsally across the base of the trigeminal nerve. The stump of the **vestibulocochlear nerve (VIII)** lies dorsal to the cut end of the facial nerve. Facial and vestibulocochlear nerves leave the cranial cavity together through the **internal acoustic meatus** (p. 90). The vestibulocochlear nerve begins, of course, at the inner ear, which is lodged within the petrosal part of the temporal bone, but the facial nerve continues through the temporal bone to emerge on the skull surface at the **stylomastoid foramen** (p. 90).

The **glossopharyngeal nerve (IX)** and the **vagus nerve (X)** are represented by a number of fine rootlets caudal to, but in line with, the eighth nerve. In most specimens, these rootlets have been cut so short that they cannot be traced into the peripheral parts of the nerves. It is impossible to say more than that the rostral rootlets belong to the glossopharyngeal nerve and the caudal rootlets to the vagus nerve. The **accessory nerve (XI)** is the large, longitudinal nerve caudal to the vagus. It arises by a number of fine rootlets from the caudal end of the medulla and the rostral end of the spinal cord. The rostral end of the nerve lies close to the vagus nerve; indeed, the accessory nerve evolved from a part of the vagus nerve of anamniotes. Glossopharyngeal, vagus, and accessory nerves leave the cranial cavity together through the **jugular foramen** (p. 90) in company with the internal jugular vein. The **hypoglossal nerve (XII)** is represented by the rootlets on the caudoventral portion of the medulla. Notice how close these are to the level of the foramen magnum; such proximity is to be expected, because the nerve evolved from the spinooccipital nerves of fishes. The hypoglossal nerve leaves through the **hypoglossal canal** (p. 90).

Now strip off the meninges on one side of the medulla, and the **tela choroidea** with its **choroid plexus** that forms much of the roof of the medulla. Pull the cerebellum forward and note the large **fourth ventricle** extending forward into the metencephalon. The caudal part of the roof of the ventricle was formed by the tela choroidea, but the rostral part of the roof is formed by a thin layer of fibers termed the **rostral medullary velum,** which you can see by pushing the cerebellum caudally. The trochlear nerve decussates in the velum.

Note the enlargement on the dorsal rim of the medulla just caudal to the point at which the restiform body turns into the cerebellum, and dorsal to the vestibulocochlear nerve. Observe that it extends medially to an oval enlargement in the ventrolateral part of the floor of the fourth ventricle. This enlargement constitutes the **acoustic area.**

Examine the dorsal surface of the caudal end of the medulla. (The medulla extends as far caudad as the **first spinal nerve.**) The prominent **dorsal median sulcus** (Fig. 9-14) continues onto the medulla nearly to the fourth ventricle. A less distinct **dorsal intermediate sulcus** lies about 0.5 centimeter lateral to the dorsal sulcus, and a **dorsal lateral sulcus** lies slightly lateral to the dorsal intermediate sulcus. These grooves outline two longitudinal fiber tracts, a dorsal **fasciculus gracilis** and a more lateral **fasciculus cuneatus.** The rostral end of the former tract expands slightly to form a structure called the **gracile tubercle.** The fasciculus cuneatus has a comparable enlargement known as the **cuneate tubercle.** The tubercles are the outward manifestations of nuclei. As already stated, the restiform body forms the dorsolateral rim of the medulla rostral to these tubercles. It then turns dorsally to enter the cerebellum.

Turn the brain over and examine the ventral surface of the medulla. The narrow, transverse band of fibers immediately caudal to the pons is the **trapezoid body** (Fig. 9-13). Notice that its fibers can be followed dorsolaterally into the acoustic area. The midventral groove extending the length of the medulla oblongata is the **ventral median fissure.** The longitudinal bands of tissue on either side of it that are approximately 0.5 centimeter wide are known as the **pyramids.** Note that some of the pyramidal fibers lie superficial to the trapezoid body.

The medulla is a transitional region between the rostral parts of the brain and the spinal cord. As in fishes (p. 257), it contains some of the reticular formation and the sensory and motor nuclei of cranial nerves that attach in this region. Numerous reflex activities occur in this region, and many important visceral activities are controlled here: respiratory movements, salivation, swallowing, rate of heart beat, and blood pressure. Most of the nuclei cannot be seen grossly, but some form bulges on the surface. The acoustic area contains vestibular and cochlear nuclei, where many of the sensory neurons in the vestibulocochlear nerve terminate and relay. Some of the relays extend vestibular impulses to the cerebellum, and some extend auditory impulses through the trapezoid body. Much of the white matter of the medulla represents fiber tracts passing through the region.

You have now seen the surface manifestations for important sensory and motor pathways to and from the cerebrum (Fig. 9-17). Certain of the deeper parts will be exposed later. General cutaneous sensations, such as pain, heat or cold, and light touch, enter the spinal cord or brainstem on somatic sensory neurons. Some of these neurons ascend without interruption in dorsal portions of the white matter and terminate in the gracile and cuneate tubercles of the medulla oblongata. Here they can be relayed to the cerebellum (Fig. 9-16) or continue on afferent interneurons in a tract,

272

Chapter 9
The Nervous
System

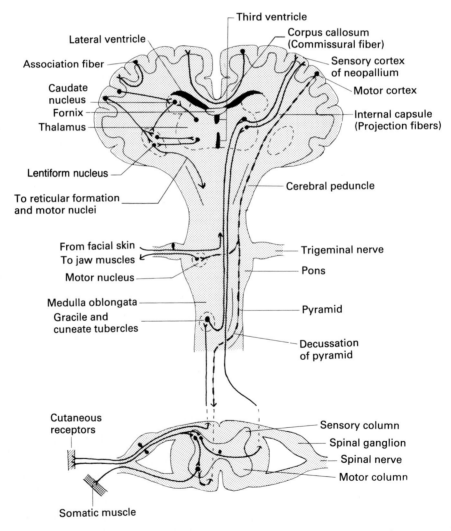

Figure 9-17
Diagram of major cutaneous sensory pathways and of the pyramidal motor system in the mammalian brain. Types of interconnections within the cerebrum and some of the connections of the lentiform and caudate nuclei of the corpus striatum are also shown. Pathways through the brain are diagrammed partly in transverse section and partly in a dorsal projection.

not visible grossly, that decussates and ascends to various dorsal thalamic nuclei (Fig. 9-17). Other neurons continue from the thalamus through the internal capsule, which you will see later deep within the cerebrum, to specific areas of the sensory cortex of the neopallium. Other somatic sensory neurons entering the central nervous system relay on afferent interneurons at the level of entry. Some of these interneurons make reflex connections with motor neurons; others decussate and ascend in lateral portions of the white matter to the dorsal thalamus.

The primary voluntary motor pathway from the motor cortex of mammals is the **pyramidal system.** Motor neurons whose cell bodies lie in the motor cortex have long axons that descend all the way to the cell bodies of the motor neurons of cranial and spinal nerves. The pyramidal neurons extend through the internal capsule, the cerebral peduncles, and the deep parts of the pons, emerging on the ventral surface of the medulla as the pyramids. Fibers continuing down the cord decussate near the caudal end of the medulla. The pyramidal system is a direct and rapid pathway to the motor nuclei and columns. It is unique to mammals. Less direct and slower motor pathways that involve ventral thalamic nuclei and the reticular formation are also present, as they are in nonmammalian vertebrates. They constitute the **extrapyramidal system.**

Sagittal Section of the Brain

Cut the brain in half as close as possible to the sagittal plane. If you deviate from the plane, take the larger half and dissect away enough tissue to be able to see clearly the median cavities of the brain (Fig. 9-18). Many of the features just described can also be seen in this view. Note in particular the way in which the cerebral hemispheres extend back over the diencephalon and mesencephalon. The **corpus callosum** also shows particularly well. Its expanded rostral end is known as the **genu;** its expanded caudal end is the **splenium.** The thinner region between is called the **trunk of the corpus callosum.** A thin, vertical septum of tissue, the **septum pellucidum,** lies ventral to the rostral part of the corpus callosum. It consists of two thin plates of gray matter. The lateral ventricle lies lateral to the septum; you may observe this by breaking it. A band of fibers called the fornix lies caudal to the septum pellucidum. The **body of the fornix** begins near the splenium, and then the band curves forward and ventrally as the **column of the fornix.** It passes out of the plane of the section caudal to a small, round bundle of fibers that is a cross section of the **anterior commissure,** an olfactory decussation. The thin ridge of tissue extending ventrally from the anterior commissure to the optic chiasma is the **lamina terminalis—** a landmark representing the rostral end of the embryonic neural tube. The cerebral hemispheres are evaginations that extend laterally and forward from the rostral end of the neural tube.

The third ventricle and diencephalon lie caudal to the column of the fornix, rostral commissure, and lamina terminalis. Note that the **third ventricle** is very narrow but has a considerable dorsoventral and rostrocaudal extent. It is lined by a shiny epithelial membrane (the **ependymal epithelium**), as are all the cavities within the neural tube. The thalamus lies lateral to the third ventricle, but a portion of it, the **interthalamic adhesion,** extends across the third ventricle and will appear as a dull, circular area not covered by the shiny ependymal epithelium. The **interventricular foramen** (foramen of Monro), through which each lateral ventricle communicates with the third ventricle, lies in the depression rostral to the adhesion. The hypothalamus lies ventral to the third

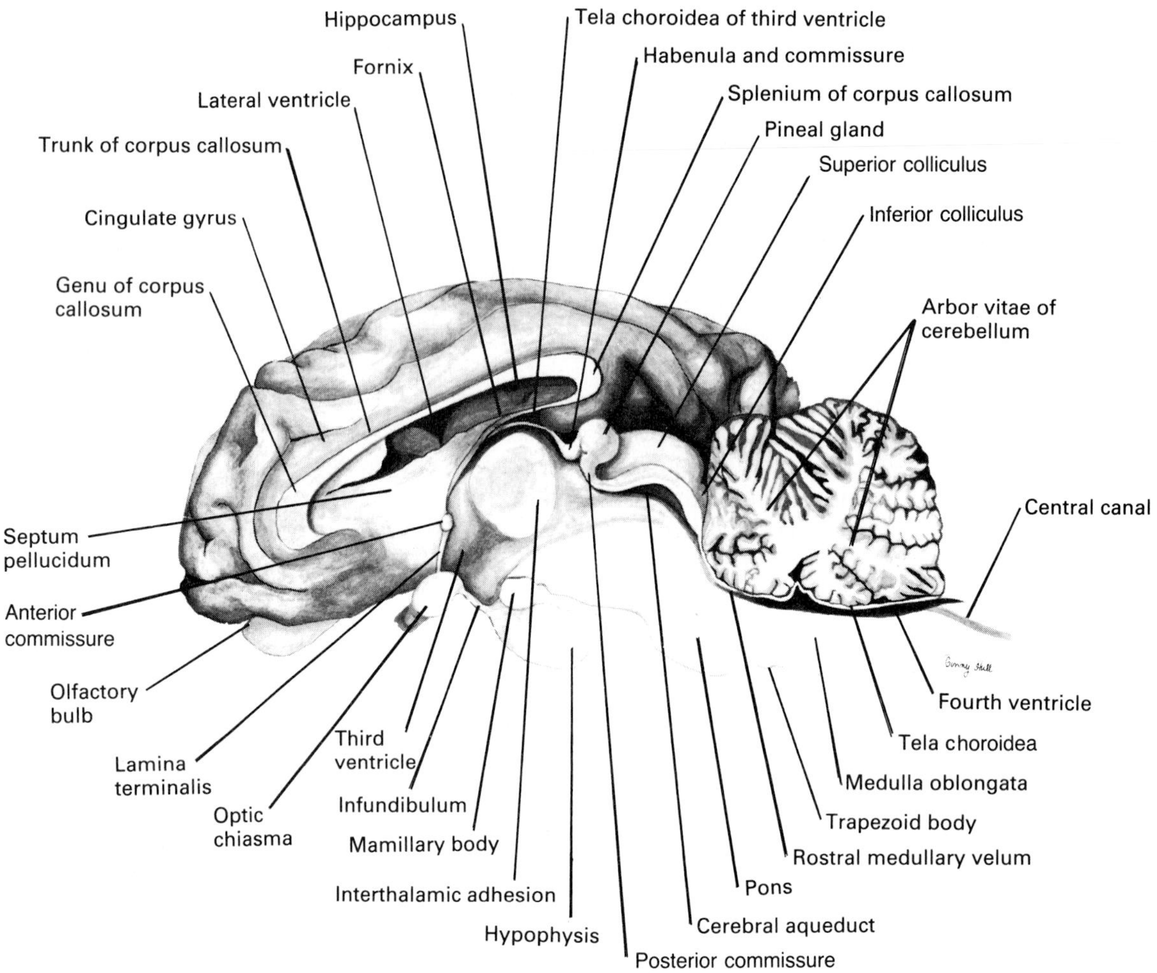

Figure 9-18
Sagittal section of the sheep brain.

Labels (clockwise from top): Hippocampus, Fornix, Lateral ventricle, Trunk of corpus callosum, Cingulate gyrus, Genu of corpus callosum, Tela choroidea of third ventricle, Habenula and commissure, Splenium of corpus callosum, Pineal gland, Superior colliculus, Inferior colliculus, Arbor vitae of cerebellum, Central canal, Fourth ventricle, Tela choroidea, Medulla oblongata, Trapezoid body, Rostral medullary velum, Pons, Cerebral aqueduct, Posterior commissure, Hypophysis, Interthalamic adhesion, Mamillary body, Infundibulum, Third ventricle, Optic chiasma, Lamina terminalis, Olfactory bulb, Anterior commissure, Septum pellucidum.

ventricle, and the epithalamus lies dorsal to it. Note again the **pineal gland,** the **habenular commissure** (which will show in cross section), and the **habenula.** These features show very well in the section. In addition, the epithalamus includes a **posterior commissure.** This is the tissue ventral to the attachment of the pineal gland and anterior to the corpora quadrigemina.

Carefully dissect away tissue between the anterior commissure and the mamillary bodies. You will soon find a direct band of fibers, the **postcommissural fornix,** which is continuous with the column of the fornix and leads to the mamillary bodies (Fig. 9-19). This is one of the main connections between the cerebrum and the hypothalamus. Dissecting just rostral to the anterior commissure, you will find a thin layer of fibers leading from the fornix into the area of gray matter just under the septum pellucidum. This is the **septal region,** and these fibers in front of the anterior commissure constitute the **precommissural fornix.**

A narrow **cerebral aqueduct** (aqueduct of Sylvius) leads through the mesencephalon to the fourth ventricle of the hindbrain (Fig. 9-18). The cerebellum lies above the fourth ventricle. Note that most of its **gray matter** is in the form of a gray **cortex** over the surface of the folia, while the **white matter** is centrally located. The white matter, which

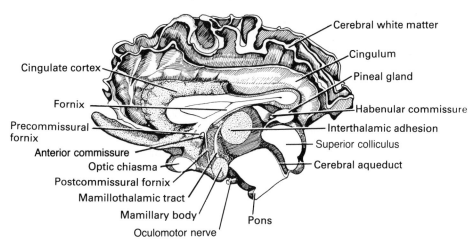

Cerebral white matter
Cingulum
Cingulate cortex
Pineal gland
Fornix
Habenular commissure
Precommissural fornix
Interthalamic adhesion
Anterior commissure
Superior colliculus
Optic chiasma
Cerebral aqueduct
Postcommissural fornix
Mamillothalamic tract
Mamillary body
Pons
Oculomotor nerve

Figure 9-19
Dissection of a sagittal section of the cerebrum and diencephalon of a sheep brain.

represents fiber tracts extending between the cerebellum and other parts of the brain, has the appearance of a tree and is called the "tree of life" (**arbor vitae**).

Dissection of the Cerebrum

◆ (A) THE CEREBRAL CORTEX

Most of the gray matter of the mammalian cerebrum forms a **gray cortex** on the surface. All of the cortex lying dorsal to the rhinal sulcus belongs to the neopallium, and it is the primary integrating area of the mammalian brain (Fig. 9-12). You can gain an appreciation of its extent by dissecting one half of your sheep brain, preferably the one in which the fornix has been dissected, in one of two ways.

(1) Using a blunt instrument, such as an orangewood manicure stick or the handle of a pair of forceps, scrape deeply into each sulcus and shuck out the cortex from the deeper white matter. The cortex is soft and scrapes out easily, whereas the underlying white fibers are quite firm. If time permits, do this over all of the neopallium except for the insular cortex, an area lying dorsal to the rhinal sulcus about the level of the optic chiasma.

(2) If less time is available, you can see cortex and white fibers by slicing off the top 1 or 2 centimeters of the cerebrum in the plane of the corpus callosum (see Fig. 9-21).

The white matter that is exposed by either approach is a tangle of three types of fibers, although these cannot be distinguished grossly (Fig. 9-17). (1) Some **association fibers,** whose cell bodies lie in the cortex, interconnect adjacent gyri; others connect more distant parts of the same cerebral hemisphere. (2) **Commissural fibers,** whose cell bodies also lie in the cortex, interconnect the two cerebral hemispheres and form pathways for the exchange of information. Those extending between the neopallium on each side form the corpus callosum. (3) The remaining fibers are **projection fibers** that interconnect the cerebrum with other parts of the brain. Projection fibers are named according to their origin (location of their cell bodies) and destination. **Thalamocortical fibers** ascend from the thalamus; **corticobulbar** and **corticospinal fibers** extend from the cortex to the brainstem and to the spinal cord, respectively.

A group of nuclei interspersed with fibers is situated deeply in the base of the cerebrum and constitutes the **corpus striatum.** Carefully scrape away the cortex lying dorsal to the central portion of the rhinal sulcus (insular cortex, Fig. 9-20) and you will come upon a thin layer of white fibers, the **external capsule.** Expose about 2 centimeters of this; then carefully peel it off to expose a large mass of gray matter beneath. This is the **lentiform nucleus.** Neuroanatomists divide this nucleus into a lateral **putamen** and a medial **globus pallidus,** but these cannot be detected grossly in the sheep brain.

Cut off the top of the cerebral hemisphere in the plane of the corpus callosum and at the level of the dorsal surface of the corpus callosum. In doing this, you will probably cut through part of the roof of the **lateral ventricle.** Continue to cut away the roof and lateral wall of this ventricle, but leave a strip of corpus callosum near the sagittal plane (Fig. 9-21). Part of the ventricle extends laterally and ventrally into the piriform lobe just posterior to the region you have been dissecting. Expose it. You can see the dark **choroid plexus** of the ventricle extending inward from the thin medioventral wall of the cerebrum. Remove it. Another major nucleus of the corpus striatum lies in the floor of the rostral part of the ventricle. It is called the **caudate nucleus** because of its thick rostral end and tapering tail, which extends caudally and ventrally.

The white matter between the lentiform and caudate nuclei constitutes the **internal capsule.** You can expose it by carefully scraping away the lentiform nucleus until you come to it (Fig. 9-22). Dorsal to the level of the caudate and lentiform nuclei, the fibers running through the internal capsule fan out to different parts of the cortex to form the **corona radiata.** This radiation has probably been cut away.

The internal capsule is a critical part of the brain, for most projection fibers pass through it (Fig. 9-17). The nuclei of the corpus striatum are a major integrating center in the brains of nonmammalian tetrapods. In mammals, the nuclei of the corpus striatum interconnect with each other and receive impulses from the neopallium, thalamus, and reticular formation. Major efferent fibers lead to the reticular formation and motor nuclei and columns. Through its interactions with other parts of the brain, the corpus striatum affects motor centers that are essential for normal locomotion, but its actions are not fully understood. Damage to these nuclei or to their connections cause movements to be jerky

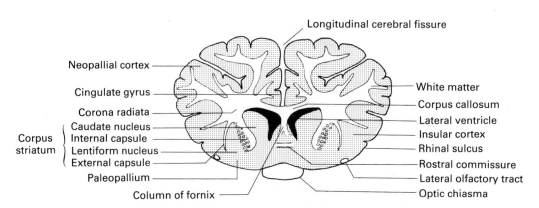

Figure 9-20
Transverse section through the cerebrum of the sheep brain at the level of the optic chiasma. White and gray matter are differentiated in this drawing.

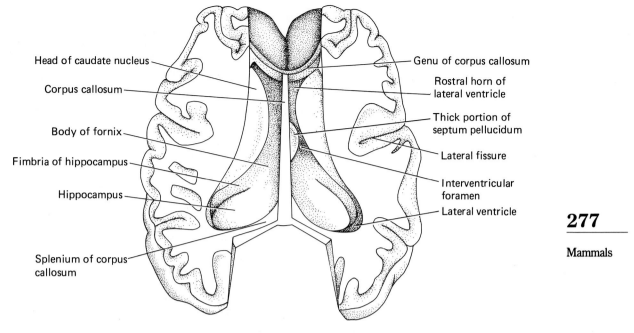

Figure 9-21
Dorsal view of a dissection of sheep cerebrum. *(From Ranson and Clark.)*

and erratic, as in cerebral palsy. Collectively, the motor pathways through the corpus striatum constitute the **extrapyramidal system.**

◆ (C) THE RHINENCEPHALON AND LIMBIC SYSTEM

The primitive olfactory portions of the cerebrum are known as the **rhinencephalon.** Primary olfactory neurons from the nasal epithelium terminate in the olfactory bulbs.

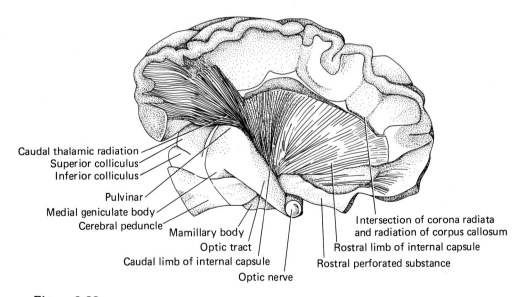

Figure 9-22
Lateral view of a dissection of the sheep brain to show the internal capsule. The lateral surface of the cerebrum and the lentiform nucleus have been removed. *(From Ranson and Clark.)*

From here, secondary olfactory neurons go to more caudal parts of the brain. One group of grossly visible fibers forms the **lateral olfactory tract** and leads to the piriform lobe (Figs. 9-12 and 9-20). Another, the **medial olfactory tract,** extends through the anterior commissure to the contralateral olfactory bulb. This tract, most of which is deeply situated, can be followed most easily by dissecting from the anterior commissure toward the olfactory bulb. Other tracts, not visible grossly, lead from the bulbs to the rostral perforated substance and make connections with the habenula, hypothalamus, and mesencephalic tegmentum.

From each piriform lobe a tertiary group of neurons extend into the **hippocampus.** The hippocampus represents the primitive archipallium, which has been rolled inward and medially (see Fig. 9-10) and now bulges into the caudal part of the floor of the lateral ventricle, where you can see it (Fig. 9-21). Nick the hippocampus with a scalpel in order to verify that it contains gray matter.

A quaternary group of neurons have their cell bodies in the hippocampus, emerge on its surface to form the thin covering of white matter—the **alveus**—that you just nicked, and flow together to form the free rostral border of the hippocampus, which is called the **fimbria of the hippocampus.** These same neurons continue into the body of the fornix, turn ventrally and caudally as the columns of the fornix and postcommissural fornix, and terminate in the mamillary bodies of the hypothalamus (Fig. 9-19).

Neurons initiating in the mamillary bodies extend dorsally as the **mamillothalamic tract** to terminate in the anterodorsal part of the thalamus (Fig. 9-19). You can expose this tract by carefully scraping away thalamic tissue caudal to the columns of the fornix. From here another group of neurons goes through the internal capsule and corona radiata to the **cingulate gyrus** of the neopallium. The cingulate gyrus is a longitudinal fold that lies just dorsal to the corpus callosum deep within the longitudinal cerebral fissure. It can be seen on the intact sagittal section (Fig. 9-18) and also shows in cross section in Figure 9-20. Fibers initiating in the cingulate cortex extend caudally in the white matter beneath it (the **cingulum**) and curve around the splenium of the corpus callosum to enter the hippocampus.

The feedback loop you have followed from the hippocampus (where olfactory information is fed in from the piriform lobe) to the fornix, mamillary bodies of the hypothalamus, thalamus, internal capsule, cingulate cortex, and back to the hippocampus constitutes the **Papez circuit** ("Papez" rhymes with "tapes"). This circuit (Fig. 9-23), which is a part of the limbic system, interconnects the paleopallium and archipallium, and it also represents a primary way in which olfactory information reaches the thalamus, hypothalamus, and neopallium, where other connections can be made. Other parts of the limbic system include the **amygdala,** a deep cerebral nucleus located lateral to the optic chiasma, and the septal region. Olfactory input into the limbic system is of prime importance in nonmammalian vertebrates, but additional sensory information reaches it in mammals: The hippocampus receives gustatory, visceral sensory, auditory, visual, and somatic sensory information as well. The limbic system influences feeding, drinking, fighting, fleeing, reproduction, and other emotional and motivational behaviors related to preservation of self and species. It acts, in part at least, by inhibiting certain activities of the hypothalamus and the reticular formation. Electrical stimulation of parts of the limbic system suppresses behavior that is occurring at the time of stimulation, and destruction of limbic centers leads to an overreaction to various stimuli. Beyond this, the limbic system appears to be involved in arousal or activation of the neopallium and in the formation of short-term memories.

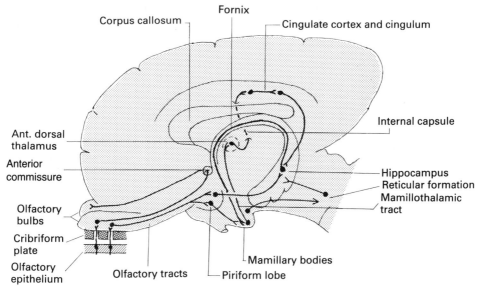

Fornix

Corpus callosum

Cingulate cortex and cingulum

Internal capsule

Ant. dorsal thalamus

Anterior commissure

Hippocampus

Reticular formation

Mamillothalamic tract

Olfactory bulbs

Cribriform plate

Olfactory epithelium

Olfactory tracts

Mamillary bodies

Piriform lobe

Figure 9-23
Diagram on a lateral projection of the sheep brain of the olfactory connections and of major parts of the limbic system.

The Lateral Surface of the Diencephalon and Mesencephalon

Remove the cerebrum from one half of the brain to expose the lateral surface of the diencephalon and mesencephalon. If you dissected the cerebrum, all you need do is to cut through the corpus callosum, fornix, and the connection between the hippocampus and piriform lobe and then lift off what remains of the cerebrum. Strip the meninges from the lateral surface of the diencephalon and mesencephalon.

Note again the optic nerve and chiasma. Optic fibers continue in the **optic tract** dorsally and caudally along the side of the thalamus to terminate primarily in a thalamic nucleus known as the **lateral geniculate body** (see Fig. 9-15). Carefully peel off the optic chiasma and tract and you can see fibers entering the gray matter of the lateral geniculate body. Continue to peel off the optic tract and notice that some of the more superficial fibers terminate in a swelling (the **pulvinar**) on the dorsolateral part of the thalamus. Others travel as the **brachium of the superior colliculus** to terminate in the superior colliculus of the mesencephalic tectum.

The structures you have exposed are parts of the optic system (Fig. 9-24A). Fibers in the optic nerve originate in ganglion cells of the retina (p. 223) and, in most mammals, undergo only a partial decussation in the optic chiasma. Fibers from the lateral half of the retina remain ipsilateral, whereas those from the medial half cross to the opposite side. Many optic fibers are relayed in the lateral geniculate body to the occipital region of the neocortex, where visual interpretation occurs. Other fibers reach this area by first going to the superior colliculus, where they are relayed to the cortex by way of the pulvinar. Some optic fibers terminate in the superior colliculus on tectal fibers that project to the reticular formation and to the motor nuclei of the nerves supplying the extrinsic ocular muscles. Tectal-mediated reflexes control pupillary size and focusing and coordinate the movements of the eyeballs.

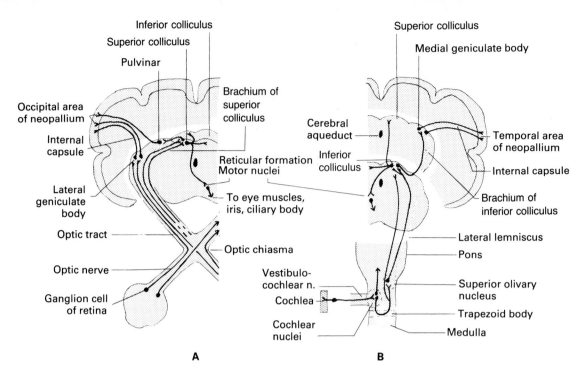

Figure 9-24
Diagrams on transverse sections and dorsal projections of sheep brains. **A**, major optic pathways,
B, major acoustic pathways.

The smaller enlargement on the side of the thalamus caudal to the lateral geniculate body is the **medial geniculate body** (Fig. 9-15). Notice that it is connected by a band of fibers, the **brachium of the inferior colliculus,** to the inferior colliculus of the mesencephalon. The lateral lemniscus, previously seen (p. 269), emerges between the brachium pontis and brachium conjunctivum of the cerebellum and also goes to the inferior colliculus. It lies dorsal to the cerebral peduncles.

All of these structures are parts of the acoustic system (Fig. 9-24**B**). Primary neurons from receptors in the cochlea terminate in cochlear nuclei beneath the acoustic area of the medulla. Some neurons beginning in these nuclei remain ipsilateral, but most fibers decussate in the trapezoid body and terminate in certain olivary nuclei in the medulla that are not grossly visible. Some fibers from the olivary nuclei extend through the lateral lemniscus to terminate in the inferior colliculus. Most fibers originating in the inferior colliculus continue via the brachium of the inferior colliculus to the medial geniculate body. The medial geniculate body is the relay station for acoustic fibers that extend through the internal capsule to the acoustic part of the neopallium, which is located in the temporal region. This cortex is essential for a mammal to interpret the biological significance of sound. Some fibers beginning in the inferior colliculus project to the reticular formation and to motor nuclei. These mediate various acoustic reflexes, including those that enable a vertebrate to position its head so as to localize the source of sound.

You have seen how specific dorsal thalamic nuclei relay olfactory, optic, and acoustic impulses. In a similar way, other, more deeply situated dorsal thalamic nuclei relay other sensory modalities to the cerebral cortex (Fig. 9-17). Ventral thalamic nuclei interconnect with the corpus striatum, and some relay efferent impulses on the way from the cerebral hemispheres. But the thalamus is more than a simple relay station. It acts as a subcortical integrating center, and its numerous interconnections with higher cortical centers implicate it in many cortical functions.

The Spinal Cord and Spinal Nerves

♦ **(A) SPINAL CORD STRUCTURE, ORIGIN OF SPINAL NERVES**

The **spinal cord** *(medulla spinalis)* is a nearly cylindrical cord lying within the vertebral canal of the vertebral column. It is not uniform in diameter, for it bears **cervical** and **lumbosacral enlargements** from which arise nerves to the appendages, and the caudal end tapers as a fine **terminal filament** to end in the base of the tail. You need to study only a section of the spinal cord. To approach it, remove the epaxial muscles from your specimen so as to expose several centimeters of the vertebral column. You should do this in the caudal thoracic region. With bone scissors, carefully cut across the pedicles of the vertebrae and remove the tops of the vertebral arches. The spinal cord will be seen lying in the vertebral canal. Continue chipping away bone and removing fat from around the cord until you have satisfactorily exposed it along with several roots of the spinal nerves.

The spinal cord and nerve roots are covered by the tough **dura mater.** Note that the dura mater is not fused with the periosteum lining the vertebral canal as it is in the cranial cavity. Leave the dura mater on for the present and examine the roots of the spinal nerves. At each segmental interval, there is a pair of dorsal and ventral roots (see Fig. 9-1). Trace a dorsal and ventral root laterally on one side. They pass into the intervertebral foramen before uniting to form a **spinal nerve.** Just before uniting, the dorsal root bears a small round enlargement—the **spinal ganglion.** If you trace the spinal nerve laterally, you may see it divide into a **dorsal ramus** to the epaxial regions of the body and a **ventral ramus** to the hypaxial regions. You will probably not see the small **communicating rami** to the sympathetic ganglia and cord.

The first spinal nerve emerges through a foramen in the vertebral arch of the atlas, and it is called the first cervical nerve. The eighth cervical nerve leaves just caudal to the seventh cervical vertebra. Thereafter, the spinal nerves carry the name and number of the vertebrae caudal to which they emerge; for example, the first thoracic nerve lies just behind the first thoracic vertebra. The number of spinal nerves varies with the number of vertebrae. Most quadrupeds have 8 cervical nerves, 12 to 14 thoracic nerves, 6 to 7 lumbar nerves, 3 to 4 sacral nerves, and 5 or more caudal nerves.

Slit open the dura mater at one end of the exposed area. This opens the **subdural space.** Observe that the roots of the spinal nerves do not have a simple attachment on the cord but unite by a spray of fine **rootlets.** Cut out a segment of the spinal cord and strip off the remaining meninges—**arachnoid** and **pia mater.** The spinal cord has a deep ventral furrow known as the **ventral median fissure,** a less distinct dorsal furrow called the **dorsal median sulcus,** and a more prominent furrow slightly lateral to the middorsal line, the **dorsal lateral sulcus.** Note that the dorsal rootlets enter along the dorsal lateral sulcus. Recall that these same grooves extend onto the medulla oblongata.

Make a fresh cross section of the cord with a sharp instrument such as a razor blade. Examine the cut surface and compare it with Figure 9-1. Also look at demonstration slides if possible. The tiny **central canal** can generally be seen grossly, and if you are fortunate you may be able to distinguish the butterfly-shaped central **gray matter** from the peripheral **white matter.** As explained in the introduction to this chapter, the gray matter consists of unmyelinated fibers and the cell bodies of motor and interneurons; the white matter consists of ascending and descending myelinated fibers. That segment of the white matter that lies between the dorsal median sulcus and the dorsal lateral sulcus is called the **dorsal funiculus.** The segment between the dorsal lateral sulcus and the line of attachment of the ventral roots is the **lateral funiculus,** and that portion between the ventral roots and the ventral median fissure is called the **ventral funiculus.**

◆   (B) THE BRACHIAL PLEXUS

There are 38 spinal nerves in a cat (8 cervical, 13 thoracic, 7 lumbar, 3 sacral, 7 caudal). The rabbit has 36 spinal nerves (8 cervical, 12 thoracic, 7 lumbar, 4 sacral, 5 caudal). The dorsal rami of these nerves extend straight out into the epaxial region, but many of the ventral rami unite in a complex manner to form networks, or **plexuses,** before being distributed to the musculature and skin. In a typical mammal, the anterior cervical nerves form a **cervical plexus** supplying the neck region; the posterior cervical and rostral thoracic nerves form a **brachial plexus** supplying the pectoral appendage; and the lumbar, sacral, and anterior caudal nerves form a **lumbosacral plexus** supplying the pelvic appendage. You may dissect the brachial and/or lumbosacral plexus as examples.

The brachial plexus lies medial to the shoulder and rostral to the first rib; you should approach it from the ventral surface. If it is still intact on the side on which the muscles were dissected, study it there; otherwise, cut through the pectoralis complex of muscles on the other side. The dissection of the plexus involves the meticulous picking away of fat and connective tissue from around the nerves and the accompanying blood vessels. If you find it necessary to cut any of the larger vessels, do so in such a way that you will be able to appose the cut surfaces when you study the circulatory system. Clean off the nerves from a point as near as you can reach to the vertebral column to the point at which they disappear into the shoulder muscles and brachium.

The brachial plexus is formed by the union of the ventral rami of the sixth to eighth cervical nerves and the first thoracic nerve in the cat. The fifth cervical also contributes to it in the rabbit. Considerable variation occurs in the details of the manner in which nerves unite to form the plexus and in the origin of peripheral branches, but a common pattern for the mammals under consideration is shown in Figure 9-25. The ventral rami of the nerves that enter the plexus are called the **roots** of the plexus. Note that each root tends to split into two **divisions** and that the divisions of different nerves unite to form **trunks** from which peripheral nerves arise. Also note that the splitting of roots into divisions, and the union of divisions to form trunks, occurs in such a way that the nerves supplying the dorsal appendicular muscles tend to segregate early from those supplying the ventral appendicular muscles. Many of the nerves are also cutaneous, being distributed to the skin, but we will describe only the major cutaneous branches.

The most ventral nerves of the plexus are several **pectoral nerves.** They are small nerves that arise from the ventral divisions of the plexus; they may or may not unite with each other, and they pass to the pectoralis complex.

A large **suprascapular nerve** leaves the front of the plexus, where it arises for the most part from the sixth cervical nerve, and passes between the subscapular and supraspinatus muscles to supply the supraspinatus and infraspinatus muscles, and some of the skin over the shoulder and brachium.

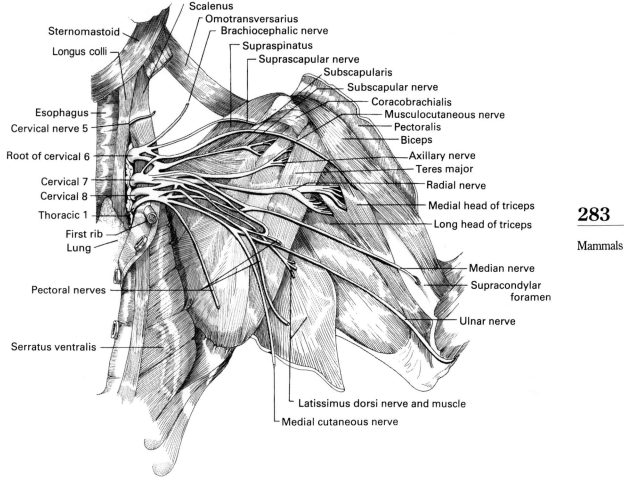

Figure 9-25
Ventral view of a dissection of the left brachial plexus of a cat. *(From Walker,* A Study of the Cat.*)*

One or more **subscapular nerves** leave caudal to the suprascapular nerve and pass to the large subscapular muscle. They arise, for the most part, from the sixth and seventh cervical nerves.

A large **axillary nerve,** which lies caudal to the subscapular nerve, arises from the seventh cervical nerve. It passes between the teres major and the subscapularis muscles to supply the teres minor and deltoid complex and often the teres major.

Smaller **latissimus dorsi nerves** to the latissimus dorsi muscle arise from the plexus near the origin of the axillary nerve. They may also supply the teres major.

The large, deep nerve caudal to the axillary nerve that is formed by the union of the dorsal divisions of the seventh and eighth cervical nerves and first thoracic nerve is the **radial nerve.** This is the largest nerve of the plexus. It passes through and between parts of the triceps, and crosses the humerus to the lateral surface of the arm and from there goes down to the extensor portion of the forearm. It supplies the tensor fasciae antebrachii, triceps, and forearm extensors. A large branch of the nerve is cutaneous.

All of these nerves go to dorsal appendicular muscles, except the pectoral and subscapular nerves, which supply parts of the ventral musculature. The remaining nerves innervate the rest of the ventral appendicular muscles. Note that all of the nerves to the ventral muscles arise from the ventral divisions of the plexus. A small **musculocutaneous nerve** springs from the sixth and seventh cervicals and enters the biceps. It generally branches before reaching the biceps, supplying the biceps, coracobrachialis, brachialis, and some of the skin over the forearm.

The ventral divisions of the seventh and eighth cervical nerves and the first thoracic nerve combine to form two prominent nerves that run down the medial side of the brachium. The more cranial of these is the **median nerve;** the more caudal nerve is the **ulnar nerve.** They are distributed to the forearm flexors and skin of the hand. The median nerve passes through the supracondylar foramen of the humerus.

A small **medial cutaneous nerve,** which arises from the first thoracic nerve, runs parallel with and caudal to the ulnar nerve. It supplies some of the skin over the forearm.

♦   **(C) THE LUMBOSACRAL PLEXUS**

The **lumbosacral plexus,** which supplies the skin and muscles of the pelvis and hind leg, is located so deep within the abdominal and pelvic cavities that you cannot dissect it until you have studied the abdominal and pelvic viscera. If you are to dissect this plexus, return to the following description after you have completed study of the urogenital system. The following description is based on the cat but is applicable to the rabbit.

The pelvic symphysis will have been cut. Push the two hind legs dorsally, thereby spreading open the pelvic canal, and push the abdominal and pelvic viscera to one side. Also cut the external iliac artery and vein shortly before they pass through the abdominal wall, and reflect them. Identify the psoas minor muscle (p. 195), and cut and reflect it. You will find a longitudinal cleft on the lateroventral part of the psoas major near the point where the deep circumflex iliac vessels cross it. Separate the psoas major along this cleft into superficial (ventral) and deep (dorsal) portions. Notice the nerves of the lumbosacral plexus emerging through this cleft, and dissect away the superficial portion of the psoas major as you trace them medially to the intervertebral foramina through which they leave the vertebral column.

The lumbosacral plexus is formed by the ventral rami of seven spinal nerves (the fourth lumbar nerve to the third sacral nerve). As with the brachial plexus, some variation occurs in the way the seven roots of the plexus split to form divisions and in the way the divisions unite to form the trunks from which the peripheral nerves arise. Figure 9-26 shows a common pattern.

Lumbar nerve 4 splits soon after emerging from the intervertebral foramen into a **genitofemoral nerve** and a branch that passes caudad to join the divisions of the fifth lumbar nerve. The genitofemoral nerve continues caudad close to the external iliac vessels and passes through the body wall with the external pudendal vessels to supply the skin in the groin and on parts of the external genitalia. In a male, small branches also go to the cremasteric muscle.

One division of **lumbar nerve 5** unites with the caudal branch of the fourth lumbar nerve to form the **lateral femoral cutaneous nerve,** which extends laterally near the deep circumflex iliac vessels, perforates the body wall, and supplies the skin over the lateral surface of the hip and thigh. The other division of lumbar nerve 5 continues caudad to unite with the divisions of the sixth lumbar nerve. Branches to the psoas muscles may arise from this division or from the lateral femoral cutaneous nerve.

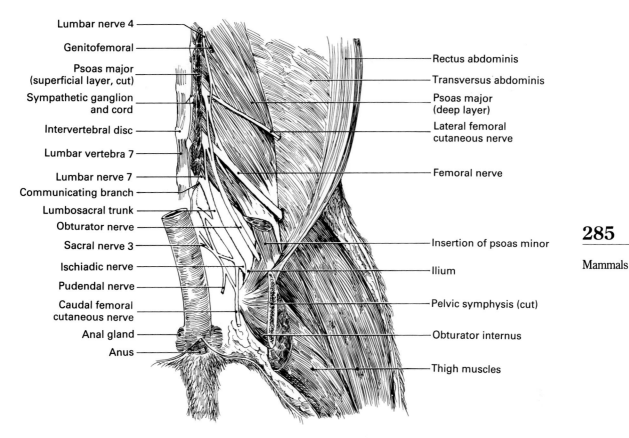

Lumbar nerve 4

Genitofemoral

Psoas major
(superficial layer, cut)

Sympathetic ganglion
and cord

Intervertebral disc

Lumbar vertebra 7

Lumbar nerve 7

Communicating branch

Lumbosacral trunk

Obturator nerve

Sacral nerve 3

Ischiadic nerve

Pudendal nerve

Caudal femoral
cutaneous nerve

Anal gland

Anus

Rectus abdominis

Transversus abdominis

Psoas major
(deep layer)

Lateral femoral
cutaneous nerve

Femoral nerve

Insertion of psoas minor

Ilium

Pelvic symphysis (cut)

Obturator internus

Thigh muscles

Figure 9-26
Ventral view of a dissection of the left lumbosacral plexus of a cat.

Lumbar nerve 6 is a large nerve. Its largest division unites with a branch of the caudal division of the fifth lumbar nerve to form the large **femoral nerve.** After perforating the body wall with the femoral artery and vein, the femoral nerve innervates the quadriceps femoris and certain other extensor thigh muscles. It also gives rise to a prominent cutaneous branch (the **saphenous nerve**) that supplies the skin on the medial side of the thigh and shin.

A second division of lumbar nerve 6 receives another division of the fifth lumbar nerve to form the **obturator nerve.** This nerve extends caudolaterally, perforates the obturator internus near the brim of the pelvis, and goes through the obturator foramen of the pelvic girdle to supply primarily the gracilis, obturator externus, and adductors of the thigh. These muscles belong to the ventral appendicular group.

A third division of lumbar nerve 6 continues caudad to join **lumbar nerve 7,** forming the large **lumbosacral trunk.** As the lumbosacral trunk continues caudolaterally, it soon receives a division from sacral nerve 1 and, a bit farther distally, a contribution from sacral nerve 2. The trunk leaves the pelvic canal by passing between the ilium and sacrum and breaks up into several branches. **Gluteal nerves** supply the gluteal and other laterodorsal hip muscles, and a large **ischiadic** (sciatic) **nerve,** which is the main continuation of the lumbosacral trunk, continues down the lateral surface of the thigh (Fig. 7-33, p. 190), innervating the biceps femoris, semimembranosus, semitendinosus, and other flexor muscles of the thigh. It bifurcates near the distal end of the thigh into **tibial** and **common peroneal** (fibular) **nerves.** These innervate the flexors and extensors of the shank, respectively. Cutaneous branches of the ischiadic, tibial, and common peroneal nerves help to supply adjacent skin.

Sacral nerves 1, 2, and 3 also interconnect with each other to form a network from which several small nerves arise, the most conspicuous being the pudendal and caudal femoral cutaneous nerves. The **pudendal nerve** contains motor fibers to striated muscles in the anal region and sensory fibers coming from the anal region and from the penis or clitoris. The **caudal femoral cutaneous nerve** helps supply the skin in the anal area and adjacent parts of the thigh.

During this dissection, you can also see a portion of the **sympathetic cord** and **ganglia** lying on the ventral surface of the lumbar vertebrae (Fig. 9-26). Each ganglion receives a **communicating branch** from adjacent lumbar spinal nerves. Pelvic viscera receive their sympathetic innervation by minute branches from the pelvic extension of the sympathetic cord that follow the blood vessels to the organs. Parasympathetic innervation is supplied by a **pelvic nerve** formed by very small branches from the sacral nerves. These nerves are seldom seen in dissections.

CHAPTER TEN

◆ ◆

THE COELOM AND THE DIGESTIVE AND RESPIRATORY SYSTEMS

We now turn from the organ systems that support, move, and integrate the body's activities to a group that sustains metabolism. The digestive system brings in water, food, minerals, and other nutrients needed by the body; the respiratory system takes care of the essential gas exchanges; the circulatory system transports materials to and from the cells; and the excretory system eliminates all, or much, of the nitrogenous waste products of cellular metabolism and helps to control the water and salt balances of the body. Additional nitrogenous excretion and salt and water exchanges occur through the gills of fishes. These systems are functionally distinct, but it is convenient to study the digestive and respiratory systems together to some extent, for they are closely associated morphologically. The respiratory system develops embryonically as outgrowths from the digestive system.

Structure and Function of the Coelomic Cavity and Its Contents

The body cavity (**coelom** or **coelomic cavity**) is filled with the inner organs (or **viscera**). Typically, organs within the coelom alter their shape and volume rhythmically (over short periods of time) or cyclically (over longer periods of time). This is true for the heart, lungs, stomach, intestine, ovaries, and uterus (although not for the liver, pancreas, and spleen, which are placed in the coelom because of their embryonic derivation). As the viscera expand and contract and shift their positions accordingly, their surfaces must be able to glide past one another and across the walls of the coelomic cavity. This mobility is ensured by the **serosa,** which lines the walls of the coelom and envelops all the viscera. The serosa consists of a thin, smooth, flat-celled epithelium, the **mesothelium,** and an underlying layer of loose connective tissue; it secretes small amounts of a serous fluid that keeps the mesothelial surfaces moist and lubricated. Each inner organ is anchored to the wall of the coelom by a **mesentery** (Fig. 10-1B and C). A mesentery consists of two layers of serosa or, in other words, of a central layer of connective tissue sandwiched between two layers of mesothelium. The serosa of the mesenteries is continuous with that lining the body cavity and enveloping the viscera. Mesenteries hold the viscera in place while allowing them some degree of movement. They also provide avenues for the nerves and blood and lymph vessels that supply the viscera and are contained within the connective tissue layer of the mesenteries.

   In vertebrates, the coelom is subdivided into at least two cavities: the **pericardial cavity** and the **pleuroperitoneal cavity.** In mammals and some reptiles, the latter is subdivided further into the **peritoneal cavity,** or **abdominal cavity,** and into the paired **pleural cavities.** It is customary to name the serosa specifically according to

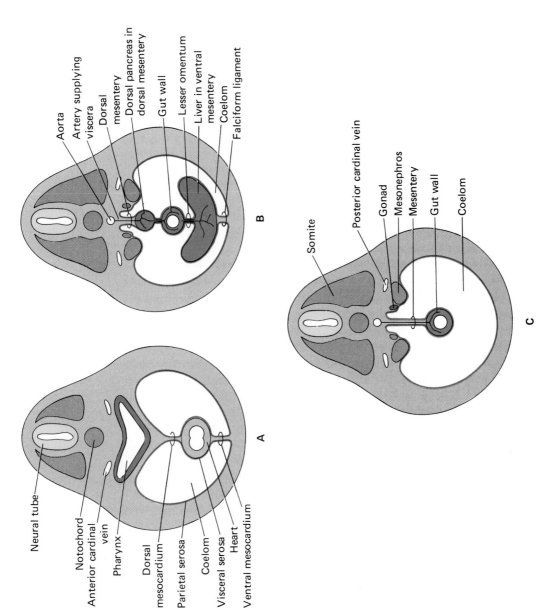

Figure 10-1

Diagrammatic cross sections through an embryo illustrating the structural relationships of the coelom and mesenteries to the visceral organs. **A**, Cross section at the level of the heart, which develops within the ventral mesentery ventrally to the pharynx and cranially to the liver; **B**, cross section at the level of the liver and pancreas, both of which are outgrowths of the primitive gut; **C**, cross section at the level of the intestine caudally to the liver, where the ventral mesentery has disappeared. (*Modified from Walker, Functional Anatomy of the Vertebrates; after Patten and Corliss.*)

Labels for B:
Aorta
Artery supplying viscera
Dorsal mesentery
Dorsal pancreas in dorsal mesentery
Gut wall
Lesser omentum
Liver in ventral mesentery
Coelom
Falciform ligament

Labels for A:
Neural tube
Notochord
Anterior cardinal vein
Pharynx
Dorsal mesocardium
Parietal serosa
Coelom
Visceral serosa
Heart
Ventral mesocardium

Labels for C:
Somite
Posterior cardinal vein
Gonad
Mesonephros
Mesentery
Gut wall
Coelom

its location in a particular subdivision of the coelom, namely **pericardium, peritoneum**[19], and **pleura.** It is also customary to distinguish the **parietal serosa,** which lines the walls of the coelom and its subdivisions, from the **visceral serosa,** which envelops the inner organs.

As we have already seen (p. 143 and Fig. 7-5), the formation of the embryonic coelom and mesenteries is a process that is relatively easily understood. However, in the course of both the embryonic development and evolutionary transformations, the inner organs grow, differentiate, and shift their positions to fit within the available space, and the coelom is subdivided by the formation of septa. The result is that the anatomy of the body cavity with its subdivisions and mesenteries can be very complex and confusing, unless one is familiar with its differentiation during embryonic development.

Embryonic Development of the Coelom and Its Subdivisions

As we have already seen, the coelom appears as paired cavities within the paired hypomeres (p. 143 and Fig. 7-5). A mesentery is formed where the expanding paired coelomic cavities meet along the midline, and the **primitive gut** (or **archenteron**) is enclosed between them. The serosa develops from the innermost mesoderm layer of the hypomere that surrounds the coelom: the parietal serosa from the somatic layer and the visceral serosa from the visceral layer (Fig. 7-5**F**).

At an early stage, the primitive gut is straight and undifferentiated and divides the mesentery into the **dorsal mesentery** and the **ventral mesentery** (Fig. 7-5**F**). As the primitive gut differentiates, the liver grows into the ventral mesentery and the pancreas into the dorsal mesentery (Figs. 10-1**B** and 10-4). The heart lies far cranially in the embryo, and ventrally to the pharynx in the branchial region (Figs. 10-1**A** and 10-4). Technically, it develops within the ventral mesentery. Caudad to the liver, most of the ventral mesentery disappears, so that the coelom becomes one confluent cavity (Fig. 10-1**C**).

The coelom surrounding the heart is separated from the rest of the body cavity by the formation of the **transverse septum** and subsequently is called the pericardial cavity (Fig. 10-2**A**). The ventral part of the transverse septum is formed in connection with the growth and differentiation of the liver (Figs. 10-2**A** to **D**). At the beginning, the heart and liver lie adjacent to each other within the ventral mesentery (Fig. 10-2**B**). As the liver expands, it reaches the wall of the coelom at the level of the pectoral girdle, where the visceral serosa of the liver fuses with the parietal serosa and thereby divides the coelom into the pericardial and pleuroperitoneal cavities (Fig. 10-2**C**). In the course of its further differentiation, the liver elongates and pulls back from the wall of the coelom. During this process, the visceral serosa of the liver doubles back towards the center, thereby forming the ventral part of the transverse septum (Fig. 10-2**D**). At the same time, the mesocardium connecting the heart to the transverse septum disappears, and the transition zone between the parietal serosa covering the caudal surface of the transverse septum and the visceral serosa enveloping the liver forms the **coronary ligament.** The dorsal part of the transverse septum is formed as the heart grows and differentiates. At an early stage, the heart is a longitudinal tube, but as it continues to elongate, its growth soon outstrips that of the surrounding body region. Because the heart is attached to the parietal pericardium cranially by the ventral aorta and caudally by the common cardinal and hepatic veins, it is folded into an S-shaped curve to fit into the given space of the pericardial cavity

---

[19]In many textbooks, the term *peritoneum* is used as a synonym for the general term *serosa.* This can be confusing unless it is specified whether the term *peritoneum* is being used in its broad or strict sense. In this manual, we will use the term *peritoneum* strictly for the serosa lining the peritoneal cavity. For the lining of the pleuroperitoneal cavity, we will use the general term *serosa* instead of the technically correct but longer term *pleuroperitoneum.*

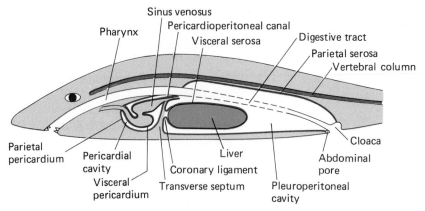

Sinus venosus
Pericardioperitoneal canal
Pharynx
Visceral serosa
Digestive tract
Parietal serosa
Vertebral column

Parietal
pericardium
Pericardial
cavity
Visceral
pericardium
Coronary ligament
Transverse septum
Liver
Cloaca
Abdominal
pore
Pleuroperitoneal
cavity

A

290

Chapter 10
The Coelom
and the Di-
gestive and
Respiratory
Systems

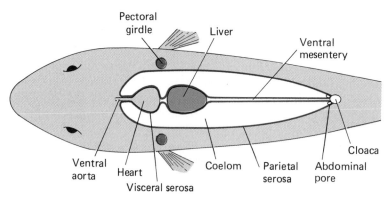

Pectoral
girdle
Liver
Ventral
mesentery

Ventral
aorta
Heart
Visceral serosa
Coelom
Parietal
serosa
Abdominal
pore
Cloaca

B

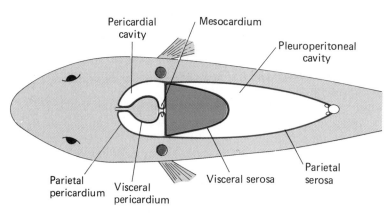

Pericardial
cavity
Mesocardium
Pleuroperitoneal
cavity

Parietal
pericardium
Visceral
pericardium
Visceral serosa
Parietal
serosa

C

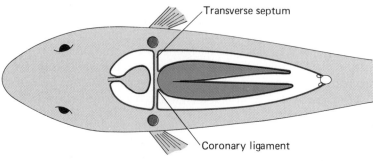

Transverse septum

Coronary ligament

D

(Fig. 10-2**A**). As a result of this folding process, the sinus venosus is displaced dorsally. At the same time, the paired, transversely oriented common cardinal veins (or ducts of Cuvier), which drain the blood from the body walls (see Fig. 11-6), are partly incorporated into the sinus venosus. As the common cardinal veins are pulled medially and dorsally in the process, they entrain the serosa, so that a pair of double-layered serosal folds forms the upper part of the transverse septum. In chondrichthyan fishes, the transverse septum does not close completely; the **pericardioperitoneal canal** remains as a communication between the two parts of the coelom (Fig. 10-2**A**).

In tetrapods, with the development of a neck, the heart migrates caudad into the thoracic region, so that the pericardial cavity is situated ventral to the pleuroperitoneal cavity. The originally dorsal part of the transverse septum is now oriented obliquely to horizontally instead of vertically as in fishes. The pericardial cavity is closed on the dorsal side by a pair of serosal folds, the **pleuropericardial membranes,** which grow towards and fuse with the ventral mesentery (Fig. 10-3**A**). The dorsal serosal layer of the pleuropericardial membrane becomes continuous with the serosa of the **mediastinal septum** (which is derived from the dorsal portion of the former ventral mesentery), whereas the ventral serosal layer becomes continuous, at least for the time being, with the serosa of the dorsal mesocardium (which is also derived from the former ventral mesentery) (Fig. 10-3**B**). The pericardial wall[20] hence consists of three layers: the parietal (or mediastinal) pleura, the parietal pericardium, and a fibrous connective tissue layer formed by and sandwiched between the two serosae. Soon, all the mesocardia disappear (see Figs. 10-2**D** and 10-3**C**). The pleuroperitoneal cavity dorsal to the pericardial cavity is partitioned by the mediastinal septum (part of the former ventral mesentery) into a pair of recesses, the **pleural recesses.** The paired lungs arise as ventrolateral evaginations from the caudal end of the pharynx and grow caudad into the pleural recesses, while their expanding surfaces remain covered by visceral serosa (Fig. 10-3**B**). Later, the ventral side of the pharynx leading to the lungs gives rise to the trachea. In amphibians, the lungs remain connected to the mediastinal septum by a pair of mesenteries, the **pulmonary ligaments** (or mesopneumonia).

In mammals and some reptiles, in contrast to amphibians and most reptiles, the pleural recesses are closed off caudally and become **pleural cavities,** which are distinct from the **peritoneal** or **abdominal cavity.** In mammals, this separation is accomplished by another pair of serosal folds, the **pleuroperitoneal membranes,** growing from the wall of the coelomic cavity toward the midline, together with some other folds. The **diaphragm** of mammals consists of the pleuroperitoneal membranes, the ventral part of the transverse septum, and a central layer of muscle. The muscle develops from the hypaxial musculature of the cervical region (remember that the heart is located far cranially in the early embryonic stages) and invades the space between the two serosal layers of the transverse septum and the various membranes. In many mammals, the pleural cavities extend ventrally (Fig. 10-3**B**), so that they eventually

---

[20]The three-layered pericardial wall of tetrapods is usually called *pericardium.* This is confusing, since the serosa lining the pericardial cavity is also called *pericardium,* of which we distinguish the parietal pericardium, which is part of the pericardial wall, and the visceral pericardium, which surrounds the heart (Fig. 10-3**C**). Our use of the term *pericardial wall* instead of the widely used term *pericardium* is an attempt at clarifying the situation for didactic purposes.

◀ **Figure 10-2**
Diagrams illustrating the formation of the transverse septum and the subdivision of the coelom in a shark. **A,** Longitudinal section showing the adult condition (the falciform ligament and other mesenteries are not shown); **B** through **D**; dorsal views of horizontal sections through the coelom of embryos at different developmental stages, at the level of the ventral mesentery; **B,** relatively early stage, in which the ventral mesentery has not yet disappeared caudally to the liver; **C,** later stage, in which the liver expands laterally and the visceral serosa of the liver fuses to the peritoneal serosa of the body wall; **D,** final stage, in which the liver contracts and elongates and the serosa doubles back towards the center to form the transversal septum, and the mesocardium disappears.

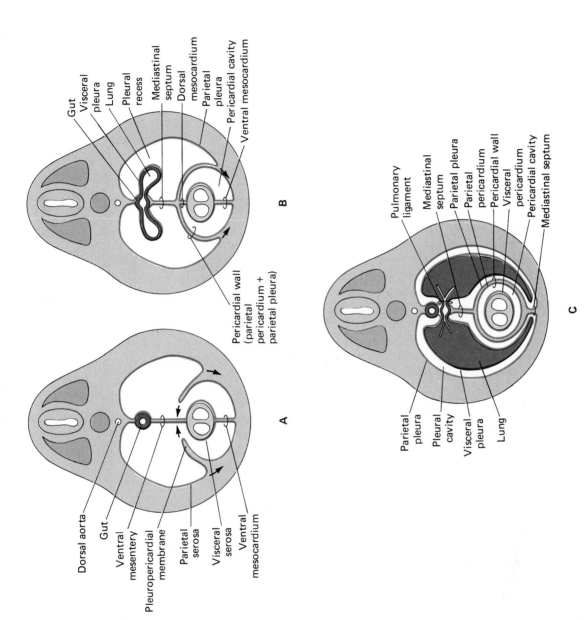

Figure 10-3

Diagrammatic cross sections through a mammalian embryo illustrating the development of the pericardial and pleural cavities. **A,** Formation of the pleuropericardial membranes from the parietal serosa; **B,** fusion of the pleuropericardial membranes with the mediastinal septum, separation of the paired pleural recesses from the pericardial cavity, ventromedial extension of the pleural recesses, and appearance of lung buds; **C,** final stage, in which the dorsal and ventral mesocardia have disappeared, the pleural cavities completely surround the pericardial cavity, and the lungs fill the pleural cavities.

meet medioventrally to the pericardial cavity and form a continuation of the mediastinal septum (Fig. 10-3C).

The organs of the reproductive and excretory systems, although located in or near the abdominal cavity, do not develop within the primitive dorsal and ventral mesenteries. The kidneys remain outside the coelomic cavity, along the dorsomedial side of the coelomic wall, a position called **retroperitoneal** (see p. 405; consult Figs. 7-5 and 10-1A and B). The gonads and their ducts, however, project into the coelomic cavity. The two layers of the serosal folds, which are pulled after these organs, adhere to each other behind the organs and form mesenteries. These mesenteries are called **subsidiary mesenteries,** because they have no developmental relationship to the primary mesenteries.

The Development of the Digestive and Respiratory Systems

Muscles and connective tissue in the walls of the digestive and respiratory tracts develop from the visceral layer of the hypomere, but most of the epithelium lining these tracts, and the secretory cells of the glandular outgrowths, develop from the embryonic archenteron (primitive gut) and, hence, are endodermal. However, variable amounts of the front and hind ends of the digestive tract are formed by ectodermal invaginations—the **stomodeum** and **proctodeum,** respectively (Fig. 10-4). The former forms the mouth, or **oral cavity;** the latter contributes to the cloacal region. At first the ectodermal invaginations are separated from the archenteron by plates of tissue, but these eventually break down. The borderlines between ectodermally and endodermally derived tissues do not remain distinct in the oral cavity and cloaca of adult vertebrates.

For purposes of description, the archenteron may be divided into a **foregut** and a **hindgut.** The foregut differentiates into the pharynx, esophagus, and, generally, a stomach; the hindgut differentiates into the intestinal region and much of the cloaca. In most mammals, the cloaca is present only in embryos. It soon becomes divided; the dorsal part contributing to the rectum and the ventral part to the urogenital passages.

In all vertebrates, a series of **pharyngeal pouches** grow out from the side of the pharynx (Fig. 10-4). There are six of these in most fishes, fewer in tetrapods. The tissues between the pouches constitute the **pharyngeal bars,** the first bar being cranial to the first pouch. The skeletal visceral arches, branchial muscles, certain

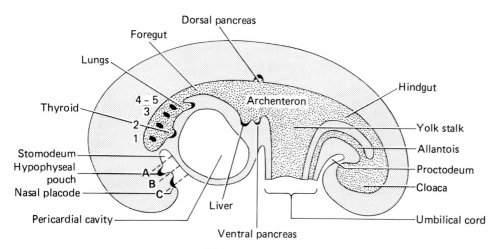

Figure 10-4
Diagrammatic sagittal section of a mammalian embryo to show the development of the digestive and respiratory systems. The points of entrance of the pharyngeal pouches are numbered. Lines **A, B,** and **C** indicate the comparative position of the mouth openings of **A,** an agnathous fish; **B,** a jawed fish not having internal nostrils; and **C,** fishes with internal nostrils and tetrapods.

nerves, and the aortic arches grow into these pharyngeal bars. In fishes, the endodermal pharyngeal pouches meet comparable ectodermal furrows, break through to the surface, and form the **gill slits** and **gill pouches** on whose walls the gills develop. The first pharyngeal pouch and furrow form the **spiracle** of a fish such as *Squalus;* the others form the five definitive gill pouches. The pharyngeal pouches do not normally break through in tetrapods (except for certain pouches in larval amphibians), but are present embryonically nonetheless. The first gives rise to the **tympanic cavity** and **auditory tube** in tetrapods having these structures; the mammalian **tonsillar fossa,** containing the palatine tonsil, develops at the site of the second pharyngeal pouch; the other pharyngeal pouches give rise to certain endocrine glands and then disappear. A **thymus** develops as epithelial thickenings of the dorsal part of most (or all) of the pharyngeal pouches in fishes and urodeles, but from the ventral part of just the third and fourth pharyngeal pouches in mammals. These primordia may remain as distinct glandlike structures, or they may coalesce into a single structure. The thymus is necessary for the normal development of parts of the body's immune system because one population of lymphocytes (the T-lymphocytes) undergoes maturation and develops the capacity to respond to certain invading antigens here. **Parathyroid glands** are absent in fishes but present in all tetrapods. They develop as epithelial thickenings from the ventral part of the third and fourth pharyngeal pouches in urodeles and from the dorsal part of these pouches in mammals. Finally, **ultimobranchial bodies** develop in all vertebrates from the posterior face of the last pharyngeal pouch. Both the parathyroid glands and ultimobranchial bodies produce hormones that regulate the level of calcium and phosphorus in the blood.

In addition to the lateral pharyngeal pouches, certain median evaginations arise from the floor of the pharynx. An endocrine **thyroid gland** grows out from the floor at the level between the first and second pharyngeal pouches. In fishes, the thyroid, although losing its connection with the pharynx, tends to remain in this cranial position, but it migrates a variable distance caudad in tetrapods. The thyroid may remain a median organ, or it may bifurcate and form a pair of glands. As discussed earlier (p. 27), the development of the thyroid of the lamprey from a part of the endostyle-like subpharyngeal gland of the ammocoetes larva points to a possible homology between the vertebrate thyroid and the endostyle of protochordates. The thyroid hormone increases an organism's metabolic rate. This is particularly important in amphibian metamorphosis and in endothermic-homoiothermic vertebrates.

The **lungs** of sarcopterygian fishes and tetrapods arise as a median bilobed evagination from the floor of the pharynx just caudal to the pharyngeal pouches. The early primordia of the lungs resemble a pair of pharyngeal pouches in the embryos of some amphibians, and it is possible that they evolved from a ventrally displaced pair of caudal pharyngeal pouches. It is believed that lungs appeared at a very early point of vertebrate evolution. The fossils of some early bony fishes show evidence of lunglike structures. Lungs also are found in the living sarcopterygian fishes and in certain actinopterygian fishes that have retained a number of primitive anatomical conditions. These lungs serve to supplement gill respiration. In most actinopterygians, however, the lungs have become transformed into the dorsally placed, gas-filled **swim bladder,** which regulates the buoyancy of these fishes (see also p. 148). In most cases, the swim bladder develops as an evagination from the roof of the posterior pharyngeal region, but in at least one fish it arises from the lateral wall. This suggests a transitional stage from the ventrally derived lungs.

No other outgrowths arise from the digestive tract of most vertebrates until the transition from foregut to hindgut. At this point are found the liver and pancreas. The **liver** arises embryonically as a prominent ventral diverticulum (Fig. 10-4), which, as mentioned earlier, grows into the ventral mesentery caudal to the heart, and by its expansion forms the ventral part of the transverse septum. Functionally, the liver is a very diverse organ. It secretes bile (a mixture of excretory products and the fat-emulsifying bile salts), and its cells come into intimate contact with blood that the hepatic portal system brings to it from the stomach and intestinal region. Many metabolic conversions occur here. Excess absorbed sugars are stored, largely in the

294

Chapter 10
The Coelom
and the Di-
gestive and
Respiratory
Systems

form of glycogen, or deficiencies are made up so that the glucose content of the blood is kept at a constant level. Amino acids are deaminated and their amino groups converted to urea. Toxins may be removed; many plasma proteins are synthesized.

The pancreas is a grossly visible organ in most vertebrates, but in agnathans it is represented only by scattered cells in the walls of the intestine and liver (p. 23). The organ arises embryonically from one or more intestinal outgrowths near the liver primordium. Frequently there are both a dorsal evagination (the **dorsal pancreas**) and a ventral one (the **ventral pancreas**). The latter is often paired and is associated with the base of the liver anlage and, hence, with the future bile duct (Fig. 10-4). The ventral pancreas grows around the intestine and fuses with the dorsal pancreas; together the two pancreatic primordia extend into the dorsal mesentery (see Fig. 10-1**B**). All the stalks of the primordia may persist as ducts, or certain stalks may be lost in the adult. A ventral pancreatic duct can be recognized by its entrance into the intestine in common with the bile duct, a dorsal pancreatic duct, by its independent entrance on the opposite side of the intestine. Most of the pancreatic cells are exocrine and secrete digestive enzymes, which are discharged into the intestine and act on proteins, carbohydrates, fats, and nucleic acids. Little islands of endocrine tissue, termed the **islets of Langerhans,** are scattered among the exocrine cells; these islets produce insulin and glucagon, which play a crucial role in carbohydrate metabolism.

More caudally along the hindgut there is, in the embryos of amniotes and certain fishes, a **yolk stalk** connecting with the **yolk sac** (Fig. 10-4). The yolk stalk and sac become smaller as the embryo grows and are lost by the adult stage.

A final major outgrowth is the **urinary bladder,** present in most tetrapods. It develops from the embryonic cloaca near the caudal end of the hindgut. In the embryos of amniotes, this structure expands considerably and extends beyond the limits of the embryo as the **allantois**—an extraembryonic membrane that serves for excretion and respiration in the embryos of reptiles and birds, and for vascularizing the fetal portion of the placenta in eutherian mammals (see p. 414).

FISHES

The coelom and the digestive and respiratory systems of *Squalus* are good examples of the condition of these structures in ancestral jawed fishes. Bony fishes ancestral to tetrapods also had lunglike outgrowths from the caudal part of the pharynx. A stomach was probably absent in very early fishes. Protochordates are (and early vertebrates may have been) filter feeders, feeding more or less continuously on minute food particles. Filter feeders need no stomach, but a change in feeding habits occurred with the evolution of jaws. Large chunks of food are taken at irregular time intervals, and the stomach serves for temporary storage and preliminary physical and chemical breakdown of food.

Pleuroperitoneal Cavity and Its Contents

♦ **(A) THE BODY WALL AND**
PLEUROPERITONEAL CAVITY

Study the caudal parts of the digestive system before the mouth and pharyngeal area. As a consequence of the method of preservation, the intestine of some specimens will have become everted, that is, turned inside out, and will be seen protruding through the cloaca. Before pushing or pulling it back into the body cavity, observe the deep, spiral fold in its lining. This is the **spiral valve** (p. 300). Now open the **pleuroperitoneal cavity** by a longitudinal incision slightly to one side of the midventral line, preferably on the right side if the muscles were dissected on the left. Extend the incision as far forward as the pectoral girdle and as far caudad as the base of the tail. In doing the latter, cut through the pelvic girdle and continue caudally on one side of the cloacal aperture. On each side, make a transverse incision that extends from about the middle of the longitudinal cut to the lateral line. You now have four flaps of the body wall, which you can fold back to expose the body cavity and its contents, but do not break tissue extending between the cranial part of the liver and the ventral body wall.

Note the layers of the body wall through which you have cut. The outermost is the **skin.** This is followed by a thin and inconspicuous layer of **connective tissue** comparable to the external fascia of mammals (p. 325), the hypaxial musculature, and finally the shiny coelomic epithelium, the **serosa**[21], lining the pleuroperitoneal cavity. The portion of the serosa lining the musculature of the body wall is the **parietal serosa;** the portion covering the inner organs, or viscera, is the **visceral serosa.** Finally, the portion that extends from the body wall to the viscera contributes to the **mesenteries.** As a result of the way they develop, mesenteries consist of a double layer of epithelium between which lie the blood and lymphatic vessels and nerves to the viscera embedded in connective tissue (see Fig. 10-1).

In most vertebrates, the coelom is a closed cavity having no direct communication with the outside. But in early vertebrates, including the dogfish, it may communicate with the exterior by a pair of **abdominal pores** (Figs. 10-2 and 10-5). You can find these by probing the most caudal recess of the pleuroperitoneal cavity beside the **cloaca** (the chamber receiving the intestine and genital ducts) on the side of the body that is still intact. Each pore opens through the lateral wall of the cloacal aperture, but sometimes the lips of the pore have grown together. The functional significance of the abdominal pores is not understood. They may serve to eliminate excess coelomic fluid, or they may be vestiges of an evolutionary stage when gametes were discharged into the environment directly from the coelom through genital pores.

♦   **(B) THE VISCERAL ORGANS**

A large **liver** with a pair of long, pointed lobes occupies most of the cranioventral portion of the pleuroperitoneal cavity. You may cut off the ends of these lobes, but do not injure blood vessels going to the liver or the bile duct going from the liver to the intestine. The **bile duct** accompanies these blood vessels for most of its length but separates from the blood vessels near the beginning of the intestine at the pylorus (Fig. 10-5). Spread the lobes of the liver apart, and you will see the **esophagus** and **stomach** in a more dorsal position. Both have about the same diameter, so there is no external line of demarcation separating them. If a constriction is seen, it represents a peristaltic contraction fixed at death and during preservation of the animal, not a line of demarcation. The caudal end of the stomach curves cranially and gives the organ a J-shape. The **pylorus** marks the end of the stomach; its wall contains a thick sphincter muscle. The digestive tract then turns caudally and forms the straight **valvular intestine** which continues to the **cloaca.** If part of the intestine has been everted through the cloaca, you must pull it back into the body cavity.

[21]See footnote 19, p. 289.

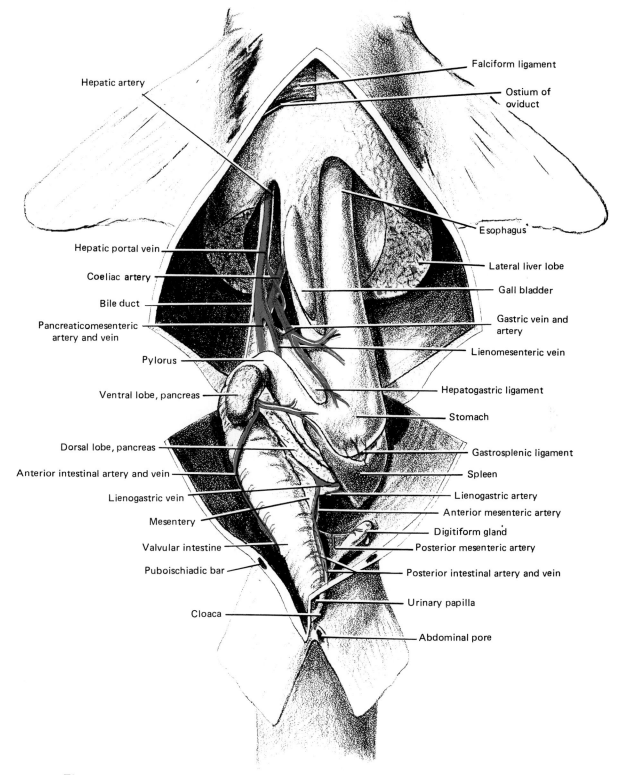

Figure 10-5
Ventral view of the abdominal viscera and blood vessels of a female *Squalus*. The distal parts of the lateral liver lobes have been cut away.

The large, triangular organ attached to the caudal end of the stomach (at the region where the stomach curves forward) is the **spleen**—an organ related to the production and storage of blood cells. The elongate dorsal lobe of the **pancreas** extends from the right side of the spleen cranially toward the beginning of the intestine. The oval ventral lobe of the pancreas is applied to the surface of the intestine near its junction with the

297

stomach. The very narrow, thin part of the pancreas between the dorsal and ventral lobes is called the **isthmus** of the pancreas. The finger-like organ extending dorsally from the caudal end of the intestine is the **digitiform gland** (or rectal gland). It is a salt-excreting gland and undergoes regressive changes in sharks that enter fresh water.

Other organs within the pleuroperitoneal cavity are parts of the urogenital system. We will consider them in more detail later, but you should identify them at this time. A pair of large **gonads**—**testes** or **ovaries**—lie one on either side of the cranial end of the esophagus and stomach. If your specimen is a mature female, a pair of prominent oviducts will be seen dorsal to the gonads and continuing to the cloaca. Their caudal ends are enlarged, greatly so in pregnant females. The paired **kidneys** are represented by the long bands of dark material located dorsal to the parietal serosa on either side of the middorsal line of the pleuroperitoneal cavity. They are very long and are not uniform in diameter, for the caudal end of each is much wider than the cranial end. If the specimen is a mature male, a large, twisted excretory duct, termed the **archinephric duct,** will be seen on the ventral surface of each kidney.

298

Chapter 10
The Coelom
and the Di-
gestive and
Respiratory
Systems

♦   **(C) THE MESENTERIES**

Pull the digestive tract ventrally and note that much of it is connected to the middorsal body wall of the pleuroperitoneal cavity by the **dorsal mesentery.** (In the embryo, the entire digestive tract is suspended by the dorsal mesentery; see Figs. 10-1**B** and **C**.) That portion of the dorsal mesentery passing from the middorsal line of the coelom to the dorsal surface of the esophagus and stomach is called the **greater omentum,** or **mesogaster;** that portion passing to the cranial part of the intestine is called the **mesentery** (in the limited sense)[22], and that portion passing to the digitiform gland and caudal end of the intestine is called the **mesocolon.** The spleen lies within the mesogaster, but the part of this mesentery between the spleen and stomach is given a special name—the **gastrosplenic ligament** (Fig. 10-5). Pull the caudal end of the stomach to the animal's left and the cranial end of the intestine to the right. On looking between them, you can see that the mesentery (in the limited sense) does not arise from the body wall but has shifted its attachment onto the mesogaster. You can also see that the pancreas lies in a special fold of the mesentery (Fig. 10-6). This relationship shows best on the portion of the pancreas near the spleen. The complicated relationships among the different parts of the mesentery, in the broad sense, are a result of the S-like curving of the stomach and cranial end of the intestine, together with their mesenteries. Subsequently, various parts of the mesentery, which came to lie next to each other, became fused to one another.

The ventral mesentery has disappeared except for that portion into which the liver has grown. The part of the ventral mesentery extending between the ventral surface of the front of the liver and the midventral body wall is the **falciform ligament;** the part between the liver and the digestive tract (caudal portion of stomach and front of intestine) is the **lesser omentum,** or **gastrohepatoduodenal ligament.** The latter mesentery, which also contains the bile duct and the blood vessels going to the liver, is very complex in

---

[22]Unfortunately, the term *mesentery* is used in two ways—in a broad sense for all membranes passing to the viscera, and in a limited sense for the mesentery suspending most of the intestine.

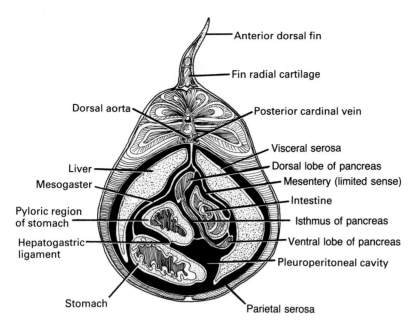

Figure 10-6
Cross section through the trunk of *Squalus* at the level of the isthmus of the pancreas, viewed from behind.

the dogfish. Near the liver, the lesser omentum is a unit, but it divides near the digestive tract. Part of it, the **hepatogastric ligament,** passes into the angle formed by the curvature of the stomach, and another part of it, the **hepatoduodenal ligament,** carries the bile duct to the beginning of the intestine. Note that a part of the dorsal mesentery suspending the pancreas extends between these two limbs of the lesser omentum and brings the larger blood vessels to the lesser omentum.

Subsidiary mesenteries (p. 293) support certain of the genital organs. Each testis is suspended by a **mesorchium,** each ovary by a **mesovarium,** and each oviduct in a mature female by a **mesotubarium.**

◆ (D) FURTHER STRUCTURE OF THE DIGESTIVE ORGANS

Cut open the esophagus and stomach by a longitudinal incision that extends all the way to the intestine. Remove the contents, if any, and wash out these organs. (You can often determine the feeding habits of animals by analysis of the stomach contents.) You can recognize the esophagus internally by the **papillae** in its lining and the stomach by the longitudinal folds called **rugae.** If the stomach was greatly distended, the rugae will have stretched out, and most of the lining will be smooth. The cranial end of the stomach, which would be adjacent to the esophagus and often near the heart, is called the **cardiac region;** the main part is called the **body;** and the portion that turns forward is called the **pyloric region.** Note that the pyloric region has a thicker muscular wall, especially just before the intestine, where it forms a muscular sphincter, the **pylorus.** It should be pointed out that these topographical regions in different vertebrates may or may not correspond to glandular regions given the same names. They do not in most nonmammalian vertebrates, for cardiac glands are absent in these. The surface of the stomach along which the mesogaster and gastrosplenic ligament attach is called the **greater curvature** of the stomach; the opposite surface is called the **lesser curvature.** The greater curvature of the adult stomach corresponds to the original dorsal side of the embryonic stomach.

The bile and pancreatic ducts, which you will see presently, enter the cranial end of the valvular intestine; the digitiform gland enters the caudal end. Aside from this slight modification at either end, the valvular intestine is undifferentiated. It contains a complex spiral fold, the **spiral valve,** which is similar to that of other ancestral fishes (Fig. 10-7). The line of attachment of the valve shows as the spiral line on the surface of the intestine. If a special preparation of the valve is not available, you can see part of it by cutting a tangential slice from the intestine wall. The spiral valve slows down the passage of food through the intestine, thus increasing digestive efficiency, and it also increases the absorptive surface area.

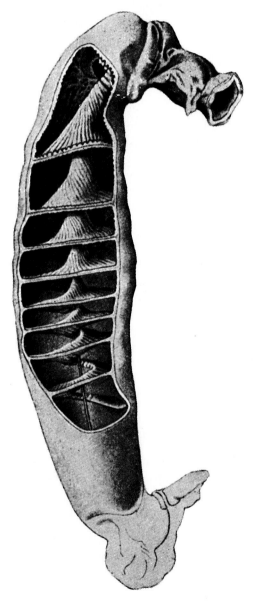

Figure 10-7
Valvular intestine of the skate, *Raja,* cut open to show the spiral valve. A bristle is shown passing through the central lumen of the intestine. *(From Mayer.)*

It is difficult to compare the intestine of ancestral fishes with the small intestine and large intestine of tetrapods. The cranial end of the fish intestine, which receives the bile and pancreatic ducts, would be roughly comparable to the duodenum. Most of the remainder would be comparable to the rest of the small intestine, and the posterior end receiving the digitiform gland would compare to at least part of the large intestine. But in the absence of clear lines of separation between these potential regions, it is best to think of the primitive fish intestine as one unit—the valvular intestine.

Examine the liver in more detail. It consists of long **right** and **left lobes** and a smaller **median lobe** containing the elongate, thin-walled **gall bladder** (Fig. 10-5). A part of the gall bladder shows on the surface, but most is imbedded within the liver lobe. Cut off the caudal end of the gall bladder and probe to find its cranial end. By scraping away superficial liver tissue from the cranial end of the gall bladder, you can find the point where the bile duct connects. In the dogfish, the bile leaves the liver through a number of inconspicuous **hepatic ducts** that enter the gall bladder; from here it flows to the intestine through the **bile duct.** The point of entrance of the bile duct into the lumen of the intestine is some distance caudad to its point of attachment on the external surface of the intestine. You can trace the duct through the intestinal wall by removing the visceral serosa and longitudinal muscle layer.

As in other vertebrates, in the dogfish the liver is an important site for the metabolism and storage of food products. Glycogen is present, but much of the food is stored as oil. In this respect, the liver of a shark is analogous to the swim bladder, which is present in most bony fishes, because the oil lowers the specific weight, or density, of a shark and makes the animal more buoyant (p. 148).

Return to the pancreas and find again the dorsal and ventral lobes and the interconnecting isthmus. Both lobes drain by a common **pancreatic duct** that leaves the ventral lobe and travels obliquely caudad in the wall of the intestine for a short distance before entering the lumen of the intestine. To see the duct, it is necessary to cut the attachment of the mesentery. Then, with the tips of a pair of forceps, slowly scrape away the soft tissue from the caudal part of the ventral lobe of the pancreas; in this way you may expose the thin whitish pancreatic duct. Follow this duct by removing the visceral serosa and longitudinal musculature from the wall of the intestine.

Relationships at the front of the intestine are confusing at first, for this portion of the digestive tract has undergone a rotation of nearly 180 degrees during development. This has brought about an apparent reversal of the relationships present in the embryo. In the adult, the bile duct enters on what would at first seem to be the dorsal surface of the intestine, and the dorsal mesentery appears to attach along the ventral surface (Fig. 10-6). The pancreas of the dogfish is entirely a dorsal pancreas; however, its duct enters on what would seem to be the original lateral surface of the intestine.

The Pericardial Cavity

The second division of the coelom, the **pericardial cavity,** is located far forward in fishes, for it lies just cranial and dorsal to the pectoral girdle and deep to the posterior hypobranchial musculature (Fig. 10-8; see also Fig. 11-13, p. 356). To expose it, continue your original ventral incision forward through the pectoral girdle and caudal hypobranchial musculature. Veer toward the midventral line as you go, but do not cut the falciform ligament. You will have to make additional transverse cuts just cranial to the pectoral girdle.

Spread open the flaps thus formed and study the pericardial cavity. The only organ within it is the **heart.** That portion of the coelomic epithelium covering the tissues surrounding the cavity is the **parietal pericardium;** that covering the heart is the **visceral pericardium.** The heart developed embryonically in the ventral mesentery beneath the pharynx (Fig. 10-1A) and at one time was suspended by parts of the ventral mesentery, termed mesocardia. The mesocardia disappear in the adult, and the heart remains attached to the wall of the pericardial cavity only at its caudal and cranial ends, where blood vessels enter and leave.

The vertical septum separating the pericardial and pleuroperitoneal cavities is the **transverse septum.** You can now see that the liver is attached to the caudal face of the transverse septum by the coronary ligament. This mesentery does not look like a typical mesentery; it is actually just the serosa bridging the gap between the liver and the transverse septum. It is continuous with the visceral serosa covering the liver and transverse septum and with the cranial part of the falciform ligament. As in many other ancestral fishes, the separation of the pericardial and pleuroperitoneal cavities is not complete in the dogfish, for a **pericardioperitoneal canal** connects the two cavities ventral to the esophagus. You will see it later after you study the heart (p. 356).

302

Chapter 10
The Coelom
and the Di-
gestive and
Respiratory
Systems

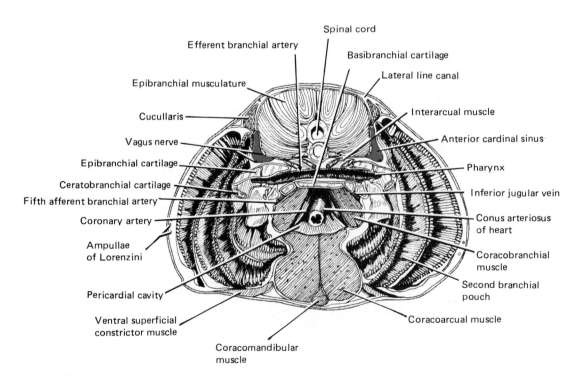

Spinal cord
Efferent branchial artery
Basibranchial cartilage
Epibranchial musculature
Lateral line canal
Cucullaris
Interarcual muscle
Vagus nerve
Anterior cardinal sinus
Epibranchial cartilage
Pharynx
Ceratobranchial cartilage
Inferior jugular vein
Fifth afferent branchial artery
Coronary artery
Conus arteriosus
of heart
Ampullae
of Lorenzini
Coracobranchial
muscle
Pericardial cavity
Second branchial
pouch
Ventral superficial
constrictor muscle
Coracoarcual muscle
Coracomandibular
muscle

Figure 10-8
Cross section through the pharynx and front of the pericardial cavity of *Squalus,* viewed from behind.

The Oral Cavity, Pharynx, and Respiratory Organs

For this part of the dissection, you need to open the pharynx (see Figs. 10-9 and 11-9). On the side on which you dissected the branchial muscles, use a strong pair of scissors and cut through the angle of the mouth and the middle level of the gill arches (where you made the cut to expose the branchial adductor, p. 155). Extend the cut through the coracoid bar and about 1 inch farther caudad through the body wall. Then proceed to cut transversely towards the midventral line. In doing so, you will first open a cavity, the posterior cardinal sinus (see Fig. 11-5). When you reach the wall of the esophagus, first make a longitudinal incision to open the pharynx, then continue the transverse cut through the ventral esophageal wall and the cranial tip of the liver. In this way, you will keep the heart and sinus venosus intact for later dissections. Continue the transverse cut across the midventral line of the body wall as far as necessary to swing open the floor of the oral cavity and pharynx. If the esophagus has everted into the pharynx, you must pull it back.

The demarcation between the **oral cavity** and the **pharynx** is not clearly definable in the adult (p. 293), but the pharynx is approximately that portion of the digestive tract into which the gill slits enter. The pharynx, of course, leads to the narrower esophagus. A tonguelike structure supported by the hyoid arch lies in the floor of the mouth and front of

x

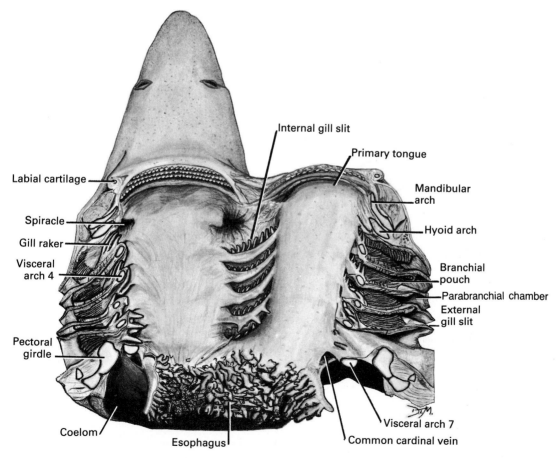

Figure 10-9
Oral cavity and pharynx of *Squalus*. The floor of the mouth and pharynx has been swung open to the right.

The Oral Cavity, Pharynx, and Respiratory Organs

For this part of the dissection, you need to open the pharynx (see Figs. 10-9 and 11-9). On the side on which you dissected the branchial muscles, use a strong pair of scissors and cut through the angle of the mouth and the middle level of the gill arches (where you made the cut to expose the branchial adductor, p. 155). Extend the cut through the coracoid bar and about 1 inch farther caudad through the body wall. Then proceed to cut transversely towards the midventral line. In doing so, you will first open a cavity, the posterior cardinal sinus (see Fig. 11-5). When you reach the wall of the esophagus, first make a longitudinal incision to open the pharynx, then continue the transverse cut through the ventral esophageal wall and the cranial tip of the liver. In this way, you will keep the heart and sinus venosus intact for later dissections. Continue the transverse cut across the midventral line of the body wall as far as necessary to swing open the floor of the oral cavity and pharynx. If the esophagus has everted into the pharynx, you must pull it back.

The demarcation between the **oral cavity** and the **pharynx** is not clearly definable in the adult (p. 293), but the pharynx is approximately that portion of the digestive tract into which the gill slits enter. The pharynx, of course, leads to the narrower esophagus. A tonguelike structure supported by the hyoid arch lies in the floor of the mouth and front of the pharynx. It is not a true tongue, as no muscle extends up beneath its epithelium, but sometimes it is called a **primary tongue,** for it is destined to contribute to the base of the true tongue of tetrapods (Fig. 10-9).

You will see the entrance of the **spiracle** at the front of the pharynx roof, and this is followed by the five definitive **internal gill slits.** The homologue of the spiracle may have been a complete gill slit in very early jawed fishes. A number of papilla-like **gill rakers** project across the internal gill slits; they act as strainers and keep food in the pharynx. Note that each internal gill slit leads into a large **branchial pouch.** The portion of the branchial pouch lateral to the gill lamellae is the **parabranchial chamber** (Fig. 10-9), and it opens on the body surface by the **external gill slit.** The tissue between each gill slit and branchial pouch constitutes the **interbranchial septum.** The outermost portion of each septum is thin and flaplike and constitutes a **flap valve** that can close and open the external gill slits. The actual gills are composed of a number of platelike **primary gill lamellae,** which are attached to the surface of the interbranchial septa. Examine the primary gill lamellae with low magnification and note that each bears many small, closely packed **secondary gill lamellae** arranged perpendicularly to the surface of each primary gill lamella. The secondary gill lamellae consist essentially of capillary beds covered with a very thin, gas-permeable epithelium; they are the site of respiratory gas exchange. Interbranchial septa that have gill lamellae on their cranial and caudal surfaces constitute a complete gill, or **holobranch.** The first interbranchial septum is a **hemibranch,** for gill lamellae are present only on its caudal surface (Fig. 10-10).

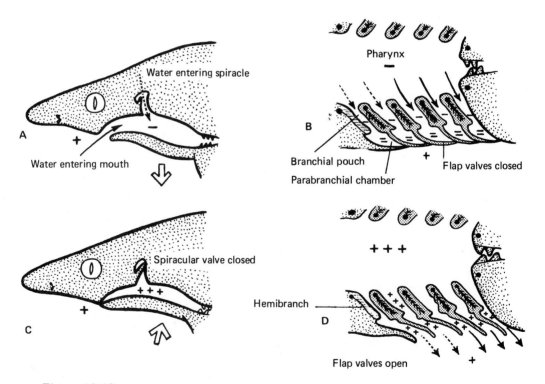

Figure 10-10
Diagrams to show the mechanics of respiration in the dogfish as seen in lateral views (**A** and **C**) and in frontal sections (**B** and **D**) of the pharynx. **A** and **B**, inspiration; **C** and **D**, expiration. Relative pressures are indicated by + and −. Open arrows indicate the direction of movement of the floor of the mouth and pharynx, solid arrows indicate the course of the current of water entering the mouth, and dotted arrows indicate the course of the water entering the spiracle. *(Slightly modified after G. M. Hughes, 1963.)*

In order to ensure the uptake of oxygen and the release of carbon dioxide by the blood in the capillaries of the gill lamellae, the organism must maintain a constant flow of water across the gill lamellae. This water flow is driven by a double-pump mechanism, in which the oropharyngeal cavity acts as a pressure pump and the parabranchial chambers act as a suction pump. As G. M. Hughes and his co-workers have shown, when the mouth and spiracle are closed and the branchial musculature contracts, the oropharyngeal cavity and branchial pouches are compressed, and water is forced across the gill lamellae and expelled through the open external gill slits (Figs. 10-10**C** and **D**). After this, the flap valves close the external gill slits, the branchial musculature relaxes, and the oropharyngeal cavity, branchial pouches, and parabranchial chambers expand passively owing to the elastic recoil of the branchial skeleton. Because the mouth is still closed at first, the internal pressure within the orobranchial cavity, branchial pouches, and parabranchial chambers is reduced relative to the external pressure, the lowest pressure being found in the parabranchial chambers. As the mouth and spiracle open, water is sucked into the oropharyngeal cavity and, following the pressure gradient, across the gill lamellae into the parabranchial chambers (Figs. 10-10**A** and **B**). Water that entered the oropharyngeal cavity through the spiracle leaves through the more cranial gill slits, while water that entered through the mouth leaves through the more caudal gill slits. Some oceanic sharks help move water through the pharynx and gill pouches by swimming with their mouths partly open; this is called ram ventilation. The spiracle is small or absent in such species. Bottom-dwelling elasmobranchs, however, have large spiracles through which most of the respiratory current enters (see also p. 42).

The double-pump mechanism of gill ventilation ensures a unidirectional flow of water across the gill lamellae and therefore makes possible an efficient countercurrent flow between respiratory water and blood in the gill capillaries. G. M. Hughes has also shown that the heartbeat in the dogfish is coupled with the respiratory movements in such a way that blood flows rapidly through the gill lamellae when water is moved across them. This increases the efficiency of gas exchange between the water and the blood in the capillaries of the gill lamellae.

Cut open the spiracle on the side you have been dissecting. You can find a minute hemibranch, known as the **pseudobranch,** on the valvelike flap (the **spiracular valve,** p. 42) on the cranial wall of the spiracle. Since oxygen-rich blood passes through the pseudobranch (see Fig. 11-9), it is often regarded as having no role in gas exchange and, therefore, as being vestigial, but it is likely that it has an accessory respiratory function. In some marine teleosts, its homologue contains secretory cells believed to be part of the salt excretory mechanism.

Examine the cut surface of a representative holobranch and note its composition (Fig. 10-11). A supporting **visceral arch** lies at its base. (Be sure that you can identify all these arches—the mandibular, hyoid, and five branchial arches.) You may also be able to see one or more cartilaginous **gill rays** extending into the interbranchial septum at right angles to the visceral arch. Cartilage also supports the **gill rakers.** Much of the septum is made up of branchial muscles (Fig. 7-12, p. 155). A **branchial adductor** lies medial to each arch, and an **interbranchial** and a **superficial constrictor** extend out into the interbranchial septum. Close examination will also reveal several blood vessels. An **afferent branchial artery** lies near the middle of the interbranchial septum just lateral to the visceral arch. It brings oxygen-depleted blood from the heart and ventral aorta to the capillaries in the gill lamellae and probably is not injected. An **efferent branchial artery,** which will be injected if the arteries have been injected, lies at the base of the primary gill lamellae on each surface of the interbranchial septum. These vessels, of which

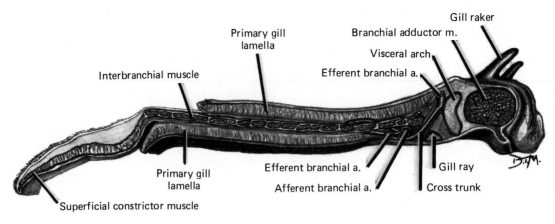

Interbranchial muscle — Primary gill lamella — Gill raker — Branchial adductor m. — Visceral arch — Efferent branchial a. — Primary gill lamella — Superficial constrictor muscle — Efferent branchial a. — Afferent branchial a. — Gill ray — Cross trunk

Figure 10-11

A frontal section through a holobranch of *Squalus*. The cranial surface is toward the top of the page. (Size shown is twice the natural size.)

there are two in a typical interbranchial septum, drain the gill capillaries and carry oxygen-rich blood to the dorsal aorta. Also recall (p. 249) that each interbranchial septum would contain a **pretrematic** and **posttrematic branch** of a cranial nerve, but these are difficult to find in a gross section. Which nerve, or nerves, are associated with the interbranchial septum you have been studying?

A thyroid gland, a series of thymus bodies, and an ultimobranchial body develop embryonically from the pharynx but are difficult to see. You may have seen the thyroid gland during the dissection of the hypobranchial muscles (p. 152).

AMPHIBIANS

The coelom of amphibians is very similar to that of fishes except that, with the beginning of the formation of the neck, the pericardial cavity has moved caudad a short distance. This causes the transverse septum to assume a somewhat oblique orientation.

More conspicuous changes are seen in the respiratory and digestive systems. In regard to respiration, tetrapods retain the internal nostrils and elaborate upon the lungs of their piscine ancestors but, excepting larval amphibians, lose the gills. Larval amphibians have gill slits and gills of two types. **External gills,** which are also present in some fish larvae, develop as outgrowths from the neck surface near the gill slits in urodele and frog larvae. In frog larvae, these are later replaced by **internal gills** that develop closer to the gill arch and hence may be homologous to the internal gills of adult fishes. At metamorphosis, the gills are lost, and the lungs become the primary site of gas exchange, supplemented in living amphibians by gas exchange through the vascularized, thin, moist skin and the epithelial lining of the oropharyngeal cavity.

Amphibians have little problem in obtaining sufficient oxygen for their relatively low metabolic needs, because air contains so much more oxygen than water (about 210 ml of oxygen per liter of air compared with 7 ml of oxygen per liter of fresh cool water). Amphibians do not need to move much air in and out of the lungs to satisfy their oxygen requirements. They ventilate their lungs by moving the floor of the mouth and pharynx, which, rather than pumping water across the gills, pumps air into the lungs. When the lungs are filled, the glottis is closed, and air is held in the lungs at greater than atmospheric pressure. When the glottis is opened, pressure in the lungs and their elastic recoil drives out the air. Ribs are very short and are sometimes (as in frogs) fused to the sides of the vertebrae, hence rib movements do not help ventilate the lungs. Eliminating carbon dioxide is more of a problem because it would require a greater rate of lung ventilation than amphibians can sustain. Contemporary amphibians supplement pulmonary respiration with cutaneous respiration and eliminate most of their carbon dioxide through the skin. Under some circumstances, such as when the animal is under water, the skin is also the site of oxygen uptake. The small size of contemporary amphibians and the resulting favorable surface-to-volume ratio facilitate cutaneous respiration.

Gas exchange may have been somewhat different in the ancestral labyrinthodont amphibians because they were larger animals (see p. 35), and some retained small bony scales in their skin. Their broad heads suggest a buccopharyngeal pump, but they also had longer ribs whose movement may have contributed to the ventilation of the lungs.

Major changes that are related to the transition from an aquatic to a terrestrial environment have also affected the feeding mechanism. G. V. Lauder and his co-workers (see references for details) have analyzed the functional and structural changes involved in this transition by comparing the feeding behavior of larval and adult *Ambystoma* salamanders in aquatic and terrestrial environments. Larvae suck food items, together with water, into the mouth cavity by using a mechanism similar to that used by fishes. They lower the pressure within the oropharyngeal cavity by closing the gill slits and the mouth and by depressing the floor of the mouth. As the mouth opens, water and food rushes into the mouth cavity; the food is swallowed and the water expelled through the opened gill slits. A metamorphosed salamander on land, however, cannot use this mechanism, because it would have to generate impossibly high suction forces to produce an air stream strong enough to carry food items into the mouth cavity. Terrestrial amphibians, in general, project their sticky tongue out of their mouth cavity onto the food item and pull the food item into the mouth cavity by retracting their tongue. Fully metamorphosed *Ambystoma* salamanders that have returned to the water (during the mating season, for example), capture prey again by suction, but because their gill slits have closed during metamorphosis, the surplus water, which is sucked in with the food items, has to leave through the mouth opening. In water, metamorphosed *Ambystoma* are less efficient at feeding than the larval animals.

Major changes in the digestive tract of amphibians include the evolution of small **oral glands** and a muscular **tongue,** both correlated with the problem of food acquisition and manipulation in a terrestrial environment. The amphibian tongue consists of the primary tongue of fishes overlying the hyoid apparatus, plus a swelling **(gland field)** that develops between the hyoid and mandibular arches. The whole structure is invaded by certain prehyoid hypobranchial muscles (p. 158). The primitive spiral valve in the intestine has been lost. The resulting reduction in internal surface area is compensated for by other folds and an increase in the length of the intestine. A differentiation into **small** and **large intestines** also occurs. Finally, a **urinary bladder** has evolved as an outgrowth from the ventral surface of the cloaca, but this is a part of the urinary system.

Necturus illustrates the early tetrapod stage well, except for the retention of many larval features, the major ones being a vertical transverse septum, the presence of gill slits and external gills, and a poorly formed tongue. Moreover, all urodeles differ from most other tetrapods in lacking the auditory tube and tympanic cavity.

The Pleuroperitoneal Cavity and Its Contents

♦ (A) THE BODY WALL AND PLEUROPERITONEAL CAVITY

If your specimen of *Necturus* has been injected, a partial incision will have been made through the body wall on one side of the midventral line. Continue this incision cranially to the pectoral girdle and caudally through the pelvic girdle and along one side of the cloacal aperture. The layers of the body wall through which you have cut are similar to those of *Squalus:* **skin, connective tissue, hypaxial muscles,** and **parietal serosa**[23]. How many muscle layers were cut through? The part of the coelom opened is, as in fishes, a **pleuroperitoneal cavity.** It is lined with the **parietal serosa,** and the inner organs, or viscera, are covered with **visceral serosa.**

♦   (B) THE VISCERAL ORGANS

The **liver** is the largest of the visceral organs and lies in a cranioventral position (Fig. 10-12). Notice that it is displaced toward the right side of the cavity. It is not as obviously subdivided into lobes as it is in *Squalus.* A portion of the ventral mesentery, the **falciform ligament,** extends from the midventral line of the pleuroperitoneal cavity to attach along the entire length of the ventral surface of the liver. You will have to cut it, but do so in such a way that you can later reconstruct the major veins that pass through it. Pull the left side of the body wall and the liver apart and you will see the elongate **stomach.** Because *Necturus* remains in the larval stage, the stomach is straight rather than J-shaped as it is in fully metamorphosed amphibia. The **esophagus** is a short connecting piece between the pharynx and stomach; you will see it more clearly later. The **spleen** is the elongate, oval organ attached to the left side of the stomach. The long, finger-like left **lung** lies dorsal to the spleen; a similarly shaped right lung lies in a comparable position on the other side of the body.

The **small intestine** begins at the pylorus at the caudal end of the stomach (Fig. 10-12) and makes one cranial loop (the **duodenum**) before continuing in a number of convolutions. An irregularly shaped **pancreas** lies near this cranial loop. Part of it is in contact with the intestine and part with the liver; one part—the **tail**—passes dorsal to the stomach and nearly reaches the caudal tip of the spleen. Shortly before entering the **cloaca,** the intestine widens and forms a short **large intestine.** The **urinary bladder,** which is usually collapsed and shriveled in preserved specimens, lies ventral to the large intestine. It enters the cloaca independently.

If your specimen is a male, a pair of large, elongate, oval **testes** lie in a dorsal position in the region of the cranial intestine. The **kidneys** lie dorsal and lateral to the testes. Each has a conspicuous, convoluted **archinephric duct** along its lateral border. If your specimen is a female, the large, paired **ovaries** may largely fill the pleuroperitoneal cavity; a long, convoluted **oviduct** lies dorsal to each ovary and ventral to the **kidney** (see Fig. 12-10, p. 411).

♦   (C) THE MESENTERIES

As in the dogfish, that portion of the dorsal mesentery passing to the stomach is called the **mesogaster.** That portion passing to the small intestine is called the **mesentery** (in the limited sense), and that portion passing to the large intestine is called the **mesocolon.**

---

[23]See footnote 19, p. 289.

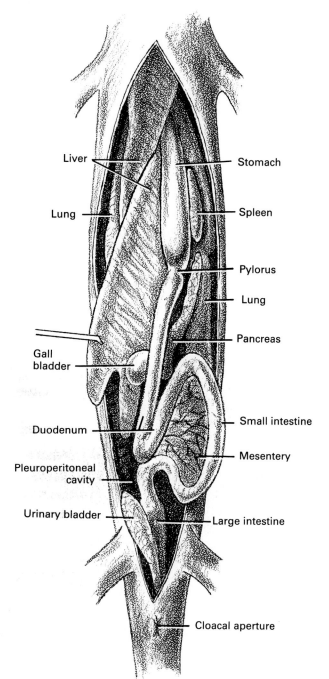

Liver — — Stomach

Lung — — Spleen

— Pylorus

— Lung

— Pancreas

Gall
bladder

— Small intestine

Duodenum

— Mesentery

Pleuroperitoneal
cavity

Urinary bladder

— Large intestine

— Cloacal aperture

Figure 10-12
Ventral view of the digestive tract and associated organs of *Necturus*. The caudal half of the liver has
been reflected and is seen in a dorsal view.

The portion of the mesogaster between the spleen and the stomach is called the
gastrosplenic ligament (Fig. 10-13). Most of the left lung is connected with the
mesogaster by a short **pulmonary ligament** (or **mesopneumonium**). The right lung
is also suspended by a comparable but wider pulmonary ligament. However, an accessory
mesentery (the **ligamentum hepatocavopulmonale**) passes from the right pulmonary ligament to the dorsal surface of the liver. A large blood vessel, the caudal vena cava,
approaches the liver through the caudal margin of this mesentery.

You have already inspected and cut the part of the ventral mesentery extending from

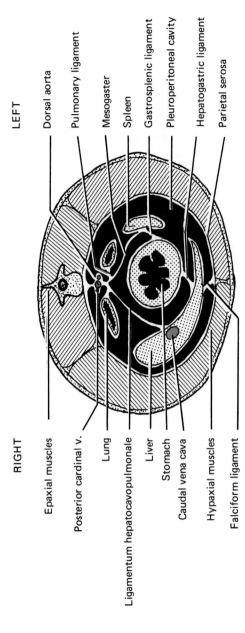

RIGHT

Epaxial muscles

Posterior cardinal v.

Lung

Ligamentum hepatocavopulmonale

Liver

Stomach

Caudal vena cava

Hypaxial muscles

Falciform ligament

LEFT

Dorsal aorta

Pulmonary ligament

Mesogaster

Spleen

Gastrosplenic ligament

Pleuroperitoneal cavity

Hepatogastric ligament

Parietal serosa

Figure 10-13
Diagrammatic transverse section through a male specimen of *Necturus* at the level of the spleen to show mesenteric relationships; viewed from the front.

the liver to the body wall (the **falciform ligament**). Two other parts extend from the liver to the digestive tract—a **hepatogastric ligament** between the cranial part of the stomach and liver and a **hepatoduodenal ligament** between the liver and the most cranial loop of the small intestine. A part of the pancreas lies in the latter ligament, but most of it extends into the dorsal mesentery. A final part of the ventral mesentery, the **median ligament of the bladder,** passes from the urinary bladder to the midventral line of the body wall.

The subsidiary genital mesenteries are the same as in *Squalus:* a **mesorchium** to the testis, a **mesovarium** to the ovary, and a **mesotubarium** to the oviduct.

♦ **(D) FURTHER STRUCTURE OF THE DIGESTIVE ORGANS**

Cut open the stomach and wash it out. Its lining is thrown into a number of irregular and longitudinal folds **(rugae).** As in other vertebrates, it can be divided into several gross regions (a **cardiac region** next to the esophagus, a central **corpus,** and a caudal **pyloric region**), but these are not sharply demarcated. The pyloric region ends in a sphincter, the **pylorus.**

Cut open the intestine at several points and notice that a spiral valve is lacking, but that its internal surface area is increased through many small, wavy, longitudinal folds **(plicae).** Compared to that of a shark, the intestine of *Necturus* is also longer and, therefore, is thrown into loops and convolutions that allow it to fit within the pleuroperitoneal cavity; this elongation of the intestine also increases the surface area available for absorption processes.

Most of the bile is drained from the liver by several **hepatic ducts;** these unite to form a **common bile duct,** which enters the duodenum (Fig. 10-14). Some of the bile backs up through a **cystic duct** into the **gall bladder,** where it is temporarily stored. You can find the gall bladder on the dorsal surface of the liver near the apex of the cranial loop of the intestine, but most of the ducts are imbedded within pancreatic tissue. Some can be found by careful dissection in this region.

The pancreas of urodeles is formed from a dorsal primordium and a pair of ventral primordia (see p. 295). All fuse in the adult, but each part retains its own duct (Fig. 10-14). The pancreatic ducts cannot be found by gross dissection.

The Pericardial Cavity

The second division of the coelom, the **pericardial cavity,** is located farther forward in *Necturus* than in most amphibians, for it still occupies the larval position cranial to the pectoral girdle and deep to the hypobranchial musculature. Expose it by removing the musculature in this region. The pericardial cavity is lined with **parietal pericardium** and contains only the **heart,** which is covered with **visceral pericardium.**

Cut through that portion of the ventral body wall lying between the pericardial cavity and the cranial end of your incision into the pleuroperitoneal cavity. The two divisions of the coelom are completely separated by a small, vertical **transverse septum.** A

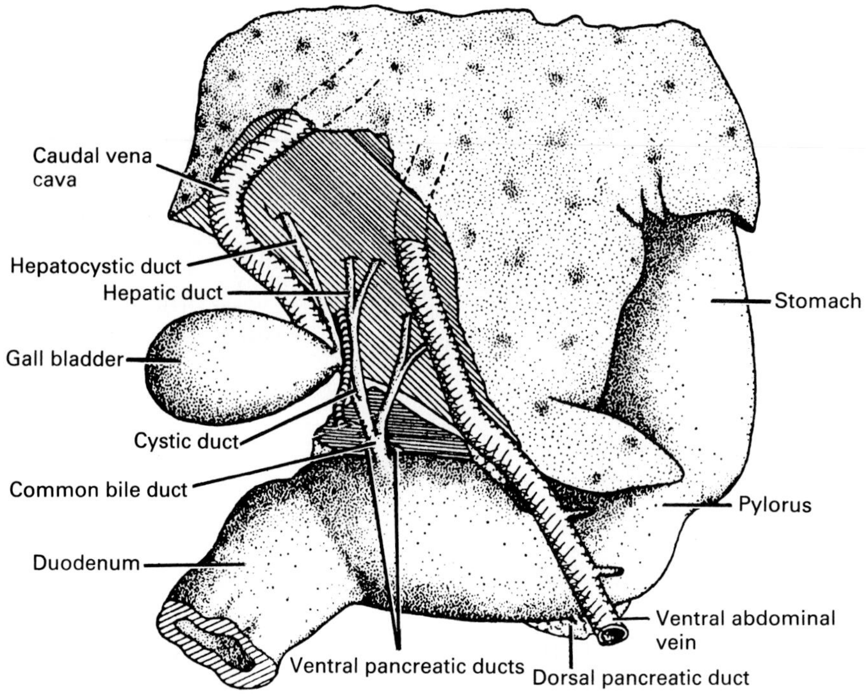

Caudal vena cava
Hepatocystic duct
Hepatic duct
Gall bladder
Cystic duct
Common bile duct
Duodenum
Ventral pancreatic ducts
Dorsal pancreatic duct
Stomach
Pylorus
Ventral abdominal vein

Figure 10-14
Ventral dissection of the bile and pancreatic ducts of *Salamandra*. Parts of the liver and pancreas have been cut away. *(Modified from Francis.)*

coronary ligament runs from the caudal surface of the transverse septum to the cranial end of the liver. A large vein (the caudal vena cava) lies within it.

The Oral Cavity, Pharynx, and Respiratory Organs

Since *Necturus* is neotenic, its major respiratory organs are not the lungs but the three pairs of larval **external gills** arising from the back of the head (Fig. 3-5, p. 44). Spread them apart and find the two **gill slits** between their bases. If a living specimen is available in an aquarium, notice the great vascularity of the numerous **gill filaments** and the way in which the gills are slowly moved back and forth through the water.

Open the oral cavity and pharynx on the side on which you have been dissecting by cutting through the angle of the mouth, caudally through the gill slits, the side of the neck lateral to the pericardial cavity, and the ventral portion of the pectoral girdle, and on to intersect the incision by which you opened the pleuroperitoneal cavity (Fig. 10-15). The deeper part of the incision should pass dorsal to the lung and into the esophagus. Swing open the floor of the mouth and pharynx.

As explained in the introduction to this chapter (p. 293), the breakdown of the plate of tissue between the stomodeum and archenteron makes it difficult to draw a sharp line between the rostral **oral cavity** and the more caudal **pharynx.** One merges with the other. Caudally, however, the pharynx leads into a somewhat constricted, short passage, the **esophagus,** which soon enters the stomach. The longitudinal folds in the lining of the esophagus tend to be smaller than those in the stomach.

The **tongue** is a fold, supported by the hyoid arch and located in the floor of the mouth

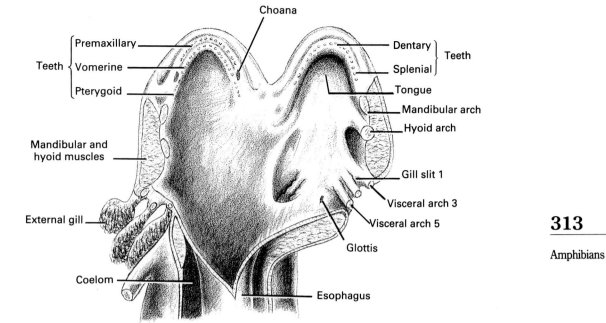

Figure 10-15
Oral cavity and pharynx of *Necturus*. The floor of the mouth and pharynx has been swung open to the right.

(Fig. 10-15). It has developed little beyond the primary tongue of fishes, but a few hypobranchial muscle fibers enter its base (p. 158). The gland field of the tongue of adult amphibians (p. 307) is barely developed. The third to fifth visceral arches lie caudal to the hyoid arch and can be palpated. The gill slits lie between the third and fourth and the fourth and fifth visceral arches. With which slits of *Squalus* are these slits homologous? Small **gill rakers** can be seen on the pharyngeal surface of the gill slits.

    Necturus seldom breathes air, but the structures needed for pulmonary respiration are present. Air can enter the oral cavity and pharynx by way of the nares, nasal cavity, and choanae. If you did not dissect the nasal cavity when you studied the sense organs (p. 214), expose it now. A choana can be seen in the roof of the mouth lateral to the most caudal and shortest of the tooth rows (pterygoid teeth). The air leaves the pharynx through the **glottis**—a short longitudinal slit in the center of the floor of the pharynx about at the level of the caudal gill slits. The glottis leads into a median **laryngotracheal chamber,** from whose caudal end a pair of openings—the **bronchi**—pass to the lungs. You can best see these structures by cutting the front of one lung, passing a blunt probe forward to the glottis, and then cutting open the floor of the esophagus and pharynx along the course of the probe. A pair of small cartilages **(lateral cartilages),** which are difficult to see but may be felt, support the wall of the laryngotracheal chamber (see Fig. 4-12). They are probably derived from the sixth and seventh visceral arches (p. 73). Cut open more of the lung and observe that it is an empty sac with a smooth internal surface. In most adult amphibians, the internal surface area is increased by pocket-like folds. If possible, examine a preparation of a lung of a frog.

    A thyroid gland and a series of parathyroid glands and thymus glands develop embryonically from the pharynx but are difficult to find. The thyroid gland may be seen during the dissection of the arteries (p. 362).

MAMMALS

The evolution from amphibians through reptiles to mammals has been one of elaboration upon terrestrial adaptations, including an increase in general body activity. Mammals are active, warm-blooded animals, most of whom maintain a high and constant rate of metabolism (they are **homoiothermic** as well as **endothermic**). Evolutionary changes in the digestive, respiratory, circulatory, and excretory systems enable mammals to sustain this high level of metabolism.

Correlated with a shift in the site of gas exchange from gills to lungs, the heart and pericardial cavity undergo a caudad migration and become situated caudal to the pectoral girdle in the thoracic region of the body. The pleuroperitoneal cavity of nonmammalian vertebrates becomes divided into a pair of **pleural cavities** containing the lungs and a **peritoneal cavity** containing most of the remaining viscera. Muscles invade the coelomic folds that separate the pleural cavities from the peritoneal cavity, and the whole complex forms a **diaphragm** whose movements, together with those of the ribs, provide an efficient means of ventilating the lungs.

314

Chapter 10
The Coelom
and the Di-
gestive and
Respiratory
Systems

With the evolution of a distinct neck, which is made possible by the caudad migration of the heart, the laryngotracheal chamber of amphibians differentiates into a **larynx,** from which a **trachea** descends to the lungs. And, as explained in connection with the skeleton, the evolution of a hard and soft palate separates a respiratory passage from the original oral cavity and rostral part of the pharynx. This permits simultaneous respiration and manipulation of food within the mouth. Food and air passages cross only in the caudal part of the pharynx, and even here, the evolution of a flaplike **epiglottis** normally prevents food from entering the larynx.

Respiratory passages branch extensively within the lungs and terminate in grapelike clusters of minute, thin-walled vascular **alveoli,** where gas exchange occurs. The greater separation of the digestive and respiratory passages, the great increase of the respiratory surface area of the lungs, and efficient pulmonary ventilation make the mammalian respiratory system far more efficient than that of amphibians and reptiles.

Many changes also occur in the digestive system. Conspicuous **salivary glands** evolve from certain of the small oral glands of amphibians. In addition to providing secretions for lubricating the food, these glands secrete a digestive enzyme, **ptyalin,** that initiates the hydrolysis of starch.

The tongue, which aids in manipulating food within the mouth and in swallowing, has become a large, muscular organ (p. 207). A pair of **lateral lingual swellings** and a gland field (**tuberculum impar** of mammalian embryology) have been added rostral to the primary tongue, and somatic muscles have continued to invade the organ. The muscles, which are of hypobranchial origin, are innervated by the hypoglossal nerve; but the tongue's sensory innervation derives from the trigeminal, facial, and glossopharyngeal nerves, because its epithelium derives from the epithelium overlying the more cranial visceral arches. General sensory fibers are supplied to approximately the anterior two thirds of the tongue by the trigeminal nerve, and taste fibers are supplied to this area by the facial nerve. The glossopharyngeal nerve supplies both general sensory and taste fibers to the posterior third of the tongue.

The most notable change in the intestinal region is a great increase in internal surface area. This is accomplished in part through an increase in the length of the intestine, and in part through the evolution of numerous minute, finger-like **villi.** Another major change is the division of the cloaca into dorsal and ventral portions. The ventral portion contributes to the urogenital passages; the dorsal portion forms the **rectum.**

Superimposed upon these general features are numerous modifications of the digestive tract correlated with the diverse diets and modes of life of the various mammalian groups. The digestive tract of the cat is a good example of the more primitive carnivore pattern, whereas the intestine of the rabbit and the stomach of the sheep illustrate two quite different patterns that have evolved in herbivorous mammals to digest cellulose with the help of symbiotic microorganisms.

The Digestive and Respiratory Organs of the Head and Neck

♦ **(A) THE SALIVARY GLANDS**

You may study the salivary glands of your specimen on the same side of the head that you used for the dissection of the muscles, provided that you did not injure the glands. If they were destroyed, carefully remove, on the opposite side of the head, the skin overlying the cheek, throat, and side of the neck ventral to the auricle. You must also take off the superficial cutaneous muscles (platysma and facial muscles, p. 166). Be especially careful in the cheek region, for one of the salivary ducts is very superficial.

Pick away the connective tissue ventral to the auricle, and you will expose the large **parotid gland.** You can recognize it by its lobulated texture. The gland is somewhat oval in the cat (Fig. 10-16) but is shaped like a dumbbell in the rabbit. In all the mammals being considered, the **parotid duct** emerges from the front of the gland, crosses the large cheek muscle (masseter), and perforates the mucous membrane of the upper lip near the molar teeth. Frequently, accessory bits of glandular tissues are found along the duct. Two branches of the facial nerve going to facial muscles emerge from beneath the parotid gland and cross the masseter, one dorsal and one ventral to the parotid duct. Do not confuse the parotid duct with them. The parotid duct can be traced into the parotid gland, whereas the facial nerve extends deep to the gland toward the stylomastoid foramen (Fig. 4-27), through which it leaves the skull (p. 90).

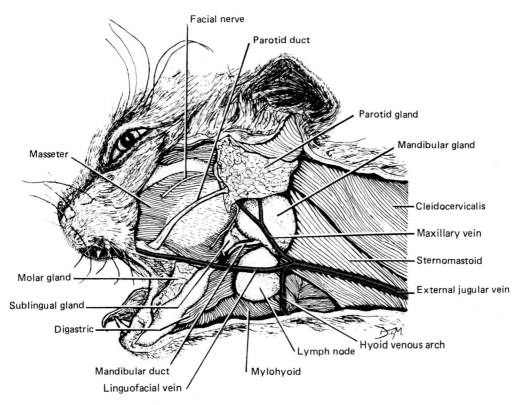

Figure 10-16
Lateral view of the salivary glands of the cat.

A **mandibular gland** lies caudal to the angular process of the jaw and deep to the ventral border of the parotid gland. It is a large oval gland with a lobulated texture, but the individual lobules are larger than those of the parotid gland. Do not confuse it with the smaller, smoother-textured **lymph nodes** of variable sizes in this region. The **mandibular duct** emerges from the front of the gland and passes forward, first going lateral to the digastric muscle. The digastric muscle is the large muscle arising from the base of the skull and inserting along the ventral border of the lower jaw (p. 205). Bisect (if you have not done so already, see p. 206) and reflect the digastric muscle; you will be able to see that the mandibular duct then passes deep to the caudal border of the mylohyoid— the thin transverse sheet of muscle lying between the paired digastric muscles. Bisect (if you have not done so already) and reflect the mylohyoid and follow the mandibular duct forward as far as you can. It is crossed rostrally by the **lingual nerve** returning sensory fibers from the tongue. The lingual nerve is considered to be a branch of the trigeminal nerve, because most of its fibers are general cutaneous fibers. Taste fibers in the lingual nerve separate and enter the skull with the facial nerve. The **hypoglossal nerve,** carrying motor fibers to the tongue musculature, lies caudal and dorsal to the mandibular duct. The mandibular ducts of opposite sides converge and enter the floor of the mouth by a pair of inconspicuous openings situated just rostral to the midventral septum of the tongue, the **lingual frenulum.** In the cat, these openings are hidden under flattened papillae, which can be lifted.

A small, elongated **sublingual gland** is located beside the mandibular duct. In the cat, the sublingual gland lies along the caudal one third of the mandibular duct and usually abuts against the mandibular gland. It lies along the rostral one third of the mandibular duct in the rabbit. The sublingual gland is drained by a minute duct that parallels the mandibular duct but is hard to distinguish grossly. The sublingual ducts enter the floor of the mouth but cannot be seen with the naked eye.

The parotid, mandibular, and sublingual glands are the most common of the salivary glands of mammals, but others are present in certain species. The cat and rabbit have a **zygomatic gland** that lies beneath the eyeball; you have already seen it if you dissected the eye (p. 225). In addition, the cat has a small, elongate **molar gland** situated between the skin and mucous membrane of the caudal half of the lower lip. Several small ducts that cannot be seen grossly lead from it to the inside of the lip. The rabbit has elongated **buccal glands** beneath the skin of the upper and lower lips.

◆ **(B) THE ORAL CAVITY**

Because it is not possible to open the jaws in preserved specimens, open the **oral cavity** by cutting through the floor of the mouth with a scalpel. Do this from the external surface, cut on each side, and keep as close to the mandible and chin as possible. Then cut through the symphysis of the mandible with bone scissors (if you have not done so already during the dissection of the tongue muscles, p. 206), spread the two halves of the lower jaw apart, and pull the tongue and the floor of the mouth ventrally through the gap, so that you can look into the oral cavity. The rostral part of the roof of the oral cavity is formed by the bony **hard palate,** the caudal part by the fleshy **soft palate** (Figs. 10-17 and 10-18). The epithelium covering the hard palate is heavily cornified and forms transversal palatine

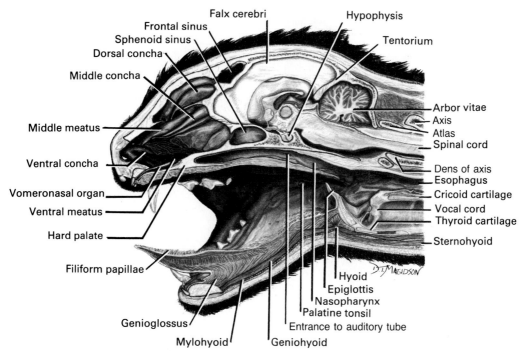

Figure 10-17
Sagittal section of the head of the cat.

ridges **(rugae)** and, in the cat, transversal rows of cornified papillae. The rugae and papillae point caudally and ensure that the food moves towards the pharynx as it is chewed and manipulated by the tongue. You will see a pair of small openings at the very front of the hard palate just caudal to the incisor teeth. These are the openings of the **incisive ducts.** These ducts pass through the palatine fissure to the vomeronasal (or Jacobson's) organs in the nasal cavities (p. 216). The lateral walls of the oral cavity are bounded by the teeth, lips, and cheeks. That portion of the cavity lying between the teeth and cheeks is called the **vestibule.** A well-developed, muscular **tongue** *(lingua)* lies in the floor of the cavity and is connected to the floor by the vertical lingual frenulum seen previously.

Pull the tongue far enough ventrally to tighten and bring into prominence a pair of lateral folds that extend from the sides of the caudal portion of the tongue to the soft palate. These folds constitute the **palatoglossal arches,** and they represent the boundary between the adult oral cavity and the **pharynx.** However, part of the back of the oral cavity defined as such probably develops from the embryonic pharynx. The passage between the palatoglossal arches is called the **isthmus faucium,** because it is the narrowest part of the pharynx. The chains of ossicles that connect the hyoid bone to the skull (p. 94) are imbedded within the soft tissues of the palatoglossal arches. Notice that the very back of the tongue lies within the pharynx.

Cut through the palatoglossal arch on one side, pull the tongue farther down, and examine its dorsal surface (Figs. 10-18 and 10-19). In the rabbit, but not in the cat, the back of the tongue is raised slightly, forming a **lingual torus.** This part has developed from the embryonic tuberculum impar (p. 314); the front of the tongue, which is divided by a **median lingual sulcus,** has developed from the paired lateral lingual swellings.

The dorsum of the tongue is covered with papillae, the most numerous of which are the pointed **filiform papillae.** The rostral filiform papillae of the cat bear spiny projections with which the animal grooms its fur or rasps flesh from bones, but the caudal

318

Chapter 10
The Coelom
and the Di-
gestive and
Respiratory
Systems

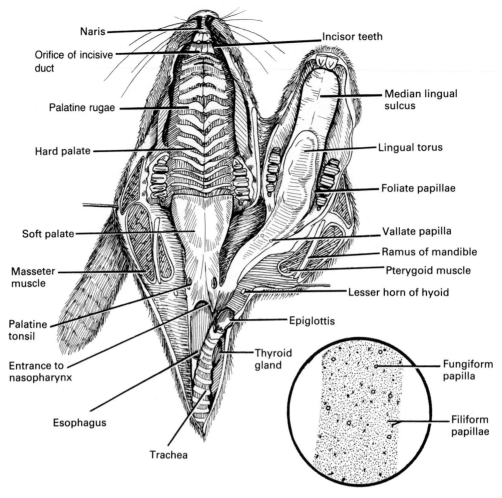

Figure 10-18
Oral cavity and pharynx of the rabbit. The floor of the mouth and pharynx has been swung open to the right.

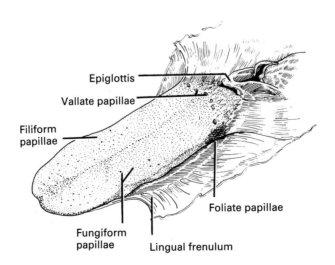

Figure 10-19
Dorsolateral view of the tongue of a cat. (*From Walker,* A Study of the Cat.)

filiform papillae are soft. Small, rounded **fungiform papillae** are interspersed among the filiform papillae, especially along the margins of the tongue. You may have to use low magnification to see them. **Vallate papillae** are located near the back of the tongue. Each vallate papilla is a relatively large, round patch set off from the rest of the tongue surface by a circular groove. There are two in the rabbit and four to six in the cat distributed in a caudally pointing, V-shaped line. Leafy **foliate papillae** can be found along the side of the caudal part of the tongue. Those of the rabbit are concentrated in a pair of patches. Microscopic taste buds are found on the sides and base of all the papillae, except for most of the filiform papillae.

♦ **(C) THE PHARYNX**

With a pair of scissors, cut caudally through the lateral wall of the pharynx on the side on which the palatoglossal arch was cut. Follow the contour of the tongue to the laryngeal region; then extend the cut dorsal to the larynx and back into the esophagus. Do not cut into the soft palate or larynx. Swing open the floor of the mouth and pharynx. The pharynx may be divided somewhat arbitrarily into oral, nasal, and laryngeal portions. The **oropharynx** lies between the palatoglossal arches and the free caudal margin of the soft palate. A pair of **palatine tonsils** lie in its laterodorsal walls. Note that each palatine tonsil is partially imbedded in a **tonsillar fossa**. The **laryngopharynx** is the space dorsal to the enlargement, the **larynx,** in the floor of the caudal part of the pharynx. It communicates caudally with the **esophagus** and ventrally with the larynx. The slitlike opening within the larynx is termed the **glottis**. A trough-shaped fold, the **epiglottis,** lies cranial to the glottis and helps to deflect food around or over the glottis. The **nasopharynx** lies dorsal to the soft palate. Open it by making a longitudinal incision through the middle of the soft palate. Spread open the incision as wide as possible and try to shine a light down into the nasopharynx. The pair of slitlike openings in its laterodorsal walls are the entrances of the **auditory tubes** (eustachian tubes) from the tympanic cavity (Fig. 10-17). Through them, air pressure on each side of the tympanum can be equalized. The choanae, or internal nostrils, enter the rostral end of the nasopharynx but cannot be seen in this view.

♦ **(D) THE LARYNX, TRACHEA, AND ESOPHAGUS**

Approach the laryngeal region from the ventral surface of the neck. You will have to remove several muscles, but do not injure any of the larger blood vessels. The **larynx** is the chamber whose walls are supported by relatively large cartilages. The **hyoid bone,** which forms a sling for anchoring the base of the tongue to the base of the skull (p. 94), is imbedded in the muscles cranial to the larynx, and its greater horn articulates with the front of the larynx. We have described its parts elsewhere (p. 94). The larynx is continued caudally as the windpipe, or **trachea,** whose walls are supported by a series of cartilaginous rings, which hold the tracheal lumen open, thus permitting the free movement of air. The **esophagus** is a collapsed, muscular tube lying dorsal to the trachea. Its lumen is pushed open as food is swallowed.

The dark **thyroid gland** lies against the cranial end of the trachea. In the cat it consists of **lobes,** one on either side of the trachea, that are connected across the ventral surface of the trachea by a very narrow band of thyroid tissue called the **isthmus.** The isthmus is frequently destroyed. The thyroid of the rabbit differs in having a wide, prominent isthmus. Two pairs of **parathyroid glands** are imbedded in the dorsomedial surface of the thyroid, but they cannot be distinguished grossly.

Return to the larynx and study it more thoroughly. Much of it is covered by **intrinsic laryngeal muscles** derived from the caudal branchiomeric musculature and hence innervated by the vagus nerve. Strip off these muscles from all surfaces of the larynx to expose the laryngeal cartilages. You may identify the muscles as you remove them by referring to Figure 10-20. They are named according to the cartilages between which they extend. The large cranial cartilage that forms much of the ventral and lateral walls of the larynx is called the **thyroid cartilage.** It is this cartilage that forms the projection known as the Adam's apple in the neck of men. The ring caudal to the thyroid cartilage is the **cricoid cartilage.** The cricoid cartilage is shaped like a signet ring, for its dorsal portion is greatly expanded and forms most of the dorsal wall of the larynx. Careful dissection will reveal a pair of small, triangular cartilages cranial to the dorsal part of the cricoid. These are the **arytenoid cartilages.** Additional minute cartilages are frequently associated with the arytenoid cartilages but are seldom seen. An **epiglottic cartilage** supports the epiglottis.

Cut open the larynx along its middorsal line. The pair of whitish, lateral folds that extend from the arytenoid cartilages to the thyroid cartilage are the **vocal cords.** They are set in vibration by the movement of air across them, and their tension is controlled by the movement of the arytenoid cartilages, the action of muscles within them, and slight changes in the shape of the larynx. The **glottis** is the space between them. In the cat, an accessory pair of folds, sometimes called the **false vocal cords,** extend from the arytenoid cartilages to the base of the epiglottis. In the rabbit, a small pair of bumps, the **epiglottic hamuli,** lie at the base of the epiglottis.

> Most of the laryngeal cartilages develop from certain of the visceral arches, but there is some doubt as to the precise homologies. The arytenoid and cricoid cartilages are the first to appear phylogenetically, being represented by the lateral cartilages of the laryngotracheal chamber of amphibians (p. 313). They appear to develop from the sixth or from the sixth and seventh visceral arches. The fourth and fifth visceral arches are incorporated in the hyoid apparatus in amphibians. In mammals, the hyoid apparatus involves only the second and third visceral arches, and in most mammals the fourth and fifth arches form the newly evolved thyroid cartilage. Some anatomists believe that the tracheal rings may have evolved from the splitting and multiplication of the seventh visceral arch, but this is doubtful. The epiglottic cartilage is apparently a new structure.

The Thorax and Its Contents

♦ (A) THE PLEURAL CAVITIES

Open the thorax by making a longitudinal incision about 2 centimeters to the right of the midventral line of the cat or rabbit and by extending it the length of the sternum. Use a pair

320

Chapter 10
The Coelom
and the Di-
gestive and
Respiratory
Systems

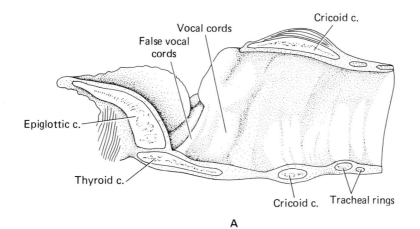

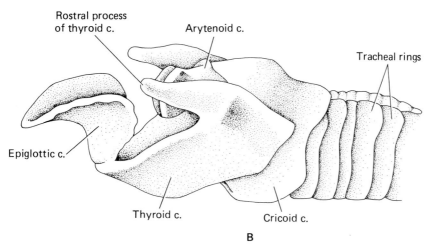

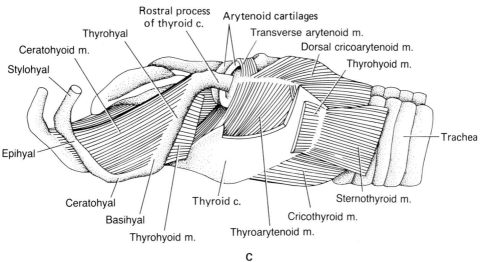

Figure 10-20
The larynx of the cat. **A**, Longitudinal section; **B**, lateral view of the laryngeal and tracheal cartilages
(the paired rostral processes of the thyroid cartilage articulate with the thyrohyoid bones of the
greater horns of the hyoid); **C**, lateral view of the laryngeal muscles (parts of the thyroid cartilage
and thyrohyoid muscle have been cut away to reveal the underlying thyroarytenoideus muscle).
(Redrawn from R. Nickel, A. Schummer, and S. Seiferle. The Viscera of the Domestic Mammals, 2nd ed. *Berlin,
Verlag Paul Parey, 1979.)*

of strong scissors. Spread open the incision and look into the **right pleural cavity.** You will see a dome-shaped transverse, muscular partition (the **diaphragm**) at the caudal end of the cavity. Just cranial to the diaphragm, make another cut that extends laterally and dorsally to the back. Follow the line of attachment of the diaphragm but keep on the pleural side. Feel for the individual ribs along the inner surface of the thoracic wall. With a strong pair of scissors, cut through each rib near its attachment to the vertebrae, but leave the ribs in place. In this way, you can open the right pleural cavity by spreading the thoracic wall laterally. (If you simply do this with force without cutting through the ribs first, the ribs will break and splinter and, subsequently, may damage the thoracic structures and your fingers.)

322

Chapter 10
The Coelom
and the Di-
gestive and
Respiratory
Systems

The right pleural cavity and its **lung** *(pulmo)* are now well exposed. The coelomic epithelium lining the walls of the pleural cavity is called the **parietal pleura;** that covering the surface of the lung is called the **visceral** *(pulmonary)* **pleura.** The right lung is divided into four **lobes—cranial, middle, caudal,** and **accessory** (Fig. 10-21). The accessory lobe extends dorsal to a large vein, the **caudal vena cava,** and then ventrally into a pocket on the medial side of the mesentery—the **caval fold**—attaching to the ventral surface of the vein (Fig. 10-22). Tear the caval fold near the vein and you can see this lobe. The lobes of the lung are attached to the medial wall of the pleural cavity by a pleural fold known as the **pulmonary ligament** (Figs. 10-3C and 10-22). The **pulmonary arteries** and **veins** connected with the heart and the **bronchi** from the trachea to the lung pass through part of this ligament (Fig. 10-22), but do not dissect them until you have studied and removed the heart. These structures constitute the **root of the lung.** Cut into a part of the lung and notice that it is not an empty organ but a very spongy one. The numerous, thin-walled, terminal pockets of the respiratory passages **(alveoli),** in which gas exchange occurs, are not visible grossly.

The medial wall of each pleural cavity (right and left) is formed only of a layer of parietal pleura. The space, or potential space, between the medial walls of the two pleural cavities constitutes the **mediastinum.** This space, however, is filled with connective tissue and structures that lie between the two cavities. For example, the pericardial cavity and heart, which form the large bulge medial and ventral to the lung, lie within the mediastinum (Fig. 10-3C). In places, the medial walls of the pleural cavities meet and form a mesentery-like structure termed the **mediastinal septum.** The mediastinal septum can be seen caudal to the heart and medial to the accessory lobe of the right lung. The caval fold is an evagination from this portion of the septum. In the cat, the mediastinal septum continues cranially ventral to the heart, but the medial walls of the pleural cavities do not meet in this region in the rabbit.

For the time being, leave the mediastinal septum attached to the ventral thoracic wall intact. Make a longitudinal incision about 2 centimeters from the midventral line on the left side of the cat or rabbit and open the left pleural cavity in the same way you opened the right pleural cavity. Examine the **left pleural cavity.** (This will give a good view of the third muscle layer of the thorax, the transversus thoracis, mentioned on p. 200.) The left lung does not have an accessory lobe. In the rabbit, the middle lobe is not well demarcated from the cranial lobe.

Pull the left lung ventrally and examine the region dorsal to it. You can see a large artery—the **aorta**—and the **esophagus** passing through the dorsal portion of the

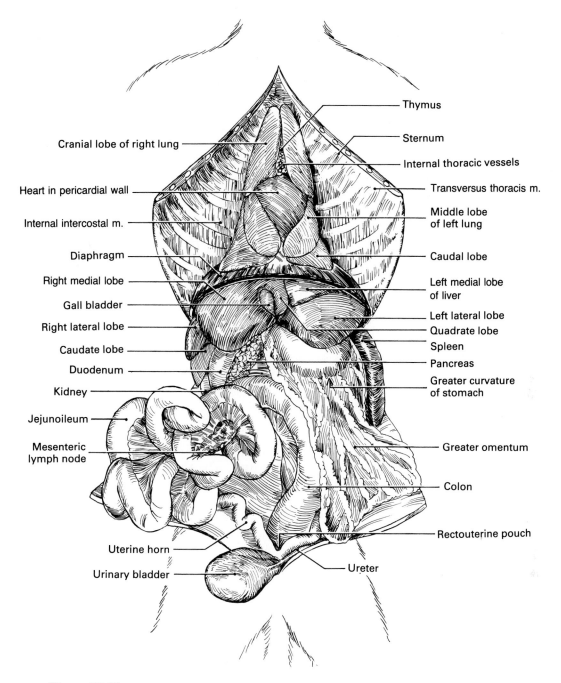

Cranial lobe of right lung

Heart in pericardial wall

Internal intercostal m.

Diaphragm

Right medial lobe

Gall bladder

Right lateral lobe

Caudate lobe

Duodenum

Kidney

Jejunoileum

Mesenteric
lymph node

Uterine horn

Urinary bladder

Thymus

Sternum

Internal thoracic vessels

Transversus thoracis m.

Middle lobe
of left lung

Caudal lobe

Left medial lobe
of liver

Left lateral lobe

Quadrate lobe

Spleen

Pancreas

Greater curvature
of stomach

Greater omentum

Colon

Rectouterine pouch

Ureter

Figure 10-21
Ventral view of the major organs of the thoracic and peritoneal cavities of a cat. *(From Walker,* A Study
of the Cat.*)*

mediastinum. The aorta lies to the left of the vertebral column; the esophagus lies more
ventrally. Let the lungs fall back into place. A pair of white strands, the **phrenic nerves,**
can be seen in the central portion of the mediastinum on each side of the pericardial cavity
and heart. They lie ventral to the roots of the lung and pass caudad to supply the
diaphragm. The right phrenic nerve follows the caudal vena cava closely; the left phrenic
nerve passes through the caudal part of the mediastinal septum (Fig. 10-22). The fact that
these nerves originate from the ventral rami of some combination of the fourth, fifth, and
sixth cervical nerves indicates that the diaphragmatic muscles derive from cervical muscle

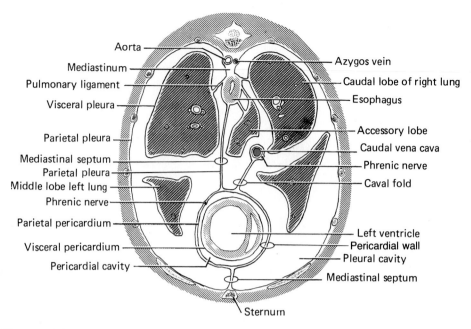

Aorta
Mediastinum
Pulmonary ligament
Visceral pleura
Parietal pleura
Mediastinal septum
Parietal pleura
Middle lobe left lung
Phrenic nerve
Parietal pericardium
Visceral pericardium
Pericardial cavity

Azygos vein
Caudal lobe of right lung
Esophagus
Accessory lobe
Caudal vena cava
Phrenic nerve
Caval fold
Left ventricle
Pericardial wall
Pleural cavity
Mediastinal septum

Sternum

Figure 10-22
A diagrammatic transverse section through the thorax of the cat at the level of the ventricles of the heart to show the serosa and its relation to the thoracic viscera. The section is viewed from behind so that left and right sides of the animal and drawing coincide.

precursors (p. 291). The portion of the mediastinum ventral and cranial to the heart is occupied by the dark, irregularly lobulated **thymus.** The thymus varies considerably in size, being best developed in young individuals.

◆   **(B) THE PERICARDIAL CAVITY**

The **pericardial cavity** and **heart** are the largest structures within the mediastinum. In order to open the pericardial cavity, first break the mediastinal septum. Then cut across the still intact midventral part of the thoracic wall by extending the incisions that you made along the attachment of the diaphragm to open the left and right pleural cavities. You can now reflect the midventral part of the thoracic wall cranially and expose the pericardium. (In this way, you will preserve the internal thoracic blood vessels for future study.)

Cut open the pericardial cavity by a midventral incision. Its saclike wall, the **pericardial wall**[24], is formed of three layers: an outer layer of parietal pleura continuing from the mediastinal septum, a middle layer of fibrous connective tissue, and an inner layer of **parietal pericardium** (see also Fig. 10-3C and p. 291). A **visceral pericardium** covers the surface of the heart. The parietal and visceral pericardia are continuous with each other over the large blood vessels at the front of the heart.

---

[24]See footnote 20, p. 291.

The Peritoneal Cavity and Its Contents

♦   **(A) THE BODY WALL AND PERITONEAL CAVITY**

Make a longitudinal incision through the abdominal wall slightly to the right of the midventral line. Extend the cut from the diaphragm to the pelvic girdle. Then cut laterally and dorsally along the attachment of the diaphragm to the body wall. Do this on both sides, thereby freeing the diaphragm as far as the back. Reflect the flaps of the abdominal wall. Notice that, apart from the skin, you have cut through the external fascia of the trunk (the fascia trunci), three layers of muscle or their aponeuroses (external oblique, internal oblique, and transversus abdominis), possibly the rectus abdominis near the midventral line, the internal fascia of the trunk (the fascia transversalis), and the parietal peritoneum. The portion of the coelom exposed is the **peritoneal cavity.** Its walls are lined with **parietal peritoneum,** and its viscera are covered with **visceral peritoneum.** Wash out the cavity if necessary.

♦   **(B) THE ABDOMINAL VISCERA AND MESENTERIES**

The concave surface of the dome-shaped diaphragm forms the cranial wall of the peritoneal cavity, and the large **liver** *(hepar)* lies just caudal to it and is shaped to fit into the dome. Pull the liver and diaphragm apart. You will now see that the central portion of the diaphragm is formed by the **central tendon,** onto which its muscle fibers insert. A vertical **falciform ligament** extends between the diaphragm and the liver and ventral abdominal wall. Sometimes a thickening, which represents a vestige of the embryonic umbilical vein, can be seen in its free edge. It is known as the **round ligament of the liver.** The diaphragm and liver are closely apposed dorsal to the falciform ligament, and the peritoneum bridging the gap constitutes the **coronary ligament.**

The liver can be divided into right and left halves at the cleft into which the falciform ligament passes. Each half is divided into a lateral and a medial lobe, thus making **left lateral, left medial, right medial,** and **right lateral lobes** (Figs. 10-21 and 10-23). A small **quadrate lobe** is interposed between the left medial and right medial lobes. It is partly united with the latter, being separated from it by the **gall bladder** *(vesica fellea).* A **caudate lobe** lies caudal to the right lateral lobe and abuts the right kidney. In rabbits it is usually united with the right lateral lobe. Part of the caudate lobe extends toward the left side of the body going deep to a mesentery, the **lesser omentum,** which extends from the liver to the stomach and duodenum (see Fig. 10-21). A **hepatorenal ligament** extends from the caudate lobe of the liver to the parietal peritoneum near the right **kidney.** The left kidney lies in a slightly more caudal position on the opposite side of the body.

As in the majority of vertebrates, most of the **stomach** *(ventriculus)* lies on the left side of the peritoneal cavity and is more or less J-shaped. Cut through the left side of the diaphragm to find the point at which the **esophagus** enters. There is an abrupt change in

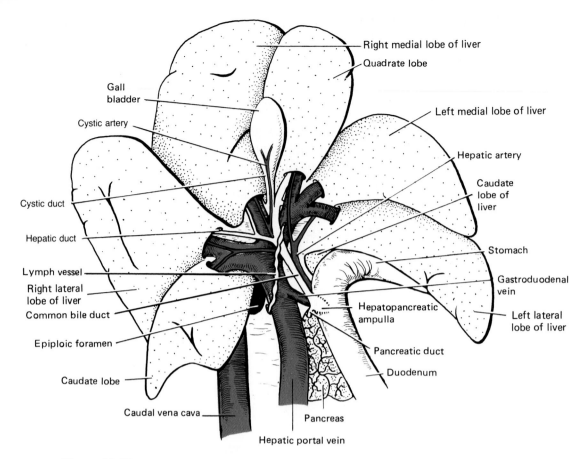

Right medial lobe of liver
Quadrate lobe
Gall bladder
Cystic artery
Left medial lobe of liver
Hepatic artery
Caudate lobe of liver
Cystic duct
Hepatic duct
Stomach
Lymph vessel
Gastroduodenal vein
Right lateral lobe of liver
Common bile duct
Left lateral lobe of liver
Epiploic foramen
Hepatopancreatic ampulla
Pancreatic duct
Caudate lobe
Duodenum
Caudal vena cava
Pancreas
Hepatic portal vein

Figure 10-23
Ventral view of the blood and lymphatic vessels and ducts of the liver in the lesser omentum in the cat. The right median, quadrate, and part of the right lateral lobes of the liver have been turned forward.

the diameter of the digestive tract at this point. The portion of the stomach adjacent to the esophagus is the **cardiac region;** the dome-shaped portion extending cranially to the left of the cardiac region is the **fundus;** the main part of the stomach is the **body;** and the narrow caudal portion is the **pyloric region.** But, as pointed out elsewhere, these gross regions do not necessarily correspond with glandular regions bearing the same names. The stomach ends in a thick muscular sphincter, the **pylorus,** which you can see if you cut open the stomach. Do not injure any mesenteries. In the cat, you will also see longitudinal ridges in the lining that are called **rugae.** The long left and caudal margin of the stomach, which represents its embryonic dorsal surface, constitutes its **greater curvature;** the shorter right and cranial margin, which represents its embryonic ventral surface, constitutes the **lesser curvature.**

Notice that the lesser omentum, which represents a part of the ventral mesentery, attaches along the lesser curvature of the stomach. The mesentery that attaches along the greater curvature is the **greater omentum,** or **mesogaster**—a part of the dorsal mesentery (Fig. 10-24). The greater omentum of mammals does not extend directly to the middorsal line of the peritoneal cavity, as it does in other vertebrates, but is modified to form a saclike structure called the **omental bursa,** which drapes down over the intestines and lies directly under the ventral abdominal wall (Figs. 10-21 and 10-24**E**). The omental bursa of the cat is very large, contains considerable fat in its wall, and extends over the intestine nearly to the pelvic region. It is often entwined with the intestines and must be untangled carefully. The rabbit's bursa is much smaller. In both animals, the greater omentum extends caudad from its attachment along the greater curvature of the stomach as the descending **ventral sheet** (or superficial sheet) and then turns upon itself and extends craniad as the ascending **dorsal sheet** (or deep sheet) of the greater omentum (Fig. 10-24**E**). After incorporating the pancreas, the greater omentum continues dorsad to attach to the dorsal wall of the peritoneal cavity. The **spleen** *(lien),* which is enormous in the cat, lies within the ventral sheet of the greater omentum on the left side of the stomach (Figs. 10-21 and 10-24**D** and **E**). The portion of the greater omentum between the stomach and spleen is known as the **gastrolienic ligament.** The duodenum, which arises from the pylorus, lies within the dorsal sheet of the greater omentum on the right side of the stomach. A small, triangular mesentery, the **gastrocolic ligament,** passes from the part of the greater omentum lying dorsal to the spleen over to the mesentery of the large intestine.

The space enclosed between the ventral and dorsal sheets of the greater omentum, though compressed, is also known as the **lesser peritoneal cavity,** because it communicates with the rest of the peritoneal cavity through the **epiploic foramen** (Figs. 10-23 and 10-24**D**). Try to grasp only one sheet of the greater omentum between the tips of a pair of forceps and cut open the omental bursa. If you do not succeed, simply tear the omental bursa and verify that it contains a space. The epiploic foramen lies dorsal to the lesser omentum and between the caudate lobe of the liver and the mesentery of the duodenum. You can pass a probe through the epiploic foramen and extend it dorsal to the stomach and into the omental bursa.

The topography of the greater omentum and other mesenteries in the cranial part of the peritoneal cavity is one of the more challenging parts of mammalian anatomy. The best way to understand the adult condition is to recapitulate the complexities of embryonic development. At a relatively early stage, the digestive tube is suspended along the middle of the coelom by the dorsal mesentery (Figs. 10-24A to C). The spleen develops within the dorsal mesentery dorsally to the stomach. The duodenal part of the intestine gives rise to the pancreas, which grows into the dorsal mesentery, and to the liver, which grows into the ventral mesentery.

Subsequently, the stomach embarks on a complex rotational movement about its longitudinal and vertical axes. For didactic purposes, we will consider the rotation about each axis separately, although the rotational movement of the stomach is a composite of both. First, the stomach rotates counterclockwise about its longitudinal axis, so that its originally dorsal side (i.e., its greater curvature) becomes now its left side (Fig. 10-24D). Simultaneously, the spleen is pulled into the left side of the peritoneal cavity, while the dorsal mesentery suspending it becomes disproportionately elongated and is called the greater omentum. The more caudal part of the dorsal mesentery, which suspends the duodenum, however, does not elongate to the same extent as the greater omentum (Figs. 10-24A and C). As a result, a pouch, the omental bursa, is formed as the greater omentum continues to elongate and expands caudad. The second rotation of the stomach proceeds counterclockwise about its vertical axis (as seen from dorsal), so that the greater curvature becomes the caudal side of the stomach, and the originally caudally located pylorus and entrance to the duodenum move into the right side of the peritoneal cavity (Figs. 10-21 and 10-26).

At first, there is a wide communication between the omental bursa and the rest of the peritoneal cavity (Fig. 10-24D), but subsequent adhesions of the liver to the dorsal part of the diaphragm and adjacent body wall reduce this to a relatively small epiploic foramen (Fig. 10-23).

328

Chapter 10
The Coelom
and the Di-
gestive and
Respiratory
Systems

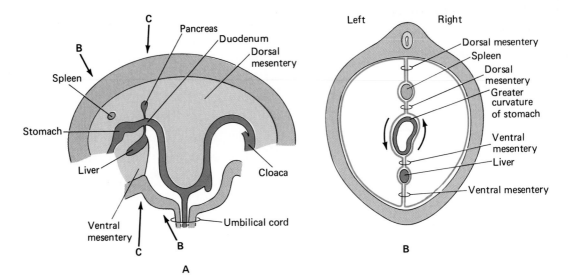

A

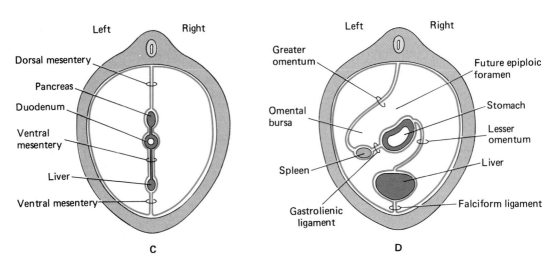

C

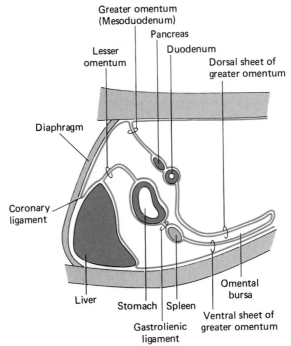

E

Carefully dissect that portion of the lesser omentum lying near the caudate lobe of the liver and the epiploic foramen in order to expose the system of bile ducts extending from the liver and gall bladder to the beginning of the duodenum. Some lymphatic vessels, which look like chains of small nodules, may have to be removed. A **cystic duct** comes down from the gall bladder and unites with several **hepatic ducts** from various parts of the liver to form a **common bile duct** *(ductus choledochus),* which passes to the duodenum. One particularly prominent hepatic duct comes in from the left lobes of the liver and another from the right lateral lobe.

The **small intestine** extends caudad from the stomach, passes through numerous convolutions, and eventually enters a **large intestine** (Fig. 10-25). Certain features of the intestinal region and the pancreas differ sufficiently in the cat and the rabbit to warrant separate treatment, but a few common features may be observed at this time. The small intestine of mammals has differentiated into a cranial **duodenum** and a more caudal **jejunum** and **ileum.** The duodenum, as will be seen presently, is the first, approximately U-shaped loop of the intestine, but there is no sharp transition between the jejunum and ileum. About all that can be said with respect to their gross anatomy is that the jejunum is the cranial portion of the postduodenal small intestine and the ileum the caudal portion. The small intestine is suspended by a part of the dorsal mesentery—that portion passing to the duodenum being the **mesoduodenum,** and that portion passing to the jejunum and ileum being the **mesentery** (in the strict sense).

Cut open a part of the small intestine and examine its lumen. The lining has a velvety appearance that results from the presence of numerous minute, finger-like projections called **villi.** These greatly increase the internal surface area. Parasitic roundworms and tapeworms are often found in the intestine of the cat.

The large intestine of mammals is much longer than that of other vertebrates and generally has a greater diameter than the small intestine. Most of it consists of the **colon,** and it is suspended by a portion of the dorsal mesentery termed the **mesocolon.** Skip the pattern that the cranial part of the colon assumes for a moment and examine the caudal part. This portion extends caudad against the dorsal wall of the peritoneal cavity and enters the pelvic canal (Fig. 10-21). This portion of the colon also lies dorsal to the pear-shaped **urinary bladder** and, if your specimen is a female, to the Y-shaped **uterus.** Notice that the urinary bladder is anchored by a vertical **median vesical ligament,**

Figure 10-24

Diagrams illustrating the embryonic development of the greater omentum and omental bursa. **A,** Longitudinal section through the middle part of an embryo, showing the position of the spleen and pancreas within the dorsal mesentery and of the liver within the ventral mesentery. (The arrows indicate the level of the cross sections shown in **B** and **C.**) **B,** Cross section through an embryo at the level of the spleen and stomach. (The arrows indicate the direction of the stomach rotation shown in **D.**) **C,** Cross section through an embryo at the level of the pancreas and duodenum. **D,** Cross section through an embryo after completed rotation of the stomach and at the beginning of the formation of the omental bursa. **E,** Longitudinal section through the cranial part of the peritoneal cavity, as seen from the left side, to show the final structural relationships of the greater omentum with the stomach, spleen, duodenum, and pancreas. (Note that in reality the spleen is found in the left side of the peritoneal cavity and the duodenum in the right side, and that, therefore, they would not be seen in the same longitudinal section.) *(A, Adapted from R. Nickel, A. Schummer, and S. Seiferle, The Viscera of the Domestic Mammals, 2nd ed. Berlin, Verlag Paul Parey, 1979; B and D, adapted from Nickel, Schummer, and Seiferle, 1979, and from K. M. Dyce, W. Sack, and C. J. G. Wensing, Textbook of Veterinary Anatomy, Philadelphia, W. B. Saunders Company, 1987.)*

which, being a part of the ventral mesentery, extends to the midventral body wall, and by a pair of **lateral vesical ligaments.** The latter often contain wads of fat. Cut open the caudal part of the colon, clean it out, and note that it lacks villi. Also note the extension of the coelom into the pelvic canal. That portion of the coelom in the male that extends caudally between the large intestine and the urinary bladder is called the **rectovesical pouch.** The comparable coelomic extension in the female is divided by the uterus into a shallow **vesicogenital pouch** between the urinary bladder and uterus and a deep **rectouterine pouch** between the uterus and large intestine.

Deep within the pelvic canal, the colon passes into the terminal segment of the large intestine, the **rectum,** which in turn opens on the body surface through the **anus.** You will see the rectal region later when you open the pelvic canal.

330

Chapter 10
The Coelom
and the Di-
gestive and
Respiratory
Systems

The numerous convolutions of the intestine are very confusing at first view. It is somewhat easier to learn the different parts of the intestine and their arrangement within the peritoneal cavity if you consider the embryonic development of the digestive tract. As was already mentioned, the primitive gut is at first straight and is suspended along the middle of the coelom by the dorsal mesentery. As it continues to grow, it elongates at a faster rate than the embryo itself and forms a ventral loop (Fig. 10-25**A**). Subsequently, the intestine is thrown into additional loops by rotating

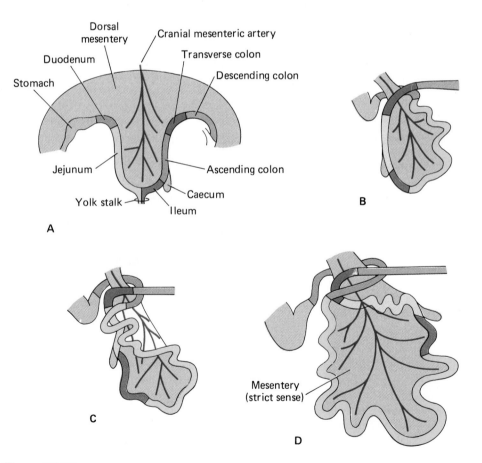

Figure 10-25
Formation of the intestinal loops during embryonic development in a mammal, as seen from the left side of the embryo. **A,** Early stage with a straight primitive gut; **B,** first rotation of the intestine around the cranial mesenteric artery; **C,** beginning of the second rotation of the intestine; **D,** final stage of the loop formation of the intestine. *(Redrawn from R. Nickel, A. Schummer, and S. Seiferle,* The Viscera of the Domestic Mammals, *2nd ed., Berlin, Verlag Paul Parey, 1979, after Zietsc hmann, 1955.)*

clockwise (as seen from dorsal) about the part of the mesentery that contains the cranial mesenteric artery (Figs. 10-25**B** to **D**). This coiling process of the intestine results in a compact, orderly packaging of the very long intestine within the relatively short peritoneal cavity. (Remember the cochlea of the mammalian ear, which is the result of a similar "packaging problem" for the greatly elongated lagena, see p. 234).

♦   **(C) FURTHER STRUCTURE OF THE DIGESTIVE ORGANS**

Keeping these general features in mind, examine the intestine and pancreas in more detail.

Cat   The intestine and pancreas of the cat are relatively simple. The **duodenum** curves caudad from the pylorus on the right side of the body; then it bends toward the left and ascends toward the stomach (Figs. 10-25**D** and 10-26**A**). It is arbitrarily considered to end at its next major bend. A small, triangular peritoneal fold, the **duodenocolic fold,** extends between the caudal end of the duodenum and the mesocolon.

You will recognize the **pancreas** by its lobulated texture. Its head lies against the descending portion of the duodenum, and its tail extends transversely across the body to the spleen. This portion lies in the dorsal sheet of the omental bursa (Fig. 10-24**E**). Two pancreatic ducts are present, for the ducts of both the dorsal and the ventral primordia are retained. The main **pancreatic duct,** which is the duct of the ventral primordium, unites with the common bile duct as the latter enters the duodenum. You can find it by carefully picking away pancreatic tissue in this region. The enlargement on the duodenum where

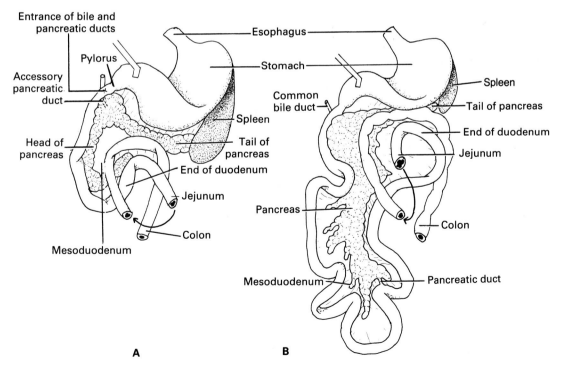

Figure 10-26
Ventral views of the duodenum, pancreas, and adjacent organs of **A**, a carnivore (mink), and **B**, a rabbit. The pyloric region of the stomach has been pulled forward.

the two ducts unite is known as the **hepatopancreatic ampulla** (Fig. 10-23). An **accessory pancreatic duct,** which is the duct of the dorsal pancreas, enters the duodenum about 1 centimeter caudal to the main duct, but it is small and very hard to find.

Follow the coils of the rest of the small intestine (jejunum and ileum) until it enters the colon (Fig. 10-27). In the cat, a short, blind diverticulum, the **caecum,** extends caudally from the beginning of the colon. The vermiform appendix of human beings is located at the end of the caecum, but an appendix is absent in the cat. Cut open the wall of the caecum and colon opposite the entrance of the ileum. Notice how the ileum projects slightly into the lumen of the colon, forming an **ileal papilla,** which helps prevent the backing up of colic material into the small intestine. The colon itself extends forward on the right side of the body for a short distance as the **ascending colon,** crosses to the left side as the **transverse colon,** and extends back into the pelvic canal as the **descending colon** to the rectum and anus. Because of rotations that occur during embryonic development, the colon loops around the beginning of the jejunum (Figs. 10-25 and 10-26A).

Rabbit The **duodenum** of the rabbit extends caudally in small convolutions nearly to the pelvic region (Fig. 10-26B). It then twists to the left and, continuing to convolute, ascends nearly to the stomach. It is arbitrarily considered to end where it next turns caudad. The duodenum is approximately 50 centimeters long in an adult rabbit.

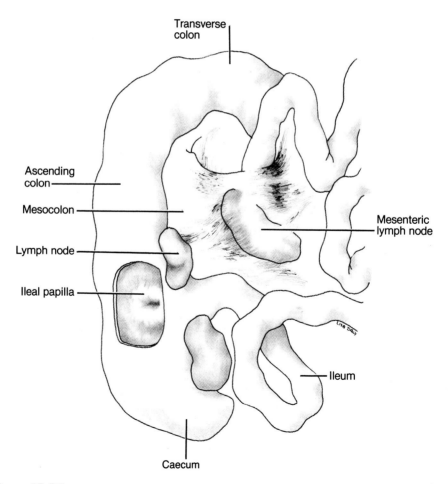

Figure 10-27
Ventral view of the ileocolic junction and caecum of the cat. The colon has been cut open to show the ileal orifice and papilla. *(From Walker, A Study of the Cat.)*

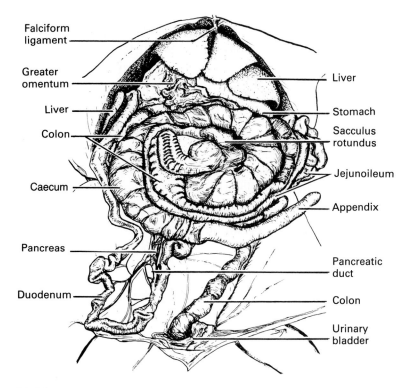

Falciform ligament

Greater omentum

Liver

Colon

Caecum

Pancreas

Duodenum

Liver

Stomach

Sacculus rotundus

Jejunoileum

Appendix

Pancreatic duct

Colon

Urinary bladder

Figure 10-28
Ventral view of the abdominal viscera of the rabbit, showing in particular the large caecum and appendix characteristic of many herbivores.

Stretch out the mesentery (mesoduodenum) that lies between the descending and ascending limbs of the duodenum. The dark, lobulated tissue that you see is the **pancreas.** It is very thin and diffuse in the rabbit. The single **pancreatic duct,** which is the duct of the embryonic dorsal pancreas, enters the ascending limb about 5 centimeters cranial to the most caudal loop of the duodenum.

Follow the coils of the rest of the small intestine (jejunum and ileum) until it enters the large intestine. The caudal end of the ileum is modified to form a round, muscular enlargement known as the **sacculus rotundus** (Fig. 10-28). Cut open the large intestine opposite the entrance of the ileum, clean out its lumen, and find the orifice of the ileum. It is on an **ileal papilla,** which prevents material in the large intestine from backing up into the small intestine.

A large caecum extends in one direction from the sacculus rotundus; the colon, which can be distinguished by its more wrinkled appearance, extends in the other direction. First follow the **caecum.** It is a wide, thin-walled, blind sac that extends for about 35 centimeters in a circular course and ends in a thicker-walled **vermiform appendix.** The appendix is about 12 centimeters long. Cut open the caecum and note that the spiral line that can be seen on its surface marks the point of attachment of a **spiral valve.** The caecum contains a colony of bacteria and other microorganisms, some of which produce the enzyme cellulase, which breaks down cellulose.[25] Fermentation occurs here, and many of the microorganisms, which reproduce rapidly, are themselves digested. Many of the organic acids that result from cecal digestion are absorbed here. Rabbits also engage in

---

[25]Metazoan animals cannot digest cellulose (also called "fiber" in everyday language), because they do not produce cellulase.

coprophagy; that is, much of the cecal contents, which are discharged as soft, mucus-covered fecal pellets, are eaten. This material is then further acted upon and absorbed in the stomach and small intestine. Cut open the appendix and note the relative thickness of its wall. Large amounts of lymphoid tissue accumulate in the wall of the appendix and may help protect the body against the effects of bacterial toxins.

Return to the **colon** and follow it. You may have to tear parts of the mesentery. The longitudinal muscle layer of the first part of the colon is limited to two or three bands, called **taeniae coli,** one of which lies along the line of attachment of the mesocolon. The wall of the colon between the taeniae protrudes to form little sacculations called **haustra coli.** More caudally, the wall of the colon is smooth, bulging only where there are pellet-shaped feces within it. After a very circuitous course, the colon descends into the pelvic canal to the rectum and anus.

The Ruminant Stomach If you have obtained the stomach of the sheep from a biological supply house, you can study an alternative method for digesting cellulose. The stomach of a sheep represents that of cud-chewing mammals, or ruminants, a group that also includes goats, antelopes, and cattle. The fundus region of the stomach of mammals with a generalized stomach has differentiated into three chambers: rumen, reticulum, and omasum (Fig. 10-29). These are lined with a stratified squamous epithelium and lack gastric glands. Only the fourth chamber, the abomasum, has the characteristic simple gastric epithelium and secretes gastric juice.

When grass or other cellulose food is ingested, it is not chewed but enters the **rumen,** where it is stored temporarily. This is by far the largest of the chambers and may have a volume of 13 liters in a large sheep. Periodically, portions of the rumen contents are regurgitated, masticated, and then returned to the rumen. As the food is being broken

334

Chapter 10
The Coelom
and the Di-
gestive and
Respiratory
Systems

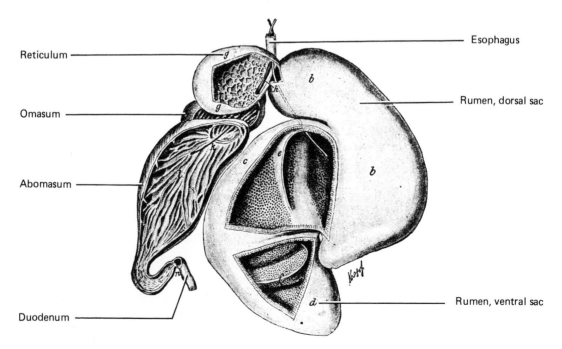

Figure 10-29
The stomach of a sheep cut open to show the characteristic internal folds of the four chambers. *(After Wiedersheim.)*

down mechanically, it is also fermented chemically by cellulase-producing bacteria that live and multiply in the anaerobic environment of the rumen.[26] Cellulose is gradually reduced to organic acids and, especially, to short-chain fatty acids such as acetic acid. Many of these are absorbed directly from the rumen through the large surface area provided by thousands of small, "leaf-shaped" papillae.

The second chamber, the **reticulum,** has a broad connection with the rumen. Its reticulated lining helps to sort out fine particles, while rejecting larger ones. Fine particles of ingested food, together with a certain amount of harvested bacteria, pass through the reticulum and into the third chamber, the **omasum.** The omasum has deep, platelike folds that further filter the food. Contraction of this chamber propels the food into the final chamber, the **abomasum,** where protein digestion is initiated. Further digestion and absorption occur in the small intestine in the usual way.

The ruminant stomach has an advantage over the cecal system, because cellulose is broken down and bacteria are harvested before food enters the main digestive and absorptive regions of the digestive tract. Food need only pass through once, and ruminants normally do not engage in coprophagy. The large amount of material being fermented in the rumen, however, also produces considerable gas. Most is discharged through the mouth, but if a ruminant eats certain types of plants, more gas is generated than can be eliminated easily. The animal may then begin to bloat. Farmers keep a tube sharpened at one end with which they can pierce the abdominal and rumen walls if this occurs.

[26]See footnote 25, p. 333.

CHAPTER ELEVEN

♦♦♦♦♦♦♦♦♦♦♦♦♦♦♦♦♦♦♦♦♦♦♦♦♦♦♦♦♦♦♦

THE CIRCULATORY SYSTEM

Functions of the Circulatory System

Continuing with the organ systems that provide for the metabolic needs of the body, we will next consider the circulatory system. This system is primarily the great transport system of the body. Oxygen and food are carried from the respiratory and digestive organs to all the tissues and cells; carbon dioxide and other excretory products are carried from the tissues to sites of removal; hormones are transported from the endocrine glands to the tissues; and heat is distributed throughout the body. But the system also has other functions: It aids in combating disease and in repairing tissues, and it helps to maintain the constancy of the internal environment (homeostasis) in many ways.

Parts of the Circulatory System and the Course of the Circulation

The blood and lymph and their cells perform the functions of the circulatory system, but we will study only the vessels that propel and carry these fluids. These vessels may be classified in most vertebrates into (1) a **cardiovascular system,** consisting of the **heart, arteries, blood capillaries,** and **veins;** and (2) a **lymphatic system,** consisting of closed **lymphatic capillaries** and **lymphatic vessels.** In addition, the lymphatic system of mammals includes many **lymph nodes** located at strategic junctions along the course of the blood and lymphatic vessels. These nodes act as sites for the production of certain white blood cells called **lymphocytes,** some of which are transformed into antibody-synthesizing cells. The **spleen** is similar to a lymph node but is interposed in the cardiovascular system. It is, at different times and in different vertebrates, a site for the production, storage, and elimination of blood cells.

Briefly, the course of the circulation through these two systems of vessels is as follows. **Blood** leaves the heart and travels through the arteries to the capillaries in the tissues. At this point, some of the blood plasma leaves the capillaries to circulate among the cells as **interstitial fluid.** Much of the interstitial fluid reenters the capillary bed and returns to the heart through veins. However, in most vertebrates, some of the fluid enters the lymphatic capillaries and is carried as **lymph** by the lymphatic vessels to the larger veins. Jawless fishes and cartilaginous fishes lack a true lymphatic system; hydrostatic pressure in their veins is very low, and many of the veins are large sinuses. There is a good drainage of the tissues and return of any plasma proteins that left the cardiovascular capillaries. As vertebrates evolved an increased blood pressure, adequate tissue drainage would become a problem if it were not for the development of the lymphatic system. Lymphatic capillaries have a very low hydrostatic pressure and walls permeable enough to pick up any plasma proteins that escape from the cardiovascular capillaries and return them to the veins. One important function of the plasma proteins is to contribute to the blood's osmotic pressure, which draws water into the blood from the interstitial fluid and counterbalances the hydrostatic pressure that drives water out. A proper balance between these forces is necessary for normal water exchanges.

The **heart** is the major pump that develops the hydrostatic pressure that circulates the blood. By the time the blood reaches the veins, the pressure has become relatively low. Pressures created by the heart do not directly affect the lymphatic system; thus, the return of lymph and, to some extent, the return of blood are implemented by other forces. Among these, the contraction and tonus of the surrounding skeletal muscles play a major role. The veins and lymphatic vessels also contain **valves** that prevent a backflow of fluids.

The **lymphatic system** is rather obscure and not commonly seen in gross dissection. We will study parts of it in the mammal, but our emphasis throughout will be on the more conspicuous cardiovascular system.

Development of the Cardiovascular System

Early in the development of a fish embryo, a system of blood vessels is established that provides for the metabolic needs of the embryo. Tetrapod vertebrates have inherited this pattern of development, but it has been modified somewhat to fit their particular requirements. Since the early basic pattern is repeated to a large extent in all vertebrates during the ontogeny of their diverse adult patterns, it forms a necessary basis for understanding the adult cardiovascular system.

The first vessels to take definite form in the embryo of any vertebrate are a pair of **vitelline veins** (Fig. 11-1), which lie beneath the embryonic gut and carry blood and food from the yolk-laden archenteron, or from the yolk sac, if such a sac is present. The anterior parts of these vessels fuse to form the **heart** and **ventral aorta.** Caudal to the heart, they are engulfed and broken up into a capillary network by the enlarging liver. The portion of the vitelline veins lying caudal to the liver becomes the **hepatic portal system;** the portion from the liver to the heart becomes the **hepatic veins. Portal veins** are simply veins that, after draining one capillary

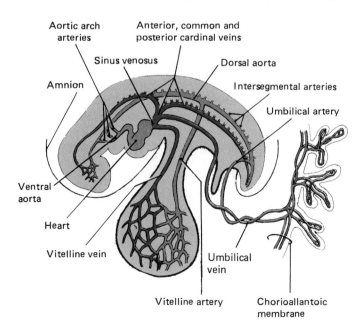

Figure 11-1
Diagrammatic lateral view of the major blood vessels of a 26-day-old human embryo. Most vessels are paired, and only those on the left side are shown. *(From K. L. Moore.)*

bed, pass to another capillary bed in a different organ. Veins going directly to the heart are called **systemic veins.**

Meanwhile, a series of six **aortic arches** develop in the first six branchial bars, which lie between the gill pouches, to carry blood from the heart and ventral aorta up to the **dorsal aortae.** The dorsal aortae, which are first paired but later fuse, carry the blood caudally and, by **vitelline arteries,** back to the archenteron and yolk sac. This completes one circuit.

It will be noted that the early circulation is largely a visceral circulation to and from the "inner tube" of the body, but a somatic circulation to the "outer tube" soon appears. Other branches from the dorsal aorta carry blood, by **umbilical** or **allantoic arteries,** into the body wall and out to an allantois if one is present. Blood from the dorsal portions of the body returns by way of **anterior** and **posterior cardinal veins.** The cardinal veins of each side unite anterior to the liver and turn ventrally to the heart as **common cardinal veins,** or ducts of Cuvier. Recall that the common cardinal veins pass through the dorsal part of the transverse septum, which they helped to form. Blood from the more lateral and ventral portions of the body wall, and from the allantois, returns by a pair of more ventrally situated vessels. These are called **lateral abdominal veins** in anamniotic vertebrates and **umbilical** or **allantoic veins** in amniotes. These vessels enter the base of the common cardinal veins in early embryos, but in the later embryos of the amniotic vertebrates they acquire a connection with the hepatic portal system and are drained through the liver.

As development proceeds and sites of nutrition, gas exchange, and excretion change, the pattern of vessel changes: New channels appear, and some old ones atrophy. Thus, in the development of an embryo—especially among amniotic vertebrates, which have had a complex phylogenetic history—we see a succession of vessels. Much of the variation seen in the vessels of the adult can be attributed to the persistence of embryonic channels that normally atrophy and to the failure of certain later channels to develop. Other variation results from the enlargement of one channel over another in a primordial capillary plexus that exists in many parts of the embryo (Fig. 11-2). The relative rate of blood flow, as well as hereditary and other factors, are important factors in determining which channels will enlarge.

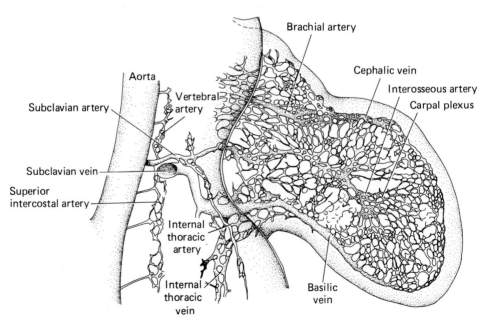

Figure 11-2
An early stage in the development of the pectoral limb bud of a pig embryo, to show the early plexus of vessels. The definitive vessels arise by the enlargement of certain channels. *(From Woollard.)*

The Study of Blood Vessels

While dissecting the vessels, you should be aware that they are subject to considerable variation and may not be exactly as described here. Since the veins of amniotic vertebrates have had a more complex ontogeny than the arteries, and since the venous blood has a more sluggish flow, it is to be expected that more variations will be found in the venous system of these animals than in the arterial system. Odd as it may at first seem, the peripheral parts of the vessels are subject to less variation than are certain of the more central and larger channels. For example, the left ovarian vein of a mammal always drains the ovary, but it may enter either the left renal vein or the caudal vena cava. If a vessel cannot be identified from its point of connection with a major vessel, it can be identified if its peripheral distribution can be established.

Another fact to bear in mind is that arteries and veins tend to follow each other, especially in amniotic vertebrates and at the periphery. For this reason, the arteries and veins of a given part of the body will often be described together. Valves frequently prevent the injection mass in veins from reaching the peripheral parts of the vessel, but the vein can usually be seen as a translucent, fluid-filled tube beside the corresponding artery.

Arteries can be recognized by their round cross sections and thick walls, veins by their flattened cross sections and thinner walls. It is important to keep in mind that arteries and veins are defined by the direction of their blood flow relative to the heart and not by the oxygen content of the blood flowing in them. Hence, **arteries** are blood vessels that conduct blood away from the heart. They may contain oxygen-rich blood (e.g., the aorta) or oxygen-depleted blood (e.g., the pulmonary artery, ventral aorta).[27] **Veins** are blood vessels that conduct blood toward the heart. They may contain oxygen-depleted blood (e.g., the cardinal veins) or oxygen-rich blood (e.g., the pulmonary veins). As already mentioned, portal veins are veins that conduct blood from one capillary bed to another (e.g., the hepatic, renal, and hypophyseal portal veins). A **sinus** is an inflated, saclike section of a vein (e.g., the anterior cardinal sinus). An **anastomosis** is a blood vessel that is larger than a capillary and connects two blood vessels, such as two arteries, two veins, or one artery and one vein.

The blood vessels of vertebrates are exceedingly numerous, and you would not be able to study them all in a course of this scope. We have emphasized the major channels in the axis of the body and the vessels connecting with them. The blood vessels of the head and the appendages and the more distal vessels in the intestinal region are treated superficially.

In order to learn the circulatory system, make your own chart of all the blood vessels mentioned in the dissection instructions for each animal. Start with the heart in the center of the chart and proceed by incorporating the arteries and veins that are mentioned in the text and figures. Finally, add capillary beds to connect the arteries and veins where appropriate. See Figure 11-3 as an example for such a chart, although yours will be much more complex, if complete, and need not look as polished in order to be useful.

FISHES

The pattern of the cardiovascular system in lungless fishes is shown diagrammatically in Figure 11-3. It is close to the pattern described for the early embryos of all vertebrates. A major determinant of the pattern in any vertebrate is the site of gas

[27]Oxygen-rich blood is sometimes also called "arterial" blood and, conversely, oxygen-depleted blood is called "venous" blood. This practice causes confusion, since it leads one to say that the pulmonary artery contains "venous" blood and the pulmonary vein "arterial" blood. The terms *arterial* and *venous* are therefore best avoided in this context.

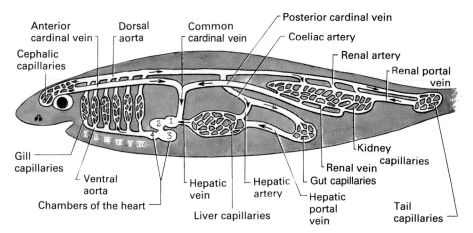

Figure 11-3

The major parts of the circulatory system of a primitive fish. *1,* Sinus venosus; *2,* atrium; *3,* ventricle; *4,* conus arteriosus of the heart. The aortic arches are numbered with Roman numerals. Only traces of the first aortic arch remain in the adults of most fishes. *(From Villee, Walker, and Barnes.)*

exchange—gills or lungs. The heart is located close to this site because it must develop enough pressure to distribute oxygen-rich blood to all tissues of the body. The fish heart is located far forward, just behind and beneath the gills. It receives only oxygen-depleted blood from the body and increases the hydrostatic pressure sufficiently to drive the blood through the gill capillaries, which are interposed in the aortic arches, and on to the body. The circulatory system of fishes is inherently a low-pressure system. The blood pressure is reduced almost immediately as it passes through the gill capillaries, since the sum of the cross-sectional areas of all the gill capillaries is much larger than the cross-sectional area of the ventral aorta. (For a given amount of flowing liquid, the hydrostatic pressure of the liquid is inversely correlated with the size of the cross-sectional area of the tube.) The blood pressure drops again sharply in the capillaries of the body tissues and, for certain circuits, still again in the liver (hepatic portal system) or kidney (renal portal system).

Measurements reported by Satchell (1971) show that blood pressure in the ventral aorta of the dogfish ranges from 25 to 39 cm H_2O (depending on whether the ventricle is relaxed or contracting). This drops about 23 percent as blood goes through the gill capillaries to a range of 20 to 29 cm H_2O in the dorsal aorta. There is a further drop as blood passes through capillaries in the tissues of the tail, and the blood pressure in the caudal vein can be as low as 2.4 cm H_2O when the animal is at rest. However, activity of caudal muscles secondarily increases the blood pressure (see "tail pump" on p. 344) enough to drive blood through the capillaries in the kidneys, where the blood pressure is reduced again. Pressures in the posterior cardinal veins, which drain the kidneys, may be as low as -0.6 to $+0.2$ cm H_2O. Venous return to the heart is facilitated by large venous sinuses near the heart, for these offer little resistance to blood flow. Blood can accumulate in them before it enters the heart.

The piscine vascular pattern also determines the thermal physiology of fishes. On its way from the capillary beds of the inner organs and body musculature, the blood picks up heat released by metabolic processes. At the gills, where the blood is separated from the surrounding respiratory water only by very thin epithelia (see p. 304), not only gas exchange but also heat exchange takes place. An equilibrium is established between the temperature of the blood in the gill lamellae and the temperature of the surrounding respiratory water. Theoretically, the blood loses heat and its temperature drops, while the surrounding respiratory water gains the heat released by the blood; the temperature of the water therefore should rise. In reality, however, the temperature rise of the water is insignificant, because there is so much more respiratory water in the gill pouches than blood in the gill lamellae. The water acts as a heat sink, with the effect that the blood temperature drops to the level of the surrounding water every time it passes through the gills, while the surrounding water maintains essentially the same temperature. In this way, a fish is, by necessity, poikilothermic as well as ectothermic.

In certain fishes, such as the tuna and mako shark, however, certain centrally located body muscles maintain a higher temperature than the rest of the body. Francis G. Carey, who studied such fishes, found that these warmer muscles are capable of working continuously over long periods of time without fatiguing. These muscles make it possible for the fishes to cruise over vast distances at relatively high velocities without tiring. Although tunas and mako sharks are poikilothermic-ectothermic like all fishes, they nevertheless manage to keep certain body muscles at a higher temperature by preventing metabolic heat, which is released by the contracting muscles, from leaving them. This is accomplished by retia mirabilia that surround the muscles and act as a thermal insulator by being a countercurrent heat exchanger.

The blood vessels of *Squalus* are a good example of those of most fishes, but remember that fishes ancestral to tetrapods had lungs as well as gills and hence had a pulmonary as well as a branchial circulation. You should study the cardiovascular system on specimens in which at least the arteries and hepatic portal system have been injected. To study all the veins, you will need triply injected specimens.

External Structure of the Heart

The pericardial cavity of the dogfish has already been opened and the heart observed (p. 302). Return to this cavity and examine the external features of the heart (Fig. 11-4). The heart of fishes is essentially an S-shaped tube (Figs. 10-2**A** and 11-12) that receives oxygen-depleted blood at its caudodorsal end, increases the blood pressure, and sends the blood out to the gills and body at its cranioventral end. Four chambers have differentiated in linear sequence along this tube. These are, from caudal to cranial, the **sinus venosus, atrium**[28], **ventricle**, and **conus arteriosus.** The last two lie along the ventral half of the S and so are the first to be seen. The ventricle is the thick-walled, muscular, oval structure lying in the caudoventral portion of the pericardial cavity; the conus arteriosus is the tubelike chamber extending from the front of the ventricle to the cranial end of the cavity. Lift up the caudal end of the ventricle (the **apex** of the heart). The thin-walled, bilobed chamber dorsal to it is the atrium. The sinus venosus is the triangular chamber lying caudal to the atrium and extending between its two lobes. The caudal surface of the sinus venosus adheres to the transverse septum and receives through it the various veins draining the body.

The Venous System

♦ **(A) THE HEPATIC PORTAL SYSTEM AND HEPATIC VEINS**

The principal vein of the hepatic portal system is the **hepatic portal vein** going to the liver (Fig. 10-5, p. 297). It can be found lying beside the bile duct in the lesser omentum. The hepatic portal vein receives small **choledochal veins** from the bile duct but is

---

[28]The terms *atrium* and *auricle* are variably used; sometimes they are used synonymously, and sometimes a subtle difference is made between them. We will follow the *Nomina Anatomica* convention, using *atrium* for the entire chamber (undivided in fish, divided in mammals), and *auricle* for the ear-shaped part of the atrium of mammals.

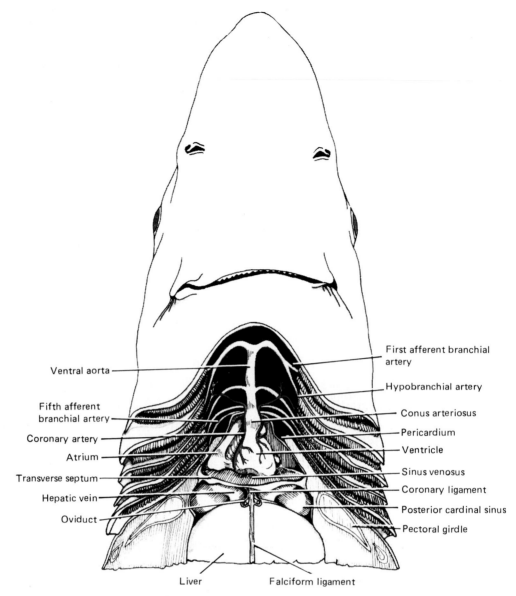

Ventral aorta

First afferent branchial artery

Hypobranchial artery

Fifth afferent branchial artery

Conus arteriosus

Coronary artery

Pericardium

Atrium

Ventricle

Transverse septum

Sinus venosus

Hepatic vein

Coronary ligament

Oviduct

Posterior cardinal sinus

Pectoral girdle

Liver

Falciform ligament

Figure 11-4
Ventral view of the heart and afferent branchial arteries of *Squalus* after removal of the hypobranchial musculature.

formed by the confluence, near the craniodorsal tip of the dorsal lobe of the pancreas, of three large tributaries. A **gastric vein** comes in from the central part of the stomach, a **lienomesenteric vein** from along the line of attachment of the mesentery to the dorsal lobe of the pancreas, and a **pancreaticomesenteric vein** from the deep side of the ventral lobe of the pancreas. The gastric and lienomesenteric veins often come together and form a short common trunk before uniting with the pancreaticomesenteric vein, which is the conspicuous vein seen passing just dorsal to the pylorus. These tributaries parallel corresponding arteries and drain the abdominal viscera. You will see their peripheral distribution when you study the arteries.

The hepatic portal vein can be traced into the substance of the liver, where it breaks up into many branches that lead to the capillary-like sinusoids. The sinusoids, in turn, are drained by a system of hepatic veins that lead ultimately to a pair of large **hepatic veins,** or **hepatic sinuses** (Fig. 11-5). The hepatic sinuses are systemic veins. Although they are not injected, you can find one of these large sinuses by making a transverse cut through the lateral edge of the liver about 1 centimeter caudal to the transverse septum. If you cut deep enough, you will cut across one of the large hepatic sinuses. Pass a probe forward through the sinuses, and you will see it go through the coronary ligament and transverse septum into the sinus venosus. The same is true for the hepatic sinus on the other side, but it should not be dissected. If desired, you can also trace the hepatic sinus caudally well into the liver.

As the portal blood passes through the liver, it comes into intimate contact with the hepatic cells, for the sinusoids are not completely lined with endothelium. Excess food products in the blood coming from the digestive tract after a meal are stored in the hepatic cells largely in the form of glycogen and (in the dogfish) oil, and deficiencies in the food content of the blood between meals is made up from food stored in the cells. Numerous other metabolic conversions also occur here.

♦ **(B) THE RENAL PORTAL SYSTEM**

Cut transversely through the back and vertebral column of the tail just caudally to the cloacal aperture. Widen the gap by bending the tail in such a way that you can look at the cut surfaces of the vertebrae. The **caudal artery** and **vein** will be seen within the hemal arch, the artery lying dorsal to the vein and being injected (Figs. 11-5 and 11-7). The vein will not be injected, unless it happened inadvertently during the injection of the artery. Make a series of partial cross sections through the base of the tail; begin your cut dorsally and extend it ventrally into, but not through, the kidneys. Space the sections about 1 centimeter apart and continue to make them until you find the caudal vein bifurcating into the two **renal portal veins.** The renal portal veins extend cranially, lying dorsal to the medial border of each kidney, but it is impractical to trace them far. They carry blood to the capillaries associated with the kidney tubules by inconspicuous **afferent renal veins.** These capillaries are drained by **efferent renal veins** that enter the posterior cardinal veins (see p. 395).

A renal portal system is not present in agnathous fishes such as the lamprey, and in such forms the caudal vein leads directly into the two posterior cardinal veins. Such a condition is also present in the embryonic dogfish (Fig. 11-6A). During subsequent development, a pair of new veins (called **subcardinal veins** in the embryo) appear ventral to the kidneys. These veins tap into the front of the posterior cardinal veins. Meanwhile a portion of the posterior cardinal veins caudal to this union atrophies (dotted line in Fig. 11-6B). The caudal portions of the embryonic posterior cardinal veins are now renal portal veins, for all their blood necessarily goes to the kidneys. The adult posterior cardinal veins are composed of the embryonic subcardinal veins plus the cranial end of the embryonic posterior cardinal veins.

The functional significance of a renal portal system is not entirely clear, and it is lost in mammals. Clearly, a large volume of blood must pass through the kidneys to be cleared of waste products. Arterial blood goes through small, segmental branches from the dorsal aorta to the glomeruli at the beginning of the kidney tubules. This provides a needed oxygen supply and a source of blood that is filtered in the glomeruli. This blood passes from the glomeruli into a capillary bed over the rest of the kidney

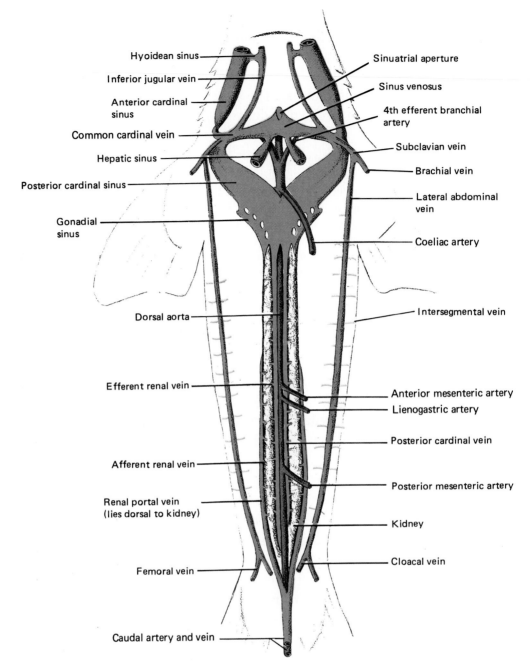

Figure 11-5
Semidiagrammatic ventral view of the renal portal and systemic veins of *Squalus*. Ventral visceral branches of the dorsal aorta are also shown.

tubules, where it is joined by oxygen-depleted blood entering via the renal portal system (see Fig. 12-1A). Processes of selective secretion of excretory products (augmentation) and reabsorption of certain materials occur in the kidney tubules (see also p. 395). A renal portal system ensures that a large volume of blood is processed by the nephrons even though the arterial pressures are low. It also enables many marine teleosts, which must conserve body water, to reduce the glomeruli, where much water is filtered from the blood, and rely upon tubular augmentation as a means of eliminating waste products from the kidneys.

The return of low-pressure, oxygen-depleted blood from the caudal vein through the kidneys is facilitated by the hemal arch, which makes a "tail pump" possible. The hemal arch protects the caudal artery and vein from the waves of muscular contraction that sweep down the tail as the fish swims. Blood flows from the caudal artery through

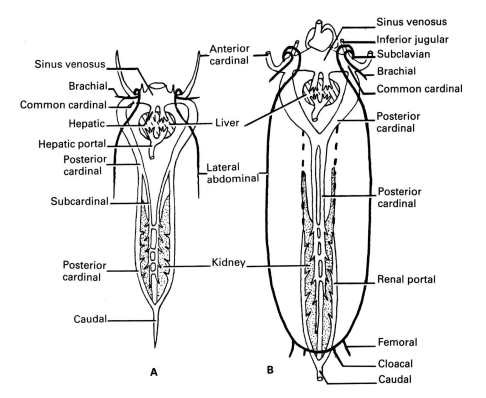

Figure 11-6
Diagrams of ventral views illustrating the development of the major veins of a dogfish: **A**, late embryo; **B**, adult. These diagrams show in particular the development of the adult renal portal system from the caudal parts of the embryonic posterior cardinal veins, and the development of the caudal parts of the adult posterior cardinal veins from the embryonic subcardinal veins. The broken lines in **B** indicate the portions of the embryonic posterior cardinal veins that disappear in the adult. *(A, Modified from Hochstetter; B, modified from Daniel.)*

intersegmental arteries during periods of muscular relaxation and is prevented from backing up when the tail muscles contract by valves present in the intersegmental arteries (Fig. 11-7). Tail muscles squeeze the peripheral veins and squirt blood from them into the caudal vein. Pressure builds up in the caudal vein because the blood vessel cannot expand within the hemal arch and because the blood cannot escape

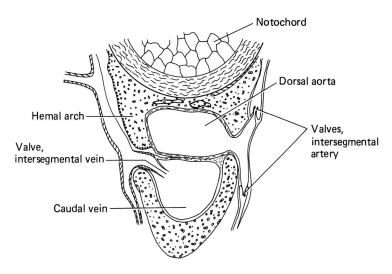

Figure 11-7
Cross section through the hemal arch and caudal blood vessels of a teleost fish *(Clupea)*. Note the valves in the intersegmental artery and vein that make possible the "tail pump." *(From G. H. Satchell.)*

through the intersegmental veins due to the valves at their entrances. Thus, the blood flows craniad into the renal portal veins.

♦ **(C) THE SYSTEMIC VEINS**

For purposes of description, the systemic veins of fishes may be arranged in five groups: (1) the hepatic veins already seen, (2) the cardinal venous system draining most of the trunk and head, (3) the inferior jugular veins draining the ventral portion of the head, (4) the lateral abdominal venous system draining the appendages and lateroventral portion of the body wall, and (5) the coronary veins draining the heart wall. You should refer to Figure 11-5 during the dissection of these veins, for many are hard to find unless you have a triply injected specimen. The following directions are based on specimens in which these veins have not been injected.

Open the sinus venosus by a transverse incision and wash it out. The small entrances of the pair of hepatic sinuses can now be seen in the central part of the sinus venosus. The large, round openings at the caudolateral angles of the sinus are the entrances of the paired **common cardinal veins,** or ducts of Cuvier (Fig. 11-5). Pass a probe into the common cardinal vein on the side of the body that has been opened and cut the vessel open along the course of the probe. The common cardinal vein passes along the lateroventral wall of the esophagus within the transverse septum and becomes continuous with the large **posterior cardinal sinus**—the large, membrane-covered space that lies against the dorsolateral surface of the esophagus and curves toward the middorsal line of the pleuroperitoneal cavity. Cut open the posterior cardinal sinus and follow it caudally. The sinuses of opposite sides are interconnected dorsally to the gonads and, in this region, receive genital vessels (**ovarian** or **testicular veins**) from **gonadal sinuses** beside the gonads, and also veins from the esophagus. It is difficult to see these blood vessels clearly. More caudally, each sinus narrows to form a posterior cardinal vein, which continues caudally along the medial border of each kidney. The posterior cardinal sinus is actually just the expanded cranial part of this vein. Each posterior cardinal vein receives numerous **efferent renal veins** from the kidneys and **intersegmental** or **parietal veins** from between the myomeres. Renal and intersegmental veins are difficult to see unless they happen to be filled with blood.

One of the paired **anterior cardinal sinuses** may have been seen during the dissection of the branchial muscles (Fig. 7-12) or cranial nerves (p. 154). If not, expose one in your specimen by making a deep longitudinal cut extending ventrally and medially from that portion of the lateral line overlying the branchial region. Open it on the side on which you have been working. Pass a probe caudally through the vessel, and you will see it turn ventrally and unite with the posterior cardinal sinus. The union of these two vessels marks the beginning of the common cardinal vein. The anterior cardinal sinuses drain the brain and all of the head except the floor of the branchial region. If desired, you can trace the anterior cardinal sinus forward by probing and cutting to a large **orbital sinus** around the eye.

The floor of the branchial region is drained by a pair of **inferior jugular veins.** The entrance of one of them can be seen in the cranial wall of the common cardinal vein just before the common cardinal vein enters the sinus venosus. The vein extends forward dorsal to the pericardial cavity. The anterior cardinal sinus and inferior jugular vein of each side are interconnected by a **hyoidean sinus.** The hyoidean sinus can be seen lying along the caudal surface of the hyoid arch on the side of the body that was cut to open the pharynx. By probing and cutting, trace it dorsally to the anterior cardinal sinus and ventrally to the inferior jugular vein. Trace the inferior jugular vein caudally to the common cardinal vein.

The appendages, and some of the lateroventral portion of the trunk, are drained by a pair of **lateral abdominal veins.** These veins are the pair of dark, longitudinal lines seen on the inside of the body wall beneath the parietal serosa. Examine the caudal end of one of these veins by probing and cutting open the vessel. After passing dorsal to the pelvic girdle, the vessel receives two tributaries. One (the **cloacal vein**) comes in from the lateral wall of the cloaca; the other (the **femoral vein**) comes in from the pelvic fin. There is also an anastomosis between the cloacal veins of the opposite sides of the body. You may not be able to find these vessels in an uninjected specimen. Cranially, the lateral abdominal vein is joined by a **brachial vein** from the pectoral fin. You can find the brachial vein by first loosening the pectoral fin from the body musculature. You will see the brachial vein on the medial surface of the fin. Trace it to its union with the lateral abdominal vein by probing and cutting along the course of the probe. You may see a **subscapular vein,** which lies along the caudal side of the scapular process under the parietal serosa, entering the proximal end of the brachial vein (Fig. 11-8). The vessel formed by the union of the brachial and lateral abdominal veins is known as the **subclavian vein.** Trace it in the same way and you will see it enter the front of the common cardinal vein beside the entrance of the inferior jugular vein. If it is necessary to find any of these veins on the opposite side, avoid injuring the falciform ligament.

Coronary veins can be seen on the surface of the heart, especially the ventricle. They enter the sinus venosus by a common aperture, which cannot be found at this stage of the dissection.

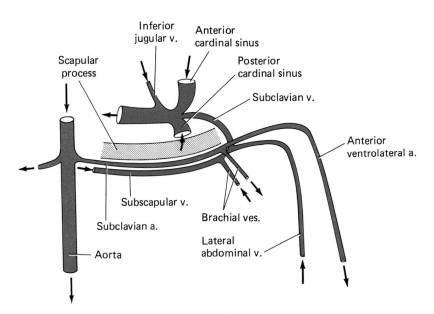

Figure 11-8
Diagram illustrating the topographical relationships among the major veins and arteries in the region of the pectoral girdle.

The Arterial System

♦ **(A) THE BRANCHIAL ARTERIES**

To dissect the branchial arteries and the ventral and dorsal aortae with their branches, swing open the floor of the mouth cavity and pharynx, as you have done while studying the mouth and pharyngeal cavities (p. 303). Remember that the heart in fishes lies cranially to the pectoral girdle (see Figs. 11-9 and 11-13). To expose the heart, the ventral aorta, and the afferent branchial arteries, first remove the epithelium covering the surface of the floor of the mouth and the primary tongue, and then carefully remove the exposed basibranchial and hypobranchial cartilages piece by piece. Start working at the rostral end of the primary tongue and work backward toward the heart. The afferent branchial arteries are structurally distinct tubes, but they are visually unremarkable and are easily overlooked and damaged because they are not injected. Use Figure 11-9 as a guide to look for them. Next, remove the epithelium from the pharyngeal roof to reveal the injected efferent branchial arteries and dorsal aortae. Finally, expose the smaller arteries in the floor of the mouth cavity and pharyngeal roof as well as those supplying and draining the gill lamellae.

As the **ventral aorta** passes forward, it gives off five pairs of **afferent branchial arteries** that continue into the interbranchial septa (Figs. 10-11, 11-4, and 11-9). The caudal two pairs of afferent branchial arteries (the **fourth** and **fifth**) leave the ventral aorta just cranial to the pericardial cavity, and from the dorsolateral side of the aorta. They come off very close together and, sometimes, by a short common trunk. The middle pair of afferent branchial arteries (the **third**) leaves slightly rostral to the caudal two pairs. The ventral aorta then passes rostrally for some distance without further branches. Slightly caudally to the level of the basihyal, it bifurcates. Trace one of the bifurcations, and you will see that it subdivides again. These subdivisions are the **first** and **second** afferent branchial arteries. Trace each of the five afferent branchial arteries far enough into the interbranchial septa on the intact side of the specimen to ascertain which interbranchial septa and gills they are supplying and to observe the numerous small branches that they send into the gill lamellae.

Oxygen-rich blood is collected from the gill lamellae by a system of **efferent branchial arteries.** The first portion of this system is a series of four and one-half **collector loops.** You can see a representative collector loop by spreading open the first definitive internal gill slit and by removing the epithelium and underlying branchial adductor muscle and branchial arch (see Fig. 10-11). A vascular circle (the first collector loop), receiving tiny vessels from the gill lamellae, will be revealed. The vessel forming the rostral half of the collector loop is the **pretrematic branch** of an efferent branchial artery; it is noticeably smaller than the vessel forming the caudal half of the collector loop, the **posttrematic branch.** Second, third, and fourth collector loops can be found in the second to fourth definitive gill pouches, but only the pretrematic half of a collector loop is present in the last branchial pouch because there are no gill lamellae on the caudal surface of this branchial pouch.

Expose the pretrematic branch of the second collector loop. In addition to receiving tiny branches from the adjacent gill lamellae, it gives off many larger branches (**cross trunks**) that pass through the interbranchial septum to the posttrematic branch of the first collector loop. This is also the case for all the other pretrematic branches, except, of

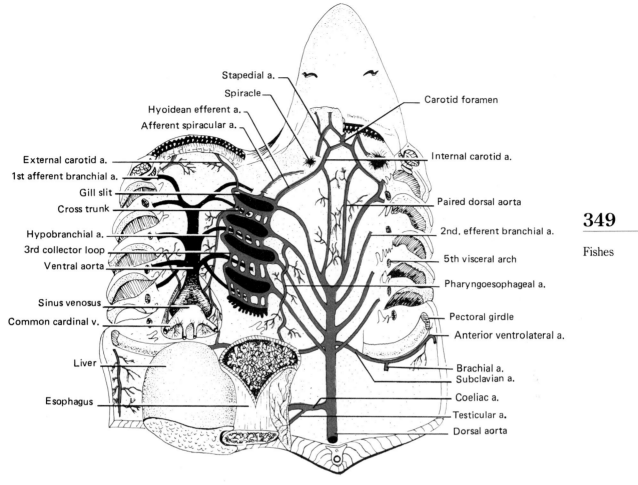

Figure 11-9
A dissection of the branchial arteries of *Squalus*. The afferent branchial arteries, which have been drawn very dark, have been approached through the floor of the pharynx and are seen in a dorsal view. The efferent branchial arteries, which are shown in red, are seen in a ventral view.

course, the first. The posttrematic branches are therefore functionally the most important parts of a collector loop, for each is the main vessel draining a particular holobranch. It receives not only the drainage from the gill lamellae adjacent to it on the cranial surface of a holobranch but also, via the cross trunks, most of the drained blood from the gill lamellae on the caudal surface of a holobranch.

Four pairs of **efferent branchial arteries** will be seen on the caudal portion of the pharyngeal roof (Fig. 11-9). They extend from the dorsal angles of the internal gill slits diagonally, medially, and caudally and converge to form a median vessel, the **dorsal aorta.** Remove connective tissue from around the efferent branchial arteries, and trace them on the intact side of the pharynx to the tops of the collector loops. The dorsal cartilages of some of the branchial arches must also be removed. (Can you name them?) The most rostral efferent branchial artery (the **first**) connects with the top of the first collector loop, and the last (the **fourth**) with the top of the fourth collector loop. These vessels receive oxygen-rich blood from the collector loops and carry it to the more caudal parts of the body. Most of the blood flows backward via the dorsal aorta, but a small **pharyngoesophageal artery** arises from the second efferent branchial artery and extends caudad in the roof of the pharynx and esophagus.

Oxygen-rich blood to the head travels in other vessels. A **hyoidean efferent artery** arises from the top of the first collector loop rostral to the first efferent branchial artery and passes forward in the roof of the pharynx. Opposite the spiracle it receives, on its medial side, a small vessel that extends forward from the medial end of the first efferent branchial artery. This artery and its mate of the opposite side represent some of the rostral portions of the embryonically paired dorsal aorta. They are called simply the **paired dorsal aortae.** The vessel rostral to the union of the hyoidean efferent artery with one of the paired dorsal aortae also develops embryonically from the front of the dorsal aorta. In the adult it is called the **internal carotid artery.** The internal carotid artery continues forward, curves toward the middorsal line, crosses or sometimes unites with its mate of the opposite side, and enters the chondrocranium through the carotid foramen. Follow the internal carotid arteries by chipping away cartilage. They soon diverge and, at the level of the hypophysis, unite with the arteries on the ventral surface of the brain. The internal carotid arteries are the major arteries supplying the brain.

At the point at which the internal carotid artery curves toward the middorsal line of the pharynx it gives off a **stapedial artery** from its rostrolateral surface. Note the proximity of this artery to the point of union of the hyomandibular cartilage (the future stapes) with the otic region of the chondrocranium. Follow the stapedial artery forward. It passes dorsally to the efferent spiracular artery extending medially from the spiracle (see later), and its blood is distributed to the orbit and snout.

An **afferent spiracular artery** arises from near the middle of the pretrematic branch of the first collector loop and continues to the pseudobranch on the spiracular valve. You can find this vessel most easily by removing skin caudal to the spiracle (Fig. 7-11, p. 151). The afferent spiracular artery will be seen just beneath the hyomandibular nerve as it crosses the lateral surface of the hyomandibular cartilage. The vessel often is not well injected, but you can trace it from the hyomandibular cartilage caudally to the pretrematic branch of the first collector loop and cranially to the pseudobranch. An **efferent spiracular artery** continues from the pseudobranch medially to unite with the internal carotid artery within the cranial cavity. This portion of the vessel can be found by removing the mucous membrane lining the front of the spiracle. Approach the spiracle from its pharyngeal entrance.

An **external carotid artery** arises from the ventral end of the first collector loop and passes forward to supply the lower jaw region. It is best approached by removing the epithelium that covers the gap between the tip of the primary tongue and the mandibular cartilage, where it is revealed as a usually injected blood vessel.

Another vessel, the **hypobranchial artery,** usually arises from the ventral end of the second collector loop, but it may receive contributions from any of the other collector loops. The vessel supplies most of the hypobranchial musculature and then bifurcates at the front of the pericardial cavity. One branch, the **coronary artery,** supplies the heart. The other, the **pericardial artery,** extends caudad in the dorsal wall of the pericardial cavity.

Ancestral fishes retain parts, at least, of all six embryonic aortic arches, but the aortic arches are obviously modified in the adult for the interposition of the gills and the supply of blood to the head. During the course of embryonic development, six

Table 11-1 Derivation of the Branchial Vessels of the Dogfish

The relation of the adult branchial vessels of a dogfish to the embryonic aortic arches is shown in this table. The derivation of each collector loop from outgrowths of two aortic arches is explained in the text.

Afferent Vessels	Efferent Vessels	Embryonic Origin
	Efferent spiracular	Dorsal half of aortic arch 1
	Afferent spiracular	Cross connection between aortic arches 1 and 2
Afferent branchial 1	Hyoidean efferent	Aortic arch 2
Afferent branchial 2	Efferent branchial 1	Aortic arch 3
Afferent branchial 3	Efferent branchial 2	Aortic arch 4
Afferent branchial 4	Efferent branchial 3	Aortic arch 5
Afferent branchial 5	Efferent branchial 4	Aortic arch 6
	Internal carotid	Rostral dorsal aorta and forward outgrowth
	Stapedial	Rostroventral outgrowth from dorsal aorta
	External carotid	Rostral outgrowth from the ventral extension of part of aortic arch 2
	Hypobranchial	Ventral outgrowth from ventral part of second collector loop (aortic arches 3 and 4)
	Pharyngoesophageal	Caudal outgrowth from dorsal part of aortic arch 4

complete aortic arches differentiate from anterior to posterior (Figs. 11-1 and 11-10**A**). The ventral portion of the first aortic arch soon disappears, but the ventral portions of the remaining five form the afferent branchial arteries (Fig. 11-10**D**). The dorsal part of the first aortic arch, together with a new connection that early develops between the first and second aortic arches (Fig. 11-10**B**), forms the spiracular artery. The dorsal part of the second aortic arch forms the hyoidean efferent artery; the dorsal parts of the remaining aortic arches form the four efferent branchial arteries. The external carotid artery, the distal part of the internal carotid artery, as well as the stapedial, hypobranchial, and pharyngoesophageal arteries represent new outgrowths from the dorsal aorta or various aortic arches, as shown in Figure 11-10 and Table 11-1.

The collector loops that drain the gill lamellae develop from the dorsal halves of the last five aortic arches. Branches extend from them ventrally, passing in front of and behind each of the definitive internal gill slits, except for the posterior surface of the last where no gill forms (Fig. 11-10**B**). These branches then connect with each other above and beneath the internal gill slit to form the collector loops (Fig. 11-10**C**). Thus, each collector loop is formed from outgrowths of two successive aortic arches—the pretrematic portion of a collector loop from the caudal branch of a bifurcation of one aortic arch, and the posttrematic portion from the cranial branch of the bifurcation of the next caudal aortic arch. At one stage of development, each collector loop would be drained by both parent aortic arches (Fig. 11-10**C**). This double drainage persists in the adult for the first collector loop (Fig. 11-10**D**). But the second and succeeding collector loops lose their original connection (**xxx** in Figure 11-10**D**) with their cranial parent aortic arch and would be drained exclusively by the caudal parent aortic arch if it were not for the development of new connections through the gills (cross trunks) with the cranial parent aortic arch.

♦ **(B) THE DORSAL AORTA, ITS BRANCHES AND ACCOMPANYING VEINS**

The basic pattern of the branches of the dorsal aorta is shown in Figure 11-11. As can be seen, there are three major categories of vessels: (1) Paired **intersegmental arteries**[29] pass between each of the body segments, and each soon bifurcates into a

---

[29]These vessels are often called segmental arteries, but the term *intersegmental* is preferable, for the vessels occupy an intersegmental position.

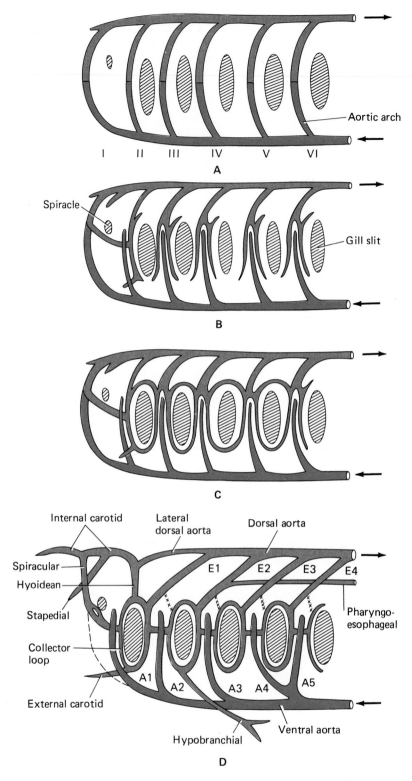

Figure 11-10
Diagrams of the aortic arches and their derivatives at different developmental stages of the dogfish, as seen in a lateral view. **A**, Early embryonic stage; **B** and **C**, successive embryonic stages; **D**, adult stage. All the vessels would be paired, except the caudal portions of the dorsal and ventral aortae. The afferent vessels are shown in blue, the efferent vessels in red. Abbreviations: *A1* to *A5*, first to fifth afferent branchial arteries; *E1* to *E4*, first to fourth efferent branchial arteries; *I* to *VI*, first to sixth aortic arches.

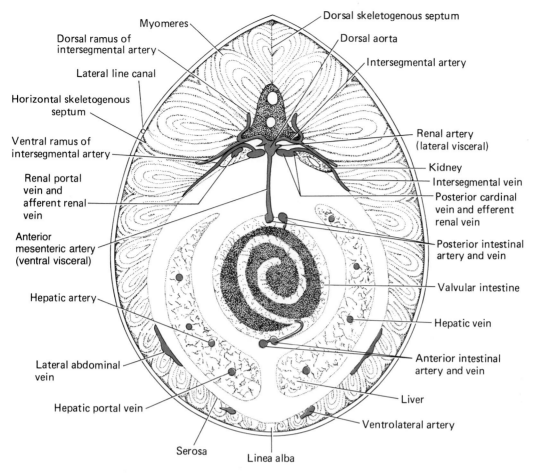

Figure 11-11
Semidiagrammatic cross section through the trunk of *Squalus* to show major arteries and veins.

dorsal ramus to the epaxial region and a **ventral ramus** to the hypaxial region. Longitudinal anastomoses may occur at various points between successive intersegmental arteries. Vessels to the appendages are simply enlarged ventral rami of intersegmental arteries. (2) Paired **lateral visceral arteries** pass at intervals to such dorsolateral organs as the kidneys, gonads, and suprarenal glands. (3) Unpaired median **ventral visceral arteries,** which develop from the embryonic vitelline arteries, pass through the dorsal mesentery to the viscera. This basic pattern of the branches of the aorta occurs in all vertebrates.

Working, if possible, from the side of the body that has already been cut open, trace the dorsal aorta caudad. You will have to separate the cranial portion of the esophagus from the body wall, and some mesenteries will have to be torn. A pair of **subclavian arteries** arise from the aorta, generally between the third and fourth efferent branchial arteries (Fig. 11-9). Follow one subclavian artery as it curves ventrally against the lateral wall of the posterior cardinal sinus. It follows the caudal margin of the scapular process and gives off several branches to the adjacent musculature (Figs. 11-8 and 11-9). The **brachial artery** branches from the dorsal side of the subclavian artery (to see the bifurcation, lift the subclavian artery and look underneath it) and enters the pectoral fin in company with the brachial vein. The **anterior ventrolateral artery** first continues cranially and ventrally along the caudal margin of the scapula and coracoid. It then curves caudally and extends caudad in the ventrolateral portion of the body wall. This part of the vessel can be seen beneath the parietal serosa between the lateral abdominal vein and the midventral line.

The subclavian arteries are modified intersegmental vessels. Other intersegmental arteries will be seen presently, but you should examine the ventral visceral arteries next. These are accompanied by tributaries of the hepatic portal vein (p. 341). The first one of these, the large **coeliac artery** (see Fig. 11-5), enters the front of the pleuroperitoneal cavity and extends ventrally and caudally along the right side of the stomach to the cranioventral tip of the dorsal lobe of the pancreas (see Fig. 10-5). Here it bifurcates. One branch, the **pancreaticomesenteric artery,** follows the pancreaticomesenteric vein dorsal to the pylorus and onto the ventral side of the intestine as the **anterior intestinal artery.** An **anterior intestinal vein,** which drains into the pancreaticomesenteric vein, lies beside the artery. In addition, the pancreaticomesenteric artery sends smaller branches to the pyloric region of the stomach and the ventral lobe of the pancreas and into the spiral valve. The other branch of the coeliac artery, the **gastrohepatic artery,** soon divides into a small **hepatic artery,** which follows the hepatic portal vein to the liver, and a **gastric artery,** which passes to the stomach and gives off a branch to the dorsal lobe of the pancreas, following closely the gastric vein and its tributaries.

The next ventral visceral branches of the aorta will be found in the free caudal edge of the dorsal mesentery. Two vessels arise closely together in this region and pass through the mesentery. The vessel going to the spleen and caudal part of the stomach, and also giving off another branch to the dorsal lobe of the pancreas, is the **lienogastric artery;** the vessel passing to the caudal part of the intestine is the **anterior mesenteric artery** (Fig. 11-5). A final ventral visceral artery is a small **posterior mesenteric artery** to the digitiform gland and caudal end of the intestine. These last three ventral visceral arteries together supply the area drained by the lienomesenteric vein (Fig. 10-5), whose tributaries are the **lienogastric vein** and the **posterior intestinal vein.** The posterior intestinal vein drains the dorsal side of the intestine and runs parallel to the **posterior intestinal artery,** which is a branch of the anterior mesenteric artery (Fig. 11-11).

More caudally, a pair of **iliac arteries** arise from the aorta. Each artery passes ventrally in the body wall lateral to the cloaca and divides into a **femoral artery** entering the pelvic fin with the **femoral vein,** and a **posterior ventrolateral artery.** The latter unites with the anterior ventrolateral artery, which was seen extending caudad from the subclavian artery. The iliac arteries, like the subclavian arteries, are modified intersegmental arteries. You can see typical **intersegmental arteries** by freeing and lifting up the lateral border of a kidney. Their most conspicuous branches pass from the aorta laterally and ventrally between the myomeres accompanied by **intersegmental veins** that drain into the posterior cardinal veins. Other branches extend into the epaxial region of the body (Fig. 11-11). Lateral visceral arteries include a number of small **renal arteries** and a pair of **gonadal arteries,** either **ovarian** or **testicular.** The renal arteries arise from the aorta close to, or in common with, the intersegmental arteries and pass to the kidneys. To see them, make a longitudinal section near the medial border of one kidney; they will be visible as cross sections of empty round tubes, because they are not injected. The gonadal arteries arise from the very base of the coeliac artery. Accompanying veins are tributaries of the posterior cardinal vein. Caudal to the iliac arteries, the dorsal aorta enters the tail as the **caudal artery,** which, as you have seen, lies dorsal to the **caudal vein** within the hemal canal.

The Internal Structure of the Heart

Preceding dissections have explored the general structure of the heart and its coronary vessels. Remove the heart by cutting the attachments of the **sinus venosus** to the transverse septum and cutting across the caudal end of the ventral aorta, last afferent branchial arteries, and coronary arteries. Wash out the sinus venosus and look inside it (Fig. 11-12). This thin-walled chamber receives oxygen-depleted blood from the body by way of the paired hepatic veins and common cardinal veins already observed. Oxygen-depleted blood then enters the **atrium** by way of the slitlike **sinuatrial aperture** at the front of the sinus venosus. This opening is guarded by a pair of lateral folds, the **sinuatrial valve,** which prevents the backflow of blood during the contraction of the atrium. One or more openings of the coronary veins may be seen in the wall of the sinus venosus near the sinuatrial aperture.

Cut forward through the dorsal end of the sinuatrial aperture and the dorsal wall of the atrium and clean out this chamber. It too is thin-walled, but muscular strands can be seen on the inside of its walls. Note that, despite its two lobes, its cavity is undivided. Find the **atrioventricular aperture** in the floor of the atrium. The opening is guarded by a pair of folds, the **atrioventricular valve.**

Insert a pair of scissors into the stump of the ventral aorta, cut back through the ventral surface of the **conus arteriosus,** and continue to the caudal end of the **ventricle.** The thicker, muscular walls of these chambers, especially of the ventricle, will be noted. The ventricular wall is also very spongy, for it is criss-crossed by numerous muscular strands. Notice that the lumen of the ventricle is U-shaped, for the entrance from the atrium and the exit to the conus arteriosus lie nearly side by side. Backflow of blood from the front of the heart into the ventricle is prevented by the presence of three rows of **semilunar valves** within the conus arteriosus. Each row consists of three pocket-shaped valves. One row lies at the very cranial end of the conus arteriosus, and the other two lie close together near the caudal end of this chamber. The entrances to the pockets face cranially, so that you can find and open them by probing caudally.

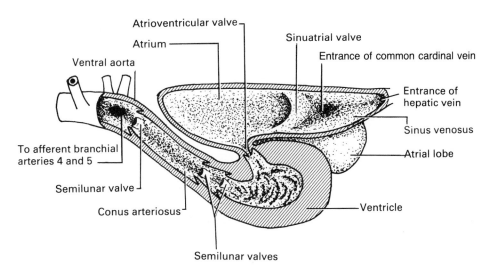

Figure 11-12
Lateral dissection of the heart of *Squalus.*

The dogfish heart lies in the pericardial cavity, which is solidly encased by the cartilaginous pectoral girdle, the large basibranchial cartilage, and the hypobranchial musculature (Figs. 10-9 and 11-13). Its location within this rigid box permits it to act alternately as a pressure pump or as a suction pump. When the highly muscular ventricle contracts, the blood in it is forced out into the conus arteriosus and ventral aorta, since the atrioventricular valve prevents a backflow into the atrium. Because the walls of the pericardial cavity are rigid and the contracted ventricle occupies less space, the pressure within the pericardial cavity drops. As a result, the flabby, less muscular walls of the atrium are pulled out, and a negative pressure builds up within the atrial chamber, so that blood is sucked into the atrium from the sinus venosus and adjacent veins. When the atrium contracts subsequently, the blood in it is forced into the ventricle; the ventricular walls are now relaxed, and the ventricle is expanded by the entering blood. The sinuatrial valve prevents a backflow of blood into the sinus venosus. The generation of a negative pressure within the atrium is crucial for ensuring the return of blood from the cardinal veins and sinuses, in which the blood pressure is exceedingly low.

Ventricular contraction drives blood out into the conus arteriosus under a relatively high pressure. The conus arteriosus expands as it receives this surge of blood and then contracts during ventricular relaxation. In this way it buffers the delicate gill capillaries against the pressure surge during ventricular contraction and tends to even out the flow of blood. However, blood flow is not perfectly even. Heart beat is synchronized with the respiratory movements of the pharynx and branchial region in such a way that the largest volume of blood flows through the gill lamellae at the time when water is discharged across them (p. 305). This, of course, increases the efficiency of gas exchange.

The Pericardioperitoneal Canal

Now that the heart has been removed, it is possible to find the communication between the pericardial and the pleuroperitoneal cavities referred to on page 302. Pass a blunt probe into the recess in the transverse septum dorsal to the line of attachment of the sinus venosus. The probe will be seen to pass into a canal beneath the visceral serosa on the ventral surface of the esophagus, and after passing a distance of about 3 centimeters, it will

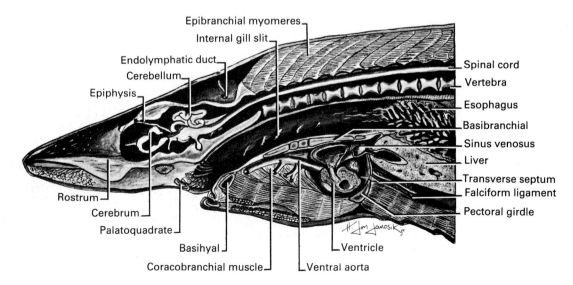

Figure 11-13
Sagittal section through the head of *Squalus*, showing the structure of the heart and its relation to surrounding structures. A bristle passes through the sinuatrial aperture.

emerge into the pleuroperitoneal cavity through a semilunar opening. This entire passage is the **pericardioperitoneal canal.** Frequently the canal bifurcates at its caudal end, in which event there would be a pair of openings into the pleuroperitoneal cavity. The canal permits liquid to escape from the pericardial cavity into the pleuroperitoneal cavity. This is important for the maintenance of the low pressure within the pericardial cavity, which is necessary for the suction action of the atrium. Fluid does not move in the other direction in the canal because its delicate walls collapse and act as a valve.

AMPHIBIANS

Important changes occur in the cardiovascular system during the transition from an aquatic to a terrestrial environment. The most conspicuous of these occur in the heart and aortic arches. Contemporary lungfish have a pulmonary as well as a branchial circulation, so that respiratory gas exchange can occur both under water and in the air. Although the lungfish heart is only partly divided into right and left sides, radiographic studies have shown that there is a good separation of oxygen-rich from oxygen-depleted blood. Blood high in oxygen content is received from the lungs by the left atrium and is directed to the more rostral aortic arches. Because gills are reduced on these aortic arches, this blood is sent directly to the head and body. Blood low in oxygen content is received from the body by the sinus venosus and right atrium and is directed to the more posterior aortic arches. These still bear gills, and the sixth aortic arch also has a branch, the **pulmonary artery,** leading to the lungs. Although lungs evolved in fish as accessory respiratory organs, permitting the fish to live in water with a low oxygen tension, the lungs are more than accessory organs in some contemporary lungfish. Even in water with a good oxygen supply, these fish must periodically surface and gulp air, because the lungs are essential for supplying the blood with enough oxygen. The gills remain necessary to unload carbon dioxide released by the cell metabolism because the lung ventilation rate is too low to accomplish this.

The crossopterygian ancestor from which amphibians evolved also had both a branchial and a pulmonary circulation. Adult amphibians lose the branchial circulation and depend on the pulmonary circulation supplemented, at least in living species, by buccopharyngeal and cutaneous gas exchange (p. 306). In these animals, the heart is only partly divided into left and right sides, but this seeming inefficiency is well adapted to their mode of life. Studies by DeLong (1962) and others indicate that the blood streams are kept separate to a large extent when the lungs are being used. The aortic arches supplying the head and body contain blood highest in oxygen content; the sixth aortic arch supplying the lungs and skin contains blood lowest in oxygen content. The lungs are of prime importance in the uptake of oxygen; the skin is of greater importance in unloading carbon dioxide. After a cycle of lung ventilation, amphibians close their glottis and hold air in their lungs at greater than atmospheric pressure; this also is the case when they stay beneath water. During prolonged breath-holding, oxygen in the lungs is depleted, pulmonary resistance increases, and blood is diverted away from the inactive lungs to the skin and body. An incomplete separation of the heart chambers permits different amounts of blood to flow to the lungs and other tissues depending on the animal's needs during different activities.

Oxygen-depleted and oxygen-rich blood streams are more completely separated morphologically in reptiles, in which there is usually a partial interventricular septum. But shunts remain that allow blood to be distributed in the most appropriate ways. During active lung ventilation, pulmonary resistance is low, and there is little mixing of oxygen-depleted blood passing through the right side of the heart to the lungs with oxygen-rich blood passing through the left side to the body. The limited mixing is a shunt from left to right that recirculates some blood through the lungs. The blood becomes highly saturated with oxygen. After a period of lung ventilation, reptiles also hold their breath for a considerable period of time. Pulmonary resistance increases as the oxygen in the lungs is depleted, and a right-to-left shunt diverts considerable blood from the lungs. Oxygen-depleted blood is recirculated, but this is not as detrimental as it may appear. In most animals, oxygen-depleted blood contains some oxygen, and the increasing accumulation of carbon dioxide causes hemoglobin to unload more of this oxygen (the Bohr effect).

As regards the aortic arches, at least the first two are lost in nonmammalian tetrapods, but those that remain are complete and are not interrupted by gill capillaries. An important corollary of the changes in the heart and aortic arches is a relative increase in the blood pressure in the dorsal aorta and, hence, in the general efficiency of circulation. A frog, for example, has a systolic pressure in the dorsal aorta of 30 mm Hg compared with a systolic pressure in the dorsal aorta of a dogfish of 17 mm Hg. The frog, moreover, is a much smaller animal, so that the relative efficiency of its circulation is even greater than these values imply.

The venous system remains somewhat fishlike in early tetrapods, but important changes are seen in the primitive lateral abdominal venous system, the right hepatic veins, and parts of the cardinal veins. Parts of the last two have been transformed into a short **caudal vena cava.**

Necturus is a good example of the early tetrapod condition, excepting certain features of its heart and aortic arches. Being neotenic, this urodele retains larval methods of respiration. The cardiovascular system should be studied on doubly or triply injected specimens.

The Heart and Associated Vessels

The pericardial cavity was noted during the study of the coelom. Open it wider, if necessary, and identify the chambers of the heart. The most conspicuous chamber in the ventral view is the large, muscular **ventricle** that occupies the caudoventral portion of the pericardial cavity. The narrow vessel that emerges from the right side of the front of the ventricle is the **conus arteriosus.** This soon expands into a wider vessel, the **bulbus arteriosus,** lying at the front of the pericardial cavity. The conus arteriosus is a chamber of the heart; the bulbus arteriosus is the modified ventral aorta, which in this case lies within the pericardial cavity. In many vertebrates that have it, the bulbus arteriosus helps to even out the pressure waves of ventricular contraction; presumably, this is its function in *Necturus.* The **atrium** lies dorsal to the conus and bulbus arteriosus and appears as the lobes on either side of these structures. It is only partially divided internally into **right** and **left atria.** Lift up the caudal end of the ventricle (the **apex** of the heart) and you will see the small, thin-walled **sinus venosus.**

The pair of large veins that enter the sinus venosus caudally are the **hepatic sinuses. Common cardinal veins** enter the caudodorsal corners of the sinus venosus just in front of the hepatic sinuses (Fig. 11-14). A pair of **pulmonary veins** from the lungs pass dorsal to the hepatic sinuses and unite to form a single vessel that enters the left atrium. This vessel may be seen between the hepatic sinuses.

Since *Necturus* is a permanent larva respiring primarily by means of external gills, its heart is not a good example of the heart of an adult amphibian. The venous drainage of the body enters the sinus venosus, which in turn opens toward the right atrium. Oxygen-rich blood from the lungs enters the left atrium. Considerable mixing occurs in the heart, for the interatrial septum is poorly developed and the two atria have a

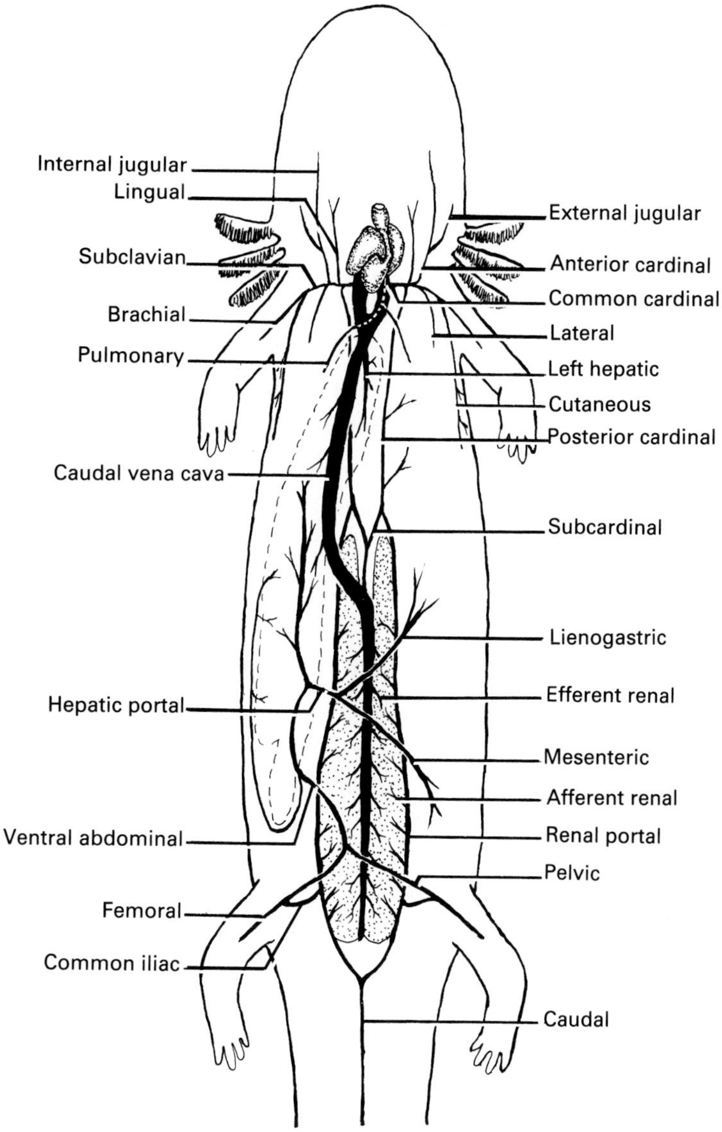

Figure 11-14
Diagrammatic ventral view of the venous system of *Necturus*. The outline of the liver is shown by broken lines.

common opening into the single ventricle. This mixing is unimportant in *Necturus*, for all the blood that leaves the heart passes through the gills before it is distributed to the body and lungs. Thus, the heart of *Necturus* is very similar to that of *Squalus* functionally, and its structure is not different enough to warrant its dissection.

The Venous System

♦ (A) THE HEPATIC PORTAL SYSTEM

Functionally, the hepatic portal system is the same in *Necturus* as in fishes, but the pattern of its tributaries differs somewhat. The major features of the pattern seen in *Necturus*, however, are very representative of those of tetrapods in general. Stretch out the mesentery of the intestine and you will see a longitudinal vessel, the **mesenteric vein,** passing forward and disappearing in the pancreas. Notice that it receives numerous

intestinal veins from the intestine. Next, look on the tail of the pancreas near the spleen. The vessel seen is the **lienogastric vein;** it is formed by the confluence of a **lienic vein** from the spleen and several **gastric veins** from the stomach. Carefully dissect away pancreatic tissue and find the point where the lienogastric and mesenteric veins unite. The common vessel that passes forward from here to the liver is the **hepatic portal vein** (Fig. 11-14).

♦ (B) THE VENTRAL ABDOMINAL VEIN

The median, longitudinal vessel that lies in the falciform ligament caudal to the liver is the **ventral abdominal vein.** This vessel has evolved from the ventral migration and fusion of the paired lateral abdominal veins of fishes, and its caudal relationships are still very similar to those of the lateral abdominal vein. It receives several small **vesical veins** from the urinary bladder (a ventral outgrowth of the embryonic cloaca) and then bifurcates into **pelvic veins.** Each pelvic vein extends laterally and caudally and after about 1 centimeter receives, on its lateral side, a **femoral vein** from the hind limb. The vessel that continues from the pelvic and femoral veins caudally and dorsally to the renal portal vein (see later) is the **common iliac vein.** Blood from the leg may pass forward on one of two routes—the ventral abdominal vein (the primitive route) or the common iliac and renal portal veins (a new route).

The cranial relationships of the ventral abdominal veins differ from those of the lateral abdominal veins of fishes, for the ventral abdominal vein has lost its primitive connection with the common cardinal vein and has developed a new one with the hepatic portal system. In this respect it resembles its homologue in late amniote embryos—the umbilical vein (p. 387).

♦ (C) THE RENAL PORTAL SYSTEM

The vessel that runs along the lateral margin of each kidney dorsal to the prominent archinephric duct in the male, or oviduct in the female, is the **renal portal vein.** As already described, a common iliac vein enters each. Trace the two renal portal veins caudad and try to find the point where they unite and receive the median **caudal vein** from the tail. The renal portal veins receive blood from the tail and some from the legs and adjacent body wall and carry most of it to the kidneys.

♦ (D) THE CAUDAL SYSTEMIC VEINS

At the very cranial end of the kidneys, the renal portal veins lead into a pair of small **posterior cardinal veins** that continue forward on either side of the dorsal aorta. This continuity of renal portal and posterior cardinal veins is not surprising, since the renal portal veins develop from the caudal end of the embryonic posterior cardinal veins (Fig. 11-6A), and since *Necturus* is an incompletely metamorphosed species. Notice the **intersegmental veins** that enter the posterior cardinal veins from the body wall. The posterior cardinal veins diverge at the level of the cranial end of the esophagus and unite with the anterior cardinal veins to form the common cardinal veins. This region will be studied presently.

The blood in the renal portal veins that passes into the kidneys leaves through numerous, small, paired, **efferent renal veins** that are located on the ventral surface of the kidneys (see also p. 395 and Fig. 12-1**B**). These, together with **gonadial** (**testicular** or **ovarian**) **veins** from the gonads, enter the **caudal vena cava** (posterior vena cava) lying between the kidneys. After an anastomosis with the caudal ends of the posterior cardinal veins, the caudal vena cava extends ventrally through the ligamentum hepatocavopulmonale and enters the liver. Trace it through the liver. It receives numerous small **hepatic veins** from various parts of the liver and a particularly large **left hepatic vein** from the front of the liver. The caudal vena cava then passes through the coronary ligament and transverse septum. After this, it bifurcates into the two **hepatic sinuses** that extend dorsal to the apex of the ventricle to enter the sinus venosus.

The caudal vena cava is a new vessel compounded largely from veins that were previously present. The portion of it cranial to the kidneys develops from the right hepatic vein and from a caudal extension of this vein. This explains why there is only one particularly prominent hepatic vein (the left one) at the front of the liver, instead of the two seen in *Squalus*. The right vein is incorporated in the caudal vena cava. The caudal extension of the right hepatic vein taps into the embryonic subcardinal veins, as it does in mammal embryos (see Fig. 11-32**B**), and these (especially the right subcardinal vein) form the segment of the caudal vena cava between the kidneys. In fishes, the embryonic subcardinal veins form the caudal portions of the adult posterior cardinal veins (p. 343). Although they now contribute to the caudal vena cava, they retain a connection with the posterior cardinal veins. Blood that has passed through the kidneys may take one of two routes forward: Either it can go through the posterior cardinal veins, its primitive route, or it can stay in the caudal vena cava. Most of it does the latter, as the posterior cardinal veins are reduced in size. The evolutionary advantages of the caudal vena cava are not clear, although it is sometimes described as a more direct route to the heart.

♦ **(E) THE CRANIAL SYSTEMIC VEINS**

The cranial systemic veins are hard to dissect unless they are filled with blood. The common cardinal veins were seen entering the sinus venosus. Each receives a number of tributaries, most of which can best be found peripherally and then traced to the common cardinal vein. You should study them on the side of the body opposite to the side on which you dissected the muscles. Do not injure arteries while dissecting the veins.

Carefully remove the skin from the lateral surface of the brachium and shoulder and you will see the **brachial vein.** It soon joins with a **cutaneous vein** from the skin to form the **subclavian vein.** Trace the subclavian vein forward. It turns into the musculature and enters the **common cardinal vein** near the latter's dorsal end. Now remove the skin ventral to the gill slits and separate the hypobranchial muscles from the brachial region. The longitudinal vessel lying on the hypobranchial musculature in this region is the **lingual vein.** Trace it caudally and you will see it enter the common cardinal vein beside, or in common with, the subclavian vein. Having located these two veins, cut through the muscles ventral to the union of these two veins with the common cardinal vein and thereby expose the common cardinal vein descending to the heart.

Next, remove the skin from the side of the trunk caudal to the shoulder. The longitudinal vessel lying between the epaxial and hypaxial muscles is the **lateral vein.** Remove the scapula and its muscles and trace the vein forward. It enters the top of the common cardinal vein slightly dorsal to the preceding two vessels. You can now trace forward the **posterior cardinal vein,** and you will see it enter the common cardinal vein caudal to the entrance of the lateral vein. The vessel entering the top of the common cardinal vein cranial to the lateral and posterior cardinal veins is the **anterior cardinal vein.** Try to trace it forward. The largest part of the vessel (the **external jugular vein**) passes dorsal to the gills, but a small branch (the **internal jugular vein**) goes into the roof of the mouth cavity.

The anterior cardinal vein, together with the internal jugular vein, represents the anterior cardinal vein of fishes. The subclavian and brachial veins represent the cranial part of the primitive lateral abdominal system and still have the same essential relationships as their homologues in *Squalus.* The lingual vein is homologous to the inferior jugular vein of fishes. Although the terminology has changed somewhat, it is obvious that the major cranial veins of *Necturus* are very similar to those of fishes.

The Arterial System

♦   **(A) THE AORTIC ARCHES AND THEIR BRANCHES**

Return to the pericardial cavity and carefully dissect away the muscular tissue (mostly rectus cervicis) that lies between the cavity and the external gills on the side of the pharynx that has not been cut open. Two arteries leave from each side of the front of the bulbus arteriosus. Trace them laterally on the intact side. They are probably not injected, so be careful. They cross the transversi ventrales muscles (p. 165) and then pass deep to the subarcual muscles (Fig. 11-15). The more cranial artery, known as the **first afferent branchial artery,** follows the first branchial arch (third visceral arch) and enters the first external gill. A small, probably well-injected artery lies just in front of the distal portion of the first afferent branchial artery. This is the **external carotid artery.** Notice that it has numerous branches supplying the muscles in the floor of the pharynx and mouth cavity. The external carotid artery is a branch of the efferent branchial system (hence its injection), but it generally has tiny anastomoses with the first afferent branchial artery. One of the paired **thyroid glands** lies in the angle formed by the meeting of the first afferent branchial and external carotid arteries. It can be recognized by its texture, for it is composed of many follicles that appear as small vesicles.

The more caudal vessel leaving the bulbus arteriosus soon bifurcates. One branch, the **second afferent branchial artery,** follows the second branchial arch and enters the second gill. The other branch, the **third afferent branchial artery,** follows the third branchial arch and enters the third gill.

Swing open the floor of the mouth cavity and pharynx and carefully remove the mucous membrane from the roof of the pharynx. A large pair of vessels will be seen

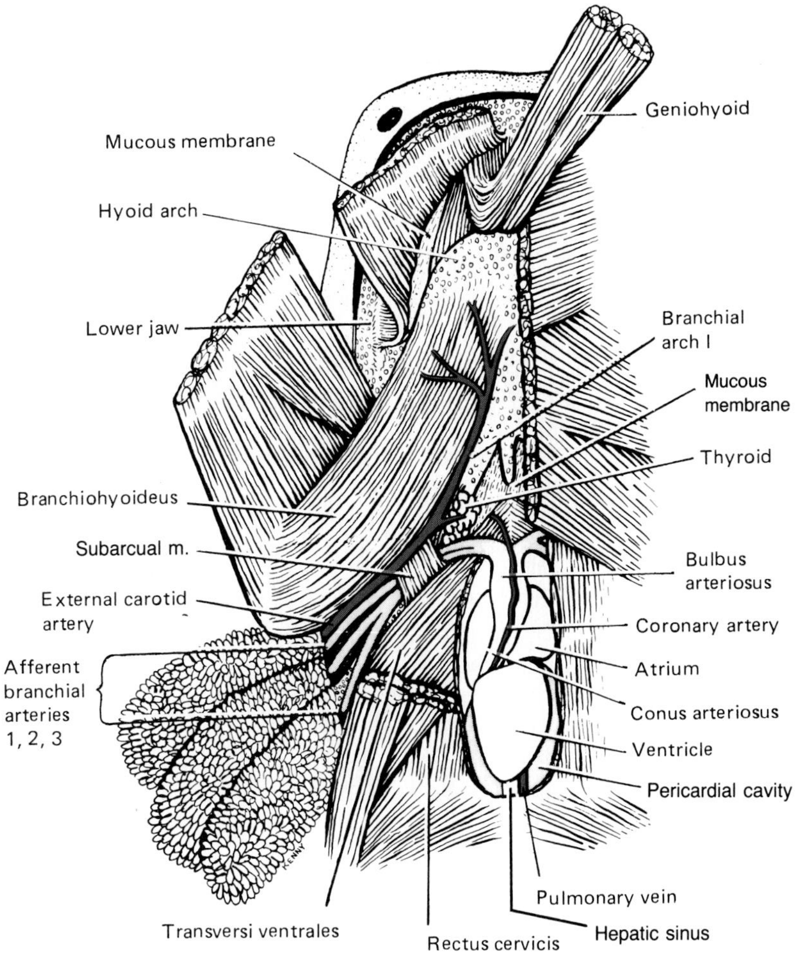

Figure 11-15
Ventral view of the afferent branchial arteries of *Necturus*. Much of the rectus cervicis muscle has been cut away.

converging toward the middorsal line, where they unite to form the **dorsal aorta.** They are the **radices** of the aorta (Fig. 11-16). Carefully trace the radix on the intact side toward the gill slits. Slightly lateral to the vertebral column it gives off a small, cranial branch, the **vertebral artery,** which soon disappears in the musculature at the base of the skull. Further laterally, the radix curves caudally. Another vessel lies cranial to this portion of the radix and is connected with the radix by a short, stout anastomosis, called the **carotid duct.** That portion of the vessel extending forward in the roof of the pharynx from the carotid ducts is the **internal carotid artery;** that portion extending laterally from the carotid duct represents the entrance of the **first efferent branchial artery.** The internal carotid artery is distributed to the facial region and enters the skull to help supply the brain. Trace the first efferent artery as far laterally as feasible, noting that it comes from the first external gill. Return to the radix and continue tracing it laterally from the carotid duct. You will soon see a caudal branch. This is the **pulmonary artery;** you should trace it to the lung. If you have difficulties, find the pulmonary artery on the lung and trace it forward. (The **pulmonary vein** lies on the opposite side of the lung. We have already noted its entrance into the heart.) Lateral to the pulmonary artery, the radix bifurcates and receives the **second** and **third efferent branchial arteries** from the respective gills.

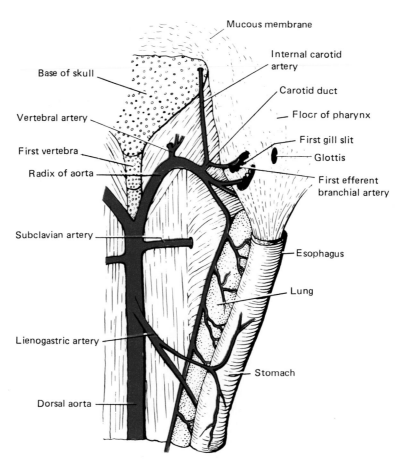

Figure 11-16
Ventral view of the efferent branchial arteries of *Necturus*.

Only the points of entrance of the efferent branchial arteries into the radix of the aorta are visible from the roof of the pharynx. In order to see these arteries more clearly, remove skin from the top of the head dorsal to the external gill slits and from the bases of the external gills. The three efferent branchial arteries lie near the dorsal edge of the gills. Trace them to the point reached in the previous dissection. You should also find the origin of the external carotid artery from the first efferent branchial artery during this dissection.

Necturus, like all tetrapods, has lost the first two aortic arches, but the dorsal aortae and the ventral aortae as far caudad as the third aortic arch persist as parts of the internal and external carotid arteries, respectively (Fig. 11-17). The third, fourth, and fifth embryonic aortic arches, which lie adjacent to visceral arches three, four, and five (or branchial arches 1, 2, and 3), give rise to the three branchial arteries going to and leading away from the external gills. In this respect, *Necturus* resembles other larval salamanders. Most larval salamanders also retain the sixth aortic arch, the ventral part of which leads from the ventral aorta to the pulmonary artery, and the dorsal part continues as the ductus arteriosus to the dorsal aorta. *Necturus* is unique, as Figge (1930) pointed out, in losing the ventral part of the sixth aortic arch (Fig. 11-17). This leaves the dorsal part of the aortic arch (the ductus arteriosus) as the origin of the pulmonary artery. Such an arrangement does no harm so long as gills are retained. If the gills should be lost, there would be no direct way of sending oxygen-depleted blood

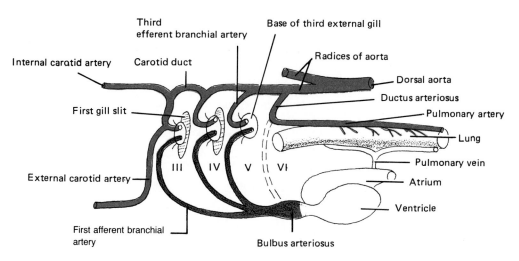

Figure 11-17
Diagrammatic lateral view of the branchial vessels of *Necturus* to show their derivation from the aortic arches. Roman numerals indicate embryonic aortic arches.

from the heart to the lungs, and the volume of blood eventually reaching the lungs would be low.

Figge postulated that the absence of the ventral part of the sixth aortic arch explains the failure to metamorphose in *Necturus,* but the reasons are probably more complex. *Necturus* is an obligate neotenic species. Metamorphosis cannot be induced, as it can be in facultative neotenic species, by giving extra amounts of thyroxine, increasing oxygen tension of the water, or manipulating other factors. The tissues of *Necturus* appear to have lost their ability to respond to thyroxine, presumably because of some modification in the biochemical receptors. Neoteny probably appeared early in the evolutionary history of *Necturus* and was followed by the loss of the ventral part of the sixth aortic arch. This would prevent the unnecessary diversion of much blood to the lungs. Had the loss of the ventral part of the sixth aortic arch preceded the ability to reproduce as a larva, *Necturus* could not have survived.

♦   **(B) THE DORSAL AORTA AND ITS BRANCHES**

A pair of **subclavian arteries** arise from the dorsal aorta just caudal to the union of the radices of the aorta (Fig. 11-16). Trace one subclavian artery laterally. At the base of the appendage it divides into a **brachial artery** that continues into the arm and a **cutaneous artery** to the skin and adjacent muscles. In amphibians with cutaneous respiration, the cutaneous artery is very large and carries blood to the skin, where some gas exchange occurs.

Continue to follow the aorta caudally. It next gives off ventrally a **lienogastric artery,** which soon branches to go to various parts of the stomach and to the spleen. The next ventral visceral branch, the **coeliacomesenteric artery,** arises some distance caudad (Fig. 11-18). It passes ventrally to the tail of the pancreas, where it divides into a **lienic artery** to the spleen, a **hepatic artery** to the liver, and a **pancreaticoduodenal artery** to the pancreas and duodenum. You will have to dissect away much of the pancreas to see all these branches. The remaining ventral visceral arteries are a number of **mesenteric arteries** to the intestine and a pair of cloacal arteries to the cloaca (see

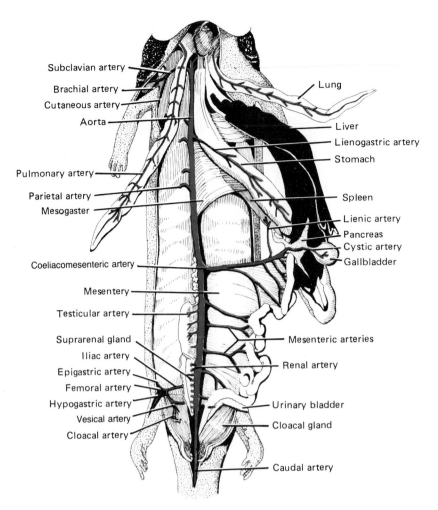

Figure 11-18
Ventral view of the arterial system of *Necturus*.

later). You will have to separate the caudal vena cava from the aorta to see the point of origin of the mesenteric arteries. Notice that the caudal vena cava lies toward the right of the aorta and the mesenteric arteries. This reflects the origin of this segment of the vena cava from the embryonic right subcardinal vein.

The lateral visceral arteries are represented by a number of paired **gonadial (testicular** or **ovarian) arteries** to the gonads and very small **renal arteries** to the kidneys. You can find the latter by dissecting away the caudal vena cava between the caudal ends of the kidneys.

Paired intersegmental arteries include the subclavian arteries already seen; a number of typical **intersegmental arteries** that arise from the dorsal surface of the aorta and pass into the body wall; and the **iliac arteries.** The iliac arteries can be found dorsal to the caudal ends of the kidneys. After traveling a short distance, each gives off cranially an **epigastric artery,** which ascends in the body wall, and caudally a **hypogastric artery,** which supplies the urinary bladder and cloaca; then the iliac artery continues as the **femoral artery** into the hind leg.

Cut through the body wall and muscles lateral and caudal to the cloaca and trace the aorta caudad. It gives off the paired **cloacal arteries** referred to previously and then enters the hemal canal of the caudal vertebrae as the **caudal artery.**

MAMMALS

Evolutionary Changes

Changes that occur in the cardiovascular system during the evolution from ancestral tetrapods to mammals correlate for the most part with the increase in activity and rate of metabolism. Since adult mammals continuously ventilate their lungs and do not hold their breath for long periods of time, there is no advantage to shunts that can alter the volume of blood going to the lungs and body. The heart becomes completely divided morphologically into a right side receiving oxygen-depleted blood from the body and sending it out to the lungs, and a left side receiving oxygen-rich blood from the lungs and sending it to the body. A complete division of the heart also makes it possible to develop quite different blood pressures in the systemic and pulmonary circuits. Except in certain reptiles, these hydrostatic pressures are nearly the same in nonmammalian terrestrial vertebrates. The left ventricular wall of the mammal heart is heavily muscularized, and the left ventricle can develop a very high systemic hydrostatic pressure—a mean of 100 mm Hg in the human dorsal aorta, compared with a mean of about 25 mm Hg in the anuran dorsal aorta. There is a rapid and efficient distribution of materials to the tissues. Such a high blood pressure would be inappropriate in the lungs, which offer far less peripheral resistance than the rest of the body, for it would cause considerable filtration of plasma-fluid into the lungs. The wall of the right ventricle is less muscularized, and the mean hydrostatic blood pressure in the human pulmonary trunk is only between 15 and 20 mm Hg.

The aortic arches are further reduced, for the fifth is lost on both sides as well as the first and second. The branches of the dorsal aorta—usually called simply the aorta, for mammals have no ventral aorta—continue to follow the basic pattern established in ancestral vertebrates.

In the cranial part of the venous system, a single cranial vena cava, or a pair of them, evolves from the anterior and common cardinal veins.

More caudally, the renal portal system is lost, and the caudal vena cava continues to the iliac and caudal veins. The loss of the renal portal system in mammals may be correlated with an increased blood pressure, for a large volume of blood now enters the kidneys directly from the aorta, and with the evolution of different mechanisms for conserving body water. It has also been suggested that the caudal migration in most mammals of a part of the embryonic kidneys with the testes (see Chapter 12) would necessitate the loss of the renal portal system. Much of the cranial portion of the posterior cardinal veins is also lost, but a part of them is transformed into an azygos system of veins.

The primitive lateral abdominal system of veins is represented embryonically by the umbilical veins, but these are lost in the adult. The veins from the appendages are not connected with the umbilical veins but enter the venae cavae directly.

Mammalian Circulation

In most parts of the body, we will describe the arteries and the veins together because they tend to parallel each other, and therefore it is convenient to dissect them at the same time. Veins are often more difficult to find than the arteries, for valves frequently prevent the injection mass from reaching the peripheral parts of these vessels, but identification of the accompanying artery will help you to find the vein. If not injected, veins appear as translucent, often fluid-filled tubes beside the corresponding arteries. If you compress the veins with a blunt object, you may see air bubbles moving in them and revealing the course of the veins.

Before studying the blood vessels on a regional basis, it is desirable to understand the overall pattern of circulation. In an adult mammal (Fig. 11-19), tributaries of the

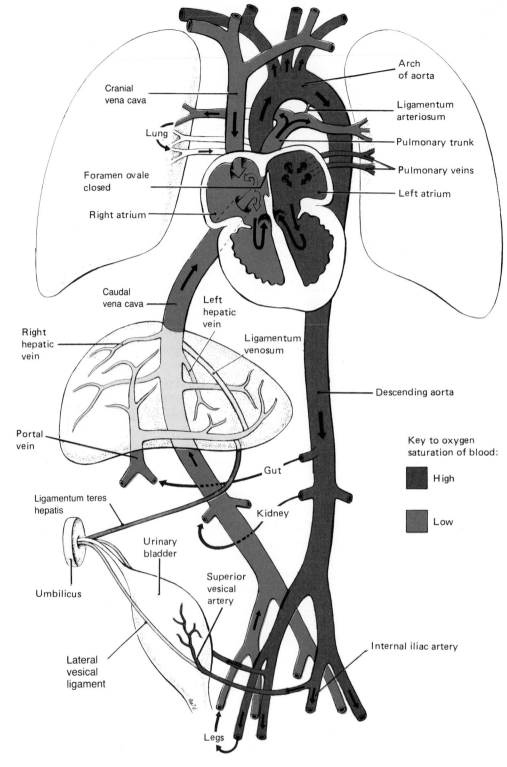

Figure 11-19
Diagram of the adult pattern of circulation in a mammal. The diagram is a ventral view, so the right side of the heart and body is on the left side of the diagram. *(From Moore.)*

caudal (inferior) **vena cava** drain the body caudal to the diaphragm, and those of the **cranial** (superior) **vena cava** drain the body cranial to the diaphragm. Both venae cavae return oxygen-depleted blood to the **right atrium** of the heart. From here the blood goes to the **right ventricle,** from which it is pumped to the lungs through the **pulmonary trunk** and **arteries.** After saturation with oxygen in the lungs, blood returns through the **pulmonary veins** to the **left atrium.** From here it goes to the

left ventricle, which pumps it into the **aorta.** Branches of the aorta carry oxygen-rich blood to nearly all parts of the body. However, blood circulating through the chambers of the heart does not supply the musculature of this organ, so a separate **coronary system** is necessary. As you will see, the coronary arteries leave the base of the aortic arch, and coronary veins return to the right atrium via a coronary sinus.

The pattern of the fetal circulation is somewhat different because the placenta, rather than the digestive tract, the kidneys, and the lungs, is the site for the intake of nutrients, the elimination of wastes, and the exchange of respiratory gases. An understanding of the fetal pattern is necessary to understand certain functionless remnants of fetal passages that persist in the adult. Blood rich in nutrients and oxygen and low in waste products enters the fetus through the **umbilical vein** (Fig. 11-20). Most of this blood passes directly through the liver in the **ductus venosus** to enter the caudal vena cava, but there is some admixture of oxygen-depleted blood in the liver. Since the entrance of the caudal vena cava into the right atrium is directed toward the interatrial septum, and since blood pressure is relatively low in the left atrium because little blood returns to this chamber from the inactive lungs, most of the oxygen-rich caudal vena caval blood passes through a valved opening in the interatrial septum, the **foramen ovale,** to enter the left side of the heart. This blood is pumped into the aorta by the left ventricle. Oxygen-depleted blood returning in the cranial vena cava, along with some admixture of blood from the caudal vena cava, passes through the right atrium into the right ventricle. This blood, which is not as rich in oxygen as the blood in the left side of the heart, is pumped toward the lungs by the right ventricle. Some of it goes to the lungs, but because the lungs are collapsed and offer considerable resistance to blood flow, most of it bypasses the lungs via a **ductus arteriosus** and enters the aorta distal to the origin of the major arteries to the head and the arms. The locations of the ductus arteriosus and the foramen ovale are such that the head, the brain, and the upper part of the trunk of the fetus receive the blood that is richer in oxygen, and a highly mixed blood is distributed to the rest of the body and, via the umbilical arteries, back to the placenta.

As development proceeds, more and more blood flows through the lungs. At birth, the lungs fill with air, and pulmonary resistance to blood flow is less than that of the rest of the body. Blood returning from the lungs increases the pressure in the left atrium, the valve in the foramen ovale is held shut, and all the blood in the right side of the heart is pumped to the lungs. As time goes on, the foramen ovale permanently closes, leaving only a depression in this region, the **fossa ovalis.** Blood in the left side of the heart, all of which has been through the lungs, is pumped into the aorta. Most blood continues to the body, but some in the aorta, because of the relatively low pulmonary resistance, flows through the ductus arteriosus into the lungs. Because of this temporary reversed flow through the ductus arteriosus, a certain fraction of blood circulates twice through the lungs. Within a few hours, the ductus arteriosus contracts. Eventually its lumen is filled in with connective tissue, and the duct is transformed into the **ligamentum arteriosum** (Fig. 11-19). With the loss of the placenta, the umbilical vein and arteries lose their function. The umbilical vein becomes the **round ligament of the liver** *(ligamentum teres hepatis),* which may have been seen in the falciform ligament. The proximal portions of the umbilical arteries remain as the proximal portions of the **internal iliac arteries.** Some of the distal parts of the umbilical arteries also persist as small **vesical arteries** supplying the urinary bladder, but the rest become the **lateral vesical ligaments.** These ligaments sometimes can be seen on either side of the urinary bladder. The adult circulatory pattern has now been established.

The Heart and Associated Vessels

Carefully cut away the pericardial sac and thymus of your specimen from around the heart and its great vessels. The **heart** *(cor)* is a large, compact organ having a pointed caudal end (its **apex**) and a somewhat flatter cranial surface (its **base**). The **right** and **left**

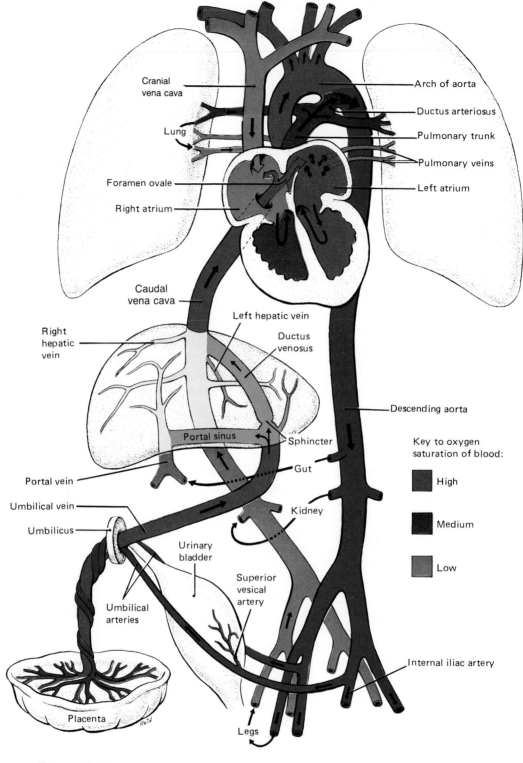

Figure 11-20
Diagram in ventral view of the fetal pattern of circulation in a mammal. Colors indicate approximate
oxygen content of the blood in the various vessels. *(From Moore.)*

ventricles form the caudal two thirds of the organ (Fig. 11-21). They are approximately
conical in shape and have thick, muscular walls. The **right** and **left atria** lie cranial to the
ventricles and are set off from them by a deep, often fat-filled groove called the **coronary
sulcus.** The atria are thinner-walled and darker than the ventricles. They are separated
from each other on the ventral surface by the great arteries leaving the front of the

ventricles. That portion of each atrium lying lateral to these arteries is called the **auricle.** The auricles are somewhat ear-shaped and tend to have scalloped margins. The separation between the ventricles appears, on the ventral surface, as a shallow groove, the **interventricular sulcus,** extending from the left auricle diagonally and toward the right.

Pick away the fat from around the large arteries leaving the cranial end of the ventricles. The more ventral vessel is the **pulmonary trunk** (Figs. 11-21 and 11-22). It arises from the right ventricle and extends dorsally to the lungs. Do not trace it until you have found the ligamentum arteriosum described below. The more dorsal vessel is the **arch of the aorta.** It arises from the left ventricle deep to the pulmonary trunk, but not much of it is visible until it emerges on the right side of the pulmonary trunk. As you continue to pick away fat and loose connective tissue from around these vessels, you will notice that they are bound together by a tough band of connective tissue, which is known as the **ligamentum arteriosum.** Try not to destroy it. Just after this connection, the pulmonary trunk bifurcates, and its branches, the **pulmonary arteries,** pass to the left and right lungs. You can see this bifurcation most easily by pushing the pulmonary trunk

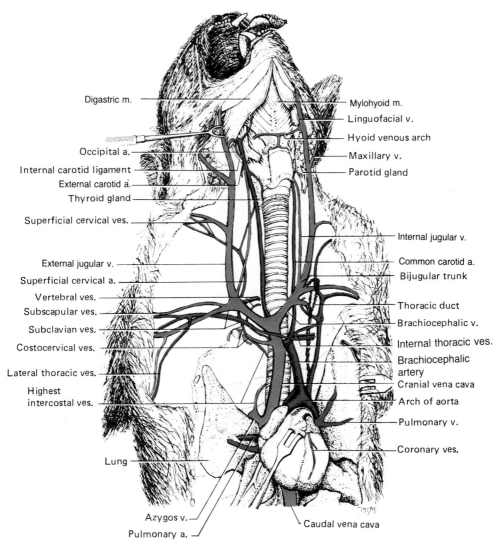

Figure 11-21
Ventrolateral view of the thoracic and cervical blood and lymphatic vessels of the cat.

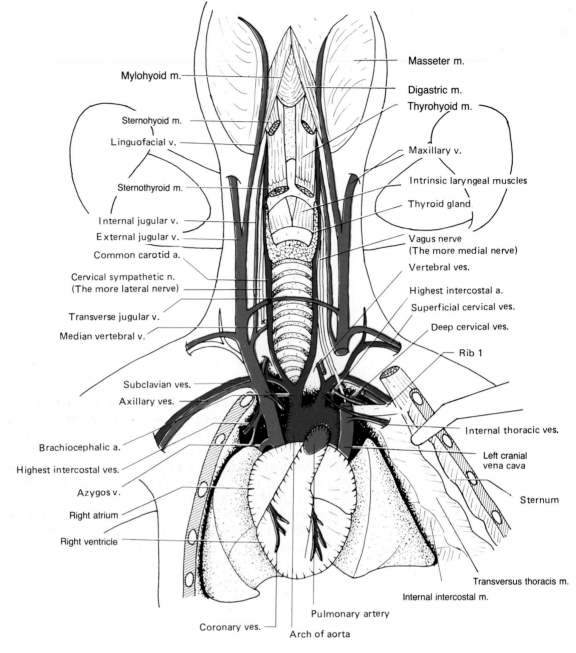

Figure 11-22
Ventral view of the thoracic and cervical blood vessels of the rabbit. Superficial veins covering the left subclavian artery and its branches have been removed.

cranially and dissecting between it and the craniodorsal portion of the heart. Two small **coronary arteries** leave the base of the arch of the aorta and pass to the heart wall. The origin of one coronary artery can be found deep between the pulmonary trunk and the left auricle; that of the other can be found deep between the pulmonary trunk and the right auricle. Distal parts of the **coronary veins** draining the heart wall parallel the arteries.

Push the heart to the left side of the thorax and you will see the **caudal vena cava** (inferior vena cava) coming through the diaphragm and entering the right atrium. You will also see a cranial vena cava entering this chamber from the right side of the neck. Adult cats normally have only the **right cranial** (superior) **vena cava,** but rabbits also have a **left cranial vena cava,** which comes down the left side of the neck, crosses the dorsal surface of the heart, and enters the right atrium. Lift up the apex of the heart and you will see this vessel. Notice that it receives the **coronary veins** from the heart wall. The coronary veins of the cat collect into a **coronary sinus,** which has a position comparable to the proximal end of the rabbit's left cranial vena cava. Carefully pick away connective tissue dorsal to the cranial venae cavae and from the roots of the lungs. You will find the **pulmonary veins** coming from the lungs and entering the left atrium. There are several veins, but those of each side generally collect into two main vessels before entering the heart.

You will note that the mammalian heart contains only two of the primitive four chambers, but that these chambers have become completely divided. The apparently missing sinus venosus and conus arteriosus are present embryonically but disappear as such in the adult. The sinus venosus is absorbed into the right atrium and forms that part of the atrium receiving the venae cavae. The conus arteriosus (bulbus cordis of mammalian embryology), together with the ventral aorta, splits and forms the bases of the pulmonary trunk and arch of the aorta.

Arteries and Veins Cranial to the Heart

◆   **(A) THE VESSELS OF THE CHEST, SHOULDER, ARM, AND NECK**

Cats have a single **cranial vena cava,** but both a left and right one are present in rabbits. Trace it (them) forward by carefully picking away surrounding portions of the thymus, connective tissue, and fat. A **subclavian vein** comes in from each shoulder and arm just in front of the first rib (a valve often prevents it from being injected) and joins the jugular veins, which receive the drainage from one side of the neck and head (Figs. 11-21 and 11-22). Usually an external and internal jugular vein unite to form a **bijugular trunk** which then joins the subclavian vein, but sometimes the jugular and subclavian veins all unite at the same point. The union of the subclavian vein and bijugular trunk forms the cranial vena cava on each side in the rabbit; in the cat it forms the **brachiocephalic veins.** Left and right brachiocephalic veins, in turn, unite to form the single cranial vena cava. The vena cava of the cat is comparable to the right one of the rabbit. A left vena cava, present in the embryo of the cat (see Fig. 11-32), disappears when the left brachiocephalic vein develops.

Next, examine certain of the tributaries of the vena cava (rabbit) or vena cava and brachiocephalic vein (cat). The most caudal tributary, entering the dorsal surface of the vena cava, is the **azygos vein,** which receives most of the **intercostal veins** from between the ribs on both sides of the body. The mammals being studied have only one azygos vein, on the right side of the body, but in some mammals a left one is also present. **Intercostal arteries** will be seen beside the veins; their origins will be seen later. A **highest intercostal vein,** which drains the cranial intercostal spaces, enters the vena cava independently and cranially to the entrance of the azygos vein. In cats, the highest intercostal vein sometimes drains into the azygos vein or deep cervical vein (see later).

The next cranial tributaries are several small veins from the thymus and a larger **internal thoracic vein,** which enters the ventral surface of the cranial vena cava. In the cat it is a single vessel at its entrance, but it bifurcates distally and drains both sides of the ventral thoracic wall. Its distal parts lie deep to the transversus thoracis muscle and are accompanied by the **internal thoracic arteries,** whose origin will be seen soon. The internal thoracic vessels continue into the cranial part of the ventral abdominal wall, where they are called the **cranial epigastric arteries** and **veins.**

Return to the arch of the aorta. After giving off the coronary arteries previously described, the arch curves dorsally and to the left, disappearing dorsal to the root of the left lung. Two vessels arise from the front of the arch—a large **brachiocephalic artery** nearest the heart and then a smaller **left subclavian artery.** Trace the brachiocephalic artery forward. It sends off small branches to the thymus and then breaks up into three vessels—two **common carotid arteries,** which ascend the neck on either side of the trachea, and a **right subclavian artery.** The common carotid arteries continue cranially deep to the vena cava and brachiocephalic veins.

Tributaries of the subclavian vein parallel branches of the subclavian artery. Since the veins often are not well injected, find the branches of the artery first, and then look beside them for the accompanying veins. Trace one of the subclavian arteries peripherally. Medial to the first rib, the subclavian artery gives rise to four branches, which are most accurately identified from their peripheral distribution (Fig. 11-23). The **internal thoracic artery,** previously identified, leaves the ventral surface of the subclavian artery and accompanies the internal thoracic vein to the ventral chest wall.

A **vertebral artery** arises from the dorsal surface of the subclavian artery nearly opposite the origin (cat), or somewhat proximal to the origin (rabbit), of the internal thoracic artery. Trace its craniomedial course and the accompanying **vertebral vein** forward. The vertebral vein normally enters the vena cava (rabbit) or brachiocephalic vein (cat). The vertebral vessels soon enter the transverse foramina of the cervical vertebrae through which they continue, finally to enter the cranial cavity and help supply the brain.

A short, caudally pointing **costocervical trunk** arises from the subclavian artery just distal to the origin of the vertebral artery and divides almost immediately into highest intercostal and deep cervical arteries. In the rabbit, and sometimes in the cat, these vessels arise independently from the subclavian artery. The **highest intercostal artery** extends caudally across the cranial ribs to supply those intercostal spaces drained by the highest intercostal vein. This vein has already been identified. The **deep cervical artery** extends dorsally to supply deep muscles of the neck. A major branch of it also passes cranial to the first rib and into the serratus ventralis muscle. The deep cervical artery is accompanied by the **deep cervical vein,** which usually drains into the vertebral vein shortly after this vein emerges from the transverse foramina. Occasionally, the deep cervical vein enters the cranial vena cava independently.

The last branch of the subclavian artery is the **superficial cervical artery.** It extends deep to the subclavian vein and follows the **external jugular vein** cranially. Trace them both. The superficial cervical artery gives off one or more small branches that extend cranially, sometimes reaching the thyroid gland, but the main part of the artery continues laterally and dorsally to supply muscles on the craniolateral surface of the shoulder. A **superficial cervical vein,** a tributary of the external jugular vein, accompanies the distal part of the artery. One tributary of the superficial cervical vein, the **cephalic vein,** is often a conspicuous vessel, draining the lateral surface of the brachium.

After giving off these vessels, the subclavian artery and the satellite subclavian vein

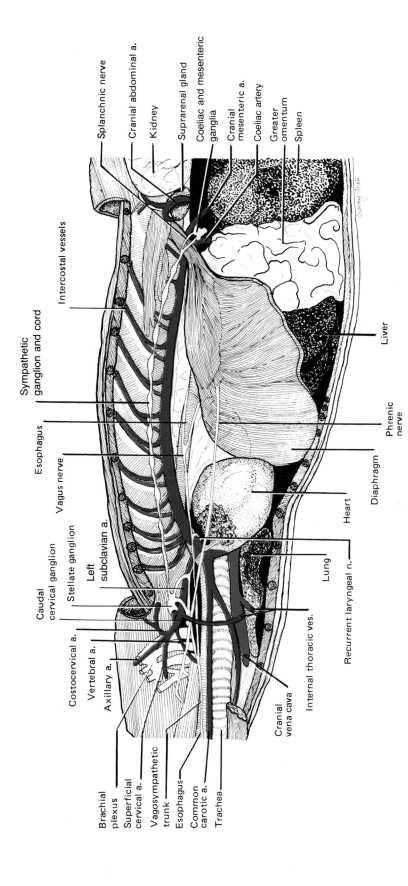

Figure 11-23

Dissection of the left side of the thorax of the cat. The left shoulder and arm, the left lung, the thoracic part of the thymus, and most of the veins have been removed. The diameter of the nerves is exaggerated slightly. (*From Walker, A Study of the Cat.*)

Brachial plexus
Superficial cervical a.
Vagosympathetic trunk
Esophagus
Common carotid a.
Trachea

Costocervical a.
Vertebral a.
Axillary a.

Caudal cervical ganglion
Stellate ganglion
Left subclavian a.

Sympathetic ganglion and cord

Esophagus
Vagus nerve

Intercostal vessels

Splanchnic nerve
Cranial abdominal a.
Kidney
Suprarenal gland
Coeliac and mesenteric ganglia
Cranial mesenteric a.
Coeliac artery
Greater omentum
Spleen

Liver

Phrenic nerve

Diaphragm

Heart

Lung

Recurrent laryngeal n.

Cranial vena cava
Internal thoracic ves.

continue laterally into the armpit (axilla). These vessels change their names at this point to the **axillary artery** and **vein.** Major branches of the axillary artery are a **lateral thoracic artery** to the pectoral muscles and a **subscapular artery.** The latter passes between the subscapularis and teres major muscles to supply deep shoulder muscles. Veins accompany the arteries but are usually not injected. When the axillary artery and vein enter the arm, they are known as the **brachial artery** and **vein.**

Return to the brachiocephalic trunk, or arch of the aorta, and trace one of the **common carotid arteries** forward. It passes deep to the brachiocephalic vein, or cranial vena cava, and continues cranially, lying lateral to the trachea and supplying the trachea, thyroid gland, and other cervical structures before reaching the head. An **internal jugular vein,** which helps drain the inside of the skull, lies lateral to the common carotid artery through most of its course. The internal and external jugular veins usually unite with each other slightly cranial to the subclavian veins to form the **bijugular trunk** previously observed. In the rabbit, a **median vertebral vein,** which courses dorsal to the esophagus, enters either jugular vein near the bijugular trunk. The rabbit also has a **transverse jugular vein,** which joins the external jugular veins slightly cranial to the bijugular trunk.

A cervical extension of the sympathetic cord and the vagus nerve can be found between the common carotid artery and the internal jugular vein. They are bound together by connective tissue to form a **vagosympathetic trunk** in the cat, but they can easily be dissected apart. The vagus is the larger nerve and passes ventrally to the brachiocephalic artery.

The major arteries described are derived from the embryonic aortic arches in the manner shown in Figure 11-24. All six of the primitive aortic arches appear during embryonic development and connect the ventral aorta (paired cranial to the fourth aortic arch) with the dorsal aorta (paired through the region of the aortic arches and for a short distance caudad). The first, second, and fifth aortic arches, the dorsal part of the right sixth aortic arch, the paired dorsal aortae between aortic arches three and four, and the right paired dorsal aortae caudad to the entrance of the right subclavian artery (an intersegmental artery) disappear during development. The dorsal part of the left sixth aortic arch persists during embryonic life as the **ductus arteriosus** and shunts blood from the pulmonary trunk directly to the dorsal aorta. For a few hours after birth it shunts some blood in the opposite direction, thereby giving this portion of the blood a double run through the lungs, but it soon becomes converted into the functionless **ligamentum arteriosum.**

The ventral portions of the sixth aortic arches persist as the pulmonary arteries. The left fourth aortic arch, together with part of the left dorsal aorta, forms the arch of the aorta. (Differential growth has the effect of shortening this arch and the adjacent dorsal aorta, so that the left subclavian artery of the adult leaves the arch of the adult aorta much closer to the common carotid arteries than it does in the embryo.) The right fourth aortic arch, plus a segment of the right dorsal aorta, forms the proximal part of the right subclavian artery. A splitting of the caudal part of the ventral aorta, and of the conus arteriosus, results in the direct origin of the arch of the aorta and the pulmonary trunk from the ventricles.

The paired ventral aortae between the fourth and third aortic arches form the common carotid arteries. The ventral aortae rostral to the third aortic arch become the external carotid arteries; the third aortic arches plus the dorsal aortae rostral to them, the internal carotid arteries. The internal carotid artery of the embryonic mammal supplies not only the intracranial part of the head but also, by its stapedial branch passing through the stapes, much of the outside of the head. However, in the adults of

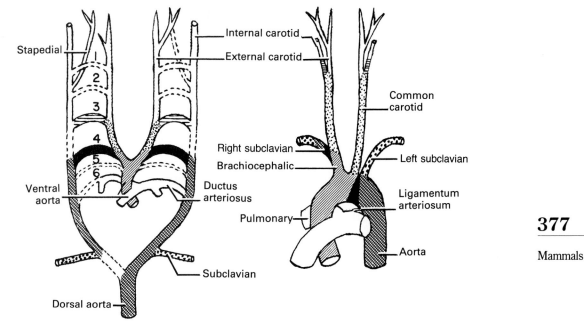

Figure 11-24
Diagrammatic ventral views of the mammalian aortic arches and their derivatives. **A,** Embryonic condition; **B,** adult condition in human beings. *(Slightly modified from Barry.)*

most mammals, the external carotid artery taps into the stapedial artery and pirates most, or all, of its peripheral distribution. If the external carotid artery takes it all over, the stapedial artery disappears.

♦ **(B) THE MAJOR VESSELS OF THE HEAD**

Remove the skin from the rest of the head on one side, if you have not done so already, and also remove the auricle. Tributaries of the **external jugular vein** are superficial to other vessels in the head and must be considered first. The external jugular vein is formed by the confluence of the linguofacial and maxillary veins (Figs. 10-16 and 11-25). Trace the **linguofacial vein** rostrally. In the cat, it soon receives on its medial side a **hyoid venous arch,** which comes from the opposite side of the body and, in turn, receives a deep vein from the larynx, the **vena laryngea impar.** The hyoid venous arch is not present in the rabbit. At the caudoventral border of the mandible, the linguofacial vein is formed by the joining of a lingual and a facial vein. The **lingual vein** enters and drains the tongue. It is accompanied by the hypoglossal nerve but has probably been cut in earlier dissections. The **facial vein** continues forward along the ventral border of the masseter muscle. Its major tributaries are a **deep facial vein** from beneath the masseter, which connects with venous plexuses in the orbit and palate; a **labial vein** from the upper lip; and a **vena angularis oculi** from the face in front of the eye.

Return to the origin of the external jugular vein and trace the **maxillary vein** dorsally toward the base of the auricle. It is formed by the confluence of a **caudal auricular vein** from behind the ear and a **superficial temporal vein** from in front of the ear. The superficial temporal vein receives tributaries from the ear, temporal muscle, and a deep branch connecting with the orbital and palatine venous plexuses. These venous plexuses, which will not be dissected, receive most of the drainage from inside the skull in

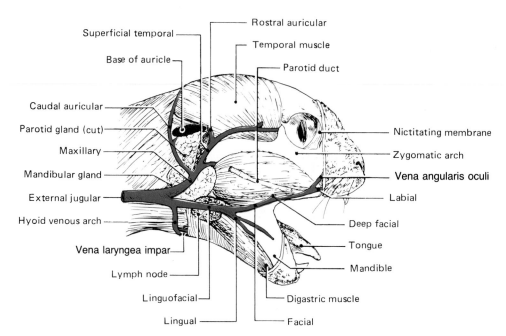

Superficial temporal
Base of auricle
Caudal auricular
Parotid gland (cut)
Maxillary
Mandibular gland
External jugular
Hyoid venous arch
Vena laryngea impar
Lymph node
Linguofacial
Lingual

Rostral auricular
Temporal muscle
Parotid duct
Nictitating membrane
Zygomatic arch
Vena angularis oculi
Labial
Deep facial
Tongue
Mandible
Digastric muscle
Facial

Figure 11-25
Lateral view of the tributaries of the external jugular vein of a cat. *(From Walker,* A Study of the Cat.*)*

the mammals being considered here, because the internal jugular veins of these mammals are small.

In order to trace the internal jugular vein and common carotid artery forward, reflect the mandibular gland and the digastric and mylohyoid muscles. At the level of the larynx, the **common carotid artery** gives off one or two **thyroid arteries** and a muscular branch (Fig. 11-26). The common carotid artery of the rabbit then divides into external and internal carotid arteries. The **internal carotid artery** goes deep toward the skull base and enters the caudal end of the carotid canal, which is located on the caudomedial side of the tympanic bulla. Together with the vertebral artery, it supplies the brain with oxygen-rich blood. An internal carotid artery is present in embryonic cats, but as development proceeds it is reduced to an **internal carotid ligament** (Fig. 11-26), which has the same topographical relationships as the artery of the same name. Much oxygen-rich blood reaches the brain of the cat by way of a small ascending pharyngeal artery (see later) and a larger anastomotic branch of the external carotid artery.

The reduction of the internal carotid artery in the cat is related to the development of a network of small arteries, known as the **carotid rete mirabile** (Fig. 11-27B), associated with the anastomotic branch of the external carotid artery. These small arteries lie within a pool of venous blood within an expanded portion of a vein, called a **cavernous sinus** (Baker 1979). The blood within the cavernous sinus is cooler than that within the carotid rete mirabile, because it is returning from the walls of the nasal cavity, where it has lost heat through evaporative cooling of the nasal epithelia. The cooler venous blood absorbs heat from the blood in the carotid rete mirabile. As a result, the temperature of the blood that the brain receives is several degrees lower than the core temperature of the body. The cooling effect is a function of the intensity of evaporative cooling in the nasal cavity, which is directly correlated with the rate of breathing.

This brain-cooling mechanism is found in many carnivores and some of their mammalian prey, such as sheep and antelopes, but not in rodents, rabbits, or primates.

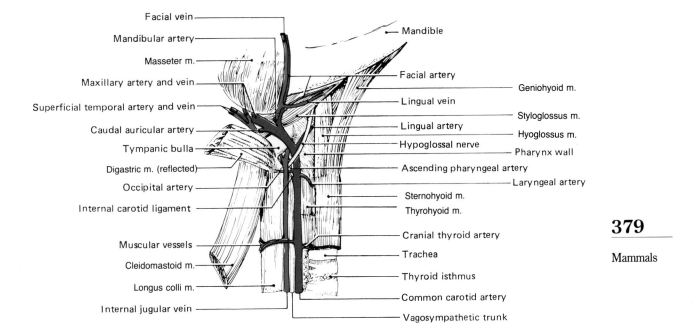

Figure 11-26
Lateroventral view of the internal jugular vein and branches of the carotid artery of the cat. Although the external jugular vein has been removed, certain of its peripheral tributaries are shown. *(From Walker,* A Study of the Cat.*)*

Since the body temperature can rise tremendously during strenuous exercise, such as chasing after a prey or being chased by a predator, this mechanism has a crucial function in protecting the brain, which is very sensitive to elevated temperatures. During prolonged hunting chases, hares, which do not have a carotid rete, may suddenly drop dead in their tracks, whereas dogs can run for a long time without apparent harm.

Cats have a second rete mirabile, the maxillary rete mirabile (Fig. 11-27**A**), whose function is not well understood.

After the origin of the internal carotid artery or ligament, the common carotid artery is known as the **external carotid artery.** Its first branches arise so close to the origin of the internal carotid artery or ligament that they can be confused with it unless their peripheral distribution is established. The **occipital artery** arises opposite to the **laryngeal artery** (see Fig. 11-26) and extends dorsally to supply neck muscles in the occipital region. The small **ascending pharyngeal artery** arises from the external carotid artery slightly cranially to the occipital artery and extends obliquely dorsocranially and deeply toward the skull base close to the internal carotid artery or ligament. This vessel is somewhat larger and more important in the cat than in the rabbit because functionally it partially replaces the internal carotid artery. The ascending pharyngeal artery follows along the ventral surface of the tympanic bulla (Fig. 11-27**A**) and finally enters the skull through the canal for the auditory tube and through the rostral portion of the carotid canal.

After the origin of these vessels, the external carotid artery gives rise to a number of branches supplying different parts external to the skull. These branches accompany the corresponding veins already observed. Lingual and facial arteries arise from the ventral surface of the external carotid artery (Fig. 11-26). They have a common origin in the

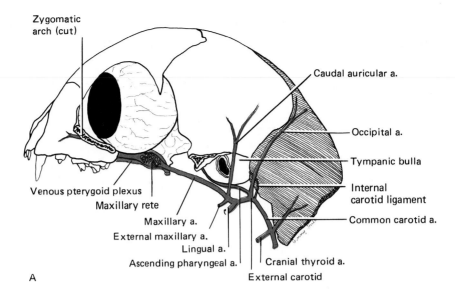

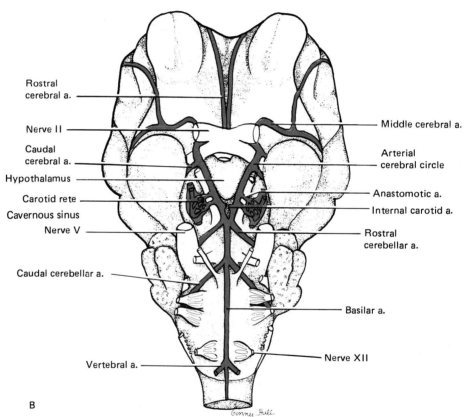

Figure 11-27

Diagrams of the circulation to the maxillary and carotid retia mirabilia and brain in a cat. **A**, Lateral view of the head; **B**, ventral view of the brain. Blood reaches the caudal parts of the brain by the vertebral arteries and a small ascending pharyngeal artery that taps into the distal, intracranial end of the internal carotid artery. Most blood reaches the cerebrum via the carotid rete mirabile on the external carotid artery, which is bathed by cool venous blood, and anastomotic arteries connecting it to the arterial cerebral circle. *(From Walker,* A Study of the Cat.*)*

rabbit, but not in the cat. The **lingual artery** enters the tongue. The **facial artery** follows the ventral border of the masseter muscle and supplies the jaws and facial structures.

Dorsal branches of the external carotid artery are (1) a **caudal auricular artery,** which extends dorsally behind the ear; (2) a **superficial temporal artery,** which extends dorsally in front of the ear; and (3) a **maxillary artery** that goes deep to the caudal border of the masseter to supply structures in the orbital and palatal regions. The caudal auricular and superficial temporal arteries have a common origin in the rabbit.

Return to the **internal jugular vein** and trace it forward. It receives small tributaries from muscles at the base of the head and then enters the skull through the jugular foramen to help drain the brain.

Arteries and Veins Caudal to the Heart

◆ **(A) THE VESSELS OF THE DORSAL THORACIC AND ABDOMINAL WALLS**

After curving to the dorsal side of the body, the arch of the aorta is known as the **descending aorta.** Trace it caudally. As it passes through the thorax along the left side of the vertebral column, it gives off paired **intercostal arteries** to those intercostal spaces not supplied by the highest intercostal arteries (Fig. 11-23), small median ventral visceral branches to the esophagus, and also small branches to the bronchi, since the tissue of lungs, like the wall of the heart, need a separate arterial supply. The **thoracic portion** of the left **sympathetic cord** can be found at this time by carefully dissecting in the connective tissue near the heads of the ribs dorsal to the aorta. Enlargements along the sympathetic cord are **sympathetic ganglia;** delicate strands passing dorsally are **communicating rami.** The left **vagus nerve** crosses the ventral surface of the arch of the aorta and passes dorsal to the root of the lung and caudally along the esophagus. **Phrenic arteries** to the diaphragm may arise from the aorta before the aorta passes through the diaphragm, or from the last intercostal arteries, or they may arise from vessels posterior to the diaphragm (first lumbar, cranial abdominal, or coeliac arteries).

The **caudal vena cava** (inferior vena cava) was seen entering the heart. Trace it caudad. As it passes through the diaphragm, it receives several small **phrenic veins** and then disappears in the liver. Scrape away tissue from the cranial surface of the right medial lobe of the liver and find the entrance of several large **hepatic veins.** The major part of the caudal vena cava, however, passes through the right lateral and caudate lobes; expose it by scraping away liver tissue. Other hepatic veins, most very small, will be seen entering.

Push the abdominal viscera to the right and find the aorta emerging from the diaphragm. Just after emerging, it gives rise to two ventral vessels—first a **coeliac artery** and then a **cranial mesenteric artery** (Figs. 11-28 and 11-29), which supply most of the abdominal viscera. Trace them when you study the vessels of the abdominal viscera (p. 383). **Coeliac** and **mesenteric ganglia,** which should not be destroyed, lie at the base of the cranial mesenteric artery. They receive one or more **splanchnic nerves** from the sympathetic cord and send out minute nerve branches that travel along the vessels to the viscera.

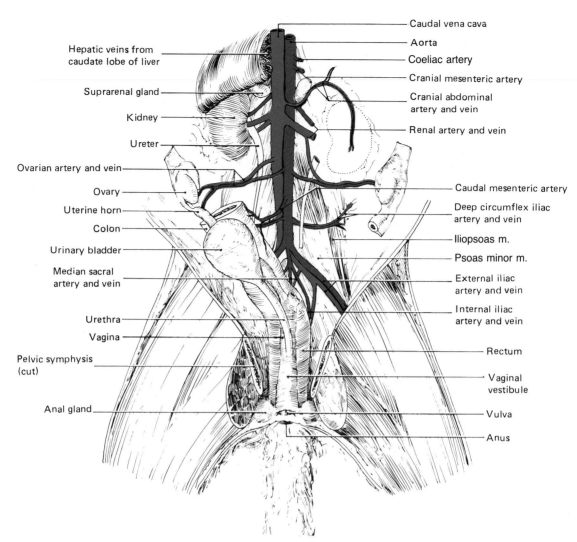

Caudal vena cava

Aorta

Coeliac artery

Cranial mesenteric artery

Cranial abdominal artery and vein

Renal artery and vein

Caudal mesenteric artery

Deep circumflex iliac artery and vein

Iliopsoas m.

Psoas minor m.

External iliac artery and vein

Internal iliac artery and vein

Rectum

Vaginal vestibule

Vulva

Anus

Hepatic veins from caudate lobe of liver

Suprarenal gland

Kidney

Ureter

Ovarian artery and vein

Ovary

Uterine horn

Colon

Urinary bladder

Median sacral artery and vein

Urethra

Vagina

Pelvic symphysis (cut)

Anal gland

Figure 11-28
Ventral view of the abdominal portion of the aorta and caudal vena cava of a female cat. The pelvic canal has been cut open, and the left kidney and uterine horn have been omitted to show deep vessels. *(From Walker,* A Study of the Cat.*)*

Slightly caudal to the cranial mesenteric artery, the aorta lies to the left side of the caudal vena cava. Trace these two vessels to the pelvic region. Their most conspicuous paired branches are the large **renal arteries** and **veins,** which supply the kidneys. Those of the right side of the body lie slightly cranial to those of the left side, since the right kidney is more cranially situated than the left one. Carefully dissect away fat from around each kidney so that you can lift up the lateral edge and look at the muscles dorsal to it. The vessels you see supplying the abdominal wall are the **cranial abdominal artery** and **vein.** Trace them toward the aorta and vena cava. The cranial abdominal vessels usually join the vena cava and aorta just cranial to the renal vessels, but they sometimes join the renal vessels directly. Before joining these vessels, they pass and supply a small, hard, oval nodule embedded in the fat between the cranial end of the kidney and the aorta and vena cava. This nodule is the **suprarenal** (adrenal) **gland.**

The suprarenal gland is an endocrine gland of dual origin. Its medullary portion, derived from postganglionic sympathetic cells of neural crest origin, secretes hormones, norepinephrine and epinephrine, that assist sympathetic stimulation in adjusting the body to meet conditions of stress. Its cortical portion, of mesodermal origin, secretes three groups of steroid hormones: mineralocorticoids, affecting sodium and potassium metabolism; glucocorticoids, affecting carbohydrate and protein metabolism; and androgens, which are male sex hormones.

The next paired branches of the aorta are the small **testicular** or **ovarian arteries** (see Figs. 12-12, 12-16, and 12-18; pp. 416, 422, and 425). They pass to the gonads accompanied by the **testicular** or **ovarian veins.** The ovaries are small, oval bodies lying near the cranial ends of the Y-shaped uterus. The testes have descended into the scrotum, and, in doing so, each has made an apparent hole (the **inguinal canal**) through the abdominal wall in the region of the groin. The testicular vessels and the sperm duct **(ductus deferens)** can be seen passing through the inguinal canals (see Figs. 12-12 and 12-16). The right gonadal vein enters the caudal vena cava; the left one may too, but it normally enters the left renal vein in the cat. It normally enters the vena cava in the rabbit.

A **caudal mesenteric artery** leaves the ventral surface of the aorta caudal to the gonadal arteries. Again, postpone tracing it until you study the vessels of the abdominal viscera. Caudal to this vessel, the aorta of the cat gives rise to a pair of **deep circumflex iliac arteries,** which pass laterally to the musculature and body wall lying ventral to the ilia of the pelvic girdle. The deep circumflex iliac arteries of the rabbit arise from a terminal branch of the aorta, the common iliac artery. Satellite **deep circumflex iliac veins** accompany the arteries and enter the caudal vena cava. The rest of the lumbar musculature is supplied by several **lumbar arteries** and **veins,** which you can find by dissecting along the dorsal surface of the aorta and vena cava between the renal and deep circumflex vessels. The lumbar vessels are single vessels where they attach to the aorta and vena cava, but they bifurcate distally. Caudal to the deep circumflex vessels, the aorta and caudal vena cava give rise to the iliac vessels supplying the pelvic region and leg. Trace them later when you study the vessels of the pelvic region and hind leg (p. 385).

♦ **(B) THE VESSELS OF THE ABDOMINAL VISCERA**

Return to the coeliac artery and mesenteric arteries where they leave the aorta. Remove surrounding connective tissue, but not the sympathetic ganglia, and trace the coeliac artery a short distance until it divides into three branches (Fig. 11-30)—a **lienic artery** to the spleen, a **left gastric artery** to the lesser curvature of the stomach, and a **hepatic artery** to the liver, pancreas, duodenum, and part of the stomach. More distal parts of these vessels will be seen with the veins described later. You can see the distribution of the **cranial mesenteric artery** to most of the small intestine and adjacent parts of the colon by stretching the mesentery. The **caudal mesenteric artery** supplies the descending colon and rectum (Figs. 11-28 and 11-29).

Although not injected, the **portal vein** can be found in the lesser omentum, where it lies dorsal to the bile duct and forms the ventral border of the epiploic foramen. Trace it caudad (Fig. 11-30). As it passes dorsal to the pylorus, it receives a small and often inconspicuous **right gastric vein** from the pyloric region of the stomach, and a larger

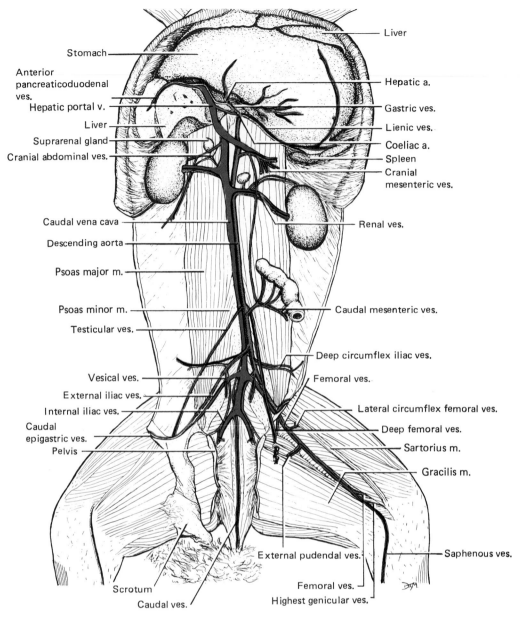

Figure 11-29
Ventral views of the posthepatic blood vessels of the rabbit. The pelvic canal has been cut open.

Labels (clockwise from top): Liver · Stomach · Anterior pancreaticoduodenal ves. · Hepatic portal v. · Liver · Suprarenal gland · Cranial abdominal ves. · Hepatic a. · Gastric ves. · Lienic ves. · Coeliac a. · Spleen · Cranial mesenteric ves. · Caudal vena cava · Descending aorta · Psoas major m. · Renal ves. · Psoas minor m. · Testicular ves. · Caudal mesenteric ves. · Vesical ves. · Deep circumflex iliac ves. · External iliac ves. · Femoral ves. · Internal iliac ves. · Caudal epigastric ves. · Lateral circumflex femoral ves. · Pelvis · Deep femoral ves. · Sartorius m. · Gracilis m. · External pudendal ves. · Saphenous ves. · Scrotum · Femoral ves. · Caudal ves. · Highest genicular ves.

gastroduodenal vein. The latter is formed by the confluence of a **cranial pancre-aticoduodenal vein** draining much of the duodenum and pancreas and a **right gas-troepiploic vein** from the greater curvature of the stomach and greater omentum. **Cranial pancreaticoduodenal, right gastroepiploic, gastroduodenal,** and **right gastric arteries** accompany the veins. All are branches of the hepatic artery, which can be seen on the left side of the epiploic foramen. After giving rise to these arteries, the hepatic artery follows the portal vein to the liver.

Push the stomach cranially and tear through the part of the greater omentum going to the spleen and dorsal body wall. Carefully dissect away the tail of the pancreas, which extends toward the spleen, and notice that the portal vein is formed by the confluence of two tributaries—a lienogastric vein, entering from the left side of the animal, and a much larger cranial mesenteric vein. Trace the **lienogastric vein** by continuing to dissect away pancreatic tissue. Its tributaries are a **left gastric vein,** which accompanies the left gastric artery and drains the lesser curvature of the stomach, and a **lienic vein,** which

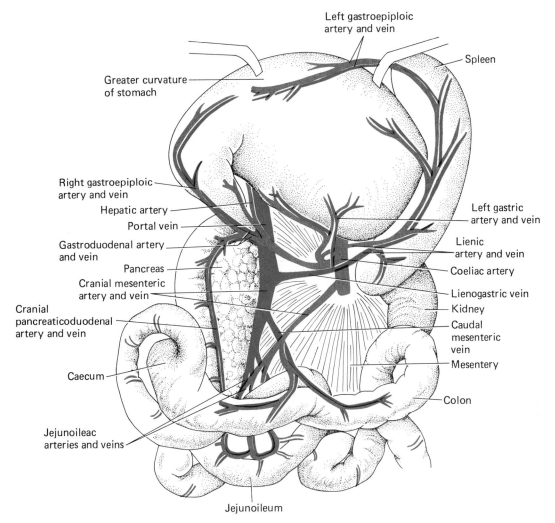

Figure 11-30
Ventral view of the hepatic portal system of veins and accompanying arteries of the cat. The stomach has been pulled forward and the tail of the pancreas dissected away.

accompanies the lienic artery to the spleen. A **left gastroepiploic artery** and **vein** can be found on the greater curvature of the stomach. They join the lienic vessels at several points.

Now trace the **cranial mesenteric vein.** One of its tributaries is the **caudal mesenteric vein** from the large intestine. Parts of this vein are accompanied by branches of the cranial mesenteric artery and parts by branches of the caudal mesenteric artery. Other tributaries of the cranial mesenteric vein accompany branches of the cranial mesenteric artery to the caudal parts of the pancreas and duodenum (**caudal pancreaticoduodenal vein** and **artery**), and to the numerous coils of the small intestine (**jejunoileac veins** and **arteries**).

♦ **(C) THE VESSELS OF THE PELVIC REGION AND HIND LEG**

Return to the caudal ends of the caudal vena cava and aorta. The terminal branches of the aorta pass superficially to the tributaries of the vena cava as they enter the pelvic cavity. In order to see the pelvic vessels clearly, open the pelvic canal. This is a simple procedure in the female. Cut the ventral ligament of the bladder and push it away from the cranioventral border of the pelvic girdle. Then take a scalpel, cut through the muscles on the ventral face of the girdle, and continue to cut right through the midventral symphysis. You may use bone scissors, but this is not necessary if you keep in the midventral line. Now take a firm grip on the thighs and bend them as far dorsally as you can. The procedure for the male is the same, but more caution is required to avoid damaging the reproductive ducts. First locate the cremasteric pouches that extend from the inguinal canals, across the ventral surface of the pelvic girdle, and into the skin of the scrotum (see Figs. 12-12 and 12-16, pp. 416 and 422). They are very narrow in the cat, but quite wide in the rabbit. They should be pushed aside before cutting. Also locate the penis emerging from the caudal end of the pelvic canal; avoid cutting it. After the canal is opened, carefully pick away fat and connective tissue from around the vessels, bladder, and rectum. Insofar as possible, confine your dissection to one side so as not to injure parts of the urogenital system.

An **external iliac artery** extends laterally and caudally toward the body wall and hind leg. It is accompanied distally by the **external iliac vein** (Figs. 11-28 and 11-29). An **internal iliac artery** and **vein** enter the pelvic cavity. The iliac arteries arise independently from the aorta in the cat but from a **common iliac artery** in the rabbit. The external and internal iliac veins of the cat unite to form a **common iliac vein** before entering the caudal vena cava; they enter the vena cava independently in the rabbit.

Trace the external iliac vessels. Usually just inside the abdominal wall, the external iliac artery and vein give off from their caudomedial surface a **deep femoral artery** and **vein,** which extend deep into the thigh (Fig. 11-31). A **caudal epigastric artery** and **vein** can be seen on the peritoneal surface of the rectus abdominis muscle. They anastomose cranially with the cranial epigastric vessels previously seen. The caudal epigastric artery is usually a branch of the deep femoral artery, but it may arise directly from the external iliac artery near the deep femoral artery. An **external pudendal artery** and **vein** can be found in the mass of fat outside the body wall in the region of the groin. They continue through the fat and supply the external genitalia. The external pudendal artery may be a branch of the caudal epigastric artery or of the deep femoral artery. The caudal epigastric and external pudendal veins normally form a short, common **pudendoepigastric trunk** before they join the deep femoral vein. After giving rise to these vessels, the external iliac vessels perforate the abdominal wall and enter the leg as the **femoral artery** and **vein.** Additional major branches of these vessels are shown in Figure 11-31.

Now trace the internal iliac vessels. Near its origin from the aorta, the internal iliac artery gives rise to a **vesical artery** to the urinary bladder. This artery is a remnant of the large umbilical artery of the embryo, which goes to the placenta (Fig. 11-20). The proximal part of the embryonic umbilical artery persists as the vessel leading to the bladder, but the portion from the bladder to the umbilicus atrophies. Deeper within the pelvic cavity, the internal iliac artery gives rise to one or two **gluteal arteries** to deep pelvic muscles and to an **internal pudendal artery** to remaining pelvic viscera. **Gluteal** and **internal pudendal veins** accompany the arteries and drain into the internal iliac vein. A small **vesical vein** normally joins the internal pudendal vein.

After the iliac arteries have branched off, the aorta continues caudad across the sacrum as a very small vessel **(median sacral artery),** and into the tail **(caudal artery).** A **caudal vein** leads to a **median sacral vein,** which normally enters a common iliac vein (cat), or an internal iliac vein (rabbit).

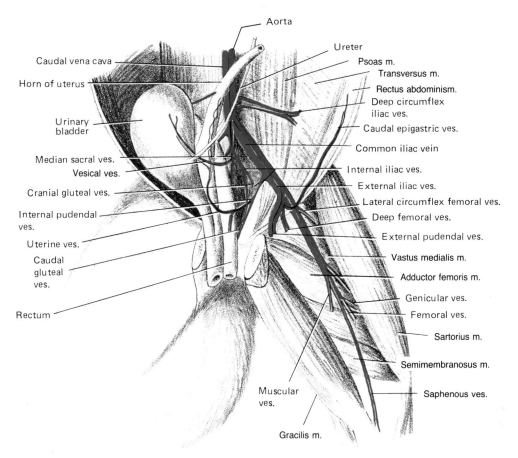

Aorta

Caudal vena cava

Horn of uterus

Urinary bladder

Median sacral ves.

Vesical ves.

Cranial gluteal ves.

Internal pudendal ves.

Uterine ves.

Caudal gluteal ves.

Rectum

Ureter

Psoas m.

Transversus m.

Rectus abdominism.

Deep circumflex iliac ves.

Caudal epigastric ves.

Common iliac vein

Internal iliac ves.

External iliac ves.

Lateral circumflex femoral ves.

Deep femoral ves.

External pudendal ves.

Vastus medialis m.

Adductor femoris m.

Genicular ves.

Femoral ves.

Sartorius m.

Semimembranosus m.

Saphenous ves.

Muscular ves.

Gracilis m.

Figure 11-31
Ventral view of the distribution of the left external and internal iliac arteries and veins in a female cat. The pelvic canal has been opened and the pelvic viscera pushed to the specimen's right side.

You have doubtless noticed during the foregoing dissections that some parts of the mammalian venous system resemble parts of the venous system in other vertebrates, but that some other parts have changed considerably. The hepatic portal system is substantially the same as in nonmammalian tetrapods, and the primitive lateral abdominal veins are represented by the umbilical veins of the embryo (p. 360). The major change is the conversion of parts of the hepatic veins and the primitive cardinal and renal portal systems into the caval and azygos system. The best way to understand how this comes about is to study the embryonic development of the veins in a mammal.

An early mammalian embryo (Fig. 11-32**A**) has a cardinal system and an incipient renal portal system, for some of the blood in the caudal part of the posterior cardinal veins passes through the kidneys to a pair of **subcardinal veins.** In this stage, the mammalian embryo is similar to a fish, except that in an adult fish the middle portion of the posterior cardinal veins atrophies, and the flow of renal portal blood through the kidneys and into the subcardinal veins is mandatory.

Later in development (Fig. 11-32**B**), the right hepatic vein enlarges, and a caudal extension of the vessel unites with the right subcardinal vein to form the proximal part of the caudal vena cava. The two subcardinal veins also unite with each other. This stage is not unlike that in the urodele.

Still later (Fig. 11-32**C**), most of the cranial portion of the posterior cardinal veins atrophies, but the caudal portion on each side forms a large vessel connecting with the subcardinal veins. The essentially new feature, regarding the trunk veins of mammals, is the subsequent formation of a pair of **supracardinal** veins connecting cranially and caudally with the remnants of the posterior cardinal veins. The supracardinal veins also become connected with the subcardinal veins by a pair of **subsupracardinal anastomoses.** This connection makes possible the elimination of most of the caudal portion of the posterior cardinal veins (the renal portal system of lower vertebrates).

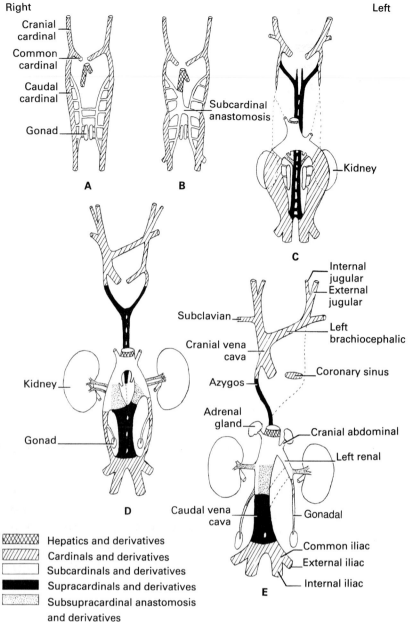

Right
- Cranial cardinal
- Common cardinal
- Caudal cardinal
- Gonad

Left

Subcardinal anastomosis

Kidney

A   **B**   **C**

Internal jugular
External jugular
Subclavian
Left brachiocephalic
Cranial vena cava
Coronary sinus
Azygos
Adrenal gland
Cranial abdominal
Kidney
Left renal
Gonad
Caudal vena cava
Gonadal
Common iliac
External iliac
Internal iliac

D   **E**

Hepatics and derivatives
Cardinals and derivatives
Subcardinals and derivatives
Supracardinals and derivatives
Subsupracardinal anastomosis and derivatives

Figure 11-32
A series of diagrams ranging from a young embryo, **A**, to an adult, **E**, to show the development of the major veins of the cat from the primitive cardinal and renal portal system. All are ventral views. For explanation, see text. *(Slightly modified after G. S. Huntington and C. F. W. McClure.)*

During subsequent development, the supracardinal veins become divided into a cranial thoracic portion and a caudal lumbar portion (Fig. 11-32**D**). The right subsupracardinal anastomosis and lumbar portion of the supracardinal vein enlarge, but those of the left side do not. Renal veins grow out from the subsupracardinal anastomosis to the definitive kidneys, which have migrated cranially.

By the adult stage (Fig. 11-32**E**), all but the most caudal segments of the posterior cardinal vein are lost, the left subsupracardinal anastomosis is lost, and the posterior vena cava is extended caudad by the enlargement of the right subsupracardinal anastomosis and lumbar portions of the supracardinal veins. In some mammals, only the right supracardinal vein is involved, but in the cat, the right supracardinal vein enlarges and absorbs the lumbar portion of the left supracardinal vein. Thus, the adult caudal vena cava is formed of the right hepatic vein, a caudal outgrowth from the right hepatic vein, the middle section of the right subcardinal vein, the right subsupracardinal anastomosis, the lumbar portion of the supracardinal veins (especially the right supracardinal vein), and a small segment of the posterior cardinal veins. The renal veins are formed primarily by outgrowths from the subsupracardinal anastomoses, but the left subcardinal vein contributes to the left renal vein. The gonadal veins are formed from the subcardinal veins plus a small segment of the posterior cardinal veins; the cranial abdominal veins are formed from the subcardinal veins.

While these changes are taking place, the thoracic portion of the left supracardinal vein disappears. But the thoracic portion of the right supracardinal vein, together with the proximal end of the right posterior cardinal vein, forms the azygos vein. Cats do not have a left highest intercostal vein, but the vessel develops in mammals that have it (rabbit) from the stump of the left posterior cardinal vein.

Transformations in the anterior veins are somewhat simpler. Internal and external jugular veins develop from the coalescence of deep and superficial tributaries of the primitive anterior cardinal vein, and they replace this part of the anterior cardinal vein. The caudal portions of both anterior cardinal veins, together with both common cardinal veins, persist in some mammals, including the rabbit, as the cranial venae cavae. This condition is close to that shown in Figure 11-32C.

In other mammals, including the cat and human beings, a cross anastomosis develops between the caudal parts of the anterior cardinal veins (Fig. 11-32D). The left anterior cardinal vein caudal to the cross trunk atrophies, whereas the right anterior and the right common cardinal veins enlarge to form the single cranial vena cava of these mammals (Fig. 11-32E). The cross anastomosis is known as the left brachiocephalic vein. The portion of the right anterior cardinal vein between the cranial vena cava and the right subclavian vein constitutes the right brachiocephalic vein.

In all mammals, the coronary veins draining the heart enter the embryonic left common cardinal vein. This becomes the base of the left cranial vena cava in the rabbit, but it forms a separate coronary sinus in the other mammals.

The Bronchi and Internal Structure of the Heart

Cut the great vessels near the heart of your specimen, remove the heart, and examine the roots of the lungs. You can now expose the bifurcation of the trachea into bronchi referred to earlier (p. 322). Trace a bronchus into a lung and notice that it subdivides repeatedly into smaller and smaller passages that terminate in clusters of thin-walled, microscopic sacs (the **alveoli**) where gas exchange occurs. This entire complex of passages is called the **respiratory tree.**

Again identify the chambers of the heart and the great vessels entering and leaving it as they appear in a ventral view (Fig. 11-33A). Carefully clean the dorsal surface of the heart and identify the chambers and vessels in this view (Fig. 11-33B and C). You can see internal features by dissecting either the heart of your own specimen or a separate sheep heart. The latter is preferable, if material is available, for the structures are larger and the chambers are not clogged with the injection mass. If you are using a sheep heart, you will have to remove the pericardial sac and clean and identify the great vessels. They are similar to those of the mammal you have studied, except that both subclavian and common carotid arteries leave the arch of the aorta by a common brachiocephalic trunk. A small left cranial vena cava is also present in the sheep, and the ligamentum arteriosum is conspicuous.

Open the heart by making the incisions shown in Figure 11-33A. To open the right atrium, make an incision that extends from the auricle into the caudal vena cava; to open

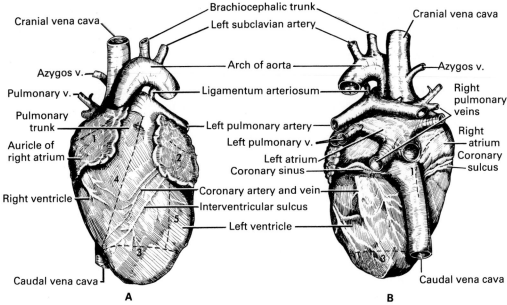

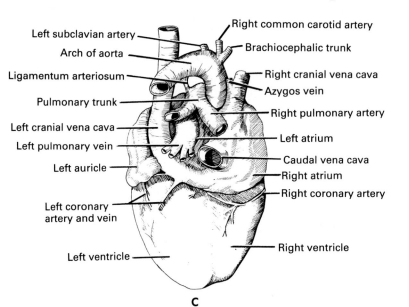

Figure 11-33
Ventral, **A**, and dorsal, **B**, views of the heart and great blood vessels of the cat. **C**, Dorsal view of the heart and great blood vessels of the rabbit. The incisions to be made in opening the heart are shown in **A**. *(A and B, from Walker,* A Study of the Cat.*)*

the left atrium, make an incision extending from its auricle through one of the pulmonary veins. To open the ventricles, first cut off the apex of the heart in the transverse plane. Cut off enough to expose the cavities of both ventricles. Then make a cut through the ventral wall of the right ventricle and extend it from the cut surface made by removing the apex into the pulmonary artery. This will be a diagonal incision. Open the left ventricle by making an incision through its ventral wall that extends from the cut surface as far forward as the base of the arch of the aorta. This will be a longitudinal incision. Clean out the chambers of the heart if necessary.

Find the entrance of the **right cranial vena cava** and **caudal vena cava** into the **right atrium.** The entrance of the **coronary sinus** (cat) or **left cranial vena cava** (rabbit) lies just caudal to the entrance of the caudal vena cava. You can determine the extent of the coronary sinus by probing. Also find the entrances of the **pulmonary veins** into the **left atrium.** The atria have relatively thin muscular walls; however, the muscles in their **auricles** form prominent bands known as **pectinate muscles** because they resemble a comb, or pecten.

The two atria are separated by an **interatrial septum.** Examine the septum from the right atrium and you will find an oval depression, the **fossa ovalis,** beside the point at which the caudal vena cava enters. Put your thumb in one atrium and forefinger in the other, and palpate this region. You will feel that the septum is unusually thin here. During embryonic life, there is an opening, the **foramen ovale,** through the septum at this point, and much of the blood in the right atrium (mostly blood coming in by the caudal vena cava) is sent directly to the left side of the heart and out to the rest of the body. This opening closes at birth (see p. 369).

The **atrioventricular openings** will be seen in the floor of the atria. The right one is guarded by the **right atrioventricular,** or **tricuspid, valve,** which consists of three flaps; the left one by the **left atrioventricular,** or **bicuspid, valve,** which consists of two flaps (Fig. 11-34). Since these flaps extend into the ventricles, you can see them better from that aspect. Note that narrow tendinous cords **(chordae tendineae)** connect the margins of the flaps with the walls of the ventricles. Many of the chordae tendineae attach onto papilla-like extensions of the ventricular muscles, the **papillary muscles.** The chordae tendineae may help to open the valves, but in any case they prevent the valves from everting into the atria during ventricular contraction.

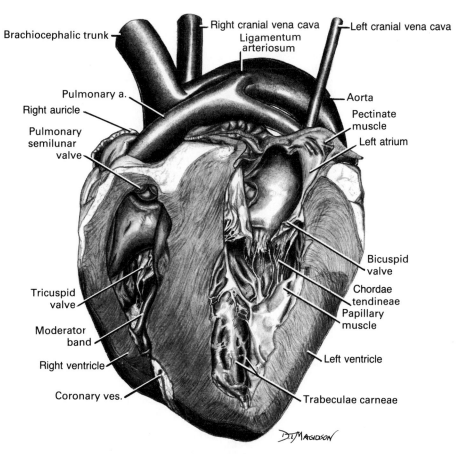

Figure 11-34
Ventral view of a dissection of the sheep heart.

Notice that the ventricles are separated from each other by an **interventricular septum** and that the walls of the ventricles are much thicker than those of the atria. As explained earlier (p. 367), the left ventricular wall is also much thicker than the right one. In addition to the papillary muscles, the inside of the ventricular walls bears irregular bands **(trabeculae carneae)** and, sometimes, bands that cross the lumen **(moderator bands)**. There is a particularly prominent moderator band in the right ventricle of the sheep. Moderator bands are believed to prevent the overdistention of the ventricle.

Notice where the pulmonary trunk and arch of the aorta leave the ventricles. Three pocket-shaped, **semilunar valves** are located in the base of each vessel, for this part of each vessel developed from a splitting of the conus arteriosus. Those in the pulmonary artery are known as the **pulmonary valve;** those in the aorta are called the **aortic valve.** The two coronary arteries leave from behind two of the semilunar valves in the aorta. One has probably been cut through.

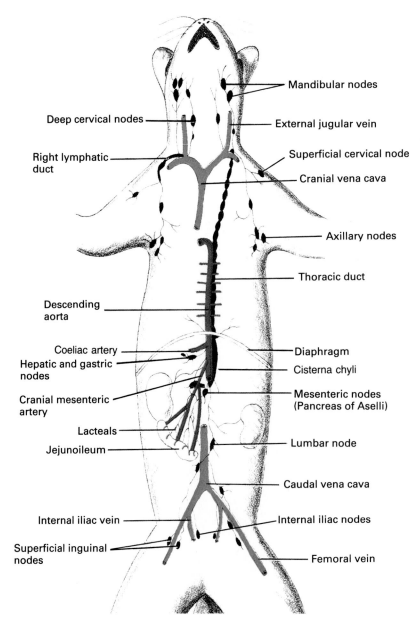

Figure 11-35
Diagrammatic ventral view of the major lymphatic vessels and groups of lymph nodes in the cat.

The Lymphatic System

The relation of the lymphatic to the cardiovascular system was considered in the introduction to this chapter (p. 336). Although the lymphatic system is not conspicuous enough in most vertebrates to be studied easily, parts, at least, of the system can be seen in mammals even though it has not been specially injected. Although the following directions are based on the cat because the lymphatics are easier to find in that animal than in the rabbit, they apply to other mammals as well.

The major lymphatic vessel of the body is the **thoracic duct** (Fig. 11-35). This is a brownish vessel that can be found in the left pleural cavity just dorsal to the aorta. Sometimes the vessel is divided into two or more channels. Trace it forward. It passes deep to most of the arteries and veins at the front of the thorax and base of the neck and then curves around to enter the left bijugular trunk beside the entrance of the subclavian vein (see Fig. 11-21). Now trace it caudad. It passes through the diaphragm dorsal to the aorta and, dorsal to the origin of the coeliac and anterior mesenteric arteries, expands into a sac called the **cisterna chyli.** The cisterna chyli is located on the left side of the body, dorsally to and between the origins of the left renal and coeliac arteries from the aorta.

Next stretch out a section of the mesentery supporting the small intestine and hold it up to the light. Very small lymphatic vessels, in this case called **lacteals** because absorbed fat passes through them, can be seen outlined by little streaks of fat. These ultimately lead into an aggregation of **mesenteric lymph nodes** (pancreas of Aselli) located at the base of the mesentery. The mesenteric lymph nodes are drained by one or more larger lymphatic vessels that pass along the cranial mesenteric artery to the cisterna chyli. These vessels have probably been destroyed. Lymphatic vessels from the stomach, liver, pelvic canal, and hind legs also pass to the cisterna chyli, but they are difficult to see. Thus, the cisterna chyli receives all the lymphatic drainage of the body caudal to the diaphragm and passes it on to the thoracic duct. The thoracic duct receives the lymphatic drainage of the thorax as it ascends through this region.

Other lymphatic vessels, which parallel the larger veins, drain the arms, neck, and head. Those of the left side enter the thoracic duct, or the left bijugular trunk close to the entrance of the thoracic duct. Those of the right side enter the right bijugular trunk near its union with the subclavian vein, either independently or by a short common trunk, the **right lymphatic duct.**

CHAPTER TWELVE

◆ ◆

THE EXCRETORY AND REPRODUCTIVE SYSTEMS

In the previous chapters, we studied the systems that serve to interact with the environment (the skeletal, muscular, and nervous systems) and to acquire food and transform it into energy that can be used by the body (the digestive, respiratory, and circulatory systems). This last chapter deals with the systems that are involved with the final aspects of the various life processes, namely, with the elimination of metabolic waste products (the excretory system) and with the generation of new life that transcends the death of the individual (the reproductive system). These two systems are treated in the same chapter because some of their parts are intimately associated and, especially in the male, are shared by both systems.

In order to understand the intimate structural relationship of the excretory system with the circulatory system on the one hand and with the reproductive system on the other hand, one must be familiar not only with the structure and function of these systems, but also with their embryonic development.

General Structure and Function of the Excretory System

The main organs of the excretory system are the kidneys.[30] Their functions include the removal of nitrogenous waste products and the maintenance of the proper composition of the body fluids (**homeostasis**) through differential secretion and reabsorption of water, ions, and other molecules.

Nitrogenous waste products are generated when amino acids are metabolically broken down and their amino groups released. In vertebrates, these amino groups are excreted as ammonia (NH_3), urea, or uric acid. Most teleosts and permanently aquatic amphibians excrete their nitrogenous waste products mainly as **ammonia,** which is highly toxic and soluble in water. It is excreted by the cells of the gill epithelium (teleosts) or by the kidneys (permanently aquatic amphibians). Because large quantities of water are needed to dissolve and carry away ammonia, terrestrial vertebrates excrete their nitrogenous waste to a greater or lesser part as urea or uric acid, depending on how much water is available for the production of urine by the kidneys. Chondrichthyan fishes, lungfishes, most amphibians, some reptiles, and mammals in general excrete mostly **urea,** which is quite soluble in water; however, it is much less

---

[30]Remember that the kidneys are not the only organs that can remove metabolic waste products and surplus substances from the body. For example, carbon dioxide, water, salt ions, and other substances can be excreted by a variety of organs, such as the gills, lungs, skin, or special salt-excreting glands (e.g., the digitiform gland of sharks).

toxic than ammonia and requires ten times less water to be washed out of the body than does ammonia. As an additional advantage, one molecule of urea contains two atoms of nitrogen, thus absorbing two amino groups per molecule. Birds and most reptiles excrete mostly **uric acid,** which is less toxic than ammonia and poorly soluble in water. It precipitates easily and is excreted as a paste, requiring 50 times less water to be removed from the body than does ammonia. Per molecule, uric acid incorporates four atoms of nitrogen.

The structure and physiology of the kidneys vary greatly among vertebrates. The vertebrate kidney that is best understood is the mammalian kidney. In general, the kidneys of other vertebrates are not as well understood and differ considerably in their gross morphology, histology, vascularization, and physiology. All vertebrate kidneys, however, consist of the same type of functional unit, the **nephron.** A nephron is composed of two parts: the renal corpuscle and the renal tubule (Figs. 12-1 and 12-4). The **renal corpuscle** consists of a capillary tuft, the **glomerulus,** which is surrounded by a **renal capsule.** The fluid that filters out of the glomerular capillaries into the renal capsule is called the **glomerular filtrate.** Its composition is essentially the same as that of blood, except that it lacks blood cells and large molecules, which do not pass through the capillary walls. The quantity of filtrated fluid is proportional to the size of the glomerulus. The **renal tubule** is a direct extension of the renal capsule; it transforms the glomerular filtrate into the final **urine** and leads it to the collecting tubules of the kidney. The structure and function of the renal tubules vary greatly among vertebrates, but basically they are all involved in secreting waste products and excess ions into the urine, or in reabsorbing water, ions, and nutrients from the glomerular filtrate. Hence, the nephrons produce a urine that removes a maximum of waste products and surplus substances and a minimum of the substances the body needs.

The blood supply to the nephrons varies among vertebrates. In order to ensure the dual function of filtration on the one hand, and of secretion and absorption on the other hand, each nephron is provided with two separate capillary networks: one—the glomerulus—to the renal corpuscle, and another—the **peritubular capillary bed**—to the renal tubule. The glomerulus receives its blood supply always through the renal artery (Fig. 12-1), but the blood supply to the peritubular capillary bed varies among vertebrates. In chondrichthyan fishes, reptiles, and birds, the peritubular capillaries receive oxygen-rich blood from the efferent renal arteriole as well as oxygen-depleted blood from the renal portal vein (Fig. 12-1**A**; see also pp. 343 and 354, and Figs. 11-5 and 11-11). In osteichthyan fishes and amphibians, the peritubular capillaries receive only oxygen-depleted blood from the renal portal vein; the efferent renal arteriole bypasses the peritubular capillary bed and joins the efferent renal venule to form the efferent renal vein (Fig. 12-1**B**; see also p. 360, and Figs. 11-14 and 11-18). In mammals, the renal portal vein has disappeared; the peritubular capillaries are supplied entirely by the efferent renal arteriole (Fig. 12-1**C**; see also p. 382, and Figs. 11-28, 11-29, and 11-32).

General Structure and Function of the Reproductive System

The "maleness" or "femaleness" of vertebrate individuals is determined by a combination of genetic and physiological factors. At fertilization, the genetic sex of a zygote is determined by the particular combination of sex chromosomes received from the oocyte and sperm cell. Nevertheless, vertebrate embryos, at least in their early stages, do not show any sexual differences, even in their reproductive organs. The development of female or male characteristics of the reproductive organs and other parts of the body is determined physiologically by hormones and, sometimes in anamniotes, by environmental cues such as temperature. Usually, the genetic and physiological factors complement one another in determining the sex of an individual. If they do not, the chromosomally determined sex of an individual may be contrary to its phenotypic sex expressed in the structure and function of the reproductive organs and

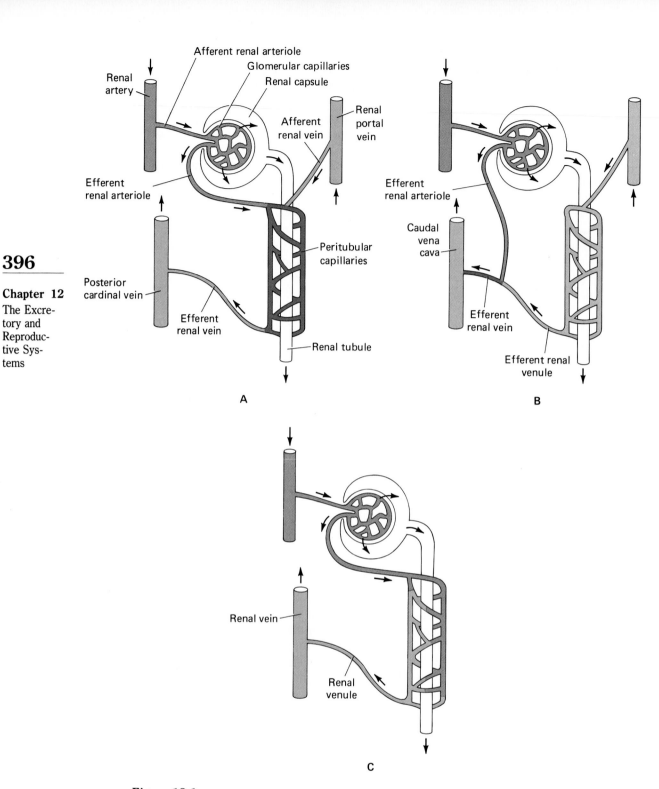

Figure 12-1
Diagrams illustrating the blood supply to a nephron in various vertebrates. (The arrows indicate the blood flow in the blood vessels and the flow of the urine in the nephron.) **A**, Chondrichthyan fishes, reptiles, and birds; **B**, osteichthyan fishes and amphibians; **C**, mammals. (Red = oxygen-rich blood; blue = oxygen-depleted blood; purple = mixed blood.)

body characteristics. Many teleost fishes can change their phenotypic sex during their lifetime, thus being **sequential hermaphrodites;** some can even produce eggs and sperms at the same time and thus are **synchronous hermaphrodites.** The phenotypic sex of certain reptiles, such as alligators and turtles, is determined by the temperature at which they are incubated. In birds and mammals, individuals with discrepancies between their genetic and phenotypic sex are rare and are usually sterile.

Embryonic Development and Evolutionary History of the Kidneys

The excretory and reproductive systems are mesodermal structures; they develop from the **mesomere,** the intermediate mesoderm portion between the epimere and hypomere (Fig. 7-5, p. 143, and Fig. 12-2). The part of the coelom that lies within the mesomere is called the **nephrocoel** (Fig. 12-4A). The mesomere soon differentiates into the **nephric ridge,** which contains the nephrocoel, and lies laterally to the genital ridge, which appears later and does not extend along the entire length of the mesomere. Starting cranially, the nephric ridge becomes segmented into **nephrotomes** (Fig. 12-2), but farther caudally, the segmentation of the nephric ridge becomes less and less obvious. The nephrotomes are thin-walled structures containing a nephrocoel that is, at first, in broad communication with the coelom (Figs. 12-4A and **B**); the unsegmented caudal part of the nephric ridge remains a more massive tissue. **Renal tubules** grow laterally out of the nephrotomes; the distal ends of the renal tubules turn caudad, forming the **archinephric duct** (also called the **wolffian duct**), into which all renal tubules tap (Fig. 12-2). As the archinephric duct reaches the caudal parts of the nephric ridge, it induces the formation of renal tubules within the tissue of the nephric ridge.

If all the renal tubules, which originate segmentally along the entire nephric ridge, remained functional, the resulting kidney would be a **holonephros,** also called **archinephros** because it is likely that this was the primitive condition of the kidney in ancestral vertebrates. In vertebrates living today, such a kidney occurs only in larval hagfishes and caecilians (Fig. 12-3A).

The first renal tubules developing at the anterior end of the nephric ridge above the pericardial cavity in all vertebrate embryos form the **pronephros** (Fig. 12-3B). It functions as a kidney in embryonic and larval cyclostomes, many osteichthyan fishes, and amphibians but serves only to initiate the formation of the archinephric duct in chondrichthyan fishes and amniotes.

Caudally to the pronephros, a few renal tubules fail to develop, creating a gap

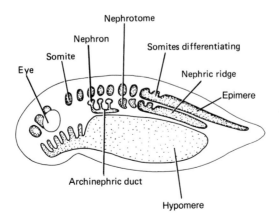

Figure 12-2
Diagram illustrating the embryonic differentiation of the nephric ridge into segmental nephrotomes and nephrons. The process starts cranially and proceeds caudad. *(From W. F. Walker,* Functional Anatomy of the Vertebrates; *after Williams and Warwick, 1980.)*

398

Chapter 12
The Excre-
tory and
Reproduc-
tive Sys-
tems

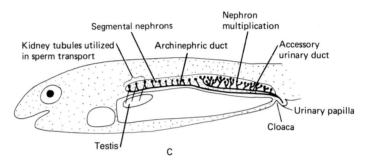

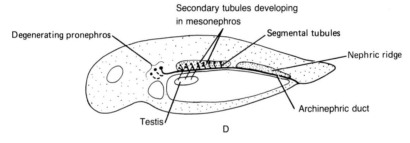

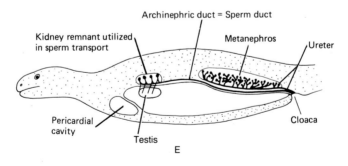

Figure 12-3
Diagrams illustrating the embryonic and evolutionary development of kidneys in vertebrates. **A,** The primitive holonephros or archinephros as exemplified by the kidneys of larval hagfishes and caecilians; **B,** the pronephros and primitive opisthonephros as exemplified by the kidneys of adult hagfishes; **C,** the derived opisthonephros as exemplified by the kidneys of most adult anamniotes; **D,** the mesonephros found in all embryonic amniotes; **E,** the metanephros found in all adult amniotes. *(From Walker,* Functional Anatomy of the Vertebrates.*)*

between the pronephros and the **opisthonephros,** which becomes the functional kidney of adult cyclostomes and anamniotes. In adult cyclostomes, the renal tubules of almost the entire opisthonephros are segmentally arranged, which is probably the primitive condition of the opisthonephros (Fig. 12-3**B**). In adult fishes and amphibians, the renal tubules proliferate, especially toward the caudal end of the opisthonephros, so that they are no longer segmentally arranged (Fig. 12-3**C**). As a consequence, the caudal part of the opisthonephros is enlarged and is the site of urine production, whereas the cranial part is reduced and, in males, is involved in sperm transport.[31] **Accessory urinary ducts** grow out of the archinephric duct in male elasmobranchs and some male urodeles to drain the caudal part of the opisthonephros.

In amniotes, the embryonic development of the kidneys differs somewhat from that in anamniotes. The renal tubules that develop in the cranial half of the nephric ridge caudal to the pronephros form a **mesonephros** (Fig. 12-3**D**). The mesonephros is a transitional organ and functions as a kidney only during the early stages of embryonic development. The archinephric duct—near its entrance into the cloaca—forms a **ureteric bud,** which grows into the caudal half of the nephric ridge. There, the ureteric bud branches and induces the formation of a large number of renal tubules, which tap into renal **collecting tubules.** The collecting tubules in turn are drained by the **ureter,** which develops out of the ureteric bud. The renal tubules that develop through induction by the ureteric bud in the caudal part of the nephric ridge form the **metanephros,** the definitive kidney of adult amniotes (Fig. 12-3**E**). The ureter of amniotes forms in the same way as the accessory urinary ducts of anamniotes; the amniote ureter, however, induces the formation of a functional kidney, while the accessory urinary ducts of the anamniote tap an already formed kidney.

Among the vertebrates living today, various types of renal tubules and nephrons are found; they are distinguished mainly on the basis of their relationship to the coelom. These various types are assumed to represent different stages in a formerly complete embryonic and evolutionary sequence (Fig. 12-4). According to this theory, capillary networks, which are supplied by branches of the aorta and drained by tributaries of the subcardinal veins, at first filter fluid into the nephrocoel and coelom. Surplus fluid that collects in the coelomic cavities enters the **nephrostomes** and is transported by ciliary action into the renal tubules and archinephric duct.[32] In a later stage, the capillary network protrudes as a glomerulus (Latin for "little ball of yarn") into the nephrocoel, thereby enlarging the surface available for filtration. This type of glomerulus is called an **external glomerulus,** because it lies outside the renal tubule. In a next stage, the passage between the nephrocoel and the coelom narrows to a ciliated **coelomic funnel**[33]. Presumably, the coelomic funnel ensures that the fluid filtered from the glomerulus enters the renal tubule directly, while still allowing surplus coelomic fluid to be drained by the kidneys. Because the glomerulus is now separated from the coelom, it is called an **internal glomerulus.** It is surrounded by a double-layered **renal capsule** (or **Bowman's capsule**), the inner layer of which is closely apposed to the capillaries of the glomerulus. The glomerulus and the renal capsule, together, form a **renal corpuscle, or malpighian corpuscle.** The renal corpuscle and renal tubule together form a **nephron**[34], the functional unit of the vertebrate kidney. In a final stage, the coelomic funnel is obliterated, and the nephron becomes independent from the coelom.

---

[31]The cranial part of the anamniote opisthonephros is homologous to the mesonephros of embryonic amniotes and often goes by the same name.

[32]Before the evolution of renal tubules, surplus fluid may have drained from the coelom directly to the outside through pores in the body wall, similar to the genital pores of lampreys or abdominal pores of sharks (see pp. 25 and 296).

[33]The terms *coelomic funnel* and *nephrostome* are often used synonymously. For didactic reasons, we will use these terms in their respective specific sense as explained here.

[34]The terms *renal tubule* and *nephron* are often used synonymously. For didactic reasons, we will use these terms as defined here.

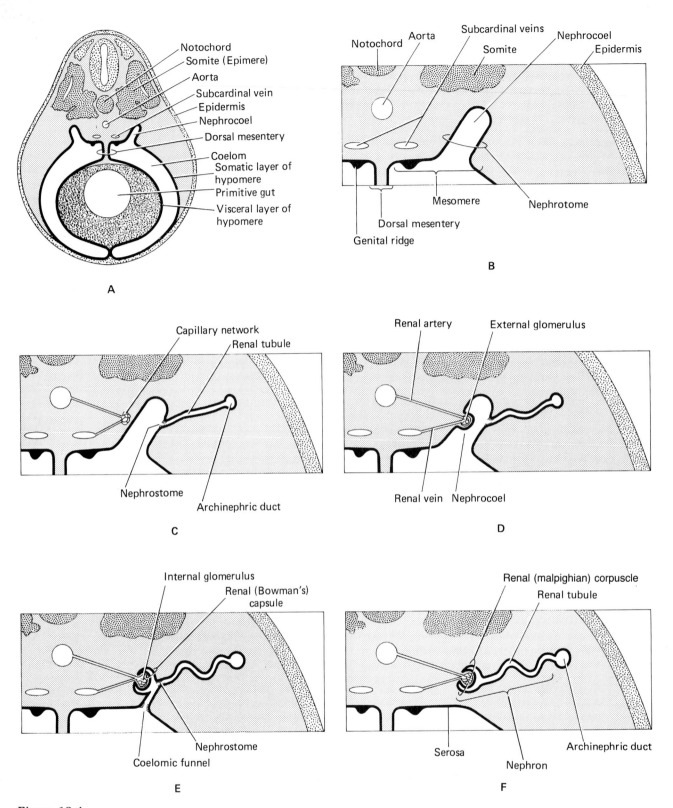

Figure 12-4

Diagrams illustrating the probable sequence of embryonic and evolutionary stages of renal tubules and nephrons. **A** and **B**, Early embryonic stage of the nephric ridge (**A**, cross section through an embryo showing the location of the nephric and genital ridges; **B**, detail from **A**); **C** and **D**, larval stages of renal tubules in the cranial end of the kidneys in hagfishes and caecilians (**C**, renal tubule with nephrostome and capillary network in the wall of the nephrocoel; **D**, renal tubule with an external glomerulus protruding into the nephrocoel); **E**, adult stage of many nephrons in elasmobranchs, some actinopterygian fishes, and some amphibians, with a persistent coelomic funnel; **F**, adult stage of nephrons in most adult vertebrates, without connection to the coelom. *(A, Adapted with permission from Hildebrand,* Analysis of Vertebrate Structure, *Copyright © 1974 and 1982 by John Wiley & Sons, Inc.)*

Embryonic Development and Evolutionary History of the Reproductive System

The early stage of the developing reproductive organs is the same in both sexes of vertebrates and therefore called the **indifferent stage.** The gonads appear first as longitudinal ridges on the medioventral surface of the mesonephros (Fig. 12-5). These **genital ridges** consist of mesenchyme (embryonic connective tissue) forming the **medulla,** and of an overlying section of thickened coelomic epithelium forming the **cortex** of the gonad. The **primordial germ cells,** which will give rise to the actual gametes, migrate from the yolk sac via the wall of the primitive gut and the dorsal mesentery into the cortex. The epithelium of the cortex is now called the **germinal epithelium.** As the germinal epithelium proliferates, it sends rodlike extensions, the **primary sex cords,** into the mesenchyme of the medulla. At the same time, the **müllerian sulcus,** which forms ventrolaterally to the archinephric duct, continues to invaginate and forms the **müllerian duct,** also called the **paranephric duct.** In elasmobranchs, the müllerian duct forms through an invagination of the archinephric duct and opens into the coelom through a modified mesonephric coelomic funnel.

In a male embryo, a layer of connective tissue invades the gonad and separates the primary sex cords from the cortical germinal epithelium (Fig. 12-5**D**). The severed primary sex cords continue to proliferate and form a mass of sex cords. Their tips in the center of the medulla form a network of **cords of the urogenital union,** or **rete cords** *(rete testis),* which connects the primary sex cords to the renal tubules of the mesonephros and thereby to the archinephric duct[35]. The testes gradually pull away from the body wall, to which they remain anchored by a subsidiary mesentery, the **mesorchium** (see also p. 293).

In most anamniotes, the sex cords and rete cords subsequently hollow out, and the mesonephric renal tubules and the archinephric duct become the permanent channels for sperm transport. In teleosts, the rete cords form a separate sperm duct, which completely bypasses the opisthonephros. In amniotes, the primary sex cords hollow out at sexual maturity and become the **seminiferous tubules** of the testis. The primary sex cord cells of epithelial origin become the **Sertoli cells.** These surround the **spermatogonia,** which develop from the primordial germ cells. The mesenchymal cells between the sex cords mature into the **interstitial cells of Leydig,** which secrete the male sex hormone testosterone. The mesonephric renal tubules become the **efferent ductules** of the testis. Their highly convoluted tips form the **head of the epididymis,** or **caput epididymidis.** The convoluted cranial part of the archinephric duct becomes the **ductus epididymidis** in the **body** (or **corpus epididymidis**) and **tail** (or **cauda epididymidis**) of the epididymis. (The amniote epididymis is homologous to the cranial part of the opisthonephros and archinephric duct of anamniotes). The rest of the amniote archinephric duct is now called the **ductus deferens** (plural, *ductus deferentes*) (Fig. 12-6**B**). The müllerian duct regresses and degenerates. In amniotes, furthermore, the caudal end of the mesonephros sometimes forms a vestigial adult structure, the **paradidymis.**

In a female embryo, the primary sex cords and traces of a connective tissue layer degenerate. At the same time, the germinal epithelium experiences a second proliferative spurt and extends **secondary sex cords** into the medulla (Fig. 12-5**E**). The secondary sex cords remain close to the cortex and break up into clusters of cells, or **primordial follicles,** which each consist of an **oogonium** developed from a primordial germ cell and surrounded by **granulosa cells,** or **follicle cells,** derived from epithelial cells of the germinal epithelium. The ovary gradually pulls away from the body wall, to which it remains anchored by a subsidiary mesentery, the **mesovarium.**

In female anamniotes, the anterior opisthonephric renal tubules do not form any direct connection with the ovary and remain—with the archinephric duct—a functioning part of the excretory system. In amniotes, the mesonephric renal tubules and the

---

[35]The archinephric duct is often called the **mesonephric duct** when it drains the mesonephros.

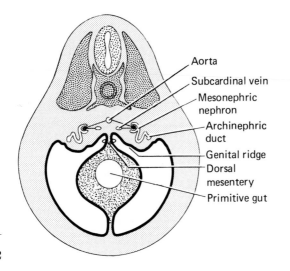

Aorta
Subcardinal vein
Mesonephric nephron
Archinephric duct
Genital ridge
Dorsal mesentery
Primitive gut

A

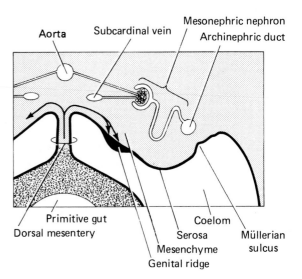

Aorta
Subcardinal vein
Mesonephric nephron
Archinephric duct

Primitive gut
Dorsal mesentery
Coelom
Serosa
Mesenchyme
Genital ridge
Müllerian sulcus

B

Renal corpuscle
Renal tubule
Archinephric duct

Müllerian duct
Mesenchyme
Primary sex cords
Germinal epithelium

C

Subcardinal vein
Aorta
Degenerating renal corpuscle
Archinephric duct

Serosa
Primary sex cords
Tunica albuginea
Müllerian duct
Renal tubule

D

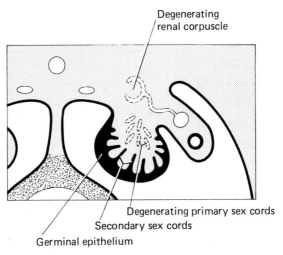

Degenerating renal corpuscle

Degenerating primary sex cords
Secondary sex cords
Germinal epithelium

E

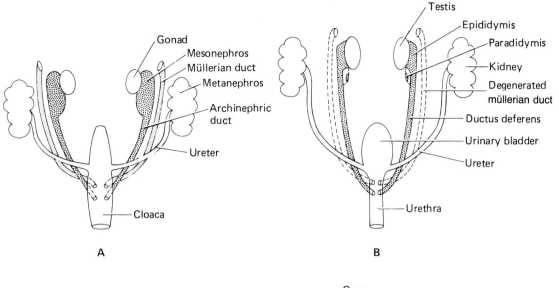

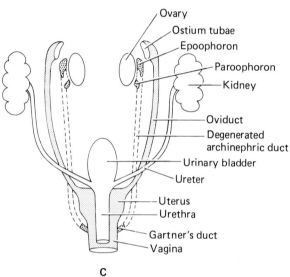

Figure 12-6
Diagrams illustrating the embryonic development of the genital ducts in amniotes. **A,** Indifferent stage; **B,** male condition; **C,** female condition. *(Redrawn from Browder, Erickson, and Jeffery, after Balinsky, 1981.)*

archinephric duct regress and degenerate, as the metanephros and the ureter become the definitive excretory organs (Fig. 12-6**C**). Some vestiges of the mesonephros and archinephric duct may remain visible in the adult amniote female, such as the **epoophoron** (cranial mesonephric renal tubules), the **paroophoron** (caudal mesonephric renal tubules), and **Gartner's duct** (caudal end of the archinephric duct).

Because the follicles remain near the surface of the ovary, and no direct connection

Figure 12-5
Diagrams illustrating the embryonic development of the gonads and genital ducts. **A,** Cross section through an embryo at the level of the mesonephros to show the location of the genital ridge; **B,** enlarged detail from **A** showing the topographical relationships among the genital ridge, mesonephros, and müllerian sulcus (the arrows indicate the path of the primordial germ cells); **C,** the formation of the primary sex cords and müllerian duct in the indifferent stage of the gonad; **D,** the separation of the primary sex cords from the peripheral germinal epithelium in the male gonad; **E,** the formation of the secondary sex cords and degeneration of the primary sex cords in the female gonad. *(A, Adapted from Hildebrand, 1982; **B-E,** redrawn from Browder, Erickson, and Jeffery, after Sadler, 1985.)*

is established between the ovary and any duct, the mature oocytes of most vertebrates are released into the coelom after breaking through the serosa enveloping the ovary. In cyclostomes, the oocytes are released directly to the outside through the genital pores (p. 25), but in most vertebrates they are transported toward the **ostium tubae** and released to the outside through the **oviduct,** the former müllerian duct.

Study of the Excretory and Reproductive Systems

We assume that you have observed the major parts of the excretory and reproductive systems during previous dissections. In the following exercises, the finer aspects will be examined and related to the more conspicuous parts. In studying these systems, you should not only dissect your own specimen, but you should also examine the dissection of a specimen of the opposite sex. Since someone else, in turn, will have to examine your specimen, make a particularly careful dissection. If possible, study sexually mature specimens.

FISHES

The excretory and reproductive systems of the cartilaginous fishes are a good example of a primitive vertebrate condition in most respects. The kidneys are opisthonephroi drained by archinephric ducts, which are supplemented in the male by accessory urinary ducts. The gonads are situated far forward in the body cavity. The eggs are discharged into the coelom and through a pair of oviducts; the sperm is discharged through the cranial part of the kidneys and through the archinephric ducts. A cloaca is present.

The nephrons of cartilaginous fishes have large glomeruli that remove a considerable volume of water from the blood by filtration. This is an advantage in a freshwater (hypotonic) environment, which was the probable environment of most of the early vertebrates, and may explain the widespread distribution of glomerular nephrons among vertebrates. But a glomerular nephron poses problems for most marine fishes, which must conserve body water because the concentration of osmotically active solutes in sea water is greater than in their body fluids. Cartilaginous fishes avoid this problem by retaining many nitrogenous excretory molecules, especially urea, in their body fluids, and by having evolved a greater tolerance for urea. Mechanisms of urea retention include having gills that are less permeable to its passage than those in most fishes and some urea reabsorption in the renal tubules. As a consequence, the osmolarity of their body fluids is equal to or greater than that of sea water, so water tends to diffuse into the body. Marine cartilaginous fishes produce a copious and dilute urine. In common with other marine vertebrates, cartilaginous fishes also must eliminate excess salts. Divalent ions (phosphate and sulfate) are excreted by the renal tubules, monovalent ions (chloride and sodium) by certain cells in the gills and by the digitiform or rectal gland (p. 298). Further information about osmoregulation in cartilaginous fishes can be found in a review by Pang, Griffith, and Atz (1977) and in several chapters in the book by Shuttleworth (1988).

Another specialized feature of many cartilaginous fishes, including *Squalus,* is that they retain their young within a uterus until embryonic development is complete and are therefore viviparous. Agnathans and sharks with a more primitive mode of reproduction are egg-laying (oviparous).

♦ **(A) THE KIDNEYS AND THEIR DUCTS**

The kidneys of the dogfish are a pair of bandlike organs lying dorsal to the parietal serosa, a position called **retroperitoneal**[36], on either side of the dorsal mesentery (Fig. 12-7). A conspicuous, white **caudal ligament** arises from the vertebral column between the kidneys and passes into the tail. The dogfish kidneys are opisthonephroi, for they extend nearly the length of the pleuroperitoneal cavity; they are drained, in part at least, by **archinephric ducts** leading to the **cloaca,** and they have a relatively primitive nephron structure. (Traces of microscopic coelomic funnels are associated with some of the cranial tubules.) You may have to cut the parietal serosa along the lateral border of a kidney to trace it cranially, for the cranial two thirds of each kidney is narrower and less conspicuous than the caudal one third. It is in the caudal third that most of the urine production occurs. The cranial part is related to the reproductive system in the male; it is poorly developed in the female.

The archinephric duct can easily be seen in the mature male, for it is a large, highly convoluted tube lying on the ventral surface of the opisthonephros. It is much smaller and straighter in the immature male. The archinephric duct of the female resembles that of an immature male and cannot be seen until the oviduct is studied.

Further aspects of the urogenital system must be considered separately in each sex.

♦ **(B) THE MALE UROGENITAL SYSTEM**

Notice that the paired **testes** are located near the cranial end of the pleuroperitoneal cavity adjacent to the cranial end of the kidneys (Fig. 12-7). They are particularly large in fish caught in January and February during their mating season. Each testis is suspended by a **mesorchium,** through the front of which pass several small, inconspicuous tubules called the **ductuli efferentes.** The ductuli efferentes carry the sperm from the testis to modified cranial renal tubules. After passing through these tubules, the sperm enter the **archinephric duct** and descend through it to the cloaca. The large size of the mature male archinephric duct is attributed to its role in sperm transport.

The part of the opisthonephros receiving the ductuli efferentes is homologous to the head of the epididymis of amniotes; the highly coiled portion of the archinephric duct adjacent to this region is homologous to the ductus epididymidis of amniotes, which comprises the body and tail of the epididymis; and the rest of the archinephric duct is comparable to the ductus deferens. Sometimes these terms are applied to the comparable organs of elasmobranchs.

The portion of the opisthonephros between the caudal end of the testis and the enlarged, caudal excretory region of the kidney is known as **Leydig's gland.** Most of the tubules in this region are modified to produce a secretion analogous to the seminal fluid of amniote vertebrates. This secretion is discharged into the archinephric duct.

---

[36]See footnote 19, p. 289. The term *retroperitoneal* for the position of the kidneys originated in human and veterinary anatomy, which deals with the mammalian condition.

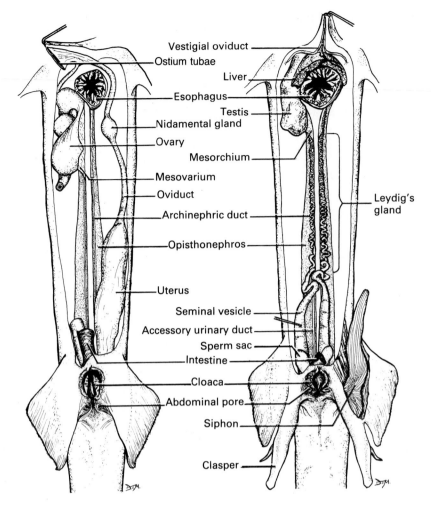

Figure 12-7
Ventral views of the urogenital system of mature specimens of *Squalus. Left,* female; *right,* male.
The siphon has been dissected free on one side of the male.

As the archinephric duct approaches the excretory portion of the kidney, it straightens and enlarges to form a **seminal vesicle.** Remove the parietal serosa from this portion of the kidney and trace the seminal vesicle caudad. You will also have to open the cloaca on the side on which you are working by cutting through the side of the cloacal aperture and into the lateral wall of the intestine. The caudal end of the seminal vesicle passes dorsal to a **sperm sac,** whose blind cranial end you should free from the seminal vesicle. The sperm sac develops as an outgrowth of the archinephric duct. Sperm are stored in the seminal vesicle and sperm sac. In addition, secretory cells in these organs contribute to the seminal fluid. The caudal end of the sperm sac and that of the seminal vesicle unite to form a **urogenital sinus,** which develops from the archinephric duct. Cut open the ventral surface of the sperm sac and urogenital sinus, clean them out, and notice the papilla that bears the opening of the seminal vesicle. The urogenital sinuses of opposite sides unite caudal to the entrances of the seminal vesicles and extend into the **urogenital papilla** located dorsally in the cloaca. Probe caudally through the urogenital sinus and notice the emergence of the probe through the tip of the urogenital papilla.

Oviducts are present in the sexually indifferent stage of a male embryo. Remnants in adult males remain as tubular folds that can be found cranially on either side of the falciform and coronary ligaments. These portions of the oviducts unite in the falciform ligament and have a common entrance, the **ostium tubae,** into the coelom. The ostium tubae can be found along the caudodorsal edge of the ligament. The rest of the oviducts are lost.

The tubules of the excretory part of the kidney do not enter the seminal vesicle but rather an **accessory urinary duct** that lies against the dorsomedial edge of the seminal vesicle. You can find this duct by freeing the lateral edge of the seminal vesicle and dissecting dorsal to the seminal vesicle. The accessory urinary duct, despite its name, carries virtually all the urine, for the cranial parts of the kidney excrete little if any urine in a mature male. Trace the accessory duct caudad; it enters the urogenital sinus caudal to the entrance of the seminal vesicle.

You can now understand that the cloaca is a sort of sewer (Latin, *cloaca,* "a sewer"), for it receives the feces from the digestive system, the urine from the excretory system, and the gametes from the reproductive system. The cloaca is not divided in the male, but the excretory and genital products are discharged more dorsally and caudally than the feces.

Unlike in most fishes, in the dogfish and other cartilaginous fishes fertilization is internal. Copulation has not been described for *Squalus,* but it is probably similar to the pattern seen in other small sharks. The female lies quietly and the male coils around the female (Fig. 12-8) in such a way that one of the pelvic claspers, which is turned forward, can be inserted into the cloaca and oviduct of the female. Spurs on the clasper help to hold it in place. Sperm pass from the cloaca into the groove on the dorsal surface of the clasper. A sac-shaped, muscular-walled **siphon** is associated with each clasper. If you have not already done so, you can find one by skinning the ventral surface of a pelvic fin (p. 153). Cut open the siphon and pass a probe through it and into the groove on the clasper. Prior to copulation, sea water is taken into the siphon. When the siphon contracts, sea water and secretions of the siphon are discharged and propel the sperm along the clasper groove and into the female. The secretion includes 5-hydroxytryptamine, which has been shown in vitro to induce contractions of the oviduct. It has been suggested that this helps to move sperm up the oviduct to the site of fertilization near the cranial end of the oviduct. (See Wourms, 1977, and Shuttleworth, 1988, for a review of chondrichthyan reproduction.)

Figure 12-8
Copulation in the spotted dogfish, *Scyliorhinus canicularis. (From Budker, after Bolau.)*

The **ovaries** are a pair of large organs located near the front of the pleuroperitoneal cavity adjacent to the cranial ends of the kidneys (Fig. 12-7). Each is supported by a **mesovarium** and contains eggs in various stages of maturity. When the eggs are mature, they attain a diameter of nearly 3 centimeters and contain an enormous amount of yolk. Cut into one of the larger eggs to see the yolk. Each egg is surrounded by a sheath of follicular cells, but this cannot be seen grossly.

When the eggs are mature, they break through the wall of the follicle and the serosa enveloping the ovary (a process called **ovulation**), fall into the coelom, and are transported (probably by the pressure of the surrounding viscera) towards the unpaired ostium tubae and into the paired oviducts. In mature females, each **oviduct** is a prominent tube suspended by a **mesotubarium** from the ventral surface of the kidney. In immature specimens, the oviducts are small tubes lying against the kidneys, and mesotubaria are lacking. Trace an oviduct cranially. It passes dorsally to the ovary and then curves ventrally and caudally in front of the liver to enter the falciform ligament. The oviducts of opposite sides unite within the falciform ligament and have a common opening, the **ostium tubae,** into the coelom. The ostium tubae is located on the caudodorsal edge of the falciform ligament; you can open the ostium tubae by spreading its lips apart. It is inconspicuous and small in immature females, but in mature females it is very large, and the falciform ligament is elongated accordingly and tears easily into shreds.

The oviduct is narrow in diameter throughout much of its length but enlarges in two regions. One enlargement lies dorsal to the ovary. This is the **nidamental gland,** or **shell gland.** Fertilization as well as shell deposition occurs here. In the dogfish, the nidamental gland secretes a thin, proteinaceous shell, composed primarily of collagen and known as the "candle," around groups of two or three fertilized eggs. Among elasmobranchs, however, the structure and composition of egg shells vary greatly. Sperm that will fertilize the eggs are stored in the nidamental gland after copulation and may remain viable for several weeks or longer. In mature females, the caudal one third to one half of the oviduct is enlarged as the **uterus.** In pregnant females, it is expanded and its wall is highly vascularized, because the embryos develop in it.

Open the cloaca by cutting through the side of the cloacal aperture and into the lateral wall of the intestine. The two oviducts enter the caudodorsal part of the cloaca just ventral to a **urinary papilla.** The urogenital portion of the cloaca, known as the **urodeum,** is partially separated by a horizontal fold from the anteroventral, fecal portion of the cloaca **(coprodeum).**

The kidneys of the female *Squalus* are drained by the **archinephric ducts,** for there are no accessory urinary ducts as there are in the male. Some other female elasmobranchs, however, have accessory urinary ducts. You can find an archinephric duct by making an incision through the parietal serosa along the lateral border of the caudal part of the kidney and very carefully reflecting the parietal serosa from the surface of the kidney. The archinephric duct lies on the ventral surface of the kidney directly dorsal to the attachment of the mesotubarium. If you do not see it on the kidney, it probably adheres to the dorsal surface of the parietal serosa and can be picked off the reflected serosa. The archinephric duct of the female is much smaller than that of the male and is not convoluted. Trace it caudad. The caudal ends of the archinephric ducts of opposite sides enlarge slightly and unite to form a small **urinary sinus,** which opens through the tip of the urinary papilla. The urinary sinus is too small to be easily dissected.

◆ (D) REPRODUCTION AND EMBRYOS

As the eggs develop within the follicles in the ovary, certain of them begin to increase in size through the accumulation of yolk. At the time of ovulation in the early winter of every second year, each ovary contains two or three ova averaging 3 centimeters in diameter. After ovulation, the follicular cells are converted into a **corpus luteum.** This is a large body 2 centimeters in diameter early in pregnancy, but it gradually regresses. The eggs enter the oviduct, which can stretch greatly in a living specimen. As they pass down the oviduct and enter the nidamental gland, they are fertilized and a thin shell is deposited around several of them. The eggs then pass to the uterus, in which they may be seen in specimens in an early stage of pregnancy.

After several months, the shell is reabsorbed, and the embryos develop within the uterus. Reproduction of sharks has been reviewed by Wourms (1977). The gestation period of *Squalus* lasts 22 months. Pups that are slightly over a year old, which is a stage often seen in pregnant specimens obtained from biological supply houses, range in length from 12 to 20 centimeters; much of the yolk is contained within an **external yolk sac** protruding from the underside of the embryo (Fig. 12-9). This is a **trilaminar yolk sac** (see Fig. 12-11), for it contains all three germ layers. The rest of the yolk is carried in an **internal yolk sac,** which can be found within the pleuroperitoneal cavity. Pups just before parturition range in length from 23 to 29 centimeters. The yolk in the external sac has been consumed by then, but a small reserve of yolk remains in the internal sac. Parturition occurs late in the autumn of the second year.

Sharks with a primitive mode of reproduction are **oviparous.** Their eggs are heavily laden with yolk, fertilization is internal, and a well-developed collagenous egg case is secreted around each fertilized egg. The eggs are laid in the surrounding sea, and the embryos develop within their protective egg cases.

In a more advanced reproduction mode, the fertilized eggs, surrounded by a remnant of the egg case, are deposited in the uterus, where development is completed and the young are born as miniature adults. Such sharks are live-bearers, or **viviparous.** Nutritional arrangements for the embryos vary in different species from complete dependence for organic nutrients upon the stored yolk, through various types of placental analogues, to the transfer of most materials needed by the embryo from the mother through a yolk sac placenta. The term *ovoviviparity* has been used to describe cases of live-bearing without the presence of a well-developed placenta, but there is such a continuous range of dependence upon the mother for nutrients that it is hard to determine where ovoviviparity shifts to viviparity. Contemporary authors speak instead of **aplacental** and **placental viviparity.**

Squalus is characterized by aplacental viviparity. Although a *Squalus* embryo derives its organic nutrients from the yolk, some exchanges of other materials occur by way of the numerous, vascular **uterine villi** that line the uterus and are applied to the surface of the vascularized yolk sac. Gas exchange and some transfer of water and minerals occur by this route early in development. Later, the uterus appears to be irrigated periodically with sea water. During development, there is a steady decrease in the organic content of the combined yolk sac and embryo, but there is a 78 percent

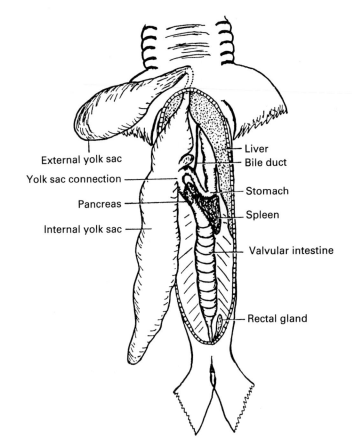

Figure 12-9
External and internal yolk sacs in a 220-mm embryo of *Squalus suckleyi*. The external yolk sac has been pulled out of the body cavity. *(From Brown, after Hoar.)*

gain in their combined weight because of the uptake of water and minerals. Early in development, the yolk is digested by endodermal cells lining the yolk sac, and the digested products are absorbed into the vitelline circulation. Later, yolk is moved by ciliary action into the embryo's intestine, where it is digested and absorbed.

The uterine villi of *Squalus* are one type of placental analogue. Others found in different species include the secretion by the mother of a uterine milk rich in organic content, which is absorbed by vessels in the yolk sac or swallowed by the embryo, and the eating by embryos of yolk-rich eggs that are ovulated later **(oophagy)**. *Mustelus canis* is an example of a small shark with placental viviparity, but even in this species the embryo uses stored yolk before the establishment of a union between the yolk sac and uterine lining. As in mammals, there is considerable variation among placental sharks in the intimacy of the union between embryonic and maternal tissues.

AMPHIBIANS

The earliest tetrapods, the amphibians, are still anamniotes, and their excretory and reproductive systems have not changed in gross morphology significantly from the condition of these systems in ancestral fishes. The only new feature is a relatively large urinary bladder formed as a ventral outgrowth of the cloaca. In many other respects, the urogenital system in amphibians is even more similar to what is believed to have been the primitive vertebrate condition than it is in the dogfish. Amphibians, with rare exceptions, are adapted strictly to freshwater, which is believed to have been the environment in which the ancestral vertebrates evolved. The nephrons in their opisthonephric kidneys are of the primitive type with a large glomerulus and therefore eliminate excess water as well as nitrogenous waste products. Amphibians also retain the primitive vertebrate mode of reproduction. They are oviparous, laying their eggs in the water or in very moist areas on land. In most cases, the eggs hatch into free-swimming larvae that later metamorphose into terrestrial adults. Thus, while amphibians have made a start in adapting to terrestrial conditions in most of their organ systems, they are limited, as a group, to moist habitats because of their inability to conserve water and reproduce under terrestrial conditions. *Necturus,* although a permanent larva, is a good example of this level of tetrapod evolution.

♦ (A) THE KIDNEYS AND THEIR DUCTS

The kidneys of *Necturus* are **opisthonephric.** They lie in the caudal half of the pleuroperitoneal cavity on either side of the dorsal mesentery (Fig. 12-10). They are easily seen in the male, but you will have to push the ovary and oviduct apart to see one in the female. The kidneys of most vertebrates have a retroperitoneal position, but in *Necturus* they protrude into the body cavity and are nearly completely surrounded by visceral serosa. Also notice that the caudal part of the kidney is much larger than the cranial part. The cranial part of the male kidney functions as part of the reproductive system, whereas that of the female kidney is degenerated. The cranial part of the kidney is best seen by lifting up the lateral border of the organ and looking at its dorsolateral surface.

The kidneys are drained in both sexes exclusively by the **archinephric ducts,** for accessory urinary ducts are absent in *Necturus.* The archinephric duct of the male is a

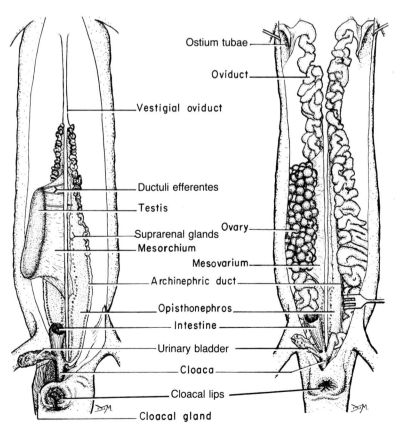

Figure 12-10
Ventral views of the urogenital system of *Necturus. Left,* male; *right,* female.

large, convoluted tube extending down the lateral border of the kidney to the cloaca. The archinephric duct of the female is similarly located but is much smaller and is not convoluted. Small **collecting tubules** may be seen entering the archinephric duct from the caudal, excretory portion of the kidney.

A **urinary bladder** lies ventral to the large intestine and is connected to the midventral body wall by a mesentery known as the **median ligament of the bladder.** The urinary bladder is not directly connected to the archinephric duct but is a cranioventral evagination of the cloaca. As the urine is discharged from the archinephric ducts into the cloaca, it collects in the urinary bladder, where it is concentrated through selective reabsorption of water by the epithelium of the urinary bladder. The urinary bladder of amphibians, unlike that of mammals, is not simply a storage organ but an organ with a crucial role in osmoregulation.

The **suprarenal glands** of amphibians consist of cortical and medullary cells clustered together in irregular patches. In well-injected specimens you will see a number of small, well-vascularized patches of suprarenal tissue along the ventral surface of the kidney of *Necturus.*

◆ **(B) THE MALE UROGENITAL SYSTEM**

The **testes** are a pair of oval organs located near the cranial ends of the kidneys. Each is suspended by a **mesorchium.** Several inconspicuous **ductuli efferentes** pass through the front of the mesorchium carrying sperm from the testes to the modified renal tubules of the cranial portion of the opisthonephros. From here, the sperm pass to the cloaca through the **archinephric duct,** which is therefore serving as a sperm duct as well as an excretory duct (Fig. 12-10). Notice that the cranial part of the archinephric duct, a part comparable to the ductus epididymidis of amniotes, is much more convoluted than the rest of the duct. The small black line along the edge of the cranial part of the archinephric duct, and extending forward along the side of the posterior cardinal vein, is a remnant of the oviduct, which has persisted from the sexually indifferent stage of the embryo. A shorter but similar remnant of the oviduct can sometimes be seen along the very caudal end of the archinephric duct.

Cut through the body musculature lateral to the cloaca and find the point at which the archinephric duct joins the dorsal wall of the cloaca just caudal to the large intestine. Open the cloaca by making an incision that extends from the cranial end of the cloacal aperture into the lateral wall of the intestine. The archinephric ducts enter the craniodorsal wall of the cloaca just caudal to a transverse ridge formed by the entrance of the intestine, but their openings probably will not be seen.

Remove the skin on one side and on the ventral surface of the cloaca, if you have not done so yet, and observe the large **cloacal gland** consisting of many small tubules. The secretions of this gland, together with the secretions of a less conspicuous **pelvic gland** in the dorsal wall of the cloaca, agglutinate the sperm into clumps called **spermatophores.** The spermatophores of salamanders generally are deposited in the water to be picked up by the cloacal lips of the female, but in a few species they are transmitted directly to the cloaca of the female. Finally, notice the numerous **papillae** on the cloacal lips, which are a characteristic feature of the male.

◆ **(C) THE FEMALE UROGENITAL SYSTEM**

The **ovaries** are a pair of large, granular organs located on either side of the dorsal mesentery adjacent to the cranial part of the kidneys (Fig. 12-10). Each is suspended by a **mesovarium** and contains eggs within follicles in various stages of maturity. The **oviducts** are a pair of large, convoluted tubes lying along the lateral side of each kidney and extending forward nearly to the front of the pleuroperitoneal cavity. Each terminates cranially in a funnel-shaped opening called the **ostium tubae,** and caudally in the cloaca. Cut through the musculature lateral to the cloaca to see that the oviduct attaches to the craniodorsal wall of the cloaca just caudal to the entrance of the large intestine. At ovulation, the eggs fall into the coelom, are carried by the action of cilia on the serosa to the ostia tubarum, and descend the oviducts to the cloaca. The oviducts are not differentiated grossly, but part of their lining is glandular. The glands of the oviduct secrete gelatinous layers around the eggs. These gelatinous layers are semipermeable and form a light-weight protective capsule when the eggs are deposited.

Open the cloaca by cutting from the cranial end of the cloacal aperture into the lateral wall of the intestine. The oviducts enter the craniodorsal wall of the cloaca through a pair of **genital papillae.** If these are not clear, slit an oviduct near its caudal end and pass a probe through it into the cloaca. The caudal end of each **archinephric duct** can be seen to leave the kidney and pass onto the wall of the oviduct, with which it becomes intimately united. However, the archinephric duct enters the cloaca independently beside the opening of the oviduct.

The **cloacal gland,** which helps to produce the spermatophores in the male, is present in a much reduced state in the female. You can find it by removing the skin on one side and on the ventral surface of the cloaca. The homologue of the pelvic gland of the male is transformed into a **spermatheca** in the female. The tubules within the spermatheca store sperm after copulation. In *Necturus,* sperm are stored from the breeding season in the fall until egg-laying in the spring. The spermatheca is located in the dorsal wall of the cloaca but cannot be seen grossly. Finally, notice that the lips of the female cloaca bear smooth folds rather than the papillae characteristic of the male.

MAMMALS

In the course of their evolution through reptiles to mammals, vertebrates have become well adapted to an active, terrestrial life characterized by a high level of metabolism. Mammals must remove a far larger volume of nitrogenous wastes from their body fluids than required in anamniote terrestrial vertebrates; at the same time, they must conserve body salts and water. This is made possible by significant increases in the number of nephrons, in the glomerular filtration rate, and in the amount of tubular reabsorption. A frog, for example, has a filtration rate of 822 ml/kg in 24 hours, which is about 100 times the volume of its body fluids. It reabsorbs about 60 percent of this, so urine flow is 317 ml/kg in 24 hours. A dog, in contrast, has a filtration rate of 6190 ml/kg in 24 hours (nearly 900 times its body fluid volume) and reabsorbs 98 to 99 percent of this, so its urine flow is only 15 to 100 ml/kg in 24 hours. This is considerably less than that of the frog! Mammals produce a very concentrated urine. Indeed, mammals and, to some extent, birds are the only vertebrates that can form a hyperosmotic urine, that is, a urine more concentrated than their body fluids.

The kidney is a **metanephros** drained by a **ureter,** and what is left of the mesonephros and archinephric duct becomes part of the male genital system. Thus, there is a more complete separation of excretory and genital functions in amniotes than in anamniotes. There is also a division of the cloaca in most mammals that separates the urogenital tract from the digestive tract.

Reptiles and mammals can also reproduce on land and do not need to return to water. Mating on land requires internal fertilization and, therefore, the evolution in the male of a **copulatory organ** and **accessory genital glands** that secrete the seminal fluid in which the sperm are carried. These organs are found in only a few anamniotes in which internal fertilization occurs.

Terrestrial reproduction makes an aquatic larval stage unnecessary. Most reptiles remain oviparous, but their eggs are **cleidoic** (Fig. 12-11). A cleidoic egg is supplied with a large store of yolk, which eventually becomes suspended in a **yolk sac.** As the egg descends the oviduct, **albumen,** or similar secretions that meet various metabolic needs, and a protective **shell** are added to the egg. The embryo itself develops **extraembryonic membranes.** A protective **chorion** and a fluid-filled **amnion** (which provides a local aquatic environment) evolve from ectodermal and somatic mesodermal layers, which at an earlier evolutionary stage covers the yolk sac in an anamniote such as the dogfish (Fig. 12-11A). The yolk sac of amniotes is therefore **bilaminar** with a wall of just splanchnic mesoderm and endoderm. Finally, an **allantois** evolves from the urinary bladder of the amphibians. Embryonic blood vessels are carried by the allantois to the chorion, which is close enough to the porous shell for gas exchange to occur. Excretory products, chiefly in the form of inert uric acid, accumulate within the allantois.

Prototherian mammals lay this type of egg, but in therian mammals this egg (minus its shell, albumen, and most of its yolk) is retained in the female reproductive tract, and a **placenta** evolves. In eutherian mammals, the placenta is simply a union of the chorion and allantois **(chorioallantoic membrane)** with the uterine lining of the female. As will be seen, the mode of reproduction necessitates changes in the female reproductive tract, notably the evolution of a uterus.

◆ **(A) THE EXCRETORY SYSTEM**

The **kidneys** (renes) of mammals, which are **metanephroi,** are located against the dorsal wall of the peritoneal cavity in a retroperitoneal position (Fig. 12-12). Each is surrounded by a mass of fat **(adipose capsule),** which should be removed, and each is closely invested by a **fibrous capsule.** The adipose capsule consists of a special type of fat tissue, called **structural fat.** It serves as a protective shield against mechanical injuries of the kidneys. Unlike the usual fat tissue, it is not used as an energy reserve for the body. Notice that the mammalian kidney is bean-shaped. The indentation on the medial border is called the **hilus.** Carefully remove connective tissue from the hilus and you will see the **renal artery** and **vein** entering the kidney and, caudal to them, the **ureter,** which drains the kidney.

Trace one of the ureters. It extends caudad and retroperitoneally and then turns ventrally into the lateral ligament of the bladder to enter the caudal part of the **urinary bladder.** As the ureter enters the lateral vesical ligament, it passes dorsal to the ductus deferens (male), or horn of the uterus (female) (Figs. 12-12 and 12-18). The urinary bladder itself is a pear-shaped organ with a broad, rounded, cranial end (its **vertex**) and a narrow, caudal part (its **body**). Cut open the urinary bladder and you may be able to see the points of entrance of the ureters in the dorsal wall of the body. Clean away connective tissue from around the urinary bladder, which gradually narrows caudally to

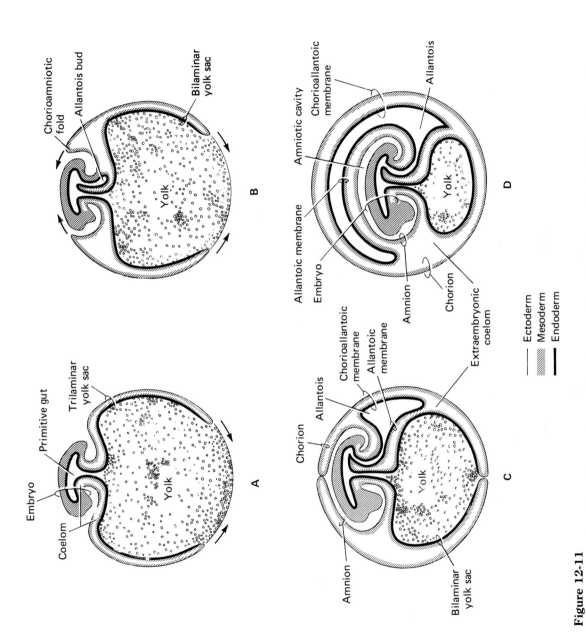

Figure 12-11

Diagrammatic sections of vertebrate embryos to show the extraembryonic membranes. The head of the embryo points to the left side of the drawing. **A**, Developing trilaminar yolk sac of a large-yolked fish embryo; **B**, hypothetical derivation of the chorioamniotic folds of an amniote embryo from the superficial layers of the trilaminar yolk sac; **C**, extraembryonic membranes of an early amniote embryo; **D**, extraembryonic membranes of a later amniote embryo. (*A through C, From Villee, Walker, and Barnes, 1984.*)

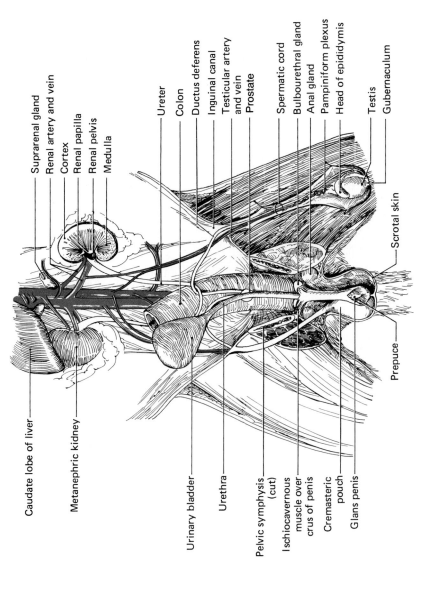

Caudate lobe of liver

Metanephric kidney

Suprarenal gland
Renal artery and vein
Cortex
Renal papilla
Renal pelvis
Medulla

Ureter
Colon
Ductus deferens
Inguinal canal
Testicular artery and vein
Prostate

Spermatic cord
Bulbourethral gland
Anal gland
Pampiniform plexus
Head of epididymis

Testis
Gubernaculum

Scrotal skin

Prepuce

Urinary bladder

Urethra

Pelvic symphysis (cut)

Ischiocavernous muscle over crus of penis

Cremasteric pouch

Glans penis

Figure 12-12
Ventral view of the urogenital system of a male cat. One kidney has been sectioned to show its internal structure. The pelvic canal has been cut open, and the left cremasteric pouch has been dissected, removed from the scrotum, and pinned to the inner surface of the left thigh. See Figure 12-15 for an enlargement of the testis and epididymis. *(From Walker, A Study of the Cat.)*

the entrance of the ureters and forms a tube. This narrow tube, which enters the pelvic canal and carries urine to the outside, is the **urethra.** The urethra begins just caudal to the entrances of the ureters. Its more caudal parts will be considered with the reproductive organs.

Leave the kidneys in place. With a long-bladed knife, cut one of them longitudinally through the hilus in the frontal plane (Figs. 12-12 and 12-16). Study the half that includes the largest portion of the ureter. The hilus expands within the kidney into a chamber called the **renal sinus.** The renal hilus and sinus are filled with the renal blood vessels, the proximal end of the ureter, and fat (Fig. 12-13A). Pick away the fat to expose the blood vessels and ureter. The portion of the ureter within the renal sinus is expanded and is called the **renal pelvis.** (If the blood vessels, nerves, ureter, and renal pelvis were now removed from the kidney, the space left would be the renal sinus). The substance of the kidney converges in the cat and in the rabbit to form a single, nipple-shaped **renal papilla** that projects into the renal pelvis. (In some mammals, humans among them, there are many renal papillae, and the proximal portion of the renal pelvis is subdivided into chambers for them, called **calyces.**) Notice in your specimen that the substance of the kidney can be subdivided into a peripheral, light **cortex** and a deeper, darker **medulla.** The medullary substance of the cat and rabbit constitutes one large **renal pyramid** whose apex is the renal papilla; in species with many papillae, there is a pyramid for each papilla.

The mammalian kidney can be divided into a cortex and a medulla because the nephrons are arranged in a particular pattern. This pattern plays a crucial role in the physiology of the mammalian kidney but is not present as such in other vertebrate kidneys (for details, consult a physiology textbook, such as Eckert, 1988). The cortex contains the renal corpuscles as well as the proximal and distal convoluted tubules, whereas the medulla contains the loops of Henle and the collecting tubules (Fig. 12-13**B**). The renal arteries branch into radially arranged **interlobar arteries.** These, in turn, branch into **arcuate arteries,** which run through the substance of the kidney along the interface between the cortex and medulla. After passing through the glomerulus and peritubular capillaries (see also Fig. 12-1**C**), the blood is collected by veins that run next to the arteries. The glomeruli can sometimes be seen in well-injected specimens under low magnification.

◆ **(B) THE MALE REPRODUCTIVE SYSTEM**

The testes of most mammals do not remain in the body cavity, where they develop embryonically, but migrate into paired, saclike extensions of the peritoneal cavity, the **processus vaginales** (singular: *processus vaginalis*). The processus vaginales lie within a cutaneous pouch, the **scrotum,** which lies below the anus. In the cat, the scrotum is situated caudally to the pelvic symphysis and penis (Fig. 12-14). But in the rabbit, the scrotum lies on the ventral surface of the pelvic girdle and cranial to the penis. This position, incidentally, occurs only in lagomorphs among placental mammals, although a comparable location is seen in marsupials.

Very carefully remove the **scrotal skin** from the scrotum. A dense layer of connective tissue containing some smooth muscle fibers, the **dartos tunic,** is closely

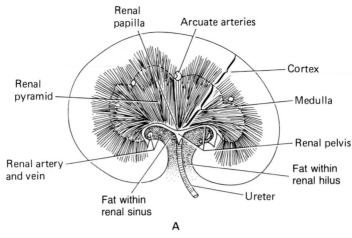

Renal papilla
Arcuate arteries
Cortex
Medulla
Renal pelvis
Fat within renal hilus
Ureter
Fat within renal sinus
Renal artery and vein
Renal pyramid

A

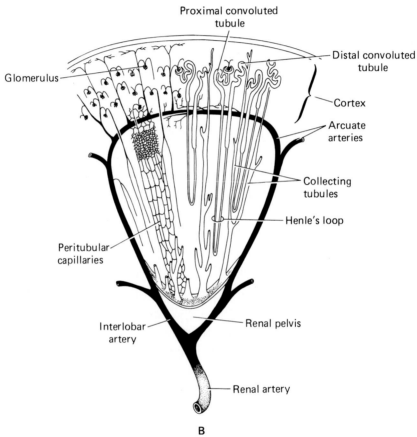

Proximal convoluted tubule
Distal convoluted tubule
Cortex
Arcuate arteries
Collecting tubules
Henle's loop
Renal pelvis
Renal artery
Interlobar artery
Peritubular capillaries
Glomerulus

B

Figure 12-13
The internal structure of the mammalian kidneys. **A,** Longitudinal section through the hilus of a kidney of the cat (the ureter is sectioned only near the renal pelvis); **B,** diagram illustrating the arrangement of the blood vessels and nephrons within a kidney. (On the left side of the drawing the nephrons are omitted; on the right side of the drawing the smaller blood vessels are omitted. The veins are completely omitted from the drawing.) (**A,** *Redrawn from R. Nickel, A. Schummer, and E. Seiferle,* The Viscera of the Domestic Mammals, 2nd ed. *Berlin, Verlag Paul Parey, 1979;* **B,** *redrawn from K. M. Dyce, W. O. Sack, and C. J. G. Wensing,* Textbook of Veterinary Anatomy, *Philadelphia, W. B. Saunders Company, 1987; after Clara, 1938.)*

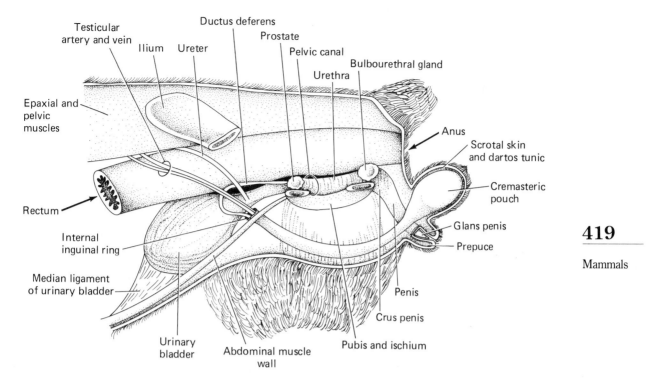

Figure 12-14
Diagrammatic lateral view of the topographical relationships of the urogenital system of the male cat. *(Adapted from R. Nickel, A. Schummer, and E. Seiferle,* The Viscera of the Domestic Mammals, *2nd ed. Berlin, Verlag Paul Parey, 1979.)*

associated with the scrotal skin and will come off with it. A pair of cordlike sacs, the **cremasteric pouches,** will be revealed. They extend from the scrotum, across the ventral surface of the pelvic symphysis, to the abdominal wall. Within the scrotum, they are separated from each other by the **septum scroti,** an extension of the dartos tunic. The proximal parts of the cremasteric pouches lie just beneath the abdominal skin and should have been seen and saved when the pelvic canal was opened (p. 385). The testes lie within the caudal ends of the cremasteric pouches. This portion of each pouch is quite large in the cat, while the rest is a constricted tube; the testes remain permanently in the scrotum. But in the rabbit, the entire pouch is wide and of nearly uniform diameter. The rabbit can pull its testes back into the peritoneal cavity during territorial combats. (Fighting male rabbits jump up against each other, striking the abdomen of their opponents with their hind legs).

The wall of the cremasteric pouches is composed of several layers that are continuous with various layers of the abdominal wall. The outermost layer, the **external spermatic fascia,** is an extension of the external fascia of the trunk, or fascia trunci, which lies directly under the skin. The next layer is the **internal spermatic fascia,** an extension of the internal fascia of the trunk, or fascia transversalis, which lies directly under the parietal peritoneum (see also p. 325). The internal and external spermatic fasciae are separated from each other by loose connective tissue and by the **cremasteric muscle,** which develops as a slip from the internal oblique and transversus abdominis muscles of the abdominal wall. The cremasteric muscle is atrophied in the cat but very large in the rabbit. The three layers of the wall of the cremasteric pouch are difficult to separate and identify in the small mammals being studied.

Leave the cremasteric pouch intact on one side, but cut open the other one along its ventral surface (Figs. 12-12, 12-15, and 12-16). Extend the cut from the caudal end of the pouch to the muscular abdominal wall, but do not cut through the wall. Notice that the cremasteric pouch contains a cavity, the **cavity of the processus vaginalis,** or **vaginal cavity** (Fig. 12-17). Probe this cavity through the constricted **vaginal canal** into the peritoneal cavity.

As already mentioned, the processus vaginalis is an outpocketing of the peritoneal cavity into which the testis descends during sexual maturation.[37] Thus, the vaginal cavity is lined with coelomic epithelium, the **parietal tunica vaginalis.** The structures within the vaginal cavity are enveloped by the **visceral tunica vaginalis** and anchored to the dorsal wall of the cremasteric pouch through a mesentery, the **mesorchium** (Fig. 12-17). A very short band of connective tissue, the **gubernaculum,** attaches the testis, epididymis, and caudal end of the mesorchium to the distal end of the cremasteric pouch (Fig. 12-15). In many mammals, humans included (but not in the cat and rabbit), the vaginal canal atrophies in the adult, so that there is no communication between the peritoneal cavity and the vaginal cavity.

The contents of the cremasteric pouch can now be examined in more detail. The **testis** is the relatively large, round body (cat; Figs. 12-12 and 12-15) or elongate body (rabbit; Fig. 12-16) lying in the caudal part of the cremasteric pouch. Testicular blood vessels and nerves attach to its cranial end. Notice that the testicular artery coils upon itself before it reaches the testis and is closely invested by a venous network, the **pampiniform plexus,** formed by the testicular vein. The **epididymis** is a band-shaped structure closely applied to the surface of the testis. It can be divided into three regions— a **head** at the cranial end of the testis, a **body** on the lateral surface of the testis, and a **tail** at the caudal end of the testis. The head of the epididymis is functionally connected with the testis and has developed from modified renal tubules (p. 401). The rest of the epididymis consists of the highly convoluted **ductus epididymidis** (former cranial end of the archinephric duct) imbedded in connective tissue.

The scrotum, dartos tunic, and pampiniform plexus are all parts of a testicular thermoregulatory system. The high body core temperature of most mammals inhibits the transformation of spermatids into tailed spermatozoa. The posterior migration, or **testicular descent,** places the testes in an environment in which the ambient temperature is several degrees lower than the core body temperature. The sparseness of the hair on the scrotum of some species of mammals further ensures low temperature, as does the countercurrent flow of blood in the pampiniform plexus. Heat is transferred from the warmer blood in the testicular artery to the cooler venous blood returning from the testes. The degree of contraction of smooth muscles within the dartos tunic helps regulate the temperature. When the muscles relax, the testes fall away from the warm body wall; when they contract, the testes are pulled closer to the body wall. Species of mammals that lack a scrotum have lower core temperatures than those with a scrotum, at least during the cooler seasons of the year when they mate.

---

[37]Strictly speaking, the wall of the processus vaginalis consists not only of the serosal tunica vaginalis but also of the internal spermatic fascia. This, however, is relevant only if the embryonic development and the testicular descent are considered in detail.

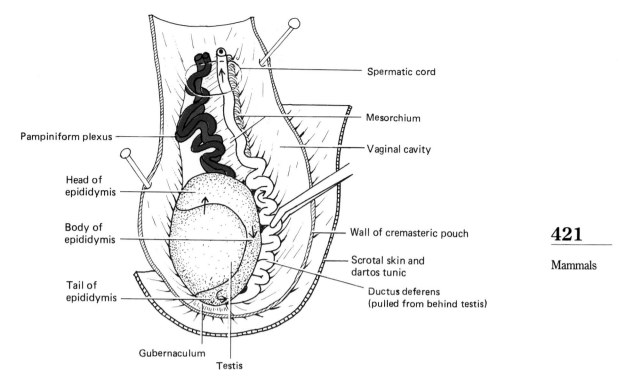

Figure 12-15
Ventral view of the left testis and epididymis of the male cat. The cremasteric pouch has been cut open; a piece of the scrotal skin and dartos tunic has been left attached. Arrows indicate the direction of the sperm movement.

The **ductus deferens,** the first part of which is convoluted, leaves the tail of the epididymis and, in company with the testicular vessels and nerves, ascends the cremasteric pouch and passes through the abdominal wall. The complex of the ductus deferens, testicular blood vessels and nerves, and their envelope of visceral tunica vaginalis is known as the **spermatic cord.** The passage through the muscular abdominal wall is known as the **inguinal canal.** The canal is very short in the mammals being considered, its cranial end **(internal inguinal ring)** being the entrance into the peritoneal cavity and its caudal end **(external inguinal ring)** being the attachment of the external spermatic fascia to the aponeurosis of the external oblique muscle. In humans, the inguinal canal passes diagonally through the abdominal wall and hence is much longer.

Continue to follow the ductus deferens. It passes craniad in the peritoneal cavity for a short distance and then loops over the ureter and extends caudad into the pelvic canal between the urethra and large intestine (Fig. 12-14). The ductus deferentes of opposite sides then converge and soon enter the urethra. The portion of the urethra distal to this union carries both sperm and urine. Various accessory genital glands, which secrete the seminal fluid, are associated with the ends of the ductus deferentes and adjacent parts of the urethra; however, these glands, and the details of the union of the ductus deferentes with the urethra, differ in the animals being considered.

In the cat, the two ductus deferentes enter the urethra independently, and a small **prostate** surrounds their point of entrance and the adjacent urethra (Figs. 12-12 and 12-14). At the caudal end of the pelvic canal, a pair of **bulbourethral glands** (Cowper's glands) enter the urethra, but you will observe these when you study the penis (see later).

In the rabbit, the two ductus deferentes pass between the urethra and a heart-shaped

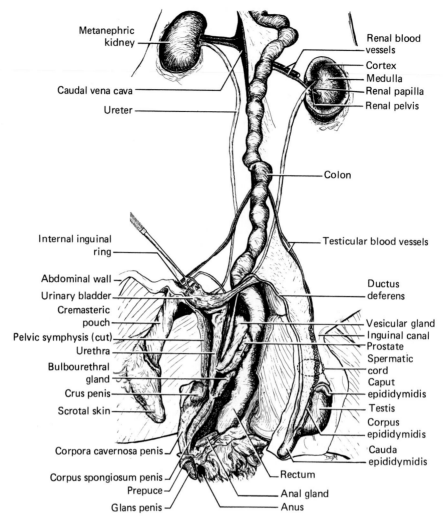

Figure 12-16
Ventral view of the urogenital system of a male rabbit. One kidney has been sectioned to show its internal structure. The pelvic canal has been opened and the urethra and penis twisted to one side to show the accessory genital glands. The scrotum is shown intact on the left side of the drawing and opened on the right side of the drawing.

vesicular gland (Fig. 12-16). Carefully separate these structures from each other. The ductus deferentes enter the narrow caudal end of the vesicular gland, which in turn enters the urethra. The dorsal wall of the vesicular gland is rather thick and includes the **prostate.** It is possible, by very careful dissection, to free the cranial end of the prostate and turn it caudad. Further dissection will reveal overlapping cranial and caudal lobes of the prostate. Both lobes enter the urethra just caudal to the entrance of the vesicular gland. A bilobed **bulbourethral gland** (Cowper's gland) enters the dorsal surface of the urethra caudal to the prostate.

The **penis** of mammals, like the claspers of sharks, is an organ that allows internal fertilization. (Recall that *Necturus* does not have a penis-like organ). The penis is attached to the pelvic ischia and encloses the part of the urethra that projects caudally beyond the pelvic canal (Fig. 12-14). The free end of the penis, the **glans penis,** lies in a pocket of

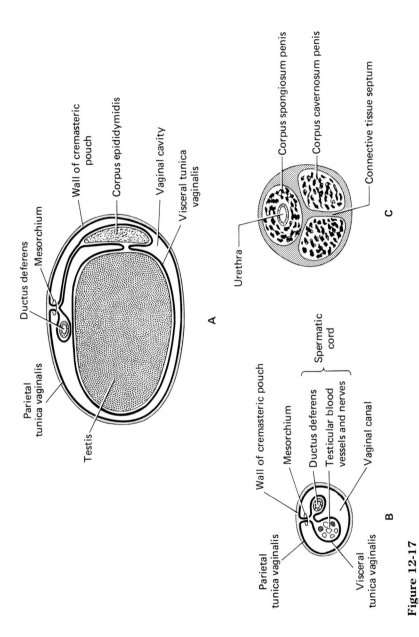

Figure 12-17
Semidiagrammatic, enlarged transverse sections through parts of the reproductive system of a male cat: **A**, through the scrotum at the level of the testis; the scrotal skin is removed; **B**, through the middle of the cremasteric pouch; **C**, through the middle of the penis; the skin is removed.

skin called the **prepuce** *(preputium)*. Cut open the prepuce to see the glans and the opening of the urethra. A number of small spines are borne on the glans penis of the cat. The rest of the penis is a firm, cylindrical structure, which you should expose by removing the skin and surrounding loose connective tissue. Make a cross section of this portion of the penis and examine it under low magnification. The urethra lies along the dorsal surface of the penis (if the organ is flaccid) imbedded in a column of spongy erectile tissue called the **corpus spongiosum penis** (Fig. 12-17C). A pair of columns of spongy erectile tissue separated by a connective tissue septum, which is often indistinct, lie along the opposite surface and are surrounded by a ring of dense connective tissue. These paired columns are the **corpora cavernosa penis.** The glans penis is simply a caplike fold of the corpus spongiosum penis that covers the distal ends of the corpora cavernosa penis. The **crura penis** (singular: *crus penis*) are the diverging proximal ends of the corpora cavernosa penis. The crura penis are anchored to the ischia, and each crus penis is covered by muscular tissue, the **ischiocavernous muscle** (Figs. 12-12, 12-14, and 12-16). If the crura penis were not torn when the pelvic canal was opened, you may have to cut one now to see the paired (cat) or bilobed (rabbit) bulbourethral gland clearly. During erection of the penis, the spaces within the spongy erectile tissue of the corpus spongiosum and corpora cavernosa become engorged with blood. Make a cross section through the glans penis and look for a small bone—the **os penis,** or baculum—which lies along one surface of the urethra and helps to stiffen this part of the penis. (The baculum of the walrus is much larger and was prized by whalers as a club.) Most mammals, including humans, lack an os penis; their penis is a purely hydrostatic skeleton that stiffens through the increased pressure of the blood accumulating within the erectile tissue.

Before leaving the urogenital system, dissect beneath the skin on either side of the rectum near the anus and find a pair of **anal glands.** These are elongate in the rabbit, round in the cat. They produce pheromones that are discharged through the anus and are believed to be of value in regulating reproductive behavior.

◆ (C) THE FEMALE REPRODUCTIVE SYSTEM

The **ovaries** are a pair of small, oval bodies (Fig. 12-18). In the adult, they lie slightly caudal to the kidneys, because they have undergone a partial descent, and the metanephroi have shifted cranially during development. The small size of the mammalian ovaries is correlated with the intrauterine development of the embryos: Fewer eggs are produced, and they contain little yolk. The eggs are microscopic, but you may see small vesicles, the **graafian follicles** *(folliculi vesiculosi),* each of which contains one egg, protruding on the surface of the ovary. The ovaries of the cat and rabbit protrude into the body cavity and are suspended by mesenteries, the mesovaria. At **ovulation,** the eggs are discharged from the follicles and break through the serosa enveloping the ovary. After ovulation, the follicles remain imbedded within the ovary and are transformed into the hormone-producing **corpora lutea,** which also may be seen protruding from the surface of the ovary, especially in pregnant specimens.

The eggs of mammals normally do not fall into the peritoneal cavity at ovulation, unlike the eggs of sharks and amphibians. In many mammals, the cranial ends of the oviducts form funnel-like expansions, **infundibula** (singular: *infundibulum*) with fringed

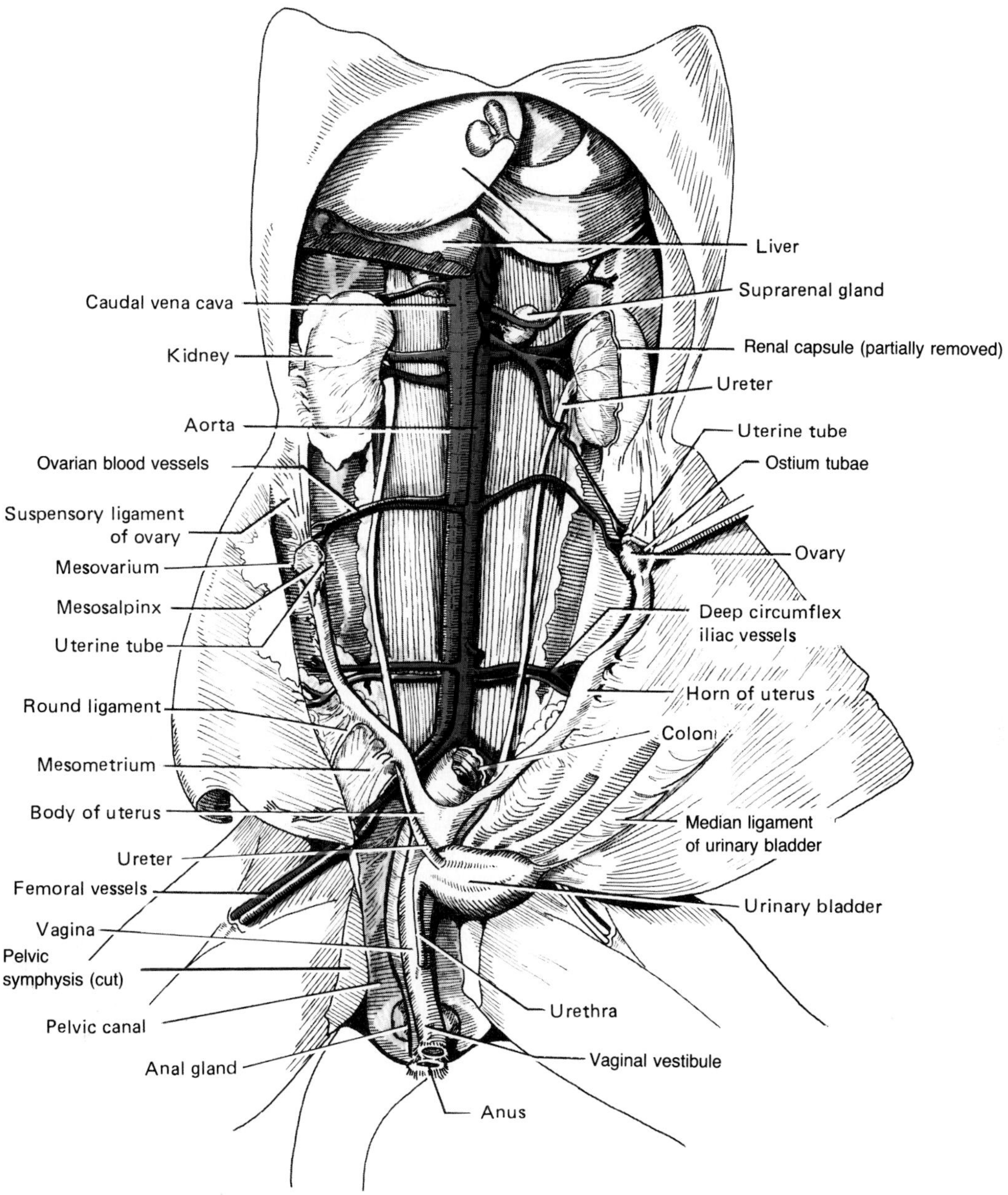
Liver
Caudal vena cava
Suprarenal gland
Kidney
Renal capsule (partially removed)
Ureter
Aorta
Uterine tube
Ovarian blood vessels
Ostium tubae
Suspensory ligament of ovary
Ovary
Mesovarium
Mesosalpinx
Deep circumflex iliac vessels
Uterine tube
Round ligament
Horn of uterus
Colon
Mesometrium
Body of uterus
Median ligament of urinary bladder
Ureter
Femoral vessels
Urinary bladder
Vagina
Pelvic symphysis (cut)
Urethra
Pelvic canal
Anal gland
Vaginal vestibule
Anus

Figure 12-18
Ventral view of the urogenital system of a female cat. The pelvic canal has been cut open.

rims. These fringes, the **fimbriae,** sweep over the surface of the ovaries and ensure that the eggs are collected into the infundibula. In other mammals, such as the cat, the hoodlike infundibula almost completely envelop the ovary; in still other mammals, the infundibula form ovarian bursae, which completely enclose the ovaries.

Typical oviducts are present in early mammalian embryos, but they differentiate into several regions during development, and their caudal ends fuse in varying degrees (Figs. 12-6 and 12-19). Thus, the adult female reproductive tract is more or less Y-shaped. The cranial part of each wing of the Y forms a narrow, convoluted **uterine tube** (fallopian tube) lying lateral to the ovary. Notice that a uterine tube curves over the front of the ovary and forms a hoodlike expansion **(infundibulum)** with fringed (fimbriated) lips. Spread open the lips and you will see the coelomic opening of the tube, the **ostium tubae.**

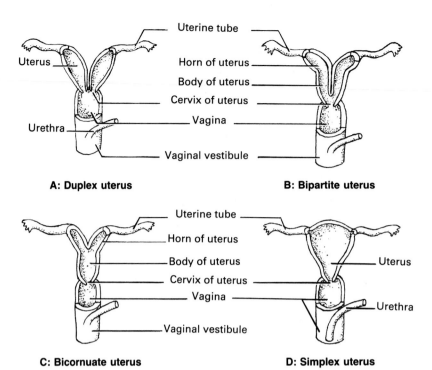

A: Duplex uterus

B: Bipartite uterus

C: Bicornuate uterus

D: Simplex uterus

Labels in figure: Uterine tube, Uterus, Horn of uterus, Body of uterus, Cervix of uterus, Vagina, Urethra, Vaginal vestibule

Figure 12-19

Diagrams to show the progressive fusion of the caudal ends of the oviducts in placental mammals. The uterus and part of the vagina have been cut open. The duplex type of uterus, in which the lower ends of the oviducts have united to form a vagina but the uteri remain distinct, is found in rodents and lagomorphs. In the bipartite uterus of carnivores, the lower ends of the uteri also have fused to form a median body from which uterine horns extend, but a partition is present in the cranial part of the body of the uterus. This partition is lost in the bicornuate uterus of ungulates. In the simplex uterus of primates, the uteri have completely united to form a large median body into which the uterine tubes open. In primates, the vaginal vestible also divides, so that the vagina and urethra open closer to the body surface. *(Modified from Wiedersheim.)*

The rest of each wing of the Y lies caudal to the ovary and forms a much wider tube—the **horn of the uterus** (cat), or **uterus** (rabbit) (see Figs. 12-19**A** and **B**). This section is straight in the cat but somewhat convoluted in the rabbit. It is very large in pregnant specimens, for the embryos develop within it.

The ovary and reproductive tract of the female are suspended by a mesentery known as the **broad ligament.** Often a great deal of fat lies within it. The portion of the broad ligament attaching to the uterus is the **mesometrium;** that attaching to the uterine tube is the **mesosalpinx;** and that attaching to the ovary is the **mesovarium** (Fig. 12-18). Pull the uterine horn (cat) or uterus (rabbit) toward the midline, thereby stretching the mesometrium. The mesenteric fold extending diagonally across the mesometrium from a point near the cranial end of the uterine horn (or uterus) to the body wall, and lying perpendicular to the broad ligament, is the **round ligament.** Notice that the round ligament attaches to the body wall at a point comparable to the location of the inguinal canal in the male. The round ligament is the female counterpart of the male gubernaculum—a connective tissue strand that plays an important role in the descent of the testis.

The two uterine horns (cat) or uteri (rabbit) converge cranial to the pelvic canal and enter a common median passage. This is the stem of the Y, and it is formed in part by the fusion of the lower ends of the oviducts and in part by the division of the cloaca (see later and Fig. 12-22). In the cat, the cranial part of this median passage is the **body of the**

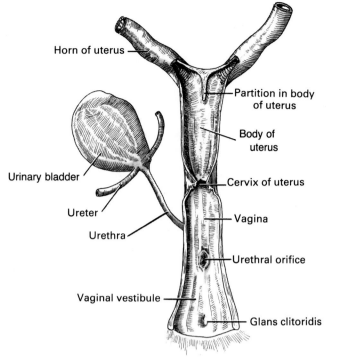

Horn of uterus

Partition in body of uterus

Body of uterus

Urinary bladder

Cervix of uterus

Ureter

Vagina

Urethra

Urethral orifice

Vaginal vestibule

Glans clitoridis

Figure 12-20
Dorsal dissection of the reproductive organs of a female cat to show the internal features of the uterus, vagina, and vaginal vestibule.

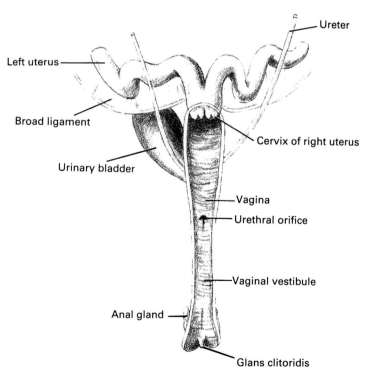

Ureter

Left uterus

Broad ligament

Cervix of right uterus

Urinary bladder

Vagina

Urethral orifice

Vaginal vestibule

Anal gland

Glans clitoridis

Figure 12-21
Reproductive tract of a female rabbit, cut open and viewed from the dorsal side.

uterus. The body of the uterus is relatively short. It is separated from the **vagina** by the cervix of the uterus, which you will see when you cut open the genital tract (see later). In the rabbit, the two uteri enter the vagina independently, each uterus having its own cervix. The vagina continues caudally through the pelvic canal, lying between the urethra and large intestine. Carefully separate these structures from one another and find the point where the vagina and urethra unite. The common passage from here to the body surface is the **vaginal vestibule,** or urogenital canal. It is a relatively long passage in quadrupeds. The comparable area in women is known as the **vulva,** but the vulva is very shallow, for vagina and urethra remain independent nearly to the body surface. The opening of the vaginal vestibule, or vulva, is flanked by skin folds, the **labia,** but these are not conspicuous in quadrupeds.

428

Chapter 12
The Excre-
tory and
Reproduc-
tive Sys-
tems

Cut through the skin around the opening of the vaginal vestibule and completely free the vestibule and vagina from the rectum. A pair of **anal glands,** which are elongate in the rabbit and round in the cat, can be found by dissecting beneath the skin on the lateral surface of the vestibule (rabbit), or of the rectum near the anus (cat). These glands produce pheromones that are discharged through the anus and probably play a role in reproductive behavior. Now open the median portion of the genital tract by making a longitudinal incision through its dorsal wall that extends from the vestibule to the horns of the uterus (cat), or uteri (rabbit). Veer away from the middorsal line toward one of the uterine horns as you open the body of the uterus in the cat. A small bump may be seen in the cat in a pocket of tissue in the midventral line of the vestibule near its orifice. This is the **glans clitoridis.** Rabbits have a larger clitoris that consists of a glans clitoridis. The clitoris can be exposed by removing the skin from the ventral surface of the vestibule. The clitoris and its glans develop from a phallus-like structure that is present in the sexually indifferent stage of the embryo. (In a male embryo, this phallus-like structure develops into the penis). Make a cross section through the organ, and it will be seen to consist of a pair of columns of spongy erectile tissue, the **corpora cavernosa clitoridis,** homologous to the corpora cavernosa penis.

More cranially in the vaginal vestibule, you will see the entrance of the urethra. The genital passage cranial to this union is the vagina, and it continues forward to the sphincter-like neck or **cervix of the uterus** (Figs. 12-20 and 12-21). In the cat, the single cervix lies about halfway between the urethral orifice and the horns of the uterus and appears as a pair of folds constricting the lumen of the reproductive tract. The body of the uterus lies between the cervix and the horns. Notice that the cranial part of the body is subdivided into right and left sides by a vertical partition. The uterus of the cat is therefore **bipartite** (Fig. 12-19). In the rabbit, the vagina extends forward to the two uteri, and each uterus has a sphincter-like **cervix** extending into the vagina. This type of uterus is called a **duplex uterus** (Fig. 12-19).

We have already discussed the evolution of most of the mammalian male and female urogenital tracts from those of their ancestors. However, we deferred a consideration of the cloacal region and its fate in mammals until we were able to study the terminal portions of the urogenital passages.

A cloaca is present in nonmammalian vertebrates and still persists in monotremes and in the embryos of marsupial and placental mammals. But it becomes divided and contributes to the formation of the intestinal and urogenital passages in the adult therian mammals. In an early, sexually indifferent eutherian embryo, the **cloaca** consists of a chamber derived from the enlargement of the caudal end of the hindgut (Fig. 12-22**A**). At first this endodermal cloaca is separated from the ectodermal proctodeum by a plate of tissue, the **cloacal membrane,** but this plate soon breaks down, and the **proctodeum** then contributes to the formation of the cloaca.

The cloaca receives the intestine dorsally and the allantois ventrally. Even at an early stage (Fig. 12-22**A**), the cranial portion of the cloaca is partly divided into a dorsal **coprodeum** receiving the intestine, and a ventral **urodeum** receiving the allantois, the ureters, and the archinephric ducts. A small **genital tubercle** is present on the ventral surface of the body cranial to the cloaca. This stage is similar to the cloaca of nonmammalian vertebrates except for the more ventral entrance of the urogenital ducts.

Later in the sexually indifferent period (Fig. 12-22**B**), the urodeum and coprodeum become completely separated from each other by a fold of tissue (**urorectal fold**) and form the urogenital sinus and rectum, respectively. Oviducts now enter the front of the urogenital sinus beside the archinephric ducts, but the attachments of the ureters shift onto the allantois and developing urinary bladder.

In the subsequent differentiation of a male (Fig. 12-22**D**), the constricted neck of the bladder (allantois) and the urogenital sinus form that portion of the urethra that is not included in the penis (segments 1 and 2). The archinephric ducts form the ductus deferens and enter the urethra. Their point of entrance is a landmark that separates the portion of the urethra derived from the allantois from the portion derived from the urogenital sinus. The oviducts disappear in the male, although their point of entrance

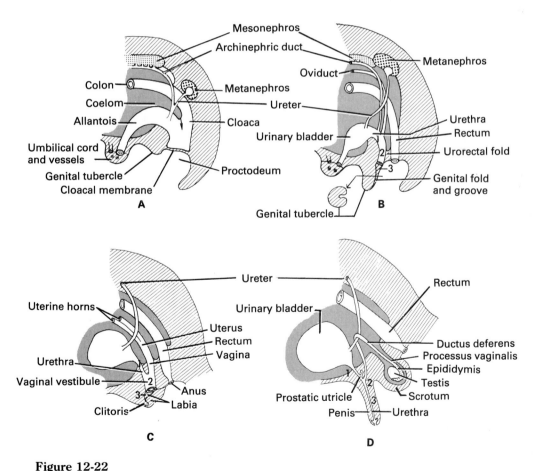

Figure 12-22
Diagrams in lateral view to show the division of the cloaca that occurs during the embryonic development of a placental mammal. **A,** An early sexually indifferent stage in which the division of the cloaca has just begun; **B,** a later sexually indifferent stage in which the cloaca has become divided into a dorsal rectum and ventral urogenital sinus; **C,** differentiation to the female condition; **D,** differentiation to the male condition. *1*, *2*, and *3* indicate comparable regions.

into the urethra may form a small sac (**prostatic utricle**) within the prostate. The genital tubercle enlarges to form the penis, and a groove on its ventral surface closes over, by the coming together of the **genital folds,** to form the penile portion of the urethra (segment 3).

In the differentiation of most female mammals subsequent to the sexually indifferent stage (Fig. 12-22C), the constricted neck of the bladder forms the entire urethra. The female urethra is thus comparable to only a small portion (the allantoic segment) of the male urethra. The urethra and the two oviducts, whose lower ends have fused to form the vagina and uterus, enter the urogenital sinus, which becomes the vaginal vestibule. In most female mammals, the vaginal vestibule remains undivided. But in primates, it too becomes divided into the urethra and vagina, which continue nearly to the surface as separate passages. This distal part of the vaginal vestibule forms the shallow vulva. The archinephric ducts disappear, and the genital tubercle forms the clitoris. The genital folds of the indifferent stage form the labia minora in those mammals that have these labia. The labia majora are skin folds comparable to the scrotum of the male.

♦   **(D) REPRODUCTION AND EMBRYOS**

Eutherian mammals are viviparous: The embryos develop within the uterus and are born as miniature adults. If any of the specimens are pregnant, cut open the uterus and examine the embryos. Mammalian embryos produce the various extraembryonic membranes characteristic of amniotes (Figs. 12-11 and 12-24). Since the outermost membrane is called the chorion, the whole complex of embryo and extraembryonic membranes is often called the **chorionic sac. Chorionic villi** arise from the surface of the chorioallantoic membrane in eutherian mammals and penetrate or unite in various ways with the uterine lining. This combination of uterine lining and villi constitutes the **placenta.** In many mammals, including those considered here, the union of villi and uterine lining is intimate, and some of the uterine lining is discharged at birth. Such a placenta is said to be **deciduous,** in contrast to a **nondeciduous** placenta, in which the union is not intimate and maternal tissue is not discharged at birth.

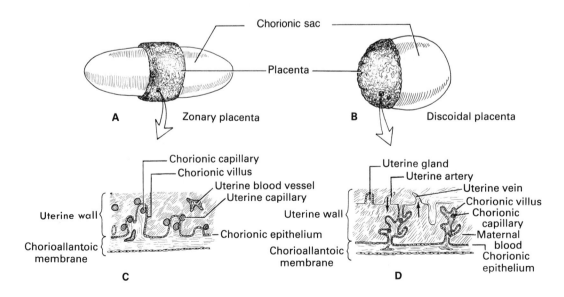

Figure 12-23
Diagrams of the cat and human placenta. **A,** Chorionic vesicle of the cat; **B,** chorionic vesicle of a human being; **C,** enlarged detail of the cat placenta; **D,** enlarged detail of the human placenta.

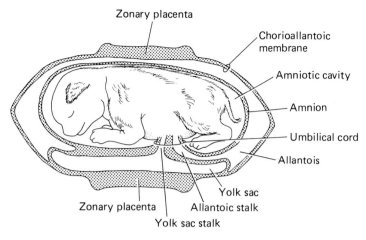

Zonary placenta

Chorioallantoic
membrane

Amniotic cavity

Amnion

Umbilical cord

Allantois

Yolk sac

Zonary placenta   Allantoic stalk

Yolk sac stalk

Figure 12-24
Diagrammatic longitudinal section through the chorionic sac and extraembryonic membranes of a cat embryo. *(Redrawn from K. M. Dyce, W. O. Sack, and C. J. G. Wensing,* Textbook of Veterinary Anatomy, *Philadelphia, W. B. Saunders Company, 1987.)*

Deciduous placentas have different shapes according to the distribution of the villi that make an intimate union with the uterine lining. In carnivores, a belt-shaped band of villi unites with the uterine lining to form a definitive placenta of the **zonary** type. This band is easily seen on the surface of the chorionic sac. In the rabbit and human beings, a disc-shaped patch of villi unites, and the placenta is said to be **discoidal** (Fig. 12-23).

Still other terms describe the microscopic details of the union. In carnivores, the surface epithelium lining the uterine portion of the placenta disappears. The chorionic villi penetrate the substance of the uterine lining and come in contact with the endothelial walls of the maternal capillaries to form an **endotheliochorial placenta.** Humans and rabbits have an even more intimate union. Maternal capillaries are eroded in humans so that the chorionic villi are bathed in pools of maternal blood **(hemochorial placenta).** In rabbits, the epithelium on the surface of the villi also disappears, so the circulation of fetus and mother is separated only by the endothelial walls of the fetal capillaries **(hemoendothelial placenta).** To summarize, carnivores have a chorioallantoic, deciduous, zonary, endotheliochorial placenta; the rabbit has a chorioallantoic, deciduous, discoidal, hemoendothelial placenta.

Cut open the chorionic sac and placenta and you will see the embryo enclosed by the allantois and the **amnion.** Open the amnion. The cord of tissue extending from the underside of the embryo to the chorionic sac and placenta is the **umbilical cord.** It contains the allantoic stalk, the umbilical or allantoic blood vessels, and sometimes a vestige of the yolk sac stalk. Its surface is covered by the amnion.

WORD ROOTS

When studying anatomy, you will be confronted with numerous unfamiliar terms that you must master because effective communication requires their use. Some understanding of the derivation of anatomical terms will help to fix their meaning and spelling in your mind. A list of the more important roots and eponyms used in anatomy, and the Greek (Gr.) and Latin (L.) words from which they have been derived, is presented below. By periodically consulting such a list, you should soon be able to see how similar roots are used in different combinations and be able to infer the meanings of unfamiliar words. The definitions of terms can be found through the index.

In the following list, usually only the nominative singular is given for a noun, but the genitive (gen.), nominative plural (pl.), or diminutive (dim.) is included if it is different enough and has formed the basis for an example cited. Only the infinitive for a verb is given, unless another form, such as the past participle (p.p.), is needed to recognize the root. Most of the examples cited are anatomical structures, but major groups of organisms mentioned in this book are included, and often a common word is cited to help fix the root in mind. This list of roots is far from inclusive; many familiar words for which the classical and the modern terms and meanings are very similar have been omitted. If this list whets your appetite, you should consult standard dictionaries or such references as Jaeger (1978) for further study.

a— Gr. *a-* or *an-*, prefix meaning without, not. (Acelous, Agnatha, anapsid)

ab— L. *ab-*, prefix meaning away from. (Abducens nerve, abduction)

abdom— L. *abdomen*, abdomen (probably from *abdo*, to conceal). (Abdomen, abdominal)

acanth— Gr. *akantha*, spine, thorn. (Acanthodii)

acetabul— L. *acetabulum*, vinegar cup (from *acetum*, vinegar). (Acetabulum, acetic acid)

acro— Gr. *akros*, topmost, extreme. (Acromiodeltoid, acromion)

actin— Gr. *aktis*, ray. (Actinopterygii)

ad, af— L. *ad-* (may be changed to *af-* before certain words), prefix meaning motion toward. (Adduction, afferent)

aden— Gr. *aden*, gland. (Adenohypophysis)

af— See *ad*.

al— L. *ala*, wing. (Alar foramen, aliform, alisphenoid bone)

all— Gr. *allos*, different, other. (Allotheria)

allant— Gr. *allas*, gen., *allantos*, sausage. (Allantoic artery, allantois)

alveol— L. *alveolus*, small pit. (Alveolus)

amelo— Old Fr. *amel*, enamel. (Amelogenins)

amn— Gr. *amnion*, fetal membrane. (Amnion, anamniote)

amphi— Gr. *amphi*, on both sides of, double. (Amphibian, amphicoelous, Amphioxus)

ampul— L. *ampulla*, a small jug. (Ampulla of Lorenzini)

an— See *a-*.

anastomos— Gr. *anastomosis*, an opening, a connection between two seas. (Subsupracardinal anastomosis)

anc— Gr. *ankon*, elbow, bend of the arm. (Anconeus muscle)

ante— L. *ante-*, prefix meaning in front of, previous to. (Antebrachium, anterior)

anus— L. *anus*, anus. (Anus)

aort— Gr. *aorte*, aorta (from *aeirein*, to raise up). (Aortic arch)

apo— Gr. *apo-*, prefix meaning away from. (Apophysis)

apsid— Gr. *apsis*, gen., *apsidos*, a loop. (Anapsid skull, synapsid)

aqu— L. *aqua*, water. (Aqueous humor)

arachn— Gr. *arachne*, spider. (Arachnid, arachnoid membrane)

arbor— L. *arbor*, tree. (Arbor vitae)

arch— Gr. *arche-*, prefix meaning primitive. (Archenteron, archinephros, archipallium)

arcu— L. *arcuare*, to bend like a bow. (Arcualia, coracoarcual muscle)

arter— L. *arteria*, artery. (Artery)

aryten— Gr. *arytaina*, ladle. (Arytenoid cartilage)

atla— Gr. mythology *Atlas*, a god condemned to support the heavens upon his shoulders. (Atlantoscapularis muscle, atlas)

atri— L. *atrium*, entrance room. (Atrium)

aur— L. *auris*, dim., *auricula*, ear. (Auricle)

auto— Gr. *auto-*, prefix meaning self. (Autostylic)

av— L. *avis*, pl., *aves*, bird. (Aves, aviary)

ax— Gr. *axon*, axle. (Axon)

axill— L. *axilla*, armpit (Axillary artery)

bas— Gr. *basis*, base, bottom. (Basal nuclei, basapophysis)

bi, bin— L. *bi-, bin-*, prefix meaning two. (Biceps muscle, binocular, biped)

bi, bio— L. *bios*, life. (Amphibian, biology)

brachi— Gr. *brachion*, upper arm. (Antebrachium, brachialis muscle, brachium pontis)

branchi— Gr. *branchion*, gill. (Branchial, Elasmobranchii)

bronch— Gr. *bronchos*, windpipe. (Bronchus)

bull— Gr. *bulla*, a bubble, seal. (Tympanic bulla)

bucc— L. *bucca*, cheek. (Buccal cavity)

burs— Gr. *bursa*, hide, wineskin. (Bursa)

caec— L. *caecus*, blind, ending blind. (Caecum, ileocaecal valve)

calc— L. *calx*, gen., *calcis*, lime, heel. (Calcaneus, calcify, calcium)

call— L. *callum*, hard skin. (Callous, corpus callosum)

caly— Gr. *kalyx*, cup. (Renal calyx)

can— L. *canis*, dog. (Canine teeth)

capill— L. *capillus*, hair. (Capillary)

capit— L. *caput*, gen., *capitis*, dim., *capitulum*, head. (Capitulum, caput)

carap— French, *carapace*, a shell. (Carapace)

cardi— Gr. *kardia*, heart. (Cardiac)

cardinal— L. *cardinalis*, principal, red. (Cardinal vein)

carn— L. *caro*, gen., *carnis*, flesh. (Carnal, carnivore, trabeculae carneae)

carotid— Gr. *karotides*, large neck arteries (from *karoun*, to stupefy). (Carotid artery)

carp— Gr. *karpos*, wrist. (Carpal, carpus)

caud— L. *cauda*, tail. (Caudal, caudofemoralis muscle)

cav— L. *cavus*, hollow. (Cavity, vena cava)

cel— See *coel*.

ceph— Gr. *kephale*, head. (Cephalic, cephalochordate, holocephalic, triceps)

cera— Gr. *keras*, horn. (Ceratobranchial, ceratotrichia, keratin)

cerc— Gr. *kerkos*, tail. (Cercaria, heterocercal fin)

cereb— L. *cerebrum*, dim., *cerebellum*, brain. (Cerebrum, cerebellum)

cervic— L. *cervix*, gen., *cervicis*, neck. (Cervical, cleidocervicalis muscle, uterine cervix)

chiasm— Gr. *chiasma*, cross. (Optic chiasma)

choan— Gr. *choane*, funnel. (Choana)

chol— Gr. *chole*, bile. (Ductus choledochus)

chondr— Gr. *chondros*, cartilage. (Chondrichthyes, chondrify, chondrocranium)

chord— L. *chorda*, string. (Chordae tendineae, Chordata, notochord)

chorio— Gr. *chorion*, skin, afterbirth. (Chorion, choroid layer of eyeball, choroid plexus)

choro— See *chorio*.

chromato— Gr. *chroma*, color. (Chromatophore)

ciner— L. *cinereus*, the color of ashes (from *cinis*, ashes). (Incinerate, tuber cinereum)

cingul— L. *cingulum*, girdle. (Cingulate cortex)

circ— L. *circum*, about, around. (Circumvallate papilla)

cistern— L. *cisterna*, reservoir, cistern. (Cisterna chyli)

clav— L. *clavis*, dim., *clavicula*, key; also, the clavicle. (Clavicle, subclavian artery)

cleid— Gr. *kleis*, gen., *kleidos*, key; also, the clavicle. (Cleidoic egg, cleidomastoid muscle)

cleithr— Gr. *kleithron*, a key, bar. (Cleithrum)

clitor— Gr. *kleitoris*, gen., *kleitoridos*, small hill. (Clitoris, glans clitoridis)

cloac— L. *cloaca*, sewer. (Cloaca)

cnem— Gr. *kneme*, shin or tibia. (Gastrocnemius muscle)

coccy— Gr. *kokkyx*, cuckoo, bone in the shape of cuckoo's bill. (Coccyx)

coch— Gr. *kokhlos*, land snail. (Cochlea, vestibulocochlear nerve)

coel— Gr. *koilos*, hollow. (Acoela, coeliac artery, coelom, neurocoele)

coll— L. *collis*, dim., *colliculus*, hill. (Superior colliculus)

coll— L. *collum*, neck. (Collar, longus colli muscle)

colon— Gr. *kolon*, colon. (Colon)

commissur— L. *commissura*, point of union of two things. (Anterior commissure of brain)

con— L. *conus*, cone. (Conus arteriosus)

conch— Gr. *konkhe*, sea shell. (Concha, conchology)

condyl— Gr. *kondylos*, articulatory prominence at

a joint, knuckle. (Epicondyle, occipital condyle)

conju— L. *conjungere*, p.p., *conjunctus*, to join together. (Conjugal, brachium conjunctivum, conjunctiva of eye)

copro— Gr. *kopros*, dung. (Coprodeum, coprophagy)

corac— Gr. *korax*, gen., *korakos*, a raven or crow. (Coracobrachialis muscle, coracoid [resembles a crow's beak])

corn— L. *cornu*, horn. (Cornea, cornify, hyoid cornua)

coron— L. *corona*, crown, or something curved. (Coronary arteries, coronation, coronoid process)

corp— L. *corpus*, pl., *corpora*, body. (Corpora cavernosa, corpse, corpus callosum)

cort— L. *cortex*, gen., *corticis*, bark. (Cerebral cortex, corticospinal tract)

cost— L. *costa*, rib. (Costal cartilage, intercostal space)

cox— L. *coxa*, hip bone. (Coxal)

cran— Gr. *kranion*, skull. (Chondrocranium, cranial)

cre— Gr. *kreas*, flesh. (Cremasteric muscle, cremate, pancreas)

cribr— L. *cribrum*, sieve. (Cribriform plate)

cric— Gr. *krikos*, ring. (Cricoid cartilage)

cross— Gr. *krossoi*, fringe. (Crossopterygian)

crus— L. *crus*, pl., *crura*, lower leg shank. (Abductor cruris muscle, crural fascia, crus penis)

cucull— L. *cucullus*, cap, hood. (Cucullaris muscle)

cun— L. *cuneus*, wedge. (Cuneiform bone, funiculus cuneatus)

cup— L. *cupa*, dim., *cupula*, tub. (Cupula, cupule)

cut— L. *cutis*, skin. (Cutaneous, cuticle, cutis)

Cuvier— Baron Georges Cuvier, eighteenth-century French scientist. (Duct of Cuvier)

cyclo— Gr. *kyklos*, circle. (Cyclic, Cyclostomata)

cyst— Gr. *kystis*, bladder. (Cyst, cystic duct)

cyt— Gr. *kytos*, hollow vessel, cell. (Lymphocyte, osteocyte)

daeum— See *deum*.

decid— L. *decidere*, to fall off. (Deciduous placenta)

delt— Gr. *delta*, fourth letter of the Greek alphabet, Δ. (Deltoid muscle)

dendr— Gr. *dendron*, tree. (Dendrite of a nerve cell)

dens, dent— L. *dens*, gen., *dentis*, tooth. (Dens, dentary bone, dentine, dentist)

derm— Gr. *derma*, skin, leather. (Dermal bone, dermis, ostracoderm)

deum— Gr. *hodaios*, on the way. (Proctodeum, stomodeum, urodeum)

di— Gr. *di-*, prefix meaning two. (Diapsid reptile, digastric muscle)

dia— Gr. *dia-*, prefix meaning across, between. (Di[a]encephalon, diapophysis, diaphragm)

didym— Gr. *didymos*, testicle. (Epididymis)

digit— L. *digitus*, finger. (Digit, digitigrade, extensor digitorum muscle)

dont— See *odont*.

dors— L. *dorsum*, back. (Dorsal, longissimus dorsi)

duct— L. *ductus*, a leading (from *ducere*, to lead). (Abduction, duct, oviduct)

duoden— New L. *duodenum*, the first part of the intestine, which is about 12 fingerbreadths long (a contraction of *intestinum duodenum digitorum*, from *duodeni*, 12 each). (Duodenal, duodenum)

dur— L. *durus*, hard. (Durable, dura mater)

ef— See *ex*.

elasm— Gr. *elasmos*, a thin plate. (Elasmobranchii)

encephal— Gr. *enkephalos*, brain (from *en* + *kephale*). (Diencephalon, encephalitis, mesencephalon)

endo, ento— Gr. *endon, entos*, within. (Endoderm, endolymph, endostyle, endotherm, entotympanic bone)

enter— Gr. *enteron*, intestine. (Enteron, mesentery)

eo— Gr. *eos*, dawn. (Eocene, Eotheria)

epi— Gr. *epi-*, prefix meaning upon, above. (Epicondyle, epidermis, epididymis, epithalamus)

epiplo— Gr. *epiploon*, the greater omentum. (Epiploic foramen)

esoph— Gr. *oisophagos*, gullet. (Esophagus)

ethm— Gr. *ethmos*, sieve. (Ethmoid bone)

eu— Gr. *eu-*, prefix meaning good, true. (Eutheria)

Eustachio— Bartolommeo Eustachio, sixteenth-century Italian anatomist. (Eustachian tube)

ex— L. *ex-, ef-*, prefix meaning out of or away from. (Ductuli efferentes, efferent neuron, extrinsic muscle)

faci— L. *facies*, form, face. (Facial nerve)

falc— L. *falx*, gen., *falcis*, sickle. (Falciform ligament, falx cerebri)

Fallopio— Gabriele Fallopio, sixteenth-century Italian anatomist. (Fallopian tube)

fasc— L. *fascia*, band, bandage. (Fascia)

fasci— L. *fascis*, dim., *fasciculus*, bundle. (Muscle fascicle) *fauc*— L. *fauces*, throat. (Fauces)

femur— L. *femur*, thigh, femur. (Femur)

fenestr— L. *fenestra*, window. (Fenestra cochleae, temporal fenestra)

fer— L. *ferre*, to carry. (Afferent neuron, deferent duct, efferent artery)

fibul— L. *fibula*, buckle, pin. (Fibula)

fid— See *fiss*.

fil— L. *filum*, dim., *filamentum*, thread. (Filamentous, filiform papillae)

fimbr— L. *fimbria*, fringe. (Fimbria of hippocampus)

fiss— L. *findere*, p.p., *fissus*, to split. (Fissure, multifidus muscle)

flex— L. *flectere*, p.p., *flexus*, to bend. (Circumflex iliac artery, flexion)

flocc— L. *floccus*, tuft of wool. (Flocculonodular lobe)

foli— L. *folium*, leaf. (Cerebellar folia, foliage, foliate papillae)

foram— L. *foramen*, an opening. (Foramen magnum)

form— L. *forma*, shape, rule. (Cribriform plate, digitiform gland, falciform ligament, formula, lentiform nucleus)

forn— L. *fornix*, a vault or arch; also, a brothel. (Fornicate, hippocampal fornix)

foss— L. *fossa*, a ditch (from *fodere*, p.p., *fossus*, to dig). (Fossil, fossorial, mandibular fossa)

fove— L. *fovea*, small pit. (Fovea in retina)

fron— L. *frons*, gen., *frontis*, forehead. (Frontal bone)

fun— L. *funis*, dim., *funiculus*, cord, rope. (Funiculi of spinal cord, funicular)

fund— L. *fundus*, bottom. (Gastric fundus)

fung— L. *fungus*, mushroom. (Fungiform papilla, fungus)

gang— Gr. *ganglion*, a swelling, tumor. (Ganglion)

Gartner— Hermann Gartner, eighteenth-to-nineteenth-century Danish anatomist. (Gartner's duct)

gastr— Gr. *gaster*, stomach. (Digastric muscle, gastric glands, gastrocnemius muscle)

gem— L. *geminus*, twin-born. (Gemellus superior muscle, trigeminal nerve)

gen— L. *genus*, race or kind (from *genere*, to beget). (Genesis, genital, genus)

gen— L. *genu*, knee. (Geniculate bodies, genioglossus muscle)

glans— L. *glans*, dim., *glandula*, acorn. (Gland, glans penis)

glen— Gr. *glene*, cavity or socket. (Glenoid cavity)

glia— Gr. *glia*, glue. (Neuroglia)

gloss, *glott*— Gr. *glossa*, or *glotta*, tongue. (Genioglossus muscle, glottis, hypoglossal nerve)

glut— Gr. *gloutos*, buttock. (Gluteus muscle)

gnath— Gr. *gnathos*, jaw. (Agnatha, gnathostome)

gon— Gr. *gonos*, seed. (Gonad)

gracil— L. *gracilis*, slender. (Fasciculus gracilis, gracile, gracilis muscle)

gubernacul— L. *gubernaculum*, a rudder (from *gubernare*, to steer). (Gubernaculum of testis, gubernatorial)

gul— L. *gula*, throat. (Gular)

gustat— L. *gusto*, to taste, p.p., *gustatus*, tasted. (Gustatory)

gyr— Gr. *gyros*, circle, round. (Cerebral gyri, gyrate)

haben— L. *haben*, dim., *habenula*, a strap. (Habenula of epithalamus)

hallu— L. *hallux*, gen. *hallucis*, big toe. (Flexor hallucis longus muscle)

ham— L. *hamus*, dim., *hamulus*, a hook. (Hamate bone, hamulus of pterygoid bone)

Harder— Johann Harder, seventeenth-century Swiss anatomist. (Harderian gland)

hem— Gr. *haima*, blood. (Hemal arch, hemoglobin, hemorrhage)

hemi— Gr. *hemi*, half. (Hemibranch, hemichordate)

hepa— Gr. *hepar*, gen., *hepatos*, liver. (Hepatic duct, hepatic portal vein)

heter— Gr. *heteros*, other, different. (Heterocercal, heterogeneity)

hippo— Gr. *hippos*, horse. (Hippocampus)

hol— Gr. *holos*, whole, entire. (Holocephalic, holonephros, Holostei)

homo— Gr. *homos*, alike, the same. (Homodont, homology)

humeru— L. *humerus*, humerus. (Humerus)

humo— L. *humor*, liquid; specifically, one of the body fluids thought in ancient times to affect disposition. (Aqueous humor, humor)

hyo— Gr. *hyoeides*, in the form of the letter upsilon (U-shaped). (Hyoglossus muscle, hyoid bone)

hyper— Gr. *hyper-*, prefix meaning above, over. (Hyperactive, hypertrophy)

hypo— Gr. *hypo-*, prefix meaning under, beneath. (Hypaxial, hypoglossal nerve, hypophysis, hypothalamus)

ichthy— Gr. *ichthys*, gen., *ichthyos*, fish. (Osteichthyes)

ile, *ili*— L. *ileum* or *ilium*, the groin. (Ileum, iliac, ilium) *incis*— L. *incidio*, to cut into, p.p., *incisus*, cut. (Incisor teeth)

incu— L. *incus*, an anvil. (Incus of ear)

ineum— Gr. *inan*, to excrete. (Perineum)

infra— L. *infra*, below. (Infraorbital foramen, infraspinatus muscle)

infundibul— L. *infundibulum*, a funnel. (Infundibulum of the uterine tube)

inguin— L. *inguen*, groin, *inguinalis*, pertaining to the groin. (Inguinal canal)

integument— L. *integumentum*, covering. (Integument)

inter— L. *inter*, between. (Intercalary plate, intercostal muscle, internuncial neuron, interparietal bone)

irid— Gr. *iris*, gen., *iridos*, rainbow. (Iridescent, iris of the eye)

ischi— Gr. *ischion*, hip. (Ischial, ischium)

Jacobson— Ludwig L. Jacobson, nineteenth-century Danish surgeon and anatomist. (Jacobson's organ)

jejun— L. *jejunus*, fasting, empty. (Jejunum [so called because it is usually found to be empty in dissection])

jug— L. *jugum*, dim., *jugulum*, collarbone. (Jugular vein)

kerat— Gr. *keras*, gen., *keratos*, horn. (Keratin)

lab— L. *labium*, lip. (Labial cartilages, labia majora)

labyrinth— Gr. mythology *Labyrinthos*, the maze in which the Minotaur was confined. (Labyrinthodontia)

lachry, lacrim— L. *lacrima*, tear. (Lacrimal bone, nasolacrimal duct)

lagen— Gr. *lagenos*, flask. (Lagena)

lamin— L. *lamina*, dim., *lamella*, layer or thin plate. (Lamellar bone, lamina terminalis, laminate)

laryng— Gr. *larynx*, gullet. (Larynx)

lat— L. *latus*, gen., *lateris*, side, broad. (Fascia lata, lateral, latissimus dorsi muscle)

lemnisc— L. *lemniscus*, ribbon. (Lateral lemniscus)

len— L. *lens*, gen., *lentis*, a lentil. (Lens, lentiform nucleus)

lien— L. *lien*, spleen. (Gastrolienic ligament, lienogastric artery)

lig— L. *ligamentum*, band. (Ligament)

limb— L. *limbus*, edge, border. (Limbic lobe)

lingu— L. *lingua*, dim., *lingula*, tongue. (Lingualis muscle, lingual nerve, sublingual gland)

liss— Gr. *lissos*, smooth. (Lissamphibia)

lith— Gr. *lithos*, stone. (Lithosphere, otolith)

lumb— L. *lumbus*, loin. (Lumbar artery, thoracolumbar fascia)

lun— L. *luna*, moon. (Lunate bone, lunatic, semilunar valve)

lymph— L. *lympha*, water. (Endolymph, lymphocyte)

macul— L. *macula*, spot. (Maculae in inner ear)

magn— L. *magnus*, great. (Foramen magnum)

malle— L. *malleus*, hammer. (Mallet, malleus)

mam— L. *mamilla*, nipple. (Mamillary body)

mamma— L. *mamma*, gen., *mamme*, breast. (Mammal)

man— L. *manus*, hand. (Manual, manubrium, manus)

mandibul— L. *mandibula*, lower jaw. (Mandible, mandibular gland)

marsup— L. *marsupium*, purse, pouch. (Marsupialia, marsupium)

mass— Gr. *masseter*, a chewer. (Masseter muscle)

mast— Gr. *mastos*, breast. (Mastoid process, neuromast)

mat— L. *mater*, mother; *maternus*, motherly. (Dura mater, maternal)

maxill— L. *maxilla*, upper jaw. (Maxilla)

meat— L. *meatus*, passage. (Acoustic meatus)

Meckel— Johann F. Meckel, eighteenth-century German anatomist. (Meckel's cartilage)

mediastin— L. *mediastinus*, median (from *medius*, middle). (Mediastinum)

medull— L. *medulla*, marrow. (Medulla oblongata, renal medulla)

melano— Gr. *melas*, black. (Melancholy, melanin)

menin— Gr. *meninx*, pl., *meninges*, membrane. (Meninges, meninx)

ment— L. *mentum*, chin. (Mental foramen)

mer— Gr. *meros*, part. (Branchiomeric)

mes— Gr. *mesos*, middle. (Mesencephalon, mesentery, mesoderm, mesonephros)

met, meta— Gr. *meta*, beside, after. (Metacarpal, metacromion, metanephros, Metatheria)

metr— Gr. *metra*, uterus. (Endometrium, mesometrium)

mol— L. *mola*, millstone. (Molar teeth)

mon— Gr. *monas*, single. (Monarch, Monotremata)

Monro— Alexander Monro, eighteenth-century Scottish anatomist. (Foramen of Monro)

morph— Gr. *morphe*, shape. (Elasmobranchiomorphi, morphology)

muc— L. *mucosus*, slimy. (Mucus, mucous membrane)

Müller— Johannes P. Müller, nineteenth-century German anatomist and physiologist. (Müllerian duct) *mult*— L. *multus*, many. (Multifidus muscle, multiply)

myel— Gr. *myelos*, marrow, spinal cord. (Myelencephalon, myelin)

myl— Gr. *myle*, millstone, molar. (Mylohyoid muscle)

myo— Gr. *mys*, gen., *myos,* muscle. (Myology, myomere, myotome)

nar— L. *naris*, pl., *nares*, external nostril. (Naris)

nas— L. *nasus*, nose. (Nasal bone)

nav— L. *navis*, dim., *navicula*, ship. (Navicular bone, navy)

neo— Gr. *neos*, new, young. (Neo-Darwinism, neopallium)

neph— Gr. *nephros*, kidney. (Mesonephros, nephron, protonephridia)

neur— Gr. *neuron*, tendon, nerve. (Neurocoele, neuromast, neuron)

nict— L. *nictare*, to wink. (Nictitating membrane)

nidament— L. *nidamentum*, nesting material. (Nidamental gland)

noto— Gr. *notos*, back. (Notochord)

nuc— L. *nux*, dim., *nucella*, nut, kernel. (Nucleus)

nuch— L. *nucha*, nape of neck. (Nuchal crest)

obturat— L. *obturare*, p.p., *obturatus*, to close by stopping up. (Obturator foramen)

occipit— L. *occiput*, gen., *occipitis*, the back of the head. (Occipital bone)

octav— L. *octavus*, eighth. (Octavolateralis system)

ocul— L. *oculus*, eye. (Bulbus oculi, orbicularis oculi muscle)

odont— Gr. *odous*, gen., *odontos,* tooth. (Labyrinthodont, thecodont)

oid— New L. -*oid* (from Gr. *o* + *eidos*, form), suffix that indicates resemblance to. (Arachnoid, diploid, pterygoid, sphenoid, xiphoid)

olecran— Gr. *olekranon*, elbow tip. (Olecranon)

olfac— L. *olfacere*, to smell. (Olfactory organ)

om— Gr. *omos*, shoulder. (Metacromion, omohyoid muscle, omotransversarius)

oment— L. *omentum*, membrane. (Greater omentum)

oo— Gr. *oon*, egg. (Epoophoron, oogenesis)

oper— L. *operire*, to cover. (Operculum)

ophthalm— Gr. *ophthalmos*, eye. (Ophthalmic nerve, ophthalmologist)

opisth— Gr. *opisthe*, behind, at the rear. (Opisthonephros, opisthotic)

opt— Gr. *optikos*, pertaining to sight. (Optic nerve)

or— L. *os*, gen., *oris*, mouth. (Oral cavity)

orb— L. *orbis*, dim., *orbiculus*, circle. (Orbicularis oculi muscle, orbit)

orch— Gr. *orchis*, testicle. (Mesorchium, orchid [from shape of root])

os, oss— L. *os*, gen., *ossis*, dim., *ossiculum*, bone. (Ossicle, ossify)

oste— Gr. *osteon*, bone. (Osteichthyes, teleost)

osti— L. *ostium*, a door or mouthlike opening. (Ostium of oviduct)

ostrac— Gr. *ostrakon*, shell. (Ostracoderm)

ot— Gr. *otikos*, pertaining to the ear. (Otic capsule, otolith, parotid gland)

ov— L. *ovum*, egg. (Ovary, oviduct, oviparous)

paedo— Gr. *pais*, gen., *paidos*, child. (Paedomorphosis)

palae, pale— Gr. *palaios*, ancient. (Paleontology, paleopallium)

palat— L. *palatum*, roof of the mouth. (Palate, palatine bone)

palli— L. *pallium*, a Roman cloak. (Neopallium)

palpebr— L. *palpebra*, eyelid. (Levator palpebrae superioris muscle)

pampin— L. *pampinus*, tendril. (Pampiniform plexus)

pan— Gr. *pan*, all. (PanAmerican, pancreas)

Papez— J. W. Papez, nineteenth- to twentieth-century American anatomist.

par— L. *pareo*, p.p., *partus*, to bring forth. (Oviparous, parturition, viviparous)

par, para— Gr. *para*, beside. (Paradidymis, parapophysis, parasympathetic nerve, parotid)

pariet— L. *paries*, gen., *parietis*, wall. (Parietal peritoneum)

pat— L. *patina*, dim., *patella*, dish or plate. (Patella bone)

path— Gr. *pathetikos*, sensitive, liable to suffer (from *pathos*, suffering). (Pathology, sympathetic nervous system)

pect— L. *pectus*, gen., *pectoris,* chest. (Pectoral girdle, pectoralis muscle)

pectin— L. *pecten*, comb. (Pectinate muscles of heart, pectineus muscle)

ped, pes— L. *pes*, gen., *pedis*, dim., *pedunculus,* foot. (Bipedal, cerebral peduncle, pes)

pelluc— L. *pellucidus*, clear, transparent. (Pellucid, septum pellucidum)

pelv— L. *pelvis*, basin. (Pelvic girdle, renal pelvis)

per, peri— Gr. *peri-*, prefix meaning around. (Periosteum, peritoneum)

perine— Gr. *perineou,* region between anus and genitals. (Perineum)

peron— Gr. *perone*, pin, fibula. (Peroneal nerve, peroneus muscle)

petr— Gr. *petros*, rock. (Petrify, petrosal bone)

phalang— Gr. *phalanx*, gen., *phalangos*, a line of soldiers. (Phalanges)

phall— Gr. *phallos*, penis. (Phallic, phallus)

pharyng— Gr. *pharynx*, gen., *pharyngos*, pharynx. (Glossopharyngeal nerve, pharynx)

phor— Gr. *phoros*, from *pherein*, to bear. (Epoophoron, pterygiophore, spermatophore)

phragm— Gr. *phragmos*, fence, partition. (Diaphragm)

phren— Gr. *phren*, diaphragm, mind. (Phrenic nerve, phrenology)

phys— Gr. *physis*, a growth. (Apophysis, epiphysis, hypophysis, symphysis)

pia— L. *pia*, tender. (Pia mater)

pin— L. *pineus*, pertaining to a pine tree. (Pineal gland)

pir— L. *pirum*, pear. (Piriform lobe, piriformis muscle)

pis— L. *pisum*, pea. (Pisiform bone)

plac— Gr. *plax*, gen., *plakos*, flat plate. (Placoderm, placoid scale)

placent— L. *placenta*, small, flat cake. (Placenta)

plant— L. *planta*, sole of the foot. (Plantaris muscle, plantigrade)

plast— contraction of Gr. *emplassein*, p.p., *emplastros*, to daub on. (Epiplastron, plaster, plastron)

platy— Gr. *platys*, flat or broad. (Duck-billed platypus, platysma muscle)

pleur— Gr. *pleura*, side, rib. (Metapleural fold, pleural cavity, pleurapophysis)

plex— L. *plexus*, an interweaving, network. (Brachial plexus)

poie— Gr. *poieo*, to make. (Hemopoietic)

poll— L. *pollex*, gen., *pollicis*, thumb. (Abductor pollicis muscle)

pon— L. *pons*, gen., *pontis*, bridge. (Brachium pontis, pons)

poplit— L. *poples*, gen., *poplitis*, knee joint. (Popliteal fossa, popliteus muscle)

post— L. *post*, after, behind. (Posterior, postorbital process)

pre— L. *prae-*, prefix meaning before, in front. (Precaval vein, premaxillary bone)

prepu— L. *praeputium*, the foreskin. (Prepuce)

prim— L. *primus*, first. (Primates, primitive)

pro— Gr. *pro-*, prefix meaning before, in front. (Pronephros, prosencephalon, protraction)

proct— Gr. *proktos*, the anus. (Proctodeum)

pron— L. *pronus*, bending, leaning forward. (Pronator teres muscle, prone)

prostat— Gr. *prostates*, one who stands before. (Prostate [stands before the bladder])

prot— Gr. *protos*, first. (Protonephridium, protoplasm, Prototheria)

pseud— Gr. pseudos, *false*. (Pseudobranch)

psoa— Gr. *psoa*, the loin. (Psoas major muscle)

pteryg— Gr. *pterygion*, gen., *pterygos*, wing, fin. (Actinopterygii, metapterygium, pterygoid process)

pub— L. *pubes*, young adult. (Puberty, pubic hair, pubis)

pudend— L. *pudendum*, external genitals (from *pudere*, to be ashamed). (Pudendal artery, pudendum)

pulmo— L. *pulmo*, gen., *pulmonis*, lung. (Pulmonary artery)

pulvi— L. *pulvinus*, cushion. (Pulvinar)

putamen— L. *putamen*, a pod. (Putamen)

pylor— Gr. *pyloros*, gate keeper. (Pyloric region of stomach, pylorus)

quadr— L. *quadrus*, fourfold. (Quadriceps femoris muscle)

quadrat— L. *quadratus*, squared. (Quadrate bone, quadratus lumborum muscle)

radi— L. *radius*, ray, spoke. (Corona radiata, radium, radius)

radi— L. *radix*, pl., *radices*, root. (Radix of aorta)

ram— L. *ramus*, branch. (Ramify, ramus)

raph— Gr. *raphe*, seam, suture. (Raphes in muscles)

re— L. *re-*, indicates backward. (Retractor muscle)

rect— L. *rectus*, straight, *rectum*, straight intestine. (Rectangular, rectum, rectus abdominis muscle)

ren— L. *ren*, kidney. (Renal artery)

rept— L. *reptare*, p.p., *reptum*, to creep. (Reptilia)

rest— L. *restis*, rope. (Restiform body)

ret— L. *rete*, dim., *reticulum*, net. (Rete mirabile, reticular formation, retina)

retinacul— L. *retinaculum*, band, holdfast. (Extensor retinaculum)

retro— L. *retro*, behind. (Retroperitoneal)

rhin— Gr. *rhis*, gen., *rhinos*, nose. (Rhinal fissure, rhinarium, rhinencephalon, rhinoceros)

rhomb— Gr. *rhombus*, parallelogram with oblique angles and unequal adjacent sides. (Rhombencephalon, rhomboideus muscle)

rostr— L. *rostrum*, beak, ship's prow. (Rostral, rostrum)

rug— L. *ruga*, fold. (Gastric rugae)

sacc— L. *saccus*, dim., *sacculus*, bag. (Sacculus, sack)

sacr— L. *sacrum*, sacred. (Sacral nerve, sacrum)

sagitt— Gr. *sagitta*, arrow. (Sagittal plane)

salp— Gr. *salpinx*, trumpet. (Mesosalpinx)

sarco— Gr. *sarx*, gen., *sarkos*, flesh. (Sarcophagus, Sarcopterygii)

sartor— L. *sartor*, tailor. (Sartorius muscle)

scal— L. *scala*, ladder. (Scala vestibuli)

scale— Gr. *skalenos*, a triangle with three unequal sides. (Scalenus muscle)

scaph— Gr. *skaphe*, bowl, boat. (Scaphoid bone)

scapul— L. *scapula*, shoulder blade. (Scapula)

scler— Gr. *skleros*, hard. (Sclera of eyeball, sclerous, sclerotome)

scrot— L. *scrotum*, pouch, scrotum. (Scrotum)

seb— L. *sebum*, grease, wax. (Sebaceous gland, sebum)

sell— L. *sella*, saddle, seat. (Sella turcica)

sem— L. *semen*, seed. (Semen, seminal vesicle)

semi— L. *semi-*, prefix meaning partly, half. (Semicircular canal, semispinalis muscle, semitendinosus muscle)

serr— L. *serra*, saw. (Serrated, serratus ventralis muscle)

sesam— Gr. *sesame*, seed of the sesame plant. (Sesamoid bone)

sin— L. *sinus*, cavity. (Sinus venosus)

sole— L. *solea*, sandal. (Sole of the foot, soleus muscle)

som— Gr. *soma*, gen., *somatos*, body. (Somatic, somatopleure)

sperm— Gr. *sperma*, gen., *spermatos*, seed, semen. (Sperm, spermatic fascia)

sphen— Gr. *sphen*, wedge. (Sphenoid bone)

sphinct— Gr. *sphinkter*, that which binds tightly. (Sphincter muscle)

spin— L. *spina*, thorn or spine. (Erector spinae muscle, spinalis muscle, supraspinatus)

spirac— L. *spiraculum*, pore, air hole. (Spiracle)

splanchn— Gr. *splanchnon*, viscus, inner organ (Splanchnic nerve)

splen— Gr. *splen*, spleen. (Spleen, splenic)

spleni— Gr. *splenion*, bandage. (Splenial bone, splenius muscle)

squam— L. *squama*, scale. (Squamata, squamosal bone, squamous epithelium)

stap— L. *stapes*, stirrup. (Stapedius muscle, stapes bone)

stern— Gr. *sternon*, breast, chest (Sternebrae, sternomastoid muscle, sternum)

stom— Gr. *stoma*, gen., *stomatos*, mouth. (Cyclostomata, stomochord, stomodeum)

stomach— Gr. *stomachos*, stomach. (Stomach)

strat— L. *stratum*, layer. (Stratum corneum)

stri— L. *striare*, p.p., *striatus*, to make furrows or stripes. (Corpus striatum, striated)

styl— Gr. *stylos*, pillar, stalk. (Endostyle, hyostylic, styloid process, stylomastoid foramen)

sub— L. *sub-*, prefix meaning beneath, below. (Subscapular fossa, subscapularis muscle)

sulc— L. *sulcus*, furrow. (Sulcus of brain)

supra— L. *supra-*, prefix meaning above. (Suprarenal gland, supraspinatus muscle)

sur— Old French *sur*, from L. *super*, above, beyond. (Surangular bone, surcharge)

sutur— L. *sutura*, seam, suture. (Suture)

Sylvius— Jacques Dubois Sylvius, fifteenth- to sixteenth- century French anatomist. (Aqueduct of Sylvius)

sym— See *syn*.

syn— Gr. *syn-*, *sym-*, prefix meaning together or with. (Symbiosis, sympathy, symphysis, synapsid skull, synotic tectum)

syst— Gr. *systema*, a composite whole. (Organ system, systemic veins)

tal— L. *talus*, ankle. (Talus bone)

tape— Gr. *tapes*, dim., *tapetion*, carpet. (Tapestry, tapetum lucidum)

tars— Gr. *tarsos*, flat surface, sole of foot. (Metatarsal bone, tarsus)

tect— L. *tectum*, roof. (Optic tectum)

tegmen— L. *tegmentum*, covering. (Tegmentum of mesencephalon)

tel— Gr. *tele*, far off, distant. (Telencephalon, teleost, television)

tela— L. *tela*, weblike membrane. (Tela choroidea)

tempor— L. *tempus*, gen., *temporis*, time. (Temple of the head, temporal bone, temporalis muscle)

ten— L. *tendere*, p.p., *tentus*, to stretch. (Tendon, tensor muscle, tent, tentorium)

teny— Gr. *teinein*, to extend. (Neoteny)

ter— L. *teres*, round, smooth. (Teres major muscle, ligamentum teres)

test— L. *testis*, a witness (originally an adult male). (Testify, testicular, testis)

tetra— Gr. *tetra*, four (Tetrapods)

thalam— Gr. *thalamos*, inner chamber. (Epithalamus, thalamus)

thec— L. *theca*, sheath. (Theca, thecodont)

thel— Gr. *thelys*, tender, delicate. (Endothelium, epithelium)

ther— Gr. *therion*, wild beast. (Eutheria, Prototheria)

thorac— Gr. *thorax*, gen., *thorakos*, chest. (Thoracic, thorax)

thym— Gr. *thymos*, the thymus (akin to *thymon*, the herb thyme or a lump resembling a bunch of thyme). (Thymine, thymus)

thyr— Gr. *thyreos*, an oblong shield. (Thyrohyal muscle, thyroid cartilage)

tibi— L. *tibia*, shin bone. (Tibia)

tom— Gr. *tomia*, a cutting or segment. (Dermatome, myotome) *ton*— Gr. *tonos*, something stretched. (Peritoneum, tone, tonus)

trab— L. *trabs*, dim., *trabecula*, beam. (Basitrabecular processes, trabeculae carnae)

trach— L. *trachia*, windpipe, from *trachys*, rough. (Trachea)

tract— L. *trahere*, p.p., *tractus*, to pull or draw. (Neuron tract, protractor muscle, retractor)

trans— L. *trans*, across, beyond. (Transcend, transverse section, transversus thoracis muscle)

trapez— Gr. *trapeza*, small table of characteristic shape. (Trapezius muscle, trapezoid bone)

trem— Gr. *trema*, hole. (Monotremata, pretrematic nerve branch)

tri— L. *tri*, three. (Triceps muscle, tricuspid valve, trigeminal nerve)

triquetr— L. *triquetrus*, triangular. (Triquetrum bone)

trochant— Gr. *trochanter*, a runner. (Femoral trochanters)

trochlea— L. *trochlea*, pulley. (Trochlea on humerus, trochlear nerve)

tuber— L. *tuber*, dim., *tuberculum*, bump, lump. (Tuber cinereum, tuberculum of a rib)

tunic— L. *tunica*, garment. (Tunicates, vascular tunic)

turb— L. *turbo*, gen., *turbinis*, a spinning thing. (Turbinate bone, turbine)

tympan— L. *tympanum*, drum. (Tympanic membrane)

ulna— L. *ulna*, elbow, lower arm. (Ulna)

ur— Gr. *oura*, tail. (Anura, Urochordata, Urodela)

ur— L. *urina*, urine. (Urea, ureter, urethra, urinary, urogenital)

uter— L. *uterus,* the womb. (Uterus)

utr— L. *utriculus*, leather bag or bottle. (Utriculus of ear)

vag— L. *vagus*, wandering, undecided. (Vague, vagus nerve)

vagin— L. *vagina*, sheath. (Vagina, vaginal ligaments, processus vaginalis)

vall— L. *vallare*, to surround with a rampart. (Vallate papilla)

vast— L. *vastus*, large area, immense. (Vast, vastus lateralis muscle)

vein— L. *vena*, vein. (Vein)

vel— L. *velum*, veil, covering. (Medullary velum)

ven— L. *vena*, vein. (Vena cava)

ventr— L. *venter*, dim., *ventriculus*, belly, womb. (Ventral, ventricle)

verm— L. *vermis*, worm. (Cerebellar vermis)

vertebr— L. *vertebra*, joint, vertebra. (Vertebrates)

vesic— L. *vesica*, dim., *vesicula*, bladder. (Vesica fellea, vesicular gland)

vestibul— L. *vestibulum*, entrance chamber. (Vestibule, vestibulocochlear nerve)

vibr— L. *vibrare*, to agitate. (Vibrate, vibrissa)

visc— L. *viscus*, pl., *viscera*, an entrail. (Visceral)

vitell— L. *vitellus*, yolk. (Vitelline veins)

vitr— L. *vitrum*, glass. (Vitreous body of eye, vitrify)

vom— L. *vomer*, plowshare (cutting blade of a plow). (Vomer bone)

vor— L. *vorare*, to devour. (Carnivore, voracious)

vulv— L. *vulva*, covering. (Vulva)

Wolff— Kaspar Friedrich Wolff, eighteenth-century German embryologist. (Wolffian duct)

xiph— Gr. *xiphos*, sword. (Xiphihumeralis muscle, xiphisternum)

zyg— Gr. *zygon*, yoke, union, pair. (Azygos vein, zygapophysis, zygomatic arch, zygote)

APPENDIX 2

◆ ◆

A NOTE ON THE HANDLING OF SPECIMENS

Specimens for a comparative anatomy course are available through biological supply houses. Those students and instructors who wish to prepare their own specimens should consult Hildebrand (1969).

There are various well-established techniques that help maintain specimens in good condition for the duration of a semester-long comparative anatomy course, but here we will describe only some newer techniques that may not be widely used yet, but that have proved highly successful in our own courses for more than a decade (see also Blaney and Johnson, 1989, and Wineski and English, 1989).

Specimens must be treated with toxic and noxious substances, such as formalin, alcohol, and phenol, in order to prevent microorganisms from attacking them. Unfortunately, these preserving agents are also volatile and quickly fill the air of a laboratory. The resulting bad air bothers most students greatly and may prevent many from even considering a teaching and research career in the anatomical sciences. Although soaking the specimens in running cold water does remove the smell, it also makes the specimens vulnerable to decay.

An excellent alternative technique that avoids virtually all problems of the older techniques involves the use of a 1% phenoxyethanol solution as a wetting agent for the preserved specimens. Phenoxyethanol (2-phenoxyethanol, practical grade; Eastman Kodak Co., Rochester, N.Y., USA) is a nontoxic liquid that has a faint but pleasant scent. It is often used as a preserving agent in cosmetic products and externally applied medications. Because it is used as a 1% solution, it is relatively inexpensive. It is nonflammable, effectively prevents the growth of microorganisms, and softens and rehydrates the tissues of specimens. European museum collections have used phenoxyethanol as a storage solution for wet specimens for at least two decades.

Phenoxyethanol is slightly lipophilic and sinks as liquid globules to the bottom, if mixed with water in a concentration that exceeds 2%. A 1% solution is best prepared by mixing exactly one part of phenoxyethanol with 99 parts of very hot tap water. The use of a magnetic stirrer for about 30 minutes ensures a homogeneous solution.

Every student should be supplied with a plastic dispensing bottle filled with the 1% phenoxyethanol solution. During the dissection, the specimen should be repeatedly and liberally sprayed or doused with the solution. Between lab sessions, the specimens should be wrapped in rags or cheesecloth soaked with the phenoxyethanol solution and then stored in tightly closed plastic bags or storage bins.

The condition of specimens with excessively dry, hardened, or brittle tissues can be greatly improved by immersing them in a 1% phenoxyethanol solution to which some fabric softener has been added. Any fabric softener commonly used for laundry may be used, preferably in the proportion of one part of fabric softener to 99 parts of the 1% phenoxyethanol solution. The specimens can be left in this conditioning solution for the duration of a semester or more.

APPENDIX 3

◆ ◆

REFERENCES

The references given below include those cited in the text, those of particular value for laboratory studies in comparative anatomy, and certain key references on the functional significance and inter-relationships of the various organs. More inclusive bibliographies can be found in many of the works cited below.

Students who are interested in current research in comparative or functional anatomy may want to look up a few journals in their library. *Scientific American* and *The American Scientist* are two journals in magazine format. They publish articles on a variety of scientific subjects, including anatomical topics, which are current, readable, and easily understood even by beginning students. *The American Zoologist* publishes excellent review articles on a variety of zoological subjects, including anatomy. They are accessible to advanced undergraduate students. Original research articles in comparative and functional anatomy are published in several scientific journals, such as *Acta Anatomica, The American Journal of Anatomy, The Anatomical Record, The Biological Journal of the Linnean Society, The Zoological Journal of the Linnean Society, The Journal of Anatomy, The Journal of Experimental Biology, The Journal of Experimental Zoology, The Journal of Morphology, The Journal of Zoology,* and *Zoomorphology*.

General

Alexander, R. McN.: *Animal Mechanics.* 2nd ed. Oxford, Blackwell Scientific Publications, 1983.

Alexander, R. McN.: *The Chordates.* 2nd ed. Cambridge, Cambridge University Press, 1981.

Baumel, J. J., King, A. S., Lucas, A. M., Breazile, J. E., and Evans, H. E. (eds.): *Nomina Anatomica Avium. An Annotated Anatomical Dictionary of Birds.* London, Academic Press, 1979.

Blaney, S. P. A., and Johnson, B.: "Technique for reconstituting fixed cadaveric tissue." *Anatomical Record,* vol. 224, pp. 550–551, 1989.

Bolis, L., Keynes, R. D., and Maddrell, S. H. P. (eds.): *Comparative Physiology of Sensory Systems.* New York, Cambridge University Press, 1984.

Bolk, L., and others: *Handbuch der vergleichenden Anatomie der Wirbeltiere.* 6 vols. Berlin and Vienna, Urban und Schwarzenberg, 1931–1938. Reprinted in 1967 by A. Asher and Co., Amsterdam.

Browder, L. W., Erickson, C. A., and Jeffrey, W. R.: *Developmental Biology.* 3rd ed. Philadelphia, Saunders College Publishing, 1991.

Carroll, R. L.: *Vertebrate Paleontology and Evolution.* New York, W. H. Freeman & Co., 1988.

DeBeer, G. R.: *The Vertebrate Skull.* Oxford, Clarendon Press, 1937.

Demski, L. S.: "Evolution of LHRH system and terminal nerve in the vertebrate reproductive brain." *American Zoologist,* vol. 24, pp. 809–839, 1984.

Dorit, R. L., Walker, W. F., and Barnes, R. D.: *Zoology.* Philadelphia, Saunders College Publishing, 1991.

Dorland, W. A. N.: *Dorland's Illustrated Medical Dictionary.* 27th ed. Philadelphia, W. B. Saunders Company, 1988.

Eckert, R., Randall, D., and Augustine, G.: *Animal Physiology—Mechanisms and Adaptations.* 3rd ed. New York, W. H. Freeman & Co., 1988.

Edgeworth, F. H.: *The Cranial Muscles of Vertebrates.* London, Cambridge University Press, 1935.

Gans, C.: *Biomechanics, An Approach to Vertebrate Biology*. Philadelphia, J. B. Lippincott Company, 1974. Reprinted in 1980 by the University of Michigan Press, Ann Arbor.

Goodrich, E. S.: *Studies on the Structure and Development of Vertebrates*. New York, Dover Publications, 1958.

Grassé, P. P. (ed.): *Traité de Zoologie*. Vols. 11–17 deal with protochordates and vertebrates. Paris, Masson et Cie., 1948–1973.

Gray, J.: *Animal Locomotion*. London, Weidenfeld and Nicolson, 1968.

Gutmann, W. F.: "Relationships between invertebrate phyla based on functional-mechanical analysis of the hydrostatic skeleton." *American Zoologist*, vol. 21, pp. 63–81, 1981.

Gutmann, W. F.: "The hydraulic principle." *American Zoologist*, vol. 28, pp. 257–266, 1988.

Hildebrand, M.: *Anatomical Preparations*. Berkeley, University of California Press, 1969.

Hildebrand, M.: *Analysis of Vertebrate Structure*. 3rd ed. New York, John Wiley and Sons, 1987.

Hughes, G. M.: *Comparative Physiology of Vertebrate Respiration*. Cambridge, Harvard University Press, 1963.

Hyman, L. H.: *Comparative Vertebrate Anatomy*. 2nd ed. Chicago, University of Chicago Press, 1942.

International Anatomical Nomenclature Committee: *Nomina Anatomica*. 5th ed. *Nomina Histologica*. 2nd ed. *Nomina Embryologica*. 2nd ed. Baltimore, Williams & Wilkins, 1983.

International Committee on Veterinary Gross Anatomical Nomenclature: *Nomina Anatomica Veterinaria*. 3rd ed. *Nomina Histologica*. 2nd ed. Ithaca, N. Y., International Committee on Veterinary Gross Anatomical Nomenclature, 1983.

Jaeger, E.: *A Source-Book of Biological Names and Terms*. 3rd ed. Springfield, Charles C Thomas, 1978.

Jollie, M.: *Chordate Morphology*. Huntington, N. Y., Robert E. Krieger Publishing Company, 1973.

Kluge, A. G., and others: *Chordate Structure and Function*. 2nd ed. New York, Macmillan Publishing Company, 1977.

Meier, S., and Tam, P. L.: "Metameric pattern development in the embryonic axis of the mouse. I. Differentiation of the cranial segments." *Differentiation*, vol. 21, pp. 95–108, 1982.

Noden, D. M.: "The embryonic origins of avian cephalic and cervical muscles and associated connective tissues." *American Journal of Anatomy*, vol. 168, pp. 257–267, 1983.

Noden, D. M.: "Craniofacial development: new views on old problems." *Anatomical Record*, vol. 208, pp. 1–13, 1984.

Nybakken, O. E.: *Greek and Latin in Scientific Terminology*. Ames, Iowa State University Press, 1959.

Pough, F. H., Heiser, J. B., and McFarland, W. N.: *Vertebrate Life*. 3rd ed. New York, Macmillan Publishing Company, 1989.

Prosser, C. L. (ed.): *Neural and Integrative Animal Physiology, Comparative Animal Physiology*. 4th ed. New York, John Wiley & Sons, Inc., 1991.

Prosser, C. L. (ed.): *Environmental and Metabolic Animal Physiology, Comparative Animal Physiology*. 4th ed. New York, John Wiley & Sons, Inc., 1991.

Romer, A. S., and Parsons, T. S.: *The Vertebrate Body*. 6th ed. Philadelphia, Saunders College Publishing, 1986.

Schmidt-Nielsen, K.: *How Animals Work*. Cambridge, Cambridge University Press, 1972.

Schmidt-Nielsen, K.: *Animal Physiology*. 4th ed. Cambridge, Cambridge University Press, 1990.

Smith, H. M.: *Evolution of Chordate Structure*. Holt, Rinehart, and Winston, Inc., 1961.

Starck, D.: *Vergleichende Anatomie der Wirbeltiere auf evolutionsbiologischer Grundlage*. Volumes 1, 2, and 3. Berlin, Springer-Verlag, 1978, 1979, 1982.

Stedman, T. L.: *Stedman's Medical Dictionary*. 25th ed. Baltimore, Williams & Wilkins, 1990.

Strother, G. K.: *Physics with Applications in Life Sciences*. Boston, Houghton Mifflin Company, 1977.

Villee, C. A., Walker, W. F., and Barnes, R. D.: *General Zoology*. 6th ed. Philadelphia, Saunders College Publishing, 1984.

Villee, C. A., Solomon, E. P., Martin, C. E., Martin, D. W., Berg, L. R., and Davis, P. W.: *Biology*. 2nd ed. Philadelphia, Saunders College Publishing, 1989.

Wake, M. H. (ed.): *Hyman's Comparative Vertebrate Anatomy*. 3rd ed. Chicago, University of Chicago Press, 1979.

Walker, W. F., Jr.: *Functional Anatomy of the Vertebrates: An Evolutionary Perspective*. Philadelphia, Saunders College Publishing, 1987.

Walls, G. L.: *The Vertebrate Eye and Its Adaptive Radiation*. Bloomfield Hills, Cranbrook Institute of Science, Bull. No. 19, 1942.

Wineski, L. E., and English, A. W.: "Phenoxyethanol as a nontoxic preservative in the dissection laboratory." *Acta Anatomica*, vol. 136, pp. 155–158, 1989.

Wolff, R. G.: *Functional Chordate Anatomy*. Lex-

444

Appendix 3

ington, MA, D. C. Heath and Company, 1991.

Young, J. Z.: *The Life of Vertebrates.* 3rd ed. London, Oxford University Press, 1981.

Protochordates and Fishes

Aleev, Yu. G.: *Function and Gross Morphology in Fish.* Washington, D.C., Smithsonian Institution, 1969.

Barrington, E. J. W.: "The supposed pancreatic organs of *Petromyzon fluviatilis* and *Myxine glutinosa.*" *Quarterly Journal of Microscopical Science,* vol. 85, pp. 391–417, 1945.

Barrington, E. J. W.: *The Biology of Hemichordata and Protochordata.* San Francisco, W. H. Freeman and Company, 1965.

Bone, Q., and Marshall, N. B.: *Biology of Fishes.* New York, Chapman and Hall, 1982.

Boord, R. L., and Campbell, C. B. G.: "Structural and functional organization of the lateral line system of sharks." *American Zoologist,* vol. 17, pp. 431–441, 1977.

Budker, P., and Whitehead, P. J.: *The Life of Sharks.* New York, Columbia University Press, 1971.

Cahn, P. H. (ed.): *Lateral Line Detectors.* Bloomington, Indiana University Press, 1967.

Carey, F. G.: "Fishes with warm bodies." *Scientific American,* vol. 228, No. 2, pp. 36–44, 1973.

Carey, F. G., and Gibson, Q. H.: "Heat and oxygen exchange in the rete mirabile of the bluefin tuna, *Thunnus thynnus.*" *Comparative Biochemistry and Physiology,* vol. 74A, pp. 333–342, 1983.

Corwin, J. T.: "Audition in elasmobranchs." In Tavolga, W. N., Popper, A. N., and Fay, R. R. (ed.): *Hearing and Sound Communication in Fishes.* New York, Springer-Verlag, 1981.

Daniel, J. F.: *The Elasmobranch Fishes.* 3rd ed. Berkeley, University of California Press, 1934.

Davis, R. E., and Northcutt, R. G. (eds.): *Fish Neurobiology. Vol. 2. Higher Brain Areas and Functions.* Ann Arbor, University of Michigan Press, 1983.

Demski, L. S.: "Electrical stimulation of the shark brain." *American Zoologist,* vol. 17, pp. 487–500, 1977.

Demski, L. S., Fields, R. D., Bullock, T. H., Schreibmann, M. P., and Margolis-Nunn, H.: "The terminal nerve of sharks and rays." *Annals of the New York Academy of Sciences,* vol. 519, pp. 15–32, 1987.

Flood, P. P.: "Fine structure of the notochord of

Amphioxus." In Barrington, E. J. W., and Jefferies, R. P. S.: *Protochordates.* Symposia of the Zoological Society of London, no. 36. London, Academic Press, 1975.

Gans, C., and Parsons, T. S.: *A Photographic Atlas of Shark Anatomy.* Chicago, University of Chicago Press, 1981.

Gilbert, S. G.: *Pictorial Anatomy of the Dogfish.* Seattle, Washington University Press, 1973.

Goodrich, E. S.: "On the development of the segments of the head of *Scyllium.*" *Quarterly Journal of Microscopical Science,* vol. 63, pp. 1–30, 1918.

Gruber, S. H., and Cohen, J. L.: "Visual system of the elasmobranchs: State of the art 1960–1975." In Hodgson, E. S., and Mathewson, R. F. (eds.): *Sensory Biology of Sharks, Skates, and Rays.* Arlington, VA, Department of the Navy, Office of Naval Research, 1978.

Heath, G. W.: "The siphon sacs of the smooth dogfish and spiny dogfish." *Anatomical Record,* vol. 125, p. 562, 1956.

Herold, R. C., Graver, H. T., and Christner, P.: "Immunohistochemical localization of amelogenins in enameloid of lower vertebrate teeth." *Science,* vol. 207, pp. 1357–1358, 1980.

Hoar, W. S., and Randall, D. J.: *Fish Physiology.* Vols. 1–10, New York, Academic Press, 1969–1984.

Hodgson, E. S., and Mathewson, R. F. (eds.): *Sensory Biology of Sharks, Skates, and Rays.* Arlington, VA, Department of the Navy, Office of Naval Research, 1978.

Hughes, G. M.: "The relationship between cardiac and respiratory rhythms in the dogfish, *Scyliorhinus canicula* L." *Journal of Experimental Biology,* vol. 57, pp. 415–434, 1972.

Hughes, G. M., and Hills, B. A.: "Oxygen tension distribution in water and blood at the secondary lamella of the dogfish gill." *Journal of Experimental Biology,* vol. 55, pp. 399–408, 1971.

Liem, K. L., and Woods, L. P.: "A probable homologue of the clavicle in the holostean fish *Amia calva.*" *Journal of Zoology (London),* vol. 170, pp. 521–532, 1973.

Marinelli, W., and Strenger, A.: *Vergleichende Anatomie und Morphologie der Wirbeltiere.* I. Lieferung. *Lampetra fluviatilis* L. III. Lieferung. *Squalus acanthias* L. Vienna, Franz Deuticke Verlag, 1954 and 1959.

Moyle, P. B., and Cech, J. J., Jr.: *Fishes: An Introduction to Ichthyology.* Englewood Cliffs, Prentice-Hall, 1982.

Nelson, J. S.: *Fishes of the World.* 2nd ed. New York, John Wiley & Sons, Inc., 1984.

Norris, H. W., and Hughes, S. P.: "The cranial, occipital, and anterior spinal nerves of the dogfish, *Squalus acanthias.*" *Journal of Comparative Neurology,* vol. 31, pp. 293–402, 1920.

Northcutt, R. G.: "Elasmobranch central nervous system organization and its possible evolutionary significance." *American Zoologist,* vol. 17, pp. 411–429, 1977.

Northcutt, R. G., and Davis, R. E.: *Fish Neurobiology. Vol 1. Brain Stem and Sense Organs.* Ann Arbor, University of Michigan Press, 1983.

Nursall, J. R.: "Swimming and the origin of paired appendages." *American Zoologist,* vol. 2, pp. 127–141, 1962.

O'Donoghue, C. H., and Abbot, E. B.: "The blood vascular system of the spiny dogfish, *Squalus acanthias* Linné, and *Squalus sucklii* Gill." *Transactions of the Royal Society of Edinburgh,* vol. 55, pp. 823–894, 1928.

Oguri, M.: "Rectal glands of marine and fresh water sharks, comparative histology." *Science,* vol. 144, pp. 1151–1152, 1964.

Pang, P. K. T., Griffith, R. W., and Atz, J. W.: "Osmoregulation in elasmobranchs." *American Zoologist,* vol. 17, pp. 365–377, 1977.

Ruppert, E. E., and Smith, P. R.: "The functional organization of filtration nephridia." *Biological Reviews,* vol. 63, pp. 231–258, 1988.

Ruppert, E. E.: "Structure, ultrastructure and function of the neural gland complex of *Ascidia interrupta* (Chordata, Ascidiacea): Clarification of hypotheses regarding the evolution of the vertebrate anterior pituitary." *Acta Zoologica* (Stockholm), vol. 71, pp. 135–149, 1990.

Satchell, G. H.: *Circulation in Fishes.* Cambridge, Cambridge University Press, 1971.

Shuttleworth, T. J. (ed.): *Physiology of Elasmobranch Fishes.* Berlin, Springer-Verlag, 1988.

Smeets, W. J. A. J., and Nieuwenhuys, R.: "Topological analysis of the brain stem of the sharks *Squalus acanthias* and *Scyliorhinus canicula.*" *Journal of Comparative Neurology,* vol. 165, pp. 333–368, 1976.

Thomson, K. S.: "The shape of a shark's tail." *American Scientist,* vol. 78, pp. 499–501, 1990.

Thomson, K. S., and Simanek, D. E.: "Body form and locomotion in sharks." *American Zoologist,* vol. 17, pp. 342–354, 1977.

Wainwright, S. A., Vosburgh, F., and Hebrank, J. H.: "Shark skin: Function in locomotion." *Science,* vol. 202, pp. 747–749, 1978.

Webb, P. W., and Weihs, D. (eds.): *Fish Biomechanics.* New York, Praeger Publishers, 1983.

Williamson, R. M., and Roberts, B. L.: "Sensory and motor interactions during movement in the spinal dogfish." *Proceedings of the Royal Society of London, Series B,* vol. 227, pp. 103–119, 1986.

Wourms, J. P.: "Reproduction and development in chondrichthyan fishes." *American Zoologist,* vol. 17, pp. 379–410, 1977.

Young, J. Z.: "The autonomic nervous system of selachians." *Quarterly Journal of Microscopical Science,* vol. 75, pp. 571–624, 1933.

Amphibians and Reptiles

Capranica, R. R.: "Morphology and physiology of the auditory system." In Llianas, R., and Precht, W. (eds.): *Frog Neurobiology.* Berlin, Springer-Verlag, 1976.

Chase, S. W.: "The mesonephros and urogenital ducts of *Necturus maculosus* Rafinesque." *Journal of Morphology,* vol. 37, pp. 457–532, 1923.

DeLong, K. T.: "Quantitative analysis of blood circulation through the frog heart." *Science,* vol. 138, pp. 693–694, 1962.

Duellman, W. E., and Trueb, L.: *Biology of Amphibians.* New York, McGraw-Hill Book Company, 1986.

Figge, F. H.: "A morphological explanation for the failure of Necturus to metamorphose." *Journal of Experimental Zoology,* vol. 56, pp. 241–265, 1930.

Francis, E. B.: *The Anatomy of the Salamander.* London, Oxford University Press, 1943.

Gans, C., and others (eds.): *Biology of the Reptilia,* vols. 1–15, London, Academic Press, 1969–1985.

Gilbert, S. G.: *Pictorial Anatomy of the Necturus.* Seattle, University of Washington Press, 1973.

Harris, J. P., Jr.: "Necturus papers: The skeleton of the arm. The pelvic musculature. The muscles of the forearm. The levator anguli scapulae. The musculus depressor mandibulae. Natural history." *Field and Laboratory,* vols. 20, 21, 22, 25, 27, 1952–1959.

Herrick, C. J.: *The Brain of the Tiger Salamander, Ambystoma tigrinum.* Chicago, University of Chicago Press, 1948.

Kinsbury, B. F.: "On the brain of *Necturus maculatus.*" *Journal of Comparative Neurology,* vol. 5, pp. 138–205, 1895.

Lauder, G. V., and Reilly, S. M.: "Metamorphosis of the feeding mechanism in tiger salamanders

(*Ambystoma tigrinum*): The ontogeny of cranical muscle mass." *Journal of Zoology, London,* vol. 222, pp. 59–74, 1990.

Lauder, G. V., and Shaffer, H. B.: "Ontogeny of functional design in tiger salamanders (*Ambystoma tigrinum*): Are motor patterns conserved during major morphological transformations?" *Journal of Morphology,* vol. 197, pp. 249–268, 1988.

Lombard, R. E., and Bolt, J.: "Evolution of the tetrapod ear: An analysis and reinterpretation." *Biological Journal of the Linnean Society,* vol. 11, pp. 19–76, 1979.

Lombard, R. E., and Straughan, I. R.: "Functional aspects of anuran middle ear structures." *Journal of Experimental Biology,* vol. 61, pp. 1–23, 1974.

Miller, W. S.: "The vascular system of *Necturus maculatus.*" *University of Wisconsin Science Series,* vol. 2, pp. 211–226, 1900.

Moore, J. A. (ed.): *Physiology of the Amphibia.* New York, Academic Press, 1964.

Noble, G. K.: *The Biology of the Amphibia.* New York, Dover Publications, 1954.

Romer, A. S.: *Osteology of Reptiles.* Chicago, University of Chicago Press, 1956.

Wilder, H. H.: "The skeletal system of *Necturus maculatus* Rafinesque." *Memoirs of the Boston Society of Natural History,* vol. 5, pp. 387–439, 1903.

Wilder, H. H.: "The appendicular muscles of *Necturus maculosus.*" *Zoologische Jahrbücher,* suppl. 15, part 2, pp. 383–424, 1912.

Mammals

Abell, N. B.: "A comparative study of the variations of the postrenal vena cava of the cat and rat and a description of two new variations." *Denison University Journal of the Science Laboratory,* vol. 40, pp. 87–117, 1947.

Allin, E. P.: "Evolution of the mammalian inner ear." *Journal of Morphology,* vol. 147, pp. 403–438, 1975.

Arey, L. B.: *Developmental Anatomy.* 7th ed., revised. Philadelphia, W. B. Saunders Company, 1974.

Baker, M. A.: "A brain-cooling system in mammals." *Scientific American,* vol. 240, No. 5, pp. 130–139, 1979.

Banks, W. J.: *Applied Veterinary Histology.* 2nd ed. Baltimore, Williams & Wilkins, 1986.

Barone, R., and others: *Atlas d'Anatomie du Lapin.* Paris, Masson et Cie, 1973.

Barry, A.: "The aortic arch derivatives in the human adult." *Anatomical Record,* vol. 111, pp. 221–238, 1951.

Bramble, D. M.: "Origin of the mammalian feeding complex, models and mechanism." *Paleobiology,* vol. 4, pp. 271–301, 1978.

Cave, A. J. E.: "The morphology of mammalian cervical pleurapophyses." *Journal of Zoology (London),* vol. 177, pp. 377–393, 1975.

Crompton, A. W., and Parker, P.: "Evolution of the mammalian masticatory apparatus." *American Scientist,* vol. 66, pp. 192–201, 1978.

Crouch, J. E.: *Text-Atlas of Cat Anatomy.* Philadelphia, Lea & Febiger, 1969.

Davis, D. D.: "The giant panda, a morphological study of evolutionary mechanisms." *Fieldiana, Zoological Memoirs,* vol. 3, pp. 1–340, 1964.

Davis, D. D., and Story, H. E.: "The carotid circulation in the domestic cat." *Zoological Series Field Museum of Natural History,* vol. 28, pp. 1–47, 1943.

Dyce, K. M., Sack, W. O., and Wensing, C. J. G.: *Textbook of Veterinary Anatomy.* Philadelphia, W. B. Saunders Company, 1987.

Fawcett, D. W.: *Bloom and Fawcett, A Textbook of Histology,* 11th ed. Philadelphia, W. B. Saunders Company, 1986.

Getty, R. (ed.): *Sisson and Grossman's Anatomy of the Domestic Animals.* 5th ed. Philadelphia, W. B. Saunders Company, 1975.

Gilbert, S. G.: *Pictorial Anatomy of the Cat.* 2nd ed. Seattle, University of Washington Press, 1975.

Griffiths, M.: *The Biology of Monotremes.* New York, Academic Press, 1978.

Huntington, G. S., and McClure, C. F. W.: "The development of the veins in the domestic cat." *Anatomical Record,* vol. 20, pp. 1–31, 1920.

Jayne, J.: *Mammalian Anatomy.* Part I. *The Skeleton of the Cat.* Philadelphia, J. B. Lippincott Company, 1898.

Jenkins, F. A., Jr.: "The evolution and development of the dens of the mammalian axis." *Anatomical Record,* vol. 164, pp. 173–184, 1969.

Jenkins, F. A., Jr.: "Limb posture and locomotion in the Virginia opossum, *Didelphis marsupialis,* and other non-cursorial mammals." *Journal of Zoology (London),* vol. 165, pp. 303–315, 1971.

Jenkins, F. A., Jr.: "The movement of the shoulder in claviculate and aclaviculate mammals." *Journal of Morphology,* vol. 144, pp. 71–84, 1974.

Lindvall, M., Edvinsson, L., and Owman, C.: "Sympathetic nervous control of cerebrospinal fluid production from the choroid plexus." *Science,* vol. 201, pp. 176–178, 1978.

Miller, M. E., Christensen, G. C., and Evans, H. E.: *Anatomy of the Dog.* Philadelphia, W. B. Saunders Company, 1964.

Mivart, St. G.: *The Cat. An Introduction to the Study of Backboned Animals, Especially Mammals.* London, John Murray, 1881.

Nickel, R., Schummer, A., and Seiferle, E.: *The Anatomy of the Domestic Animals.* Volumes 1, 2, and 3. New York, Springer-Verlag, 1986, 1979, 1981.

Northcutt, R. G., Kenneth, L. W., and Barber, R. P.: *Atlas of the Sheep Brain.* 2nd ed. Champaign, IL, Stiles Publishing Company, 1966.

Ranson, S. W.: *The Anatomy of the Nervous System.* 10th ed., revised by S. L. Clark. Philadelphia, W. B. Saunders Company, 1959.

Rasmussen, A. T.: *The Principal Nervous Pathways.* 4th ed. New York, The Macmillan Company, 1952.

Reighard, J. E., and Jennings, H. S.: *Anatomy of the Cat.* 3rd ed., Revised by R. Elliot. New York, Henry Holt and Company, 1935.

Schmidt-Nielsen, K., Bolis, L., and Taylor, C. R. (eds): *Comparative Physiology: Primitive Mammals.* Cambridge, Cambridge University Press, 1980.

Slijper, E. J.: "Comparative biological-anatomical investigations on the vertebral column and spinal musculature of mammals." *Koninklijke Nederlandse Academie Van Wetenschappen (Tweede Sectie),* vol. XLII, No. 5, pp. 1–128, 1946.

Smith, K. K., and Kier, W. M.: "Trunks, tongues, and tentacles: Moving with skeletons of muscle." *American Scientist,* vol. 77, pp. 29–35, 1989.

Thomason, J. J., and Russell, A. P.: "Mechanical factors in the evolution of the mammalian secondary palate: A theoretical analysis." *Journal of Morphology,* vol. 189, pp. 199–213, 1986.

Walker, W. F.: *A Study of the Cat with Reference to Human Beings.* 4th ed. Philadelphia, Saunders College Publishing, 1982.

Williams, P. L., Warwick, R., Dyson, M., and Bannister, L. H.: *Gray's Anatomy,* 37th ed. Edinburgh, Churchill Livingstone, 1989.

Yoshikawa, T.: *Atlas of the Brains of Domestic Animals.* University Park, Pennsylvania State University Press, 1968.

Young, J. Z., and Hobbs, M. J.: *The Life of Mammals.* 2nd ed. New York and Oxford, Oxford University Press, 1975.

◆ ◆

INDEX

Many terms are indexed under their defining noun, e.g., artery, bone, canal, cartilage, cavity, duct, fascia, fenestra, foramen, fossa, ganglion, gland, ligament, muscle, nerve, process, and vein. These lists of arteries and other major organs may also be useful for review. Organ systems are indexed by name or by major part (e.g., Skull), and under Mammal, *Necturus, Squalus*. Most figures are indexed under the name of the animal illustrated. References to figures are indicated by an (f); references to tables by a (t).